THE GLOBALIZING CITIES READER

The newly revised *Globalizing Cities Reader* reflects how the geographies of theory have recently shifted away from the western vantage points from which much of the classic work in this field was developed.

The expanded volume continues to make available many of the original and foundational works that underpin the research field, while expanding coverage to familiarize students with new theoretical and epistemological positions as well as emerging research foci and horizons. It contains 38 new chapters, including key writings on globalizing cities from leading thinkers such as John Friedmann, Michael Peter Smith, Saskia Sassen, Peter Taylor, Manuel Castells, Anthony King, Jennifer Robinson, Ananya Roy, and Fulong Wu. The new *Reader* reflects the fact that world and global city studies have evolved in exciting and wide-ranging ways, and the very notion of a distinct "global" class of cities has recently been called into question. The sections examine the foundations of the field and processes of urban restructuring and global city formation. A large number of new entries focus on the emerging urban worlds of Asia, Latin America, and Africa, including Beijing, Bogota, Cairo, Cape Town, Delhi, Istanbul, Medellín, Mumbai, Phnom Penh, Rio de Janeiro, São Paulo, and Shanghai. The book also presents cases off the conventional map of global cities research, such as smaller cities and lesser known urban regions that are undergoing processes of globalization.

The book is a key resource for students and scholars alike who seek an accessible compendium of the intellectual foundations of global urban studies as well as an overview of the emergent patterns of early 21st century urbanization and associated sociopolitical contestation around the world.

Xuefei Ren is Associate Professor of Sociology and Global Urban Studies at Michigan State University.

Roger Keil is York Research Chair in Global Sub/Urban Studies in the Faculty of Environmental Studies, York University, Toronto.

THE ROUTLEDGE URBAN READER SERIES

Series editors

Richard T. LeGates
Professor Emeritus of Urban Studies and Planning, San Francisco State University.

Frederic Stout
Lecturer in Urban Studies, Stanford University

The Routledge Urban Reader Series responds to the need for comprehensive coverage of the classic and essential texts that form the basis of intellectual work in the various academic disciplines and professional fields concerned with cities and city planning.

The readers focus on the key topics encountered by undergraduates, graduate students, and scholars in urban studies, geography, sociology, political science, anthropology, economics, culture studies, and professional fields such as city and regional planning, urban design, architecture, environmental studies, international relations and landscape architecture. They discuss the contributions of major theoreticians and practitioners and other individuals, groups, and organizations that study the city or practice in a field that directly affects the city.

As well as drawing together the best of classic and contemporary writings on the city, each reader features extensive introductions to the book, sections, and individual selections prepared by the volume editors to place the selections in context, illustrate relations among topics, provide information on the author, and point readers towards additional related bibliographic material.

Each reader contains:

- Between thirty and sixty *selections* divided into five to eight sections. Almost all of the selections are previously published works that have appeared as journal articles or portions of books.
- A *general introduction* describing the nature and purpose of the reader.
- *Section introductions* for each section of the reader to place the readings in context.
- *Selection introductions* for each selection describing the author, the intellectual background and context of the selection, competing views of the subject matter of the selection, and bibliographic references to other readings by the same author and other readings related to the topic.
- One or more plate sections and illustrations at the beginning of each section.
- An index.

The series consists of the following titles:

THE CITY READER

The City Reader, sixth edition – an interdisciplinary urban reader aimed at urban studies, urban planning, urban geography and urban sociology courses – is the *anchor urban reader*. Routledge published a first edition of *The City Reader* in 1996, a second edition in 2000, a third edition in 2003, a fourth edition in 2007, and a fifth edition in 2011. *The City Reader* has become one of the most widely used anthologies in urban studies, urban geography, urban sociology and urban planning courses in the world.

URBAN DISCIPLINARY READERS

The series contains *urban disciplinary readers* organized around social science disciplines and professional fields: urban sociology, urban geography, urban politics, urban and regional planning, and urban design. The urban disciplinary readers include both classic writings and recent, cutting-edge contributions to the respective disciplines. They are lively, high-quality, competitively priced readers which faculty can adopt as course texts and which also appeal to a wider audience.

TOPICAL URBAN ANTHOLOGIES

The urban series includes *topical urban readers* intended both as primary and supplemental course texts and for the trade and professional market. The topical titles include readers related to sustainable urban development, global cities, cybercities, and city cultures.

INTERDISCIPLINARY ANCHOR TITLE

The City Reader, sixth edition
Richard T. LeGates and Frederic Stout (eds)

URBAN DISCIPLINARY READERS

The Urban Geography Reader
Nick Fyfe and Judith Kenny (eds)

The Urban Politics Reader
Elizabeth Strom and John Mollenkopf (eds)

The Urban and Regional Planning Reader
Eugenie Birch (ed.)

The Urban Sociology Reader, second edition
Jan Lin and Christopher Mele (eds)

The Urban Design Reader, second edition
Michael Larice and Elizabeth Macdonald (eds)

TOPICAL URBAN READERS

The City Cultures Reader, second edition
Malcolm Miles, Tim Hall with Iain Borden (eds)

The Cybercities Reader
Stephen Graham (ed.)

The Sustainable Urban Development Reader, third edition
Stephen M. Wheeler and Timothy Beatley (eds)

Cities of the Global South Reader
Faranak Miraftab and Neema Kudva (eds)

The Globalizing Cities Reader
Xuefei Ren and Roger Keil (eds)

For further Information on The Routledge Urban Reader Series please visit our website:
www.routledge.com/Routledge-Urban-Reader-Series/book-series/SE0529
or contact

Andrew Mould
Routledge
2 Park Square, Milton Park,
Abingdon, Oxon, OX14 4RN
England
Andrew.Mould@tandf.co.uk

Richard T. LeGates
Department of Urban Studies and
Planning
San Francisco State University
1600 Holloway Avenue
San Francisco, CA 94132
(510) 642-3256
dlegates@sfsu.edu

Frederic Stout
Urban Studies Program
Stanford University
Stanford, California 94305-2048
fstout@stanford.edu

The Globalizing Cities Reader

Second Edition

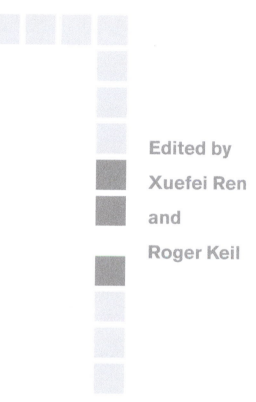

Edited by

Xuefei Ren

and

Roger Keil

Routledge
Taylor & Francis Group

LONDON AND NEW YORK

Second edition published 2018
by Routledge
2 Park Square, Milton Park, Abingdon, Oxon, OX14 4RN

and by Routledge
711 Third Avenue, New York, NY 10017

Routledge is an imprint of the Taylor & Francis Group, an informa business

© 2018 selection and editorial matter, Xuefei Ren and Roger Keil; individual chapters, the contributors

The right of Xuefei Ren and Roger Keil to be identified as the authors of the editorial material, and of the authors for their individual chapters, has been asserted in accordance with sections 77 and 78 of the Copyright, Designs and Patents Act 1988.

First edition published by Routledge 2006

British Library Cataloguing-in-Publication Data
A catalogue record for this book is available from the British Library

Library of Congress Cataloging-in-Publication Data
Names: Ren, Xuefei, editor. | Keil, Roger, 1957- editor.
Title: The globalizing cities reader / edited by Xuefei Ren and Roger Keil.
Other titles: Global cities reader.
Description: 2nd Edition. | New York : Routledge, [2018] |
Series: The Routledge urban reader series | Revised edition of The global cities reader, 2006. | Includes index.
Identifiers: LCCN 2017020625 | ISBN 9781138923683 (hardback : alk. paper) |
ISBN 9781138923690 (pbk. : alk. paper) | ISBN 9781315684871 (ebook)
Subjects: LCSH: Cities and towns. | Urbanization. | Globalization.
Classification: LCC HT119 .G64 2018 | DDC 307.76—dc23
LC record available at https://lccn.loc.gov/2017020625

ISBN: 978-1-138-92368-3 (hbk)
ISBN: 978-1-138-92369-0 (pbk)
ISBN: 978-1-315-68487-1 (ebk)

Typeset in Amasis
by Keystroke, Neville Lodge, Tettenhall, Wolverhampton

Printed in Great Britain by Ashford Colour Press Ltd

For John Friedmann (1926–2017)

Contents

Plates

Contributors

Janet Abu-Lughod (1928–2013) was Professor of Sociology at the New School for Social Research in New York City, USA.

Jean-Paul Addie is Assistant Professor at the Urban Studies Institute at Georgia State University, USA.

S. Harris Ali is Professor and Chair of Sociology at York University, Toronto, Canada.

David Bassens is Assistant Professor of Economic Geography at the Free University of Brussels, Belgium.

Jonathan V. Beaverstock is International Director (Associate Dean) for the Faculty of Social Sciences and Law, and Professor of International Management in the School of Economics, Finance and Management, University of Bristol, UK.

Bernd Belina is Professor of Human Geography at Goethe-University Frankfurt, Germany.

Fernand Braudel (1902–1985) was President of École Pratique des Hautes Études and a Professor at the Collège de France in Paris, France.

Neil Brenner is Professor of Urban Theory in the Graduate School of Design at Harvard University, USA.

Teresa Caldeira is Professor of City and Regional Planning at the University of California, Berkeley, USA.

Manuel Castells holds the Wallis Annenberg Chair in Communication, Technology, and Society at the University of Southern California, USA.

Hamid Dabashi is the Hagop Kevorkian Professor of Iranian Studies and Comparative Literature at Columbia University, USA.

Ben Derudder is Professor of Human Geography in the Department of Geography at Ghent University, Belgium, and Associate Director of the Globalization and World Cities (GaWC) research network.

Mike Douglass is Professor at the Lee Kuan Yew School of Public Policy at the National University of Singapore.

Veronique Dupont is a senior research fellow at the Institute of Research for Development in France.

Mohamed Elshahed is a Cairo-based architect, independent researcher, and writer.

James Farrer is Professor of Sociology at Sophia University, Tokyo, Japan.

Joe Feagin is Ella McFadden Professor of Liberal Arts at Texas A&M University, USA.

Andrew Field is Associate Dean of Undergraduate Studies at Duke Kunshan University, Suzhou, China.

Laurent Fourchard is a senior researcher with the French Foundation of Political Science (FNSP) at the research institute 'Les Afriques dans le Monde' at Sciences Po Bordeaux, France.

John Friedmann (1926–2017) was Professor Emeritus of Urban Planning in the Graduate School of Architecture and Planning at UCLA, USA, and Honorary Professor in the School of Community and Regional Planning at the University of British Columbia, Canada.

Frank Goetzke is Associate Professor of Urban and Public Affairs at the University of Louisville, Kentucky, USA.

Mark Graham is Professor of Internet Geography at the Oxford Internet Institute, University of Oxford, UK.

Stephen Graham is Professor of Cities and Society at the Global Urban Research Unit at Newcastle University, UK, and is based in Newcastle University's School of Architecture, Planning and Landscape.

Sir Peter Hall (1932–2014) was Bartlett Professor of Planning at University College London, UK.

David Harvey is Distinguished Professor of Anthropology and Geography at the Graduate Center of the City University of New York, USA.

Richard Child Hill is Professor Emeritus of Sociology at Michigan State University, USA.

Michael Hoyler is Senior Lecturer in Human Geography at Loughborough University, UK.

Roger Keil is York Research Chair in Global Sub/Urban Studies in the Faculty of Environmental Studies at York University, Canada.

June Woo Kim teaches at the Centre for Advanced Studies in the Faculty of Arts and Social Sciences at the National University of Singapore.

Anthony D. King is Professor Emeritus of Sociology and Art History at the State University of New York at Binghamton, USA.

Stefan Krätke is the Chair of Economic and Social Geography at the Europa-Universität Viadrina in Frankfurt (Oder), Germany.

Henri Lefebvre (1901–1991), a French urbanist and philosopher, was one of the most influential urban theorists of the 20th century.

Ute Lehrer is Associate Professor at the Faculty of Environmental Studies at York University, Canada.

Timothy W. Luke is University Distinguished Professor in Political Science at the Virginia Polytechnic Institute and State University in Blacksburg, Virginia, USA.

Christoph Mager is Research Associate in the Department of Geography at University of Heidelberg, Germany.

Warren Magnusson is Professor Emeritus of Political Science at the University of Victoria, Canada.

Peter Marcuse is Professor Emeritus of Urban Planning in the Graduate School of Architecture at Columbia University, USA.

Doreen Massey (1944–2016) was Emeritus Professor of Geography at the Open University, UK.

Margit Mayer is Senior Fellow at the Center for Metropolitan Studies in Berlin and Professor Emerita of Political Science in the John F. Kennedy Institute of the Free University Berlin, Germany.

Cameron McAuliffe is Senior Lecturer of Human Geography at the University of West Sydney, Australia.

Eugene McCann is Professor in the Department of Geography at Simon Fraser University, Canada.

Michiel van Meeteren is a postdoctoral researcher at the Free University of Brussels, Belgium.

Eduardo Mendieta is Professor of Philosophy at Pennsylvania State University, USA.

Daniel Mullis is a PhD candidate in Human Geography at Goethe-University Frankfurt, Germany.

David Murakami-Wood is Canada Research Chair (Tier II) in Surveillance Studies at Queen's University, Canada.

Cynthia Negrey is Professor of Sociology at the University of Louisville, USA.

Jeffery L. Osgood is Principal Deputy to the Provost and Dean of the School of Interdisciplinary and Graduate Studies at Westchester University, USA.

Kathryn Pain is Professor and Research Director of Real Estate and Planning at the University of Reading, UK.

Susan Parnell is Professor of Environmental and Geographical Sciences at the University of Cape Town, South Africa.

Tino Petzold is a postdoctoral researcher in Human Geography at Goethe-University Frankfurt, Germany.

Edgar Pieterse is South African Research Chair in Urban Policy and Director of the African Center for Cities at the University of Cape Town, South Africa.

Lucas Pohl is a PhD candidate in Human Geography at Goethe-University Frankfurt, Germany.

Xuefei Ren is Associate Professor of Sociology and Global Urban Studies at Michigan State University, USA.

Jennifer Robinson is Professor of Geography at University College London, UK.

Nestor Rodriguez is Professor and Chair of Sociology and Co-Director of the Center for Immigration Research at the University of Houston, USA.

Andrew Ross is Professor of Social and Cultural Analysis at New York University, USA.

Ananya Roy is Professor of Urban Planning and Social Welfare and inaugural Director of the Institute on Inequality and Democracy at the University of California, Los Angeles, USA.

Saskia Sassen is the Robert S. Lynd Professor of Sociology at Columbia University, USA.

Sebastian Schipper is a postdoctoral researcher at the Department for Human Geography, Goethe-University Frankfurt, Germany.

Christian Schmid is Titular Professor of Sociology at the Department of Architecture, ETH Zurich, Switzerland.

Allen J. Scott is Distinguished Professor Emeritus of Public Policy and Geography at University of California, Los Angeles, USA.

Gavin Shatkin is Associate Professor of Urban Planning at Northeastern University, USA.

AbdouMaliq Simone is Research Professor at the Max Planck Institute for the Study of Religious and Ethnic Diversity, Visiting Professor of Sociology at Goldsmiths College, University of London, and Visiting Professor at the African Centre for Cities, University of Cape Town.

Leslie Sklair is Professor Emeritus at the London School of Economics, UK.

Michael Peter Smith is Distinguished Professor Emeritus of Community Studies and Development at the University of California, Davis, USA.

Neil Smith (1954–2012) was Distinguished Professor of Anthropology and Geography at the Graduate Center of the City University of New York, USA.

Richard G. Smith is Associate Professor of Geography at Swansea University, UK.

Edward Soja (1940–2015) was Distinguished Professor of Urban Planning at the University of California, Los Angeles, USA, and the London School of Economics, UK.

Peter J. Taylor is Emeritus Professor of Human Geography at Northumbria University in Newcastle upon Tyne, UK, and Emeritus Professor of Geography at Loughborough University, UK.

Michael Timberlake is Professor of Sociology at the University of Utah, USA.

Steven Tufts is Associate Professor of Geography at York University, Canada.

Rashimi Varma is Associate Professor of English and Comparative Literary Studies at Warwick University, UK.

Raf Verbruggen is a policy consultant for the Flemish Youth Council, Belgium.

John Walton is Professor Emeritus of Sociology at the University of California, Davis, USA.

Kevin Ward is Professor of Human Geography and Director of the Manchester Urban Institute at the University of Manchester, UK.

Allan Watson is Senior Lecturer in Human Geography at Staffordshire University, Stoke-on-Trent, UK.

Goetz Wolff is an independent researcher, teacher, and consultant in Los Angeles, USA.

Fulong Wu is Bartlett Professor of Planning at University College London, UK.

Sharon Zukin is Broeklundian Professor of Sociology at Brooklyn College and the Graduate Center of the City University of New York, USA.

Acknowledgments

This book is the new edition of the *Global Cities Reader*, which was originally edited in 2006 by Neil Brenner and Roger Keil. Neil's contribution to the project has been invaluable. He participated in preliminary editorial conversations and also worked with us on selecting new contributions. Neil generously allowed us to revise and use the editorial introductions from the first edition. But due to competing work demands, Neil could not participate in the later editorial process. We regret but respect Neil's decision. His intellectual contributions remain, and we are very grateful for his support throughout.

We thank Richard T. LeGates and Frederic Stout, the editors of Routledge's Urban Reader Series, and Andrew Mould at Routledge, for prodding us to reconceptualize the field of global urban studies. We believe we have put together a collection that reflects the changing contours of the field. We also benefitted from a round of reviews of our proposal, and we would like to thank the reviewers.

Likewise, teaching the *Global Cities Reader* to hundreds of undergraduate and graduate students over the years has given us ample experiences on how the *Reader* can be used in the classroom and on what themes and chapters capture the imagination of a new generation of students who have been born into a world of globalizing cities and regions. In revising this *Reader*, we have been inspired by the urban questions raised by this new generation of students.

Taylor & Francis has supported our work by dealing with the copyright issues that a book with more than sixty contributions entails. We thank them for that.

We appreciate the enthusiasm we encountered from the authors whose research is featured here. Nobody turned us down when we asked to use their work in this project. This, in our view, speaks to the growth and reach of the field we are trying to define with this book. This epistemic community also includes those contributions that did not find their way into this current edition. Their place in the pantheon of global cities research remains undisputed.

Roger Keil received funding to support this work through his York Research Chair in Global Sub/Urban Studies. He would like to thank two graduate assistants at the Faculty of Environmental Studies at York University. Daniel Taylor contacted the authors initially, helped to scan countless manuscripts under consideration, and provided other assistance. His professional style and friendly insistence were of great help in getting us started. Joyce Chan provided important formatting and editing input at the final stage of production. Xuefei Ren would like to thank the American Council of Learned Societies for its generous sabbatical support in 2016–2017 through the Frederick Burkhardt Fellowship program, and also the Newberry Library in Chicago where she spent her sabbatical year and worked on the new edition of this *Reader*.

Xuefei Ren and Roger Keil
Chicago and Toronto, March 2017

Editors' introduction

from *Global to Globalizing Cities*

It is possible you have been to Troy without recognising the city. The road from the airport is like many others in the world. It has a superhighway and is often blocked. You leave the airport buildings which are like space vessels never finished, you pass the packed carparks, the international hotels, a mile or two of barbed wire, broken fields, the last stray cattle, billboards that advertise cars and Coca-Cola, storage tanks, a cement plant, the first shanty town, several giant depots for big stores, ring-road flyovers, working class flats, a part of an ancient city wall, the old boroughs with trees, crammed shopping streets, new golden office blocks, a number of ancient domes and spires, and finally you arrive at the acropolis of wealth.

–John Berger (1990: 170)

At some point, most travelers have been on some version of the road into the imagined city of Troy, which British novelist John Berger (1926–2017) described in this opening quotation. We all seem to know this kind of metropolitan region, which appears to exist in one form or another around the globe. This metropolitan environment is now the place where most of us live. It is a global network of local places that are now increasingly bound together and interdependent. It is what most people in the world now call "home."

This book is about global and globalizing cities. When we hear the term "global city" today, it is often in the context of city rankings in the international marketplace. Government agencies, global think tanks, and news media routinely publish rankings of cities comparing various aspects of their performance. The Chicago Council of Global Affairs organizes an annual forum on global cities, a networking event aimed to put Chicago on the map of the world's leading cities. The international management consulting company A. T. Kearney publishes a global cities index and named London and San Francisco as the world's leading international cities in 2016 (*The A.T. Kearney*). The online publication *Atlantic CityLab* has its own global city rankings, and according to its latest media release, New York has replaced London as the top global city (Florida 2015). These rankings and ratings produced by mainstream media and think tanks are often based on cities' economic prowess, business climate, real estate prices, and human capital reserve– their ability to command talent or cultural influence. Tourism and entertainment underlie some rankings too, as do safety, family friendliness, and environmental sustainability. Clearly, the term global city has entered everyday vocabulary, referring to an elite club of the most powerful and recognizable cities around the world. While we acknowledge the popular notion of global cities, in this book, we will present a more critical view of what makes a city global.

This book is the second, and much revised, edition of the *Global Cities Reader*, first edited in 2006 by Neil Brenner and Roger Keil. This new edition builds on the general framework established in the original edition but departs from it in many ways. The global city was the focal point in the first edition. Originally coined by Saskia Sassen in the early 1990s, the analytical concept of the global city is rested on the assumption that the global economy and associated flows of culture, information, and migration are centered on large, powerful cities that are command and control centers of the global economy. While this view still holds true, it needs to be complemented by other perspectives of globalized urbanization.

As scholars who have lived in and continue to write about urban social change in Europe, North America, Asia, and Latin America, we believe that the lens of the "global city," which helped launch a progressive research agenda in the 1980s and 1990s, no longer captures the depth and wide range of scholarly work that is being undertaken on the question of cities and globalization. Since the publication of this book's first edition, the field of global urban studies has substantially advanced and diversified, and most contemporary urban scholarship is no longer confined to examining the command-and-control functions and socioeconomic polarizations of top-tier global cities, which had been a major focus of first-generation global cities research in the 1980s and 1990s. More recent scholarship is highly attuned to the divergent pathways of global city formation, the new centralities and marginalities that crystallize within global urban networks, emerging forms of regulation and governance, variegated sociopolitical contestations, and the multidimensional connectivities (not only economic) that link cities to one another. To better represent the diverse intellectual currents, analytical orientations, and research agendas that are emerging during this extraordinarily vibrant moment within the field of global urban studies, this book is titled *Globalizing Cities Reader*.

Over the past decade, the world has been shaken by significant transformation, and this book aims to capture the more recent developments in globalizing cities. A major recession in the United States and Western Europe in 2008, triggered by a crisis in real estate financing, disrupted cities and local economies in a largely uneven manner. The recession destabilized the previous global urban hierarchy, with one notable consequence being the rise of Asian cities that have supplanted some established North American and European cities as centers of global finance. Certain countries in the global South have emerged as economic superpowers in the past decade. Between 2006 and 2016, cities in the global South experienced mega-event fever, as the BRICS countries hosted three Olympic Games, two World Cups, and one Commonwealth Games. The high-profile mega-events attracted global spotlight to the host cities, helping to realize their global city ambitions, but also intensifying patterns of polarization and contestation in the local society. From the Middle East to Latin America, cities have become the central stage for large-scale social protests and mobilizations. In the same period, global connectivity has intensified among major urban centers in the world, largely due to advances in telecommunication technologies and social media. Hyper-densification in multiple cores, ubiquitous suburbanization, megacity expansion, and mega-regional constellations have transformed global urban landscapes at a speed and intensity that have outpaced even the predictions of boosters of the urban age (Harris and Keil 2017). The restructuring of the world economy, variegated contestations, and intensified connectivity have spawned transformations in major urban regions in the world and have become the subject of critical urban studies. Presenting both historical and current writings, this book maps the trajectory of global urban research from its initial stage in the 1980s to the present.

In this introduction, we outline some of the methodological foundations for, and major contours of investigation within, research on globalizing cities, while also alluding to several emergent debates and agendas that have animated and further differentiated this field during the last decade. In so doing, we hope to stimulate readers to contribute their own critical energies to the tasks of understanding and shaping the future dynamics and trajectories of worldwide urbanization, both within and beyond major urban regions.

URBANIZATION AND GLOBAL CAPITALISM

The notion of a world or global city—largely used interchangeably throughout this book—was consolidated as a core concept for urban studies during the 1980s, in the context of interdisciplinary attempts to decipher the crisis-induced restructuring of global capitalism following the collapse of the political-economic and spatial order in the second half of the 20th century. This work built on early uses of the term "world city" by famous 20th century urbanists such as Patrick Geddes and Peter Hall (Geddes 1924; Hall 1966). In their work, the cosmopolitan character of world cities was interpreted as an outgrowth of their host states' geopolitical power, especially during the imperial and industrial expansion of the late 19th and early 20th centuries. The possibility that urban development or the formation of urban hierarchies might be conditioned by supranational or global forces was not yet systematically explored. Until the middle of the 20th century, the dominant approaches to urban studies tended to presuppose that cities were neatly enclosed within national territories and nationalized central place hierarchies. Thus, for example, regional development theorists during that time viewed the national economy as the basic container of spatial polarization between core urban growth centers and peripheries. Similarly, urban geographers then generally assumed that the national territory was the primary scale on which rank-size urban hierarchies and city-systems were organized.

This nationalized vision of the urban process was challenged as of the late 1960s and early 1970s with the rise of radical approaches to urban political economy. The seminal contributions of neomarxist urban theorists such as Henri Lefebvre, David Harvey, and Manuel Castells generated a wealth of new categories and methods through which to analyze the specifically capitalist character of modern urbanization processes. From this perspective, contemporary cities were viewed as spatial materializations of the core social processes associated with the capitalist mode of production, including, in particular, capital accumulation and class struggle. While these new approaches did not, at that time, explicitly investigate the global parameters of urbanization, they did suggest that cities had to be understood within a macrogeographical context defined by the ongoing development and restless spatial expansion of capitalism. In this manner, radical urbanists elaborated an explicitly *spatialized* and reflexively *multiscalar* understanding of capitalist urbanization. These notions refer to a differentiated, layered, networked, hierarchical, and socially constructed space in which urban built and social environments take shape. Within this new conceptual framework, the spatial and scalar parameters for urban development could no longer be taken for granted, as if they were pre-given features of the social world. Instead, urbanization was now increasingly viewed as an active moment within the ongoing production and transformation of capitalist sociospatial configurations.

GLOBAL CITIES AND URBAN RESTRUCTURING

According to Peter J. Taylor (2004: 21), "the world city literature as a cumulative and collective enterprise begins only when the economic restructuring of the world-economy makes the idea of a mosaic of separate urban systems appear anachronistic and frankly irrelevant." During the course of the 1980s and 1990s, the latter assumption was widely abandoned among critical urban researchers, leading to a creative outpouring of research on the interplay between urban restructuring and various worldwide economic—and, subsequently, political, cultural, and environmental—transformations. Numerous scholars contributed key insights to this emergent research agenda, but the most influential, foundational statements were presented by John Friedmann (1986) and Saskia Sassen (1991). The global city concept was initially most closely associated with the work of these authors, who are appropriately cited as pioneers in exploring the interplay between globalization and urban development, particularly in the contemporary period.

During the course of the late 1980s and well into the 1990s, global city theory was employed extensively in studies of the role of major cities as global financial centers, as headquarters locations for transnational corporations (TNCs), and as agglomerations for advanced producer services industries, particularly in Euro-America and East Asia. During this time, much research was conducted on several broad issues.

The formation of a global urban hierarchy

Global city theory postulates the formation of a worldwide urban hierarchy in and through which transnational corporations coordinate their production and investment activities. The geography, composition, and evolutionary tendencies of this hierarchy have been a topic of intensive research and debate since the 1980s. Following the initial interventions of Sassen and Friedmann, subsequent scholarship has explored a variety of methodological strategies and empirical data sources through which to map this hierarchy. However, whatever their differences of interpretation, most studies of the global urban system have conceptualized this grid of cities simultaneously as a key spatial infrastructure for the accelerated and intensified globalization of capital, including finance capital, and as a medium and expression of the new patterns of global sociospatial polarization that have emerged during the post-1970s period.

The contested restructuring of urban space

The consolidation of global cities is understood, in this literature, not only with reference to the global scale, on which new, worldwide linkages among cities are being established. Just as importantly, researchers in this field have suggested that the process of global city formation also entails significant spatial transformations within cities themselves, as well as within their surrounding metropolitan regions. The globalization of urban development has led to drastic transformations in the built environment. For example, the intensified clustering of transnational corporate headquarters and advanced corporate services firms in the city core overburdens inherited land use patterns and infrastructures, leading to new, often speculative, real estate booms as new office towers and high-end residential, infrastructural, cultural, and entertainment spaces are constructed within and beyond established downtown areas. Meanwhile, the rising cost of office space in the global city core may generate massive spillover effects on a regional scale, as small- and medium-sized agglomerations of corporate services and back offices crystallize throughout the urban region. The consolidation of such headquarter economies may also lead to significant shifts within local housing markets as developers attempt to transform once-devalorized properties into residential space for corporate elites and other members of the "creative" professional milieux. Consequently, gentrification ensues in formerly working-class neighborhoods and deindustrialized spaces, and considerable residential and employment displacement may be caused in the wake of rising rents and housing costs (see Ch. 35 by Smith). Global cities researchers have tracked these and many other spatial transformations at some length. The urban built environment is viewed as an arena of contestation in which competing social forces and interests struggle over issues of urban design, land use, and public space.

The transformation of the urban social fabric

One of the most provocative, if also highly controversial, aspects of global cities research during its initial phase involved claims regarding the effects of global city formation upon the urban social fabric. Friedmann (1986) and Sassen (1991), in particular, suggested that the emergence of a global city hierarchy would generate a "dualized" urban labor market structure dominated, on the one hand, by a high-earning corporate elite and, on the other hand, by a large mass of workers employed in menial, low-paying, and/or informalized jobs (see also Ross and Trachte 1983). For Sassen, this "new class alignment in global cities" emerged in direct conjunction with the downgrading of traditional manufacturing industries and the emergence of the advanced producer services complex (Sassen 1991: 13). Her work on London, New York, and Tokyo suggested that broadly analogous, if place-specific, patterns of social polarization were emerging in these otherwise quite different cities, as a direct consequence of their newly consolidated roles as global command and control centers. This "polarization thesis" has attracted considerable discussion and debate. Whereas some scholars have attempted to apply the arguments by Friedmann and Sassen to a range of globalizing cities, other analysts have questioned their logical and empirical validity (see Marcuse Ch. 10).

As global city research consolidated around the above-mentioned themes, new generations of critical urban scholars began to extend the empirical scope of the theory beyond the major urban command and control centers of the world economy—that is, cities such as New York, London, Tokyo; as well as various supraregional centers in East Asia (Singapore, Seoul, Hong Kong, Shanghai), North America (Los Angeles, Chicago, Miami, Toronto), and western Europe (Paris, Frankfurt, Amsterdam, Zurich, Milan). In this important line of research, the basic methodological impulses of global city theory were applied to diverse types of cities around the world, but particularly in the global North, that were likewise undergoing accelerated processes of economic and sociospatial restructuring (Smith and Feagin 1987). Here, the central analytical agenda was to relate the dominant socioeconomic trends within particular cities—for instance, industrial restructuring, changing patterns of capital investment, labor-market segmentation, sociospatial polarization, and class and ethnic conflict—to the emergence of a worldwide urban hierarchy and the geoeconomic and geopolitical forces that underlie it. In this manner, analysts demonstrated the usefulness of global city theory in relation to a broad range of urban transformations—now also including urban governance restructuring and the emergence of new forms of urban social protest—that were unfolding during the post-1980s period. They thus signaled a significant reorientation of the literature away from "global cities" as such, to what Peter Marcuse and Ronald van Kempen famously labeled "globalizing cities," a term intended to underscore the diversity of pathways and the place- and region-specific patterns in and through which processes of globalization and urban restructuring were being articulated (Marcuse and van Kempen 2000).

NEW FEATURES

The parameters of global city theory have been expanded considerably since the 1980s. On the one hand, a broader range of cases of globalized urban development is now being explored. At the same time, new theoretical categories and methodological strategies have been introduced in order to examine the interplay between global, national, regional, and local dynamics within rapidly changing urban spaces. This volume contains critical essays that are intended to represent this vast, multifaceted, and still evolving literature. It would be impossible in a single volume to provide a full, systematic survey of this sprawling research field. We have selected articles from a much larger set of relevant texts, and there are, no doubt, some omissions and lacunae in our presentation.

We have organized this book into seven thematic sections, each of which surveys a key set of issues in the literature on globalization and urban social change.

- Part one, "Foundations," introduces the themes of the global city literature and includes a number of classic texts on global city formation.
- Part two, "Pathways," examines place-specific trajectories of urban restructuring and global city formation in North America, West Europe, Asia, and Latin America.
- Part three, "Relations," examines emerging global urban networks, not only in economic sectors but also in telecommunication, infrastructure, logistics, security, and epidemiology.
- Part four, "Regulations," explores the political and institutional dimensions of urban restructuring in globalizing cities.
- Part five, "Contestations," examines various forms of sociopolitical mobilization emerging in globalizing cities in China, Brazil, Africa, and the Middle East, and also urban social movements in the west.
- Part six, "Culture," examines cultural strategies for global city formation, cultural networks of creative industries, and cultural politics of representation.
- Part seven, "Frontiers," identifies new directions for global urban studies, emphasizing relational and comparative approaches.

Among the 68 essays included in this second edition, 36 are new selections, and the majority were published after 2010. The essays investigate key themes in contemporary urban research, such as polarization, institutional restructuring, governance, political ecology, sociopolitical mobilization, and representation. In addition to including these more recent publications, this edition also devotes more attention to the dynamics of globalizing cities in the global South. Today more and more urban researchers come from the global South, and research on global South cities is making significant contributions to urban theory building. The book includes recently published essays on a variety of global South cities, such as Shanghai, Delhi, Medellín, and Bogotá (Part 2), various cities in Africa (Part 4), and Beijing, Guangzhou, Johannesburg, Mumbai, São Paulo, and Rio de Janeiro (Parts 5, 6, and 7.).

This book presents carefully edited versions of longer research articles—typically only one-third of the original material. Due to strict length requirements, and in the interest of practicality and accessibility, we were obliged to cut out significant sections of each reading, including (a) discussions of ongoing debates in world city theory and other relevant academic literatures; (b) various types of historical, contextual, and empirical background material; (c) most footnotes and endnotes; (d) a variety of maps, data tables, and diagrams; and (e) significant numbers of bibliographic references. Readers are strongly encouraged to consult the original, full-length publications. We feel reasonably confident, however, that the edited versions published here convey the most central arguments and insights from the original texts.

AGENDA FOR RESEARCH AND ACTION

Research on globalizing cities is intrinsically political. Like all forms of knowledge, research on globalizing cities can be used by some groups to endorse capitalist urbanization, or it can be mobilized by others for diametrically opposed, critical purposes. One of the more persistent criticisms leveled at global city researchers is that their work serves to glorify the status of particular cities in worldwide interurban competition, and thus represents an uncritical affirmation of global neoliberalism. In our view, the misunderstanding that underlies these criticisms is based on a mistaken identification of the colloquial notion of the global city with the scholarly concept developed in the literatures we have reviewed in this volume. While the former is a descriptive, affirmative notion often used by municipal power brokers to draw attention to specific places, the latter is a polysemic analytical term employed by critical urbanists to decipher the globalizing dimensions of contemporary urbanization.

Finally, how can we best take advantage of the urban knowledge produced by global city researchers? Here it is crucial to recall that Friedmann and Wolff's (1982) first foray into global cities research contained the programmatic subtitle "an agenda for research and *action*" [our emphasis]. For Friedmann and many of his colleagues, the analysis and description of the global city was meant to be a first step in actively effecting positive, progressive, and even radical social change. Thus, knowledge on the formation of global urban hierarchies and on the intensification of sociospatial polarization within global cities were clearly understood as a call to arms for progressive planners and urban scholars. Their role, in Friedmann's view, was to mobilize new public policies designed to reduce the suffering of the increasingly impoverished working classes and migrant populations in global cities and, more ambitiously still, to subject the apparently "deterritorialized" operations of transnational capital to localized, democratic political control. From our perspective, critical research on globalizing cities can offer us firm intellectual grounding as we navigate in the increasingly complex and contested urban century.

REFERENCES

Berger, J. (1990) *Lilac and Flag*, New York: Vintage.

Florida, R. (March 3, 2015) "Sorry, London, New York is the world's most economically powerful city," www.citylab.com/work/2015/03/sorry-london-new-york-is-the-worlds-most-economically-powerful-city/386315/, accessed on January 5, 2017.

Friedmann, J. (1986) The world city hypothesis, *Development and Change*, 17, 1: 69–83.

Friedmann, J. and G. Wolff (1982) World city formation: An agenda for research and action, *International Journal of Urban and Regional Research*, 6, 3: 309–44.

Geddes, P. (1924) A world league of cities, *Sociological Review*, 26: 166–67.

Hall, P. (1966) *The World Cities*, New York: McGraw-Hill.

Harris, R. and R. Keil (2017) Globalizing cities and suburbs, in A. Bain and L. Peake (eds.) *Urbanization in a Global Context*, Oxford: Oxford University Press.

Marcuse, P. and R. van Kempen (eds.) (2000) *Globalizing Cities: A New Spatial Order?* Oxford: Blackwell.

Ross, R. and K. Trachte (1983) Global cities and global classes: The peripheralization of labour in New York City, *Review*, 6, 3: 393–431.

Sassen, S. (1991) *The Global City*, Princeton, NJ: Princeton University Press.

Smith, M. P and J. Feagin (eds.) (1987) *The Capitalist City*, Cambridge, MA: Blackwell.

Taylor, P. J. (2004) *World City Network: A Global Urban Analysis*, New York: Routledge.

The A.T. Kearney Global Cities Index and Global Cities Outlook, www.atkearney.com/research-studies/global-cities-index, accessed on January 5, 2017.

PART ONE

Foundations

Plate 1 Times Square, New York City

Source: Roger Keil

INTRODUCTION TO PART ONE

While the notion of *global* or *world* cities in its contemporary usage may be new, the idea that cities are of world-historical importance—economically, militarily, politically, and culturally—has been around for some time. Cities played fundamental geostrategic roles and had long-distance, networked relationships prior to the consolidation of the modern interstate system. Whether in Mesopotamia or Egypt, the Indus Valley or in the Far East, wherever the first urban cultures appeared, their settlements were the core of territorial or maritime empires (Soja 2000). Athens and Rome are perhaps the most pervasively cited urban cores of two major world empires.

In the Middle Ages, through trade networks such as the German Hansa, cities once again came to serve as the spatial infrastructure of emerging continental and, eventually, during the early modern period, global economies (see Ch. 2 by Braudel). Byzantium, which took the mantle from Rome and remained the important buckle in a belt that tied together Occident and Orient in the Middle Ages, was certainly a type of global city, even by today's standards. But there are also many examples of smaller, less well-known cities, which fulfilled global city functions, in particular financial control, and which are now little more than regional centers and tourist destinations. One such place is Augsburg, in southern Germany, a city of impressive wealth in the late Middle Ages, when the Fugger family financed the global enterprises of the Hansa and other commercial, mining, and manufacturing projects. Located at the northern foothills of the Alps, it was the ideal connector of the Mediterranean, eastern European, North Sea, and western European economies. Another such place was the legendary Cahors in the French southwest, which is cited, alongside Sodom and Gomorrah, as a model for Hell in Dante's *Divine Comedy* (1321). At the time of Cahors' greatest power during its Golden Age in the 13th century, local and Lombard bankers transformed the city into the chief banking center of Europe and earned a reputation as usurers—a characteristic of the town that was subsequently noted by the great Italian poet.

The developmental trajectories of major cities and inter-city networks have been linked to the emergence and decline of precapitalist imperial systems and, subsequently, to the expansion of capitalism on a world scale (Abu-Lughod 1989; Chase-Dunn, Manning, and Hall 2000). In each period, the identity and character of a global city is tied into the dominant mode of production: globality is defined by the scale of the world system—the large-scale framework of material, political, and cultural life—in which that city is embedded. Cities are in turn connected in diverse, long-distance relationships that are designed to maintain the world system as a whole. However, since the mid-1970s we have witnessed the consolidation of a truly worldwide urban hierarchy that has significantly expanded the scale of major cities' command and control functions within the capitalist world system as a whole.

Peter Hall (1966: 7) attributed the term "world city" to a book by Scottish urbanist Patrick Geddes, *Cities in Evolution* (1915), which emphasized world cities' centralized economic functions. In Hall's view, world cities were sites of intensive population growth, centralized political power, and major commercial, financial, and transportation functions. These critical functions were treated not only as attributes but also as capabilities that connected world cities to one another. In Hall's view, relational properties were defining features of world cities. Crucially, however, Hall conceived world cities primarily as *national* centers that channeled international forces and influences towards national interests. Hall's conception of a world city is thus arguably a product of a period in which cities operated primarily as nodes within national urban systems. By contrast, contemporary notions of the world city emphasize the embeddedness of urban centers within an emergent system of global capitalism; this may entail their partial delinking from the territorialized economic spaces regulated by national state institutions (Ross and Trachte 1990; Sassen 1991).

The world city literature can be viewed as an implicit critique of what Ben Derudder has called "the zonal implementation of core-periphery models" (2003: 100), which were developed by world system theorists such as Wallerstein (1974) to characterize the polarization of global capitalism among core, semi-peripheral, and peripheral zones. These models are generally grounded upon territorialist assumptions in which economic space is conceived as being composed of clearly delineated, bounded geographical containers. However, by directing attention to inter-urban connections and interdependencies, which generally crosscut territorial borders in complex networked relationships, world city theorists have suggested an alternative conceptualization of capitalism's underlying economic geographies, in which economic territoriality represents only one among various possible forms of sociospatial organization (for a related perspective, see Arrighi 1995). How best to map the emergent global urban system in relation to the landscape of global capitalism remains one of the most controversial and fascinating questions within the entire field of global cities research, and it is likely to continue to stimulate energetic theoretical debate and empirical analysis in the years to come.

In terms of periodizing the emergence and development of global cities, there are basically two approaches—one which emphasizes the long-term role of cities as basing points for global economic flows; and one which emphasizes the historical specificity of contemporary patterns of global city formation. While we believe these approaches can be compatible, they have, in fact, led to quite divergent research agendas.

On the one hand, some urbanists have insisted that global cities are an age-old phenomenon. This position is strongly articulated by world city researchers such as Janet Abu-Lughod (Ch. 8), John Walton (Ch. 6), Michael Timberlake (Ch. 7), and Christopher Chase-Dunn (1985). These researchers have systematically examined the long-term structural and historical background for world city formation and have argued that cities have long served as nodal points within large-scale economic systems both prior to and throughout the history of capitalist industrialization. An important aspect of this historically based work has been an emphasis on "urban specialization," a notion that was reformulated in the 1980s to describe the development of urban centers in global systems of cities. The contribution by Rodriguez and Feagin (Ch. 5) illustrates the powerful explanatory capacity of such an approach when it is applied to the dynamics of global city formation in successive stages of capitalist development.

The second, alternative approach to the periodization of global city formation emphasizes the uniqueness of contemporary global cities due to their role as basing points for a qualitatively new formation of globalizing capitalism. Scholars who have worked in this research tradition have linked the emergence of a globalized city system to the specific forms of worldwide capitalist restructuring that began to unfold as of the 1970s. This strand of research emerged in response to two intertwined transformations—first, the end of American-Fordist-Keynesian hegemony, which entailed the crisis of the postwar framework of accumulation and state regulation across the North Atlantic zone; and second, the development of a new international division of labor in the 1970s, which entailed the increasing industrialization of formerly peripheralized states and, concomitantly, intensive processes of industrial restructuring in the former heartlands of global capitalism (Fröbel et al. 1980).

Accordingly, this second approach interprets world city formation as a key spatial expression of the new forms of capital accumulation that have been consolidated since the 1970s (Keil 1993). This means that, while specific global cities emerge as the command and control centers of the new world economy (Sassen 1991), other cities are likewise subject to closely analogous, globally induced forms of political-economic and spatial restructuring. Thus, cities that are not global command and control centers may nonetheless be transformed through, for example, globalized patterns of consumption, cultural politics, and economic restructuring. Such spaces may be most appropriately characterized, according to Marcuse and van Kempen (2000), as "globalizing cities."

This section provides a broad survey of key contributions to both of the aforementioned strands of global cities research. The section begins with a selection from Peter Hall's classic work on world cities, and then presents two seminal works by Friedmann and Wolff (Ch. 3) and Sassen (Ch. 4), who laid the foundation of the global city research. It also includes contributions that examine the role of global cities in various historical phases of capitalist development (Ch. 5 by Rodriguez and Feagin; Ch. 8 by Abu-Lughod), and chapters that situate the emergent global city research in the 1980s within the world-system perspective (Ch. 6 by Walton; Ch. 7 by Timberlake). Each of these chapters emphasizes the continuities between contemporary global cities and various types of global urban centers during the history of capitalism. We then turn to two contributions that problematize studies focusing on global city formation in developed countries. Robinson (Ch. 9) urges us to study cities in the global South by examining how global economic restructuring has shaped these cities that are "off the map". Since the early 2000s, cities in the global South have attracted much research attention, and the large scholarly output on these globalizing cities has significantly enriched our understanding of global urban transformations outside North America and west Europe. Marcuse (Ch. 10) questions the assumption of a single model of urban sociospatial organization resulting from global economic restructuring, and on this basis presents differential patterns of sociospatial fragmentation in globalizing cities as a promising avenue of research. Together, these two contributions complement earlier works on global city formation by identifying new avenues of investigation for studies of the interplay between globalization and urban restructuring.

REFERENCES AND SUGGESTIONS FOR FURTHER READING

Abu-Lughod, J. (1989) *Before European Hegemony: The World System AD 1250–1350*, New York: Oxford University Press.

Arrighi, G. (1995) *The Long Twentieth Century*, London: Verso.

Benevolo, L. (1984) *Die Geschichte der Stadt*, Frankfurt and New York: Campus.

Chase-Dunn, C. (1985) The system of cities, AD 800–1975. In M. Timberlake (ed) *Urbanization in the World-Economy*, New York: Academic Press, 269–92.

Chase-Dunn, C., Manning, S., and Hall, T. D. (2000) Rise and fall: East-West synchronicity and indic exceptionalism reexamined, *Social Science History*, 24, 4: 727–54.

Derudder, B. (2003) Beyond the state: Mapping the semi-periphery through urban networks, *Capitalism, Nature, Socialism*, 14, 4: 91–120.

Fröbel, F., Heinrichs, J., and Kreye, O. (1980) *The New International Division of Labor*, New York: Cambridge University Press.

Geddes, P. (1915) *Cities in Evolution: An Introduction to the Town Planning Movement and the Study of Civics*, London: Ernest Benn Limited.

Geddes, P. (1924) A world league of cities, *Sociological Review*, 26: 166–7.

Hall, P. G. (1966) *The World Cities*, New York: McGraw-Hill.

Hymer, S. (1979) The multinational corporation and the international division of labor. In R. B. Cohen et al. (ed), *The Multinational Corporations: A Radical Approach*, Cambridge: Cambridge University Press, 140–164.

Keil, R. (1993) *Weltstadt – Stadt der Welt*, Münster: Westfalisches Dampfboot.

Keil, R. (1998) *Los Angeles: Globalization, Urbanization and Social Struggles*, Chichester: John Wiley and Sons.

Marcuse, P. and van Kempen, R. (eds) (2000) *Globalizing Cities: A New Spatial Order?* Oxford: Blackwell.

Ross, R. and Trachte, K. (1990) *Global Capitalism: The New Leviathan*, Albany: State University of New York Press.

Sassen, S. (1991) *The Global City*, Princeton, NJ: Princeton University Press.

Soja, E. W. (1989) *Postmodern Geographies*, London: Verso.

Soja, E. W. (2000) *Postmetropolis: Critical Studies of Cities and Regions*, Oxford and Malden, MA: Blackwell Publishers.

Taylor, P. J. (2004) *World City Network: A Global Urban Analysis*, London and New York: Routledge.

Wallerstein, I. (1974) *The Modern World-System I*, New York: Academic Publishers.

1 Prologue
"The metropolitan explosion"

from *The World Cities* (1966)

Peter Hall

There are certain great cities in which a quite disproportionate part of the world's most important business is conducted. In 1915 the pioneer thinker and writer on city and regional planning, Patrick Geddes, christened them "the world cities." This book is about their growth and problems. By what characteristics do we distinguish the world cities from other great centers of population and wealth? In the first place, they are usually major centers of political power. They are the seats of the most powerful national governments and sometimes of international authorities too, of government agencies of all kinds. Round these gather a host of institutions, whose main business is with government; the big professional organizations, the trade unions, the employers' federations, the headquarters of major industrial concerns.

These cities are the national centers not merely of government but also of trade. Characteristically they are the great ports, which distribute imported goods to all parts of their countries, and in return receive goods for export to the other nations of the world. Within each country, roads and railways focus on the metropolitan city. The world cities are the sites of the great international airports: Heathrow, Charles de Gaulle, Schiphol, Sheremetyevo, Kennedy, Benito Juarez, Kai Tak. Traditionally, the world cities are the leading banking and finance centers of the countries in which they stand. Here are housed the central banks, the headquarters of the trading banks, the offices of the big insurance organizations and a whole series of specialized financial and insurance agencies.

Government and trade were invariably the original raisons d'être of the world cities. But these places early became the centers where professional talents of all kinds congregated. Each of the world cities has its great hospitals, its distinct medical quarter, its legal profession gathered around the national courts of justice. Students and teachers are drawn to the world cities: they commonly contain great universities, as well as a host of specialized institutions for teaching and research in the sciences, the technologies and the arts. The great national libraries and museums are here. Inevitably, the world cities have become the places where information is gathered and disseminated: the book publishers are found here; so are the publishers of newspapers and periodicals, and with them their journalists and regular contributors. In this century also the world cities have naturally become headquarters of the great national radio and television networks.

But not only are the world cities great centers of population: their populations, as a rule, contain a significant proportion of the richest members of the community. That early led to the development of luxury industries and shops; and in a more affluent age these have been joined by new types of more democratic trading: by the great department stores and the host of specialized shops which cater for every demand. Around them, too, the range of industry has widened: for the products of the traditional luxury trades, forged by craftsmen in the world cities, have become articles of popular consumption, and their manufacture now takes place on the assembly lines of vast factories in the suburbs of the world cities.

As manufacture and trade have come to cater for a wider market so has another of the staple businesses of the world cities – the provision of entertainment. The traditional opera houses and theatres and concert halls and luxurious restaurants, once the preserve of the aristocracy and the great merchant, are now open to a wider audience, who can increasingly pay their price. They have been joined by new and more popular forms of entertainment – the variety theatre and revue, the cinema, the nightclub and a whole gamut of eating and drinking places.

The staple trades of the world cities go, with few exceptions, from strength to strength. Here and there, a trade may wither and decay: thus shoemaking in nineteenth-century London, diamond-cutting in twentieth-century Amsterdam, shirt-making in twentieth-century New York. In the long view, even the world cities may themselves decline. Where now is Bruges – a world city of late medieval Europe? But so far in history, such cases are conspicuous by their rarity. Nothing is more notable about the world cities, taking the long historic view, than their continued economic strength. Not for them the fate of depressed regions which see their staple products decline: regions like the coalfields of Northumberland–Durham in Great Britain or Pennsylvania–West Virginia in the United States, or remote rural regions like the Massif Central of France or the south-east uplands of the Federal Republic of Germany. True, one disquieting note is that, during the 1970s, some great world city regions – London, New York – for the first time recorded declines in population, while in others – Paris, Tokyo – the rate of growth notably slowed. But this should be seen, in large measure, as the continuation of a long process of economic adaptation and of outward deconcentration; the statistical trends suggest that the official definitions of these city regions, big as they are, are no longer big enough.

As the economies of the advanced nations become steadily more sophisticated, and as those of the newly industrializing nations strain to catch them up, so in all world cities does the economic emphasis shift to those industries and trades most aptly carried on in the metropolis: industries and trades dependent on skill, on design, on fashion, on conduct with the specialized needs of the buyer. Associated with these trends, white-collar jobs grow faster than blue-collar ones; for every producer of factory goods, more and more people are needed at office desks to achieve good design, to finance and plan production, to sell the goods, to promote efficient nation-wide and worldwide distribution. So it is not surprising that, as they gain such new jobs, the world cities shed those activities that can be as readily performed elsewhere – mass production of standardized goods, space-consuming docking and warehousing, routine paper-processing in factory-like offices: such processes of economic invasion and succession are no novel event for the world cities.

2
"Divisions of space and time in Europe"

from *The Perspective of the World* (1984)

Fernand Braudel

EDITORS' INTRODUCTION

In this chapter, Fernand Braudel (1902–1985), one of the most celebrated French historians of the postwar years, depicts the large-scale processes that constitute the modern capitalist urban system. Braudel integrated economics and geography into his studies of global history and changed the way history has subsequently been written. As is apparent in his great works, *The Mediterranean and Mediterranean World in the Age of Phillip II* (1972), *Capitalism and Material Life, 1400–1800* (1973), and the three-volume *Civilization and Capitalism, 15th–18th Century* (1979), Braudel focused on the social and economic agency of people in relation to large-scale, long-term socioeconomic trends. While Braudel was most explicitly concerned to grasp the geographical dimensions of large-scale economic systems—as evidenced, for instance, in his key distinction between *world economy* and *world-economy* and in his emphasis on the polarization of economic development under capitalism—his analysis below provides a prescient account of the interplay between major cities, economic power and systems of imperial rule. Braudel emphasizes that relatively autonomous world-economies may co-exist and interact with one another. These economic systems have clear boundaries, a dominant city at the center, and also a group of cities in competition with one another. In addition, Braudel also highlights the spatial hierarchy within world-economies, which were often divided into a core, a middle zone and a vast periphery. Through his argument that a major city exists at the center of each world-economy, Braudel provides an early blueprint for the notion of the global city itself. Just as importantly, Braudel's conceptualization of spatial hierarchy anticipates the diverse types of world city models that we will encounter in subsequent chapters.

WORLD ECONOMIES

To open the discussion, I should elucidate two expressions which might lead to confusion: *the world economy* and *a world-economy*. *The world economy* is an expression applied to the whole world. It corresponds, as Sismondi (1991 [1951]) puts it, to "the market of the universe," to "the human race, or that part of the human race which is engaged in trade, and which today in a sense makes up a single market." *A*

world-economy (an expression which I have used in the past as a particular meaning of the German term *Weltwirtschaft*) only concerns a fragment of the world, an economically autonomous section of the planet able to provide for most of its own needs, a section to which its internal links and exchanges give a certain organic unity. There have been world-economies if not always, at least for a very long time – just as there have been societies, civilizations, states and even empires. If we take giant steps back through history, we could say of ancient Phoenicia that it was an early version of a world-economy, surrounded by great empires. So too was Carthage in its heyday; or the Hellenic world; or even Rome; so too was Islam after its lightning triumphs. In the ninth century, the Norman venture on the outer margins of western Europe laid down the lines of a short-lived and fragile world-economy which others would inherit. From the eleventh century, Europe began developing what was to be its first world-economy, afterwards succeeded by others down to the present day. Muscovy, connected to the East, India, China, Central Asia and Siberia, was another self-contained world-economy, at least until the eighteenth century. So was China, which from earliest times took over and harnessed to her own destiny such neighboring areas as Korea, Japan, the East Indies, Vietnam, Yunnan, Tibet and Mongolia – a garland of dependent countries. Even before this, India had turned the Indian Ocean into a sort of private sea, from the east coast of Africa to the islands of the East Indies.

Might it not in short be said that here was a process of constant renewal as each configuration gave way almost spontaneously to another, leaving plentiful traces behind – even in a case, at first sight unpromising, like the Roman Empire? The Roman economy did in fact extend beyond the imperial frontier running along the prosperous line between Rhine and Danube, or eastwards to the Red Sea and the Indian Ocean. According to Pliny the Elder, Rome had a deficit of 100 million sesterces in its trade with the Far East every year. And ancient Roman coins are still being dug up in India today.

SOME GROUND RULES

The past offers us a series of examples of world-economies then – not very many but enough to make some comparisons possible. Moreover since each world-economy lasted a very long time, it changed and developed within its own boundaries, so that its successive ages and different states also suggest some comparisons. The data available is thus sufficiently plentiful to allow us to construct a *typology* of world-economies and at the very least to formulate a set of rules or tendencies which will clarify and even define their relations with geographical space.

Our first concern, in seeking to explain any world-economy, is to identify the area it occupies. Its boundaries are usually easy to discover since they are slow

to change. The zone it covers is effectively the first condition of its existence. There is no such thing as a world-economy without its own area, one that is significant in several respects:

- it has boundaries, and the line that defines it gives it an identity, just as coastlines do a sea;
- it invariably has a center, with a city and an already-dominant type of *capitalism*, whatever form this takes. A profusion of such centers represents either immaturity or on the contrary some kind of decline or mutation. In the face of pressures both internal and external, there may be shifts of the center of gravity: cities with international destinies;
- world-cities – are in perpetual rivalry with one another and may take each other's place;
- it is marked by a hierarchy: the area is always a sum of individual economies, some poor, some modest, with a comparatively rich one in the center. As a result, there are inequalities, differences of voltage which make possible the functioning of the whole. Hence that "international division of labor," of which as P.M. Sweezy (1974) points out, Marx did not foresee that it "might harden into a pattern of development and under-development which would split mankind into haves and have-nots on a scale far wider and deeper than the bourgeois-proletarian split in the advanced countries themselves." All the same, this is not in fact a "new" division, but an ancient and no doubt an incurable divide, one that existed long before Marx's time. So there are three sets of conditions, each with general implications.

RULE ONE: THE BOUNDARIES CHANGE ONLY SLOWLY

The limits on one world-economy can be thought of as lying where those of another similar one begin: they mark a line, or rather a zone which it is only worth crossing, economically speaking, *in exceptional circumstances*. For the bulk of traffic in either direction, "the loss in exchange would outweigh the gain." So *as a general rule*, the frontiers of a world-economy are quiet zones, the scene of little activity. They are like thick shells, hard to penetrate; they are often natural barriers, no-man's lands – or no-man's-seas. The Sahara, despite its caravans, would have been one such, separating Black Africa from White Africa. The Atlantic was another, an empty expanse to the south and west of Africa, and for long centuries a barrier compared to the Indian Ocean, which was from early days the scene of much trade, at least in the north. Equally formidable was the Pacific, which European explorers had only half-opened to traffic: Magellan's voyage only unlocked one way into the southern seas, not a gateway for return journeys. To get back to Europe, the expedition had to take the Portuguese route round the Cape of Good Hope. Even the first voyages of the Manila galleon in 1572 did not really overcome the awe-inspiring obstacle posed by the South Sea.

RULE TWO: A DOMINANT CAPITALIST CITY ALWAYS LIES AT THE CENTER

A world-economy always has an urban center of gravity, a city, as the logistic heart of its activity. News, merchandise, capital, credit, people, instructions, correspondence all flow into and out of the city. Its powerful merchants lay down the law, sometimes becoming extraordinarily wealthy. At varying and respectful distances around the center will be found other towns, sometimes playing the role of associate or accomplice, but more usually resigned to their second-class role. Their activities are governed by those of the metropolis: they stand guard around it, direct the flow of business toward it, redistribute or pass on the goods it sends them, live off its credit or suffer its rule. Venice was never isolated; nor was Antwerp; nor, later, was Amsterdam. These metropolises came accompanied by a train of subordinates.

Any town of any importance, particularly if it was a seaport, was a "Noah's Ark," "a fair of masks," a "Tower of Babel," as President de Brosses described Livorno. How much more so were the real metropolises! They were the scene of fantastic mixtures, whether London or Istanbul, Isfahan or Malacca, Surat or Calcutta (the latter from the time of its very earliest successes). Under the pillars of the Amsterdam Bourse – which was a microcosm of the world of trade – one could hear every dialect in the world. In Venice, "if you are curious to see men from every part of the earth, each dressed in his own different way, go to St Mark's Square or the Rialto and you will find all manner of persons."

This colorful cosmopolitan population had to coexist and work in peace. The rule in Noah's Ark was live and let live. Of the Venetian state, Villamont thought in 1590 "that there is nowhere in all Italy where one may live in greater liberty ... firstly because the Signoria rarely condemns a man to death, secondly arms are not forbidden, thirdly there is no inquisition in matters of faith, lastly everyone lives as he pleases in freedom of conscience, which is the reason why several libertine Frenchmen reside there so as not to be pursued and controlled and so as to live wholly without constraint." I imagine that Venice's innate toleration helps to explain her "notorious anticlericalism" or as I would prefer to call it her vigilant opposition to Roman intransigence. But the miracle of toleration was to be found wherever the community of trade convened. Amsterdam kept open house, not without some merit after the religious violence between the Arminians and the Gomarists (1619–1620). In London, every religion under the sun was practiced. "There are," said a French visitor in 1725, "Jews, Protestants from Germany, Holland, Sweden, Denmark and France; Lutherans, Anabaptists, millenarians, Brownists, independents or Puritans; and Tremblers or Quakers." To these might be added the Anglicans, Presbyterians and the Catholics who, whether English or not, were in the habit of attending mass in the chapels of the French, Spanish or Portuguese embassies. Each sect or faith had its own churches and meeting-places. And each one was identifiable to the outside world. "The Quakers can be recognized a mile off by their dress: a flat hat, a small cravat, a coat buttoned up to the neck and their eyes shut most of the time."

Perhaps the most distinctive characteristic of all of these super-cities was their precocious and pronounced social diversification. They all had a proletariat, a bourgeoisie, and a patriciate, the latter controlling all wealth and power and so self-confident that before long it did not even bother, as it had in Venice or Genoa in the old days, to take the title of *nobili*. Patriciate and proletariat indeed grew further apart, as the rich became richer and the poor even poorer, since the besetting sin of these pulsating capitalist cities was their high cost of living, not to mention the constant inflation resulting from the intrinsic nature of the higher urban functions whose destiny it was to dominate adjacent economies.

Dominant cities did *not* dominate forever; they replaced each other. This was as true at the summit as it was at every level of the urban hierarchy. Such shifts, wherever they occurred (at the top or half-way down), whatever their causes (economic or otherwise) are always significant; they interrupt the calm flow of history and open up perspectives that are the more precious for being so rare. When Amsterdam replaced

Antwerp, when London took over from Amsterdam, or when in about 1929, New York overtook London, it always meant a massive historical shift of forces, revealing the precariousness of the previous equilibrium and the strengths of the one which was replacing it. The whole circle of the world-economy was affected by such changes and the repercussions were never exclusively economic.

The reference to dominant cities should not lead us to think that the successes and strengths of these urban centers were always of the same type: in the course of their history, these cities were sometimes better or worse equipped for their task, and their differences or comparative failings, when looked at closely, oblige one to make some fairly fine distinctions of interpretation.

If we take the classic sequence of dominant cities of western Europe – Venice, Antwerp, Genoa, Amsterdam, London – which we shall presently be considering at length – it will be observed that the three first-named did not possess the complete arsenal of economic domination. Venice at the end of the fourteenth century was a booming merchant city; but possessed no more than the beginnings of an industrial sector; and while she did have financial and banking institutions, this credit system operated inside the Venetian economy, as an internal mechanism only. Antwerp, which possessed very little shipping of her own, provided a haven for Europe's merchant capitalism: operating as a sort of bring and buy center for trade and business, to which everything came from outside. When Genoa's turn came, it was really only because of her banking supremacy, similar to that of Florence in the thirteenth and fourteenth centuries; if she played a leading role, it was firstly because her chief customer was the king of Spain, controller of the flow of bullion, and secondly because no one was quite sure where the center of gravity really lay between the sixteenth and seventeenth centuries: Antwerp fulfilled this role no longer and Amsterdam was not yet ready: the Genoese supremacy was no more than an interlude. By the time Amsterdam and London took the stage, the world-cities possessed the whole panoply of means of economic power: they controlled everything, from shipping to commercial and industrial expansion, as well as the whole range of credit.

Another factor which could vary from one dominant city to another was the machinery of political power. From this point of view, Venice had been a strong and independent state: early in the fifteenth century, she had taken over the *Terraferma*, a large

protective zone close at hand; since 1204 she had possessed a colonial empire. Antwerp by contrast had virtually no political power at her disposal. Genoa was a mere territorial skeleton: she had given up all claim to political independence, staking everything on that alternative form of domination, money. Amsterdam laid claim in some sense to the United Provinces, whether they agreed or not. But her "kingdom" represented little more than the *Terraferma* of Venice. With London, we move into a completely different context: this great city had at its command the English national market and later that of the entire British Isles, until the day when the world changed and this mighty combination dwindled to the dimensions of a minor power when compared to a giant like the United States (see Figures 1 and 2).

In short, the outline of the history of these successive dominant cities in Europe since the fourteenth century provides the clue to the development of their underlying world-economies: these might be more or less firmly controlled, as they oscillated between strong and weak centers of gravity. This sequence also incidentally tells us something about the variable value of the weapons of domination: shipping, trade, industry, credit and political power or violence.

RULE THREE: THERE IS ALWAYS A HIERARCHY OF ZONES WITHIN A WORLD-ECONOMY

The different zones within a world-economy all face towards one point in the center: thus "polarized," they combine to form a whole with many relationships. And once such connections were established, they lasted. At ground level and sea level so to speak, the networks of local and regional markets were built up over century after century. It was the destiny of this local economy, with its self-contained routines, to be from time to time absorbed and made part of a "rational" order in the interest of a dominant city or zone, for perhaps one or two centuries, until another "organizing center" emerged; as if the centralization and concentration of wealth and resources necessarily favored certain chosen sites of accumulation.

Every world-economy is a sort of jigsaw puzzle, a juxtaposition of zones interconnected, but *at different levels*. On the ground, *at least* three different areas or categories can be distinguished: a narrow *core*, a fairly developed middle zone and a vast *periphery*. The qualities and characteristics of the type of society, economy, technology, culture and political order necessarily alter as one moves from one zone to another. This is an explanation of very wide application, one on which Immanuel Wallerstein (1974) has based his book *The Modern World-System*.

The center or *core* contains everything that is most advanced and diversified. The next zone possesses only some of these benefits, although it has some share in them: it is the "runner-up" zone. The huge periphery, with its scattered population, represents on the contrary backwardness, archaism and exploitation

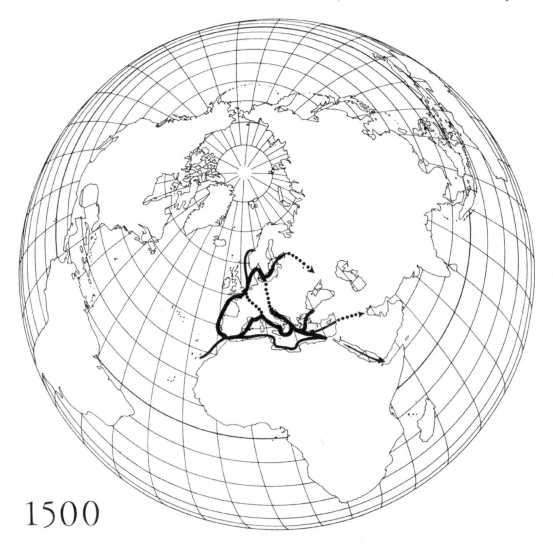

1500

EUROPEAN WORLD-ECONOMIES ON A GLOBAL SCALE
The expanding European economy, represented by its major commodity trades on a world scale.
In 1500, the world-economy with Venice at its centre was directly operating in the
Mediterranean (see Fig. 15 for the system of the *galere da mercato*) and western Europe; by way
of intermediaries, the network reached the Baltic, Norway and, through the Levant ports, the
Indian Ocean.

Figure 1 European world-economies 1550

Source: Fernand Braudel

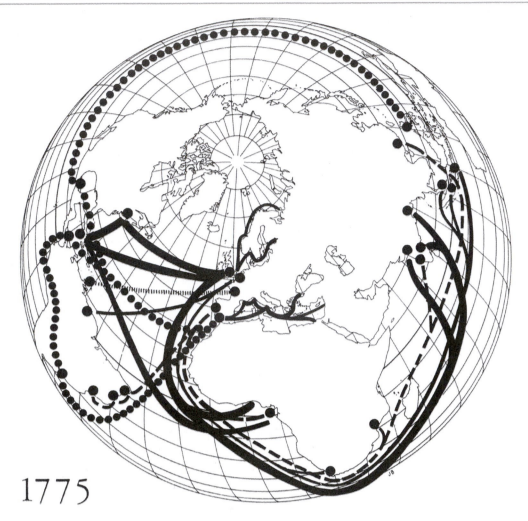

1775

In 1775, the octopus grip of European trade had extended to cover the whole world: this map shows English, Dutch, Spanish, Portuguese and French trade networks, identifiable by their point of origin. (The last-named must be imagined as operating in combination with other European trades in Africa and Asia.) The important point is the predominance of the British trade network which is difficult to represent. London had become the centre of the world. The routes shown in the Mediterranean and the Baltic simply indicate the major itineraries taken by all the ships of the various trading nations.

Figure 2: European world-economies 1775

Source: Fernand Braudel

by others. This discriminatory geography is even today both an explanation and a pitfall in the writing of world history – although the latter often creates the pitfalls itself by its connivance.

The central zone holds no mysteries: when Amsterdam was the "warehouse of the world," the United Provinces (or at any rate the most active among them) formed the central zone; when London

imposed its supremacy, England (if not the whole of the British Isles) formed the surrounding area. When Antwerp found itself in the sixteenth century the center of European trade, the Netherlands, as Henri Pirenne said, became "the suburb of Antwerp" and the rest of the world its periphery. The "suction and force of attraction of these poles of growth" were clear to see. Detailed identification is more difficult though when it comes to the regions outside this central zone, which may border on it, are inferior to it but perhaps only slightly so: seeking to join it, they put pressure on it from all directions, and there is more movement here than anywhere else. Any uncertainty evaporates on the other hand as soon as one enters the regions of the periphery. Here no confusion is possible: these are poor, backward countries where the predominant social status is often serfdom or even slavery (the only free or quasi-free peasants were to be found in the heart of the West); countries barely touched by the money economy; countries where the division of labor has hardly begun; where the peasant has to be a jack of all trades; where money prices, if they exist at all, are laughable. A low cost of living is indeed in itself a sign of under-development.

The backward regions on the fringes of Europe afford many examples of these marginal economies: "feudal" Sicily in the eighteenth century; Sardinia, in any period at all; the Turkish Balkans; Mecklenburg, Poland, Lithuania – huge expanses drained for the benefit of the western markets, doomed to adapt their production less to local needs than to the demands of foreign markets; Siberia, exploited by the Russian world-economy; but equally, the Venetian islands in the Levant, where external demand for raisins and strong wines, to be consumed as far away as England, had already by the fifteenth century imposed an intrusive monoculture, destructive of local balance.

There were *peripheries* in every quarter of the world of course. Both before and after Vasco da Gama, the black gold-diggers and hunters of the primitive countries of Monomotapa, on the east coast of Africa, were exchanging their gold and ivory for Indian cottons. China was always extending her frontiers and trespassing on to the "barbaric" lands as the Chinese texts call them – for the Chinese view of these peoples was the same as that of the classical Greeks of non-Greek-speaking populations: the inhabitants of Vietnam and the East Indies were "barbarians." In Vietnam however, the Chinese made a distinction between those barbarians who had been touched by Chinese civilization and those who had not. According to a Chinese historian of the sixteenth century, his compatriots called "those who maintained their independence and their primitive customs '*raw*' barbarians, and those who had more or less accepted Chinese ways and submitted to the empire '*cooked*' barbarians." Here politics, culture, economy and social model contributed jointly to the distinction. The raw and the cooked in this semantic

code, explains Jacques Dourbes, also signifies the contrast between culture and nature: rawness is exemplified above all by nakedness. "When the Potao ["kings" of the mountains] come to pay tribute to the Annamite court [which was Chinesified] it will cover them with clothes."

RULE THREE (CONTINUED): DO NEUTRAL ZONES EXIST?

However, the backward zones are not to be found exclusively in the really peripheral areas. They punctuate the central regions too, with local pockets of backwardness, a district or "pays," an isolated mountain valley or an area cut off from the main communication routes. *All* advanced economies have their "black holes" outside *world time.*

A world-economy is like an enormous envelope. One would expect a priori, that given the poor communications of the past, it would have to unite considerable resources in order to function properly. And yet the world-economies of the past did incontestably function, although the necessary density, concentration, strength and accompaniments only effectively existed in the core region and the area immediately surrounding it; and even the latter, whether one looks at the hinterland of Venice, Amsterdam or London, might include areas of reduced economic activity, only poorly linked to the centers of decision. Even today, the United States has pockets of under-development within its own frontiers.

REFERENCES FROM THE READING

Braudel, F. (1972) *The Mediterranean and Mediterranean World in the Age of Phillip II*, trans S. Reynolds. New York: Harper and Row.

Braudel, F. (1973) *Capitalism and Material Life, 1400–1800*, New York: Harper & Row.

Braudel, F. (1979) *Civilization and Capitalism, 15th–18th century*, Berkeley: University of California Press.

Sismondi, S. (1991) [1951] *New Principles of Political Economy: Of Wealth in its Relation to Population*, trans R. Hyse. New Brunswick, NJ: Transaction Publishers.

Sweezy, P. (1974) *Modern Capitalism*, New York: Monthly Review Press.

Wallerstein, I. (1974) *The Modern World-System*, New York: Academic Press.

3

"World city formation: an agenda for research and action"

from *International Journal of Urban and Regional Research* (1982)

John Friedmann and Goetz Wolff

EDITORS' INTRODUCTION

John Friedmann (1926–2017) was one of the pioneering urbanists of the late 20th century. Many of Friedmann's most important contributions to urban and regional studies were produced during his nearly three decades on the faculty of the Program for Urban Planning in the Graduate School of Architecture and Planning at UCLA, which he helped found in the late 1960s and where he was Professor Emeritus. Friedmann last was Honorary Professor in the School of Community and Regional Planning at the University of British Columbia. International recognition for his scholarship includes Honorary Doctorates from the Catholic University of Chile, the University of Dortmund, and York University in Toronto. Friedmann's publication record includes 14 individually authored books, 11 co-edited books, and more than 150 chapters, articles, and reviews. A dedication to his life's work can be found in Rangan et al. (2017).

Goetz Wolff, who was a PhD student at UCLA when the below article was written, is an independent researcher, teacher, and consultant in Los Angeles. His research and teaching interests center on industrial change and regional economic development issues, with particular reference to the Southern California region.

Friedmann and Wolff's study of world city formation was published in 1982 and became an instant classic: it kickstarted an impressive outburst of research on this topic. Building upon a broad range of intellectual sources—including radical international political economy, world systems theory, Marxian urban studies, urban systems theory, and radical community studies—Friedmann and Wolff's "agenda for research and action" represented a genuinely original synthesis. Reminiscent of previous arguments by Jane Jacobs (1984), Friedmann and Wolff viewed cities rather than national economies as the motors of contemporary capitalist development. World cities, they argued, represented a new breed of global command and control centers within the new international division of labor associated with post-1970s capitalism. These cities, moreover, concentrated many of the contradictions and inequalities: they had to be viewed, simultaneously, as spaces of hope and as spaces of gloom, that is, as sites in which "citadel" and "ghetto" existed in uneasy proximity.

It was by no means accidental that Friedmann and Wolff wrote this paper during a period in which Los Angeles was becoming what some would hyperbolically describe as the "capital of the 21st century" (Scott and Soja 1996). Indeed, Los Angeles in the 1980s provided an ideal backdrop for the development of global city theory. Los Angeles exemplified many of the dramatic economic changes that were unfolding in the United States during that decade, as the Fordist industrial structure of the postwar years was increasingly replaced by an internationalized, more flexible, and less stable regime of accumulation. More generally, many patterns of urban restructuring that would subsequently unfold in cities throughout the older industrialized world were already strikingly evident in Los Angeles in the late 1970s and early 1980s (Keil 1998; Soja Ch. 14).

I

Our paper concerns the spatial articulation of the emerging world system of production and markets through a global network of cities. Specifically, it is about the principal urban regions in this network, dominant in the hierarchy, in which most of the world's active capital is concentrated. As cities go, they are large in size, typically ranging from five to fifteen million inhabitants, and they are expanding rapidly. In space, they may extend outward by as much as 60 miles from an original center. These vast, highly urbanized – and urbanizing – regions play a vital part in the great capitalist undertaking to organize the world for the efficient extraction of surplus. Our basic argument is that the character of the urbanizing processes – economic, social, and spatial – which define life in these "cities" reflect, to a considerable extent, the mode of their integration into the world economy.

We propose, then, a new look at cities from the perspective of the world economic system-in-formation. The processes we will describe lead to new problem configurations. The central issue is the control of urban life. Whose interests will be served: those of the resident populations or of transnational corporations, or of the nation states that provide the political setting for world urbanization? Planners are directly engaged on this contested terrain. They are called upon to clarify the issues and to help in searching for solutions. Obviously, they will have to gain a solid, comprehensive understanding of the forces at work. And they will have to rethink their basic practices, since what is happening in world cities is in large measure brought about by forces that lie beyond the normal range of political – and policy – control. How can planners and, indeed, how can the people themselves, living in world cities, gain ascendancy over these forces? That is the basic question. Towards the end of this paper we shall venture a few observations about the tasks we face and their implications for planning.

II

Our argument is a relatively simple one. Since the second world war, the processes by which capitalist institutions have freed themselves from national constraints and have proceeded to organize global production and markets for their own intrinsic purposes have greatly accelerated. The actors principally responsible for reorganizing the economic map of the world are the transnational corporations, themselves in bitter and cannibalistic conflict for the control of economic space. The emerging global system of economic relations assumes its material form in particular, typically urban, localities that are enmeshed with the global system in a variety of ways.

The specific mode of their integration with this system gives rise to an urban hierarchy of influence and control. At the apex of this hierarchy are found a small number of massive urban regions that we shall call world cities. Tightly interconnected with each other through decision-making and finance, they constitute a worldwide system of control over production and market expansion. Examples of world-cities-in-the-making include such metropolises as Tokyo, Los Angeles, San Francisco, Miami, New York, London, Paris, Randstadt, Frankfurt, Zurich, Cairo, Bangkok, Singapore, Hong Kong, Mexico City and Sao Paulo.

To label them world cities is a matter of convenience. In each and every instance, their specific role must be determined through empirical research. Only this much we can say: their determining characteristic is not their size of population. This is more properly regarded as a consequence of their economic and political role. A more fundamental question is in what specific ways these urban regions are becoming integrated with the global system of economic relations. Two aspects need to be considered:

1 The form and strength of the city's integration (e.g. to what extent it serves as a headquarters location for transnational corporations; the extent to which it has become a safe place for the investment of "surplus" capital, as in real estate; its importance as a producer of commodities for the world market; its role as an ideological center; or its relative strength as a world market).

2 The spatial dominance assigned by capital to the city (e.g. whether its financial and/or market control is primarily global in scope, or whether it is less than global, extending over a multinational region of the world, or articulating a national economy with the world system). These criteria of world system integration must be viewed in a dynamic, historical perspective. Urban roles in the world system are not permanently fixed. Functions change; the strength of the relationship changes; spatial dominance changes. Indeed, the very concept of a world economy articulated through urban structures is as old as the ancient

empires. Rome was perhaps the first great imperial city. One may think of Venice in its Golden Age, or of nineteenth-century London. While recognizing this historical continuity, we would still argue that the present situation is substantially different. What then is new?

First, we must consider the truly global nature of the world economy. Even imperial London, ruling over an empire "where the sun never sets," controlled only portions of the world. The present transnational system of space economy, on the other hand, is in principle unlimited. It is best understood as a spatial system which has its own internal structure of dominance/subdominance. Following Immanuel Wallerstein, we may label its three major regional components as core, semi-periphery and periphery. *Core areas* include those older, already industrialized and possibly "postindustrial" regions that contain the vast majority of corporate headquarters and continue to be the major markets for world production (northwest Europe, North America, Australia, Japan). The *semi-periphery* includes rapidly industrializing areas whose economies are still dependent on core-region capital and technical knowledge. They play a significant role in extending markets into the world periphery. Mexico, Brazil, Spain, Egypt, Singapore, Taiwan and the Republic of Korea (ROK) would be examples of semi-peripheral regions. And the *world periphery* comprises what is left of market economies. Predominantly agrarian, the people of the world periphery are poor, technologically backward and politically weak.

This analytical scheme must be deftly handled. It is a first approximation to a deeper understanding of world city structure. Above all, it is an historical classification. Over the span of one or two generations, a country may change its position as it moves from periphery to semi-peripheral status (ROK, Spain, Brazil), from semi-periphery to core (Japan) and even perhaps back from core region status to the semi-periphery (Great Britain), or the ultimate decline into peripheral obscurity (Lebanon, Iran). What makes this typology attractive is the assumption that cities situated in any of the three world regions will tend to have significant features in common. As the movement of particular countries through the three-level hierarchy suggests, these features do not in any sense determine economic and other outcomes. They do, however, point to conditions that significantly influence city's growth and the quality of urban life.

The world economy is thus no longer defined by the imperial reach of a Rome, a Venice or even a London, but by a linked set of markets and production units organized and controlled by transnational capital. World cities are a material manifestation of this control, and they occur exclusively in core and semi-peripheral regions where they serve as banking and financial centers, administrative headquarters, centers of ideological control and so forth. Without them, the world-spanning system of economic relations would be unthinkable.

This conception of the world city as an instrument for the control of production and market organization implies that the world economy, spatially articulated through world cities, is dialectically related to the national economies of the countries in which these cities are situated. *It posits an inherent contradiction between the interests of transnational capital and those of particular nation states that have their own historical trajectory.* World cities are asked to play a dual role. Essential to making the world safe for capital, they also articulate given national economies with the world system. As such, they have considerable salience for national policy makers who must respond to political imperatives that are only coincidentally convergent with the interests of the transnationals. World cities lie at the junction between the world economy and the territorial nation state.

Finally, the global economy is superimposed upon an international system of states. Nation states have their own political fears and ambitions. They form alliances, and they exact tribute. They must protect their frontiers against actual and potential enemies. Wishing to ensure their continuing power in the assembly of nations, or even to enlarge their power, they must provide for a continuing flow of raw materials and food supplies.

On the other hand, although transnational capital desires maximum freedom from state intervention in the movements of finance capital, information and commodities, it is vitally interested in having the state assume as large a part as possible of the costs of production, including the reproduction of the labor force and the maintenance of "law and order." It is clear, therefore, that they would benefit from a strategy to prevent a possible collusion among nation states directed against themselves. Being essential to both transnational capital and national political interests, world cities may become bargaining counters in the ensuing struggles.

They are therefore also major arenas for political conflict. How these conflicts are resolved will shape the future of the world economy. Because many diverging interests are involved, it is a multifaceted struggle. There is, of course, the classical instance of the struggle between capital and labor. This remains. In addition, there is now a struggle between transnational capital and the national bourgeoisie; between politically organized nation states and transnational capital; and between the people of a given city and the national polity, though this may be the weakest part.

There is, then, nothing inevitable about either the world economy or its concrete materialization in world cities. Capital is in conflict with itself and with the political territorial entities where it must come to rest. There is no manifest destiny. Yet the emerging world economy is an historical event and this allows us to formulate our central hypothesis: *the mode of world system integration (form and strength of integration; spatial dominance) will affect in determinate ways the economic, social, spatial and political structure of world cities and the urbanizing processes to which they are subject.*

What we describe is not a Weberian "ideal type" of a fully formed "world city." All that we intend is to point to certain structural tendencies in the formation of those cities that appear to play a major role in the organization of world markets and production. We have in mind a heuristic for the empirical study of world city formation.

III

In making the internationalization of capital central to our analysis, we focus upon a combination of complex processes that are indeterminate, contradictory and irregular. There is little dispute about the fact of the worldwide expansion of market relations. But the cause – the driving force – of the internationalization of capital is debated: is internationalization merely a working out of the internal logic of capitalism? Or has labor in the industrialized countries created a situation where capital now finds it more profitable to locate in the periphery?

We proceed under the assumption that both "windows" – to use David Harvey's felicitous image – contribute to the needed understanding of the global order. The contradictions inherent in the capitalist economy and the basic struggle which results from the domination of labor by capital are the major forces

which account for both the spatial and temporal irregularities of the world economy. A city's mode of integration with the global economy cannot simply be understood by identifying its functional role in the articulation of the system. Rather, we suggest that the driving forces of competition, the need for accumulation and the challenges posed by political struggle make the intersection of world economy/world city a point of intense conflict and dynamic change. World city integration is not a mechanical process; it involves many interconnected changes that leave few aspects of its life untouched and create the arenas for concerted action.

IV

The world city today is in transition, which is to say it is in movement. Perhaps it has always been like this. Equilibrium is not part of the experience of large cities. Structural instability manifests itself in a variety of ways: dramatic changes in the distribution of employment, the polarization of class divisions, physical expansion and decaying older areas, political conflict. We shall have a quick look at all of these to render more specific how the formation of the world cities is affecting the quality of their life experience.

World cities are the control centers of the global economy. Their status, of course, is evolving in the measure that given regions are integrated in a dominant role with the world system. And like the golden cities of ancient empires, they draw into themselves the wealth of the world that is ruled by them. They become the major points for the accumulation of capital and "all that money can buy." They are luxurious, splendid cities whose very splendor obscures the poverty on which their wealth is based. The juxtaposition is not merely spatial; it is a functional relation: rich and poor define each other.

It is not a new story, and yet its particular features are new. As we attempt to describe the changes that occur as urban regions strive for world city status several things must be borne in mind. The characteristics we are describing are merely tendencies, not final destinations. Particular cities will exhibit particular features. Still, the account we give of conditions prevailing in urban regions as they become world cities may be regarded as the best current hypothesis. In every instance, we have tried to relate it back to specific aspects of integration with global economy.

Not only are world cities in themselves not uniform, there is no definite cut-off point with other cities that belong to the same system but are not so tightly integrated with the global economy, have only a national/subnational span of control, or are integrated primarily on a basis of dependency. In a way, the world economy is everywhere, and many of the features we will describe may be found in cities other than those we are discussing here.

The world city "approach" is, in the first instance, a methodology, a point of departure, an initial hypothesis. It is a way of asking questions and of bringing footloose facts into relation. We do not have an all-embracing theory of world city formation.

Economic restructuring

A primary fact about emerging world cities is the impact which the incipient shifts in the structure of their employment will have on the economy and on the social composition of their population. The dynamism of the world city economy results chiefly from the growth of a primary cluster of high-level business services which employs a large number of professionals – the trans-national elite – and ancillary staffs of clerical personnel. The activities are those which are coming to define the chief economic functions of the world city: *management; banking and finance, legal services, accounting, technical consulting, telecommunications and computing, international transportation, research and higher education.*

A secondary cluster of employment, also in rapid ascendancy, may be defined as essentially serving the first. Its demand is largely derived, and it employs proportionately a much smaller number of professionals: *real estate, construction activities, hotel services, restaurants, luxury shopping, entertainment, private police and domestic services.* A more varied mix than the primary cluster, its fortunes are closely tied to it. Although most jobs in this cluster are permanent and reasonably well-paid, this is not true for domestic services which is the most vulnerable employment sector and the most exploited.

A tertiary cluster of service employment centers on *international tourism.* To a considerable extent, this overlaps with the secondary cluster (hotels, restaurants, luxury shopping, entertainment), and like that cluster, it is tied to the performance of the world economy.

The growth of the first three clusters is taking place at the expense *of manufacturing* employment. Although a large cluster, its numbers are gradually declining as a proportion of all employment. Some industry serves the specialized needs of local markets, while other sectors – in Los Angeles, for example, electronics and garment industries – are choosing world city locations because of the large influx of cheap labor which helps to keep the average cost of wages down. The future of manufacturing employment in the world city is not bright, however. The next two decades will see the rapid automation and robotization of many jobs. While factories will still be producing and earning large and perhaps even rising profits, they will be largely devoid of working people.

Government services constitute a fifth cluster. They are concerned with the maintenance and reproduction of the world city, as well as the provision of certain items of collective consumption: the planning and regulation of urban land use and expansion; the provision of public housing, basic utilities and transportation services; the maintenance of public order; education; business regulation; urban parks; sanitation; and public welfare for the destitute.

Because of the uniqueness and scale of world cities, and because they are often considered national showcases, the government sector tends to be larger here than elsewhere. And because it is a political and, for the most part, technically backward sector, with uncertain criteria of adequate performance, it tends to be bloated, employing large numbers of people at relatively low levels of productivity and wages. Moreover, because the world city extends over many political jurisdictions that are contiguous with each other, there is much overlap and redundancy in employment. During periods of depression, government will often be the employer of last resort. Its internal rhythm tends to be counter-cyclical.

A sixth and, at least in some cities, numerically the largest cluster, embraces the "informal," *"floating,"* or *"street" economy* which ranges from the casual services of day laborers and shoeshine boys to fruit vendors, glaziers, rug dealers and modest artisans. Frequently an extension of the household economy, most informal activities require little or no overhead (though they do require start-up capital). They demand long hours, and the returns are low and uncertain. They offer no security to those who work in them. New arrivals to the city often find their first job in the informal sector, and many of them stay there. When times are bad, some makeshift income earning opportunities can always be found in the informal sector for people who are temporarily unemployed. Although informal sector work may be a choice between independence and security for some, for many more it is the only way to survive in the city. The cluster of informal activities takes up the slack in the "formal" economy, and thus despite its marginal character, it tends to be tolerated by the state.

Some informal activities are not as "unorganized" or "casual" as they might appear. Perhaps increasingly, small businesses are subcontracted by large, frequently multinational corporations who in this way are able to lower their costs of operation. Informal businesses are usually beyond the reach of government regulation. They don't pay minimum wages and their labor is often self-exploited. Much of it is done by women and children.

But essentially the informal sector exists because of the large influx of people into the world city from other cities and from the countryside, people who are attracted to the world city as to a honey pot. They don't all find legitimate employment. A significant number drift into illicit occupations which perhaps more than elsewhere appear to thrive in the large city: thieves, pickpockets, swindlers, pimps, prostitutes, drug peddlers, black marketeers ... the list can be extended with endless refinements.

Finally, there is the undefined category of those without a steady income: the full-time unemployed, who depend on family and public charity for support. Excepting women, who manage their households but are not paid for this, and therefore do not appear in the official statistics, their numbers are surprisingly small in the order of 5–10 percent of the labor force.

Social restructuring

The primary social fact about world city formation is the polarization of its social class divisions. Transnational elites are the dominant class in the world city, and the city is arranged to cater to their lifestyles and occupational necessities. It is a cosmopolitan world that surrounds them, corresponding to their own high energy, rootlessness and affluence. Members of this class are predominantly males between the ages of 30 and 50. Because of their importance to the city, they are a class well served.

The contrast with the third (or so) of the population who make up the permanent underclass of the world city could scarcely be more striking. The underclass are the victims of a system that holds out little hope to them in the periphery from which they came but also fails them in the very nerve centers of the world economy where they are queuing for a job. They crowd along the edges of the primary economy – the "formal" sector – or settle in its interstices, barely tolerated, yet providing personal services to the ruling class, doing the dirty work of the city. The ruling class and its dependent middle sectors enjoy permanent employment, a steady income and complete legality; they do not have to justify their existence. For all practical purposes, they *are the* city. The underclass lives at its sufferance.

Many, though not all, of the underclass are of different ethnic origin than the ruling strata; often, they have a different skin color as well, or speak a different dialect or language. These immigrant workers give to many world cities a distinctly "third world" aspect: Puerto Ricans and Haitians in New York; Mexicans in Los Angeles and San Francisco; barefoot Indians in Mexico City; "nordestinos" in Sao Paulo; Jamaicans in London; Algerians in Paris; Turks in Frankfurt; Malays in Singapore.

There is a city that serves this underclass, as suited to their own condition if not their preferences and needs, as is the city of the "upper circuit." Physically separated from and many times larger than the citadel of the ruling class, it is the ghetto of the poor. Both cities live under the constant threat of violence: the upper city is guarded by private security forces, while the lower city is the double victim of its own incipient violence and of police repression. The typical world city situation is thus for *both* the crime rate and police expenditures to rise.

Racism reinforces class contradictions, and a good deal of ethnic and racial hostility is found within the working class itself. Under conditions of tight labor markets, "foreign" workers, whether undocumented or not, frequently occasion racist outbursts, as "national" workers (particularly among the underclass) struggle to preserve their limited terrain for livelihood. Street gangs of different ethnic origin, and the pitched battles between them, especially in the United States, are a major manifestation of this violence.

Yet racial conflict is only one facet of the general increase in violence that is brought on by class polarization. Terrorism, kidnapping, street demonstrations and rioting are other common forms. There is an undeniable fascination with violence among residents in the world city which is picked up and amplified by the popular media. Yet for all the turmoil, world city conflicts are not a sign of an impending revolution. A good deal of the violence occurs within the working class itself and is a measure of its internal divisions. The world city is in any event immune to revolutionary action. Lacking a political center, it can only be rendered irrelevant, at the present time a rather unlikely occurrence.

Confronted with violence, the nation state responds in coin. Given the severity of its fiscal constraints, in the face of constantly rising costs, it resorts to the simplest, least imaginative alternative: the application of brute force. The response is acceptable to the new ruling class who generally prefer administrative to political solutions. But police repression can at best contain class violence; it cannot eliminate or significantly reduce it. Violence is here to stay.

Physical restructuring

Over the next generation, world cities can be expected to grow to unprecedented size. By the end of the century, the typical world city will have 10 million people or more. Much of the increase will have come from migration. Obviously, a population that rivals that of a medium-sized country by today's standards can no longer be considered a city in the traditional sense. It is an urbanized region or an "urban field." In the case of Los Angeles, for example, the pertinent economic region has been defined as having a radius of about 60 miles (roughly 80 minutes commuting at normal speeds); it represents the life space for more than half the population of California!

The urban field is essentially an economic concept. Although it does not respond to the traditional political concept of the city as civitas, it imposes its own logic on the vestiges of the political city which struggles to survive in this highly charged, volatile materialization of capitalist energy. The urban field is expanding, more rapidly in most cases than even the increase in population, but it is expanding unequally, regardless of whether one applies functional, social or economic criteria. Underlying its kaleidoscopic spatial form is the ever-shifting topography of land values which quickly and efficiently excludes all potential users who are unable to meet the price of a given parcel of land. Of course, this method does not at all correspond to any social need, least of all to the need of what Joan Nelson (1979) has called "access to power."

The concentration of activities and wealth on a world city scale imposes extraordinary strains on the natural resources on which the continued viability of the world city depends. To feed its voracious appetite for water and energy (which tends to grow at multiples of the increases in population), the city must reach further and further afield, sometimes for hundreds of miles, across mountains and deserts and even national boundaries. As it does so, it comes inevitably into conflict with competing interests and jurisdictions.

At the same time, and considering alone sheer volume, the world city faces enormous problems of waste management. Pollution, at levels of concentration dangerous to human health, poses a constant and growing threat. Huge areas must be set aside for low-intensity uses of the land, such as airports, water treatment facilities, solid waste disposal, agriculture and dairying and urban mass recreation.

This enormously varied complex of activities must be knitted together through high-speed transport devices and linked to the outside world through a system of international transport terminals and telecommunications capable of serving the entire region. And of course it must have a basic infrastructure of utilities that serves the housing and industrial needs of the city. At a scale of 10 or 20 million people each, and an area of several hundred square miles, these common basic needs of the world city can be satisfied only with the considered application of high technology. But high technology renders the city more vulnerable: the growing reliance on ever more sophisticated methods may be counterproductive in the end. Its costs will escalate even as the quality of its services deteriorates.

In its internal spatial structure, the world city may be divided, as we suggested in the preceding section, into the "citadel" and the "ghetto." Its geography is typically one of inequality and class domination. The citadel serves the specific needs of the transnational elites and their immediate retinues who rule the city's economic life; the ghetto is adapted to the circumstances of the permanent underclass.

With its towers of steel and glass and its fanciful shopping malls, the citadel is the city's most vulnerable symbol. Its smooth surfaces suggest the sleek impersonality of money power. Its interior spaces are ample, elegant and plush. In appropriately secluded spaces, the transnational elites have built their residences and playgrounds: country clubs and bridle paths and private beaches.

The overcrowded ghettos exist in the far shadows of the citadel, where it is further divided into racial and ethnic enclaves; some areas are shantytowns. None are well provided with public services: garbage does not get collected, only the police in their squad cars are visibly in evidence. In many places, ghetto residents are allowed outside their zones only during working hours: their appearance in the citadel after dark creates a small panic. With its dozens of apartheid Sowetos, South Africa is perhaps the extreme case of a country whose elites are gripped in fear of their underclass, but political manifestations of this fear are found to a degree in nearly all world cities. Not long ago, the papers in Los Angeles were filled with horror stories of "marauders" sallying forth from the black ghetto to the citadel on the West Side to rob, rape and kill. They called it "predatory crime." The implication was clear: at night, ghetto residents belong to the ghettos. There, isolated like a virus, they can harm only themselves.

Political conflict

Every restructuring implies conflict. And when conflict occurs in the public domain, it concerns the distribution of costs and benefits, and who shall gain advantage for the next round of the struggle. Focused on the world city, political conflict is multidimensional; it is also at the heart of the matter, as if all the world's lines of force came together here, and the contradictions in the self expansion of capital are magnified on a scale commensurate with that of the world city itself. In the absence of counter-forces, complex feedback loops tend to destabilize the system, and localized conflicts may suddenly erupt into a worldwide crisis.

The emergence of world cities sets processes into motion that restructure people's life spaces and the economic space that intersects with them. Economic space obeys the logic of capital: it is profit-motivated and individualized. Life spaces are territorial: they are the areas that people occupy in which their dreams are made, their lives unfold. They are thus the areas they really care about. For the dominant actors in economic space, life space is nothing but a hindrance, an irrational residue of a more primitive existence. Yet it can neither be denied nor circumvented. Every economic space overlaps with an existing life space, and without this, economic enterprise would perish. On the other hand, for people who are collectively the

sovereign in their respective life spaces, the space of economic logic is the basis for their physical survival.

It is thus very clear that a restructuring of these two kinds of space is bound to generate deep conflicts. In practice, because so many cross currents bear down on them, the reasons for conflicts are often hard to isolate. Conflicts persist; they have a history and a future, they interconnect in place, and the system of world cities ensures their transmittal from continent to continent. Though it may not be very enlightening, it might be more accurate to speak of them as a form of social turbulence.

There are the conflicts over livelihood, as the restructuring of economic space draws jobs away from one place to resurrect them in another, or as entirely new kinds of jobs are suddenly ascendant for which older workers are not qualified. There are racial and ethnic conflicts, as workers battle over access to the few good jobs there are, and even over jobs that are less desirable but relatively more abundant. There are struggles between world city and the national periphery over the political autonomy of peripheral regions, as the periphery sees its collective life chances systematically denied by the imperial interests of world city cores. There are conflicts over the spaces within the city, as people seek access to housing they can afford, defend their neighborhoods from the intrusion of capitalist logic, or merely struggle for turf to enjoy the freedom of following lifeways of their own. There are the political campaigns launched by concerned citizens to protect and enhance the quality of their lives as they perceive it: in struggles for the environment, against nuclear power, for child care centers, for the access of handicapped workers in public facilities ... struggles and campaigns which are incapable of being separated from the peculiar setting of world cities. Or there are the bitter and tenacious struggles of poor people, workers who belong to one class fraction or another, for greater access to the conditions for social power: the right to organize, to demonstrate, to call to account, to gain control over the conditions of their work, to keep their bodies healthy, to educate their children and themselves to higher incomes, to sources of the means for livelihood.

All these struggles are occurring simultaneously. They are centered on the restructuring process and in the contradictions that arise from the interfacing of economic and life spaces. As such, they reveal to the astute observer the true forces at work in the world city and the actual distribution of power. Also, of course, they determine the ultimate outcomes we observe ... not only the outcomes of the particular struggles being waged, but of the form and direction which world city development will take. For outcomes are not predetermined. Broad tendencies may be irreversible in the medium term, such as the integration of the world economic system under the aegis of transnational capital. But within that historical tendency, there are always opportunities for action. It is precisely at the searing points of political conflict that opportunities arise for a concerted effort to change the course of history.

Conflict between life space and economic space poses new questions for the state. More than ever, the state is faced with multiple contradictions and difficult choices. Some would say that the state itself is threatened by the dramatic appearance of transnational power. On the world periphery, the choice may be the relatively simple one between complete dependency (with all that this implies in terms of uneven development) and stone age survival. In the semi-periphery, which has already chosen a dependent development path, the problems of choice are more complex. More highly articulated than in the periphery, the state reflects within its own structure many of the conflicts within civil society and the economy, as class fraction is pitted against class fraction, territory against class, working class against capital, and citizens against the state, and as these conflicts come together in successive waves of "turbulence" in the world city, the state is merely one more actor, trying to safeguard its own specific interests.

One such interest is the political integration of its territory. Semi-peripheral states are especially concerned with regional inequalities, focal points of regional revolt, regional resources development and rural land reform. In countries such as Mexico and Brazil, regional issues have been major concerns of state action, and civil wars have been fought over them.

Regional questions recede in the territorially more integrated countries of the world core area. Here the more important conflicts tend to be between national and transnational fractions of capital, and the state becomes a major "arena" for the conduct of this struggle. Much of it is over specific legislation and the budget allocation. In all this turmoil, a major loser is the local state. Small, isolated without financial power, and encapsulated within the world economy, it is barely able to provide for even the minimal services its population needs. And yet, instead of seeking alliances with neighboring cities and organized labor, it leaves the real decisions to the higher powers on which it is itself dependent, or to the quasi-independent authorities created by state charter that manage the infrastructure of global capital-system-wide facilities such as ports, airports, rapid transit, water supply, communications and electric power.

REFERENCES FROM THE READING

Jacobs, J. (1984) *Cities and the Wealth of Nations: Principles of Economic Life*, New York: Random House.

Keil, R. (1998) *Los Angeles: Globalization, Urbanization, and Social Struggles*, Chichester, UK and New York: J. Wiley.

Nelson, J. (1979) *Access to Power: Politics and the Urban Poor in Developing Countries*, Princeton, NJ: Princeton University Press.

Rangan, H., Ng, M. K., Chase, J. and Porter, L. (eds.) (2017) *Insurgencies and Revolutions: Reflections on John Friedmann's Contributions to Planning Theory and Practice*, New York: Routledge.

Scott, A. and Soja, E. (1996) *The City: Los Angeles and Urban Theory at the End of the Twentieth Century*, Los Angeles: University of California Press.

4

"Locating cities on global circuits"

from *Environment and Urbanization* (2002)

Saskia Sassen

EDITORS' INTRODUCTION

Saskia Sassen, the Robert S. Lynd Professor of Sociology at Columbia University, developed the concept of the "global city" in her highly influential book *The Global City: New York, London, Tokyo* (1991). Her global city framework redefined urban studies and inspired scholars across the world to study the conditions that gave rise to different global cities, as well as the consequences of becoming a global city, including changes in employment relations, social polarization, and the built environment. Building upon her work in *The Global City*, Sassen here develops a more sustained analysis of specialized service industries and advanced digital technologies within global city economies. First, she reiterates her central contention that global city economies contain a broad complex of specialized service industries that enables transnational corporations to coordinate production, investment, and finance on a world scale. For Sassen, the worldwide dispersion of production is linked intrinsically to increasingly centralized key command and control capacities within the agglomeration economies of global cities. While her previous research focused on core cities such as New York, London, and Tokyo, she now identifies similar trends at work in many cities in developing countries (Sassen 2002). Second, Sassen explores one of the most perplexing paradoxes of contemporary global urbanization—the consolidation of a small number of strategically dominant global financial centers, at a time when such sites would be expected to be obsolete by the proliferation of new informational technologies. Sassen examines this paradox through a multifaceted analysis that emphasizes the various organizational and logistical advantages flowing from the intensive patterns of spatial agglomeration within global cities.

A key feature of the contemporary global system is that it contains both the capability for enormous geographic dispersal and mobility as well as pronounced territorial concentrations of resources necessary for the management and servicing of that dispersal. The management and servicing of much of the global economic system takes place in a growing network of global cities and cities that might best be described as having global city functions. The expansion of global management and servicing activities has brought with it a massive upgrading and expansion of central urban areas, even as large portions of these cities fall into deeper poverty and infrastructural decay. While this role involves only certain components of urban economies, it has contributed to a repositioning of cities both nationally and globally.

WORLDWIDE NETWORKS AND CENTRAL COMMAND FUNCTIONS

The geography of globalization contains both a dynamic of dispersal and of centralization. The massive trend towards the spatial dispersal of economic activities at the metropolitan, national and global level which we associate with globalization has contributed to a demand for new forms of territorial centralization of top-level management and control functions. Insofar as these functions benefit from agglomeration economies, even in the face of telematic integration of a firm's globally dispersed manufacturing and service operations, they tend to locate in cities. This raises a question as to why they should benefit from agglomeration economies, especially since globalized economic sectors tend to be intensive users of the new telecommunications and computer technologies. In my book, *The Global City* (Sassen, 1991), I have found that the key variable contributing to the spatial concentration of central functions and associated agglomeration economies is the extent to which this dispersal occurs under conditions of concentration of control, ownership and profit appropriation.

This dynamic of simultaneous geographic dispersal and concentration is one of the key elements in the organizational architecture of the global economic system. I will first give some empirical referents and then examine some of the implications for theorizing the impacts of globalization and the new technologies on cities.

The rapid growth of affiliates illustrates the dynamic of simultaneous geographic dispersal and concentration of a firm's operations. By 1999, firms had well over half a million affiliates outside their home countries, accounting for US$ 11 trillion in sales, a very significant figure if we consider that global trade stood at US$ 8 trillion. Firms with large numbers of geographically dispersed factories and service outlets face massive new needs for central coordination and servicing, especially when their affiliates involve foreign countries with different legal and accounting systems.

Another current instance of this negotiation between a global cross-border dynamic and territorially specific sites is that of the global financial markets. The orders of magnitude in these transactions have risen sharply, as illustrated by the US$ 65 trillion in the value of traded derivatives, a major component of the global economy. These transactions are partly embedded in electronic systems that make possible the instantaneous transmission of money and information around the globe, and much attention has been paid to these new technologies. But the other half of the story is the extent to which the global financial markets are located in an expanding network of cities, with a disproportionate concentration in cities of the global North. Indeed, the

degrees of concentration internationally and within countries are unexpectedly high for an increasingly globalized and digitized economic sector. Within countries, the leading financial centers today concentrate a greater share of national financial activity than even 10 years ago, and internationally, cities in the global North concentrate well over half of the global capital market.

The specific forms assumed by globalization over the last decade have created particular organizational requirements. The emergence of global markets for finance and specialized services, the growth of investment as a major type of international transaction, all have contributed to the expansion in command functions and in the demand for specialized services for firms. By central functions, I do not only mean top-level headquarters but, rather, all the top-level financial, legal, accounting, managerial, executive and planning functions necessary to run a corporate organization operating in more than one country, and increasingly in several countries. These central functions are partly embedded in headquarters, but also in good part in what has been called the corporate services complex, that is, the network of financial, legal, accounting and advertising firms that handle the complexities of operating in more than one national legal system, national accounting system, advertising culture, etc. and do so under conditions of rapid innovation in all these fields. Such services have become so specialized and complex, that headquarters increasingly buy them from specialized firms rather than producing them in-house. These agglomerations of firms producing central functions for the management and coordination of global economic systems are disproportionately concentrated in the highly developed countries – particularly, though not exclusively, in global cities. Such concentrations of functions represent a strategic factor in the organization of the global economy and they are situated in an expanding network of global cities.

National and global markets, as well as globally integrated organizations, require central places where the work of globalization gets done. Finance and advanced corporate services are industries producing the organizational commodities necessary for the implementation and management of global economic systems. Cities are preferred sites for the production of these services, particularly the most innovative, speculative, internationalized service sectors. Further, leading firms in information industries require a vast physical infrastructure containing strategic nodes with hyper-concentration of facilities. Finally, even the most advanced information industries have a production process that is at least partly place-bound because of the combination of resources it requires even when the outputs are hyper-mobile.

Capital mobility cannot be reduced simply to that which moves nor can it be reduced to the technologies that facilitate movement. Rather, multiple components of what we keep thinking of as capital fixity are actually components of capital mobility. This conceptualization allows us to reposition the role of cities in an increasingly globalizing world, in that they contain the resources that enable firms and markets to have global operations. The mobility of capital, whether in the form of investments, trade or overseas affiliates, needs to be managed, serviced and coordinated. These are often rather place-bound, yet are key components of capital mobility. Finally, states – place-bound institutional orders – have played an often crucial role in producing regulatory environments that facilitate the implementation of cross-border operations for their national firms and for foreign firms, investors and markets.

In brief, a focus on cities makes it possible to recognize the anchoring of multiple cross-border dynamics in a network of places, prominent among which are cities, particularly global cities or those with global city functions. This, in turn, anchors various features of globalization in the specific conditions and histories of these cities, in their variable articulations with their national economies and with various world economies across time and place. This optic on globalization contributes to identifying a complex organizational architecture which cuts across borders and is both partly de-territorialized and partly spatially concentrated in cities. Further, it creates an enormous research agenda in that every particular national or urban economy has its specific and partly inherited modes of articulating with current global circuits. Once we have more information about this variance, we may be able also to establish whether position in the global hierarchy makes a difference, and the various ways in which it might do so.

THE INTERSECTION OF SERVICE INTENSITY AND GLOBALIZATION

To understand the new or sharply expanded role of a particular kind of city in the world economy since the early 1980s, we need to focus on the intersection of two major processes. The first is the sharp growth in the globalization of economic activity, which has raised the scale and the complexity of transactions, thereby feeding the growth of top-level multinational headquarter functions and the growth of advanced corporate services. The second process we need to consider is the growing service intensity in the organization of all industries. This has contributed to a massive growth in the demand for services by firms in all industries, from mining and manufacturing to finance and consumer services industries with mixed business and consumer markets; they are insurance, banking, financial services, real estate, legal services, accounting and professional associations. Cities are key sites for the production of services for firms. Hence, the increase in service intensity in the organization of all industries has had a significant growth effect on cities, beginning in the 1980s and continuing today.

In the case of cities that are major international business centers, we are seeing the formation of a new urban economy. This is so in at least two regards. First, even though these cities have long been centers for business and finance, since the late 1970s there have been dramatic changes in the structure of the business and financial sectors, as well as sharp increases in the overall magnitude of these sectors and their weight in the urban economy. Second, the ascendance of the new finance and services complex, particularly international finance, engenders what may be regarded as a new economic regime, that is, although this sector may account for only a fraction of the economy of a city, it imposes itself on that larger economy. Most notably, the possibility for superprofits in finance has the effect of devalorizing manufacturing insofar as the latter cannot generate the superprofits typical in much financial activity.

This is not to say that everything in the economy of these cities has changed. On the contrary, they still show a great deal of continuity and many similarities with cities that are not global nodes. Rather, the implantation of global processes and markets has meant that the internationalized sector of the economy has expanded sharply and has imposed a new valorization dynamic – that is, a new set of criteria for valuing or pricing various economic activities and outcomes. This has had devastating effects on large sectors of the urban economy. High prices and profit levels in the internationalized sector and its ancillary activities, such as top-of-the-line restaurants and hotels, have made it increasingly difficult for other sectors to compete for space and investments. Many of these other sectors have experienced considerable downgrading and/or displacement as, for example, neighborhood shops tailored to local needs are replaced by upscale boutiques and restaurants catering to new high-income urban elites.

Although of a different order of magnitude, these trends also became evident, beginning in the late 1980s and early 1990s, in a number of major cities in low- and middle-income nations that have become integrated into various world markets: Sao Paulo, Buenos Aires, Bangkok, Taipei, Shanghai, Manila, Beirut and Mexico City are a few examples. Here also, the new urban core was fed by the deregulation of various economic sectors, the ascendance of finance and specialized services and integration into the world markets. The opening of stock markets to foreign investors and the privatization of what were once public sector firms have been crucial institutional arenas for this articulation. Given the vast size of some of these cities, the impact of this new core on the broader city is not always as evident as in central London or Frankfurt, but the transformation is still very real.

It is important to recognize that manufacturing remains a crucial sector in all these economies, even when it may have ceased to be a dominant sector in major cities. Indeed, several scholars have argued that the producer services sector could not exist without manufacturing (Cohen and Zysman, 1987). A key proposition for these authors is that producer services are dependent on a strong manufacturing sector in order to grow. There is considerable debate around this issue. I argue that manufacturing indeed feeds the growth of the producer services sector, but that it does so whether located in the area in question, somewhere else in the country or overseas (Sassen, 1991). Even though manufacturing plants – and mining and agriculture, for that matter – feed growth in the demand for producer services, their actual location is of secondary importance in the case of global level service firms: thus, whether manufacturing plants are located off shore or within a country may be quite irrelevant as long as it is part of a multinational corporation likely to buy the services from those top-level firms. Second, the territorial dispersal of factories, especially if international, actually raises the demand for producer services. This is yet another meaning, or consequence, of globalization: the growth of producer services firms headquartered in New York or London or Paris can be fed by manufacturing located anywhere in the world as long as it is part of a multinational corporate network. Third, a good part of the producer services sector is fed by financial and business transactions that either have nothing to do with manufacturing, as is the case in many of the global financial markets, or for which manufacturing is incidental, as in much merger and acquisition activity, which is centered on buying and selling firms rather than on buying manufacturing firms as such.

THE LOCATIONAL AND INSTITUTIONAL EMBEDDEDNESS OF GLOBAL FINANCE

Several of the issues discussed thus far assume particularly sharp forms in the emerging global network of financial centers. The global financial system has reached levels of complexity that require the existence of a cross-border network of financial centers to service the operations of global capital. This network of financial centers will increasingly differ from earlier versions of the international financial system. In a world of largely closed national financial systems, each country duplicated most of the necessary functions for its economy; collaborations among different national financial markets were often no more than the execution of a given set of operations in each of the countries involved, as in clearing and settlement. With few exceptions, such as the offshore markets and some of the large banks, the international system consisted of a string of closed domestic systems.

The global integration of markets pushes towards the elimination of various redundant systems and makes collaboration a far more complex matter, one which has the effect of raising the division of labor within the network. Rather than each country having its own center for global operations, the tendency is towards the formation of networks and strategic alliances with a measure of specialization and division of functions. This may well become a system with fewer strategic centers and more hierarchy, even as it adds centers. In this context, London and New York, with their enormous concentrations of resources and talent, continue to be powerhouses in the global network for the most strategic and complex operations for the system as a whole.

The financial centers of many countries around the world are increasingly fulfilling gateway functions for the circulation in an out of national and foreign capital. The incorporation of a growing number of these financial centers is one form through which the global financial system expands: each of these centers is the nexus between that country's wealth and the global market, and between foreign investors and that country's investment opportunities.

WHY THE NEED FOR CENTERS IN THE DIGITAL ERA?

What really stands out in the evidence for the global financial industry is the extent to which there is a sharp concentration of the shares of many financial markets in a few financial centers. London, New York, Tokyo (notwithstanding a national economic recession), Paris, Frankfurt and a few other cities regularly appear at the top and represent a large share of global transactions. London, followed by Tokyo, New York, Hong Kong and Frankfurt, account for a major share of all international banking. London, Frankfurt and New York account for an enormous world share in the export of financial services. London, New York and Tokyo account for over one-third of global institutional equity holdings, this as of the end of 1997 after a 32 per cent decline in Tokyo's value over 1996. London, New York and Tokyo account for 58 per cent of the foreign exchange market, one of the few truly global markets; and together with Singapore, Hong Kong, Zurich, Geneva, Frankfurt and Paris, they account for 85 per cent of this, the most global of markets.

Why is it that at a time of rapid growth in the network of financial centers, in overall volumes and in electronic networks, we have such high concentration of market shares in the leading global and national centers? Both globalization and electronic trading are about expansion and dispersal beyond what had been the confined realm of national economies and floor trading. Indeed, one might well ask why financial centers matter at all. The continuing weight of major centers is, in a way, counter-sensical, as is, for that matter, the existence of an expanding network of financial centers. The rapid development of electronic exchanges, the growing digitalization of much financial activity, the fact that finance has become one of the leading sectors in a growing number of countries, and that it is a sector that produces a dematerialized, hyper-mobile product, all suggest that location should not matter. In fact, geographic dispersal would seem to be a good option given the high cost of operating in major financial centers.

There are, in my view, at least three reasons that explain the trend towards consolidation in a few centers rather than massive dispersal. I developed this analysis in *The Global City* (1991), focusing on New York, London and Tokyo, and since then events have made this even clearer and more pronounced.

SOCIAL CONNECTIVITY AND CENTRAL FUNCTIONS

While the new communications technologies do indeed facilitate geographic dispersal of economic activities without losing system integration, they have also had the effect of strengthening the importance of central coordination and control functions for firms, and even markets. Indeed, for firms in any sector, operating a widely dispersed network of branches and affiliates and operating in multiple markets has made central functions far more complicated. Their execution requires access to top talent, not only inside headquarters but also, more generally, from innovative milieux – in technology, accounting, legal services, economic forecasting and all sorts of other, many new, specialized corporate services. Major centers have massive concentrations of state-of-the-art resources that allow them to maximize the benefits of the new communication technologies and to govern the new conditions for operating globally. Even electronic markets such as NASDAQ and E*Trade rely on traders and banks which are located somewhere, with at least some in a major financial center.

One fact that has become increasingly evident is that to maximize the benefits of the new information technologies, firms need not only the infrastructure but a complex mix of other resources. Most of the added value that these technologies can produce for advanced services firms lies in so-called externalities; and this means the material and human resources – state-of-the-art office buildings, top talent and the social networking infrastructure that maximizes connectivity. Any town can have fiber optic cables, but this is not sufficient.

Cross-border mergers and alliances

Global players in the financial industry need enormous resources, which is leading to rapid mergers and acquisitions of firms, and strategic alliances between markets in different countries. These are taking place on a scale and in combinations few would have foreseen just three or four years ago. There are growing numbers of mergers among, respectively, financial services firms, accounting firms, law firms, insurance brokers, in brief, firms that need to provide a global service. A similar evolution is also possible for the global telecommunications industry, which will have to consolidate in order to offer a state-of-the-art, globe-spanning service to its global clients, among which are the financial firms.

These developments may well ensure the consolidation of a stratum of select financial centers at the top of the worldwide network of 30 or 40 cities through which the global financial industry operates. We now also know that a major financial center needs to have a significant share of global operations to become such. If Tokyo does not succeed in getting more of these operations, it is going to lose standing in the global hierarchy, notwithstanding its importance as a capital exporter. It is this same capacity for global operations that will keep New York at the top levels of the hierarchy even though it is largely fed by the resources and the demands of domestic (although state-of-the-art) investors.

In brief, the need for enormous resources to handle increasingly global operations, in combination with the growth of central functions described earlier, produces strong tendencies towards concentration and hence hierarchy in an expanding network.

DENATIONALIZED ELITES AND AGENDAS

Finally, national attachments and identities are becoming weaker for these global firms and their customers. Thus, the major US and European investment banks have set up specialized offices in London to handle various aspects of their global business. Even French banks have set up some of their global specialized operations in London, inconceivable even a few years ago. Deregulation and privatization have further weakened the need for national financial centers. The nationality question simply plays differently in these sectors than it did even a decade ago. Global financial products are accessible in national markets and national investors can operate in global markets. For instance, some of the major Brazilian firms now list on the New York Stock Exchange and bypass the Sao Paulo exchange.

One way of describing this process is as an incipient denationalization of certain institutional arenas. It can be argued that such denationalization is a necessary condition for economic globalization as we know it today. The

sophistication of this system lies in the fact that it only needs to involve strategic institutional areas – most national systems can be left basically unaltered. Major international business centers produce what we could think of as a new sub-culture, a move from the "national" version of international activities to the "global" version. The longstanding resistance in Europe to mergers and acquisitions, especially hostile takeovers, or to foreign ownership and control in East Asia, signals national business cultures that are somewhat incompatible with the new global economic culture. I would posit that major cities, and the variety of so-called global business meetings (such as those of the World Economic Forum in Davos), contribute to de-nationalizing corporate elites. Whether this is good or bad is a separate issue, but it is, I would argue, one of the conditions for setting in place the systems and sub-cultures necessary for a global economic system.

CONCLUSION

Economic globalization and telecommunications have contributed to producing a spatiality for the urban which pivots on cross-border networks and territorial locations with massive concentrations of resources. This is not a completely new feature. Over the centuries, cities have been at the crossroads of major, often worldwide, processes. What is different today is the intensity, complexity and global span of these networks, the extent to which significant portions of economies are now dematerialized and digitalized and hence the extent to which they can travel at great speeds through some of these networks, and the numbers of cities that are part of cross-border networks operating on vast geographic scales.

REFERENCES FROM THE READING

Cohen, S. and Zysman, J. (1987) *Manufacturing Matters*, New York: Basic Books.

Sassen, S. (1991) *The Global City: New York, London, Tokyo*, Princeton, NJ: Princeton University Press.

Sassen, S. (ed) (2002) *Global Networks, Linked Cities*, New York: Routledge.

5

"Urban specialization in the world system: an investigation of historical cases"

from *Urban Affairs Quarterly* (1986)

Nestor P. Rodriguez and Joe R. Feagin

EDITORS' INTRODUCTION

Along with Christopher Chase-Dunn (1985), Michael Timberlake (1985), Fernand Braudel (Ch. 2) and Janet Abu-Lughod (Ch. 8), sociologists Joe Feagin (Ella McFadden Professor of Liberal Arts at Texas A&M University) and Nestor Rodriguez (Professor and Chair of Sociology at the University of Houston and Co-Director of the Center for Immigration Research) were among the first scholars to examine systematically the process of world city formation in comparative-historical perspective. Their classic study of urban specialization in world cities was published during the mid-1980s, just as a large number of North American and western European urbanists were beginning to grapple more explicitly with the problematic of globalization. Both authors have continued subsequently to make important contributions to the study of globalized urbanization.

Rodriguez and Feagin cast their net widely in this article to examine three sets of cities, in each case one financial center and one industrial center, situated within the globally hegemonic national state during three different phases of world capitalist development. Through their comparison of Amsterdam-Leiden (during the period of Dutch hegemony), London-Manchester (during the period of British hegemony) and New York-Houston (during the period of US hegemony), the authors explore the key role of "urban specialization"– the spatial concentration of particular types of economic activities within cities–in the dynamics of world city formation. Their major insight is that urban specialization assumes different forms (a) within globally hegemonic national economies and (b) during successive phases of capitalist development. In each case, Feagin and Rodriguez show how advanced economic capacities in a particular sector, be it financial or industrial, serve as an essential foundation for the development and consolidation of global cities. More generally, in investigating the political economies of urbanization from early to late capitalism, they show that urban development patterns cannot be explained adequately in regional and national terms, but must be understood, instead, with reference to cities' positions in worldwide spatial divisions of labor. In this manner, Rodriguez and Feagin criticize mainstream US urban studies, including writers in the Chicago School tradition, for an excessively localist, culturalist orientation that brackets the large-scale political-economic contexts in which urban development unfolds. In addition, through their expert use of thick historical description, the authors also demonstrate how locally powerful corporate elites have managed to transform their private economic assets not only into a basis for local political influence, but also into a means for consolidating and extending their host city's global reach.

INTRODUCTION

In the history of urban sociology in the United States, much of urban-growth theory has been limited to a national level of analysis. From classical-location to uneven-development theories, conceptualization of urban growth has been contained mainly in regional or national frameworks. But in many cases, involvement in the capitalist world-economy has been a source of major urban development. Many major cities grow during certain periods mainly because of a specific function they play in the capital-accumulation circuits of the world-system. That is, some cities grow as the world-system grows because they have specialized in some function of capital accumulation, for example, as producers or as financial markets, and thus fit a specific niche in the world-economy. Taking a world-system perspective as a necessary level of analysis for understanding major urban growth, the discussion in this article elaborates on the concept of urban specialization, offers case studies of urban growth through a specialization niche, and compares the development of specialization from the standpoint of the three stages of capitalist development (mercantile, industrial, and monopoly) in which the cases are situated.

SPECIALIZATION IN THE WORLD-SYSTEM: A THEORETICAL INTRODUCTION

What is urban specialization?

It is perhaps an empirical commonplace to observe that cities tend to specialize in certain types of economic activities. London, for example is known as a financial center, and Detroit is identified with automobile production. In the present discussion, taking a world-system perspective of urban growth, the concept of specialization is developed and advanced from the standpoint of the development of specific cities in the world-economy. Structural dimensions of world capitalism are emphasized: (a) the circuits of capital, (b) the international division of labor, and (c) the spatial organization. These three dimensions are believed to be the basic matrices of socioeconomic development in the world capitalist system. In contrast to the conventional view, the relational aspect of specialization in a world context is emphasized. A city develops specialized economic activity and a corresponding social structure as a consequence of its relationship to other cities and regions in the capitalist world-system. This development may or may not involve the subdivision of extant industries. It may involve subdivision of an old industry, but specialization may also occur through the development of a new industry.

THREE SETS OF HISTORICAL CASES

In this section, historical cases of urban specialization in the capitalist world-system are described. The cities are financial and production centers in three historical periods of capitalist development: the mercantile, industrial, and advanced-monopoly periods. The specialized cities are situated in countries that dominated the world-economy at certain periods during its development. The core countries that acquired hegemony in the world-economy were settings where urban specialization intensified, especially since superior productive efficiency substantially undergirded hegemonic status. For the period of Dutch hegemony (1625–1671), the cases of Amsterdam and Leiden are examined; for the period of British hegemony (1815–1873), London and Manchester; and for the period of U.S. hegemony (1945–1967), New York and Houston. For each period, the dominant financial/administrative center was chosen. Leiden and Manchester were dominant industrial cities in the Dutch and British periods of hegemony, respectively. Houston was the dominant industrial-technological center of the developing petrochemical industry during the U.S. period of hegemony.

In the development of the world-system, new stages of capitalism brought new structural conditions of urban specialization. The transitions from mercantile to advanced-monopoly capitalism contained changes in the systemic relations between financial and industrial specialized centers, in the relations between specialized industrial cities and the periphery of the world-economy, and, related to this, in the working-class development of the specialized industrial cities.

THE ERA OF DUTCH HEGEMONY

Amsterdam's financial specialization

Amsterdam developed into an international financial center during the period of Dutch hegemony in the seventeenth century. In the division of labor of the

Plate 2 Amsterdam historic canal

Source: Roger Keil

then young capitalist world-system, Amsterdam was a center of manufacturing, shipbuilding, and trading, a place where commodity capital was converted into money capital. But it was foremost a center of finance for business projects throughout the world. Amsterdam's Exchange Bank (established in 1609), Loan Bank (1614), and Burse (1611) made the city the focal point where money capital was converted into commodity capital and where monies were enlarged through loans and investments abroad. From a town with no special status in the mid-sixteenth century, Amsterdam grew into the "Wall Street" of the world-system in the seventeenth century.

During this growth, the population of this world city quadrupled, from about 50,000 in 1600 to about 200,000 by the end of the century. Amsterdam's specialization in finance, which stimulated the growth of its manufacturing and trading industries, involved the in-migration of impoverished peasants from northern Holland and craft workers from southern Holland.

From the perspective of the spatial organization of intra-European trade, Amsterdam was ideal. It was situated between the Baltic area countries of Russia, Poland, Sweden, and Denmark and western and southern European areas that depended on the former countries for grain imports. Amsterdam became a capital-circuit microcosm of mercantile capitalism, with "an abundance of ever-ready goods and a great mass of money in constant circulation" (Braudel, 1984: 236). This position in the circuit of capital was significant for the financial rise of Amsterdam because the city's bank was originally developed to enable merchants to settle mutual debts. Trade and finance are thus intimately linked.

The bank of Amsterdam provided security rare in seventeenth-century banking because of the interventionist role of the state. Instead of adhering to mercantilist policies, at an earlier point in time the Dutch state (funded by Amsterdam) created conditions that facilitated private enterprise, such as the free movement of bullion, a system of bills of exchange, and credit functions. In some cases, the Dutch government guaranteed loans made by private financiers. Amsterdam's government enhanced the city's capitalistic environment by promoting the immigration of business entrepreneurs. The municipal council kept the cost of citizenship, required in some businesses, low and found housing and offered other inducements for immigrant master craft workers who agreed to start industries in silk finishing, cloth making, leather gilding, glassblowing, mirror manufacturing, salt refining, or shipbuilding.

Relations between employers and workers in Amsterdam were relatively stable, in part because of social welfare measures (e.g., lodging, hospitalization, and monetary payments for some) funded by private sources and by revenues derived from the sale of confiscated properties of the Roman Catholic Church. Finance capitalists in Amsterdam invested in profit-motivated projects that greatly enhanced the development of the capitalist world-system in the seventeenth century. Examples included mining, timber exporting, and the development of sawmills in Sweden; mining, timber exporting, and salt production in Denmark; copper mining in Norway; sulfur exporting and fishing in Iceland; exporting caviar, tea, oil, and wool in Russia; grain exporting in Poland; and lake dredging and canal construction in Italy, France, and England. Amsterdam capitalists invested heavily in the Dutch East and West India companies, which through colonization helped expand the periphery of the world-economy to the West Indies, the Indian Ocean, and the Far East.

Textile specialization in Leiden

The world-trading networks that opened up with the assistance of Amsterdam finance capital stimulated the development of specialized industries in many areas in Holland. In the period of Dutch hegemony, textile production centered in Leiden. Leiden's cloth production was central to the Dutch economic ascension in the young capitalist world-system. Cloth and draperies became an important resource for the Dutch trade in the Baltic, the Mediterranean, the Levant, Africa, and the West Indies. Textile industrialization made Leiden second only to Amsterdam in population size in the United Provinces.

Within the circuits of capital of seventeenth-century Europe, Leiden was the largest single-industry setting where money capital was converted into commodity capital of greater value. In addition to large-scale manufacturing, its labor structure involved traditional craft (*fabrieken*) production that used capital mainly for the wages of the skilled workers. Many of the *fabrieken* industries produced for a consumer-market niche of new lighter-weight cloth. Textile specialization made Leiden the largest manufacturing center of the "new draperies" in seventeenth-century Europe. Leiden's textile development benefited from its geographical position at the center of an area of cloth commerce, a confined northern European area where semifinished cloths imported from England were dyed and dressed in Amsterdam, and German linens were bleached and finished in Haarlem.

Leiden's growth as a center of textile specialization involved political and class developments. The urban centers of Holland such as Leiden prospered because they dominated the countryside economically and politically. With large-scale immigration, the Leiden area became a labor-rich environment. But immigration was not the only source of labor. Leiden's labor force was supplemented with child workers, many of whom were imported. Through the use of child and immigrant labor, large manufacturers circumvented the older artisan (guild) mode of production, in which craft workers controlled the scale of operation.

In contrast to the later industrialization, in which factories operated with unorganized work forces with little power, Leiden's industrialism involved *fabrieken* industries, in which artisan guilds still had some control. The cloth guilds in Leiden and in other towns in Holland tried, albeit unsuccessfully, to resist changes in their regulation of work hours, wages, and sizes of work forces. The guilds also refused to accept alien workers.

The second half of the seventeenth century witnessed a sharp decline in Leiden's textile industry. Several factors precipitated this decline: the high wage levels of skilled cloth workers, the loss of markets in Spanish-controlled territories, English mercantile policies that restrained the export of English and Scottish wool, the relocation of textile production by Dutch capitalists to the low-wage countryside outside Holland, and the growing popularity in Europe of the cheaper calicoes imported from India.

Leiden's textiles helped put the United Provinces (especially Holland) at the top of the spatial hierarchy of the emerging world-system in the seventeenth century. With its relatively advanced industrialization, Leiden became the leading textile area in the spatial organization of the evolving world-economy. The technical superiority of Leiden's textile specialization gave Dutch merchant capitalists an important trading advantage in other core countries and in the semiperipheral and peripheral areas of the Mediterranean, the Levant, Africa, and the West Indies. For a few places in the expanding world-economy, Leiden also provided textile technology and skilled labor.

THE ERA OF BRITISH HEGEMONY

London's financial specialization

In the period of British hegemony in the world-system (1815–1873), world trade and finance centered on London. The city was a major source of finance capital for the vast British trading network that, in addition to western Europe, included cities in the Baltic area, the West Indies, India, the Far East, and

Plate 3 The Shard, London, looming

Source: Roger Keil

the Americas. Other major cities (e.g., Manchester) were also important markets for major primary commodities, but no city in England or abroad matched London's financial function in the world-economy. Capitalists from across Europe and a few more from the United States traveled to London to obtain funds for a diverse range of enterprises. Foreign national and state government officials from places such as Russia, Holland, Egypt, Massachusetts, Alabama, and Bolivia journeyed to the banking houses of London for loans, especially for the development of transportation systems.

London bankers funded operations that enhanced transportation in many regions of the world-system. A prime example of this was the financing of railroad construction in western Europe, India, and the Americas. Railroads added to economic growth by cheapening the price of transport, and consequently the cost of producing commodities. Railroads stimulated the development of supporting industries (coal mining, locomotion production); passenger traffic became a business in itself. Investment in railroad construction was an initial event that linked London's financial houses directly to industrialization. Prior to this investment, London's financiers were involved in the growth of industrialism mainly by providing funds for the buying or transportation of industrial commodities.

London developed as a world financial center even before it developed as a national financial center; this speaks for the complexity of regional and world capital circuits that underlie the growth of urban specialization. From the chartering of the Bank of England in 1695 to the end of England's industrial revolution, merchant-bankers were the core of London's financial sector. Throughout the first half of the nineteenth century, English industrial capitalists were not prominent in London's financial establishment.

London's rise to national financial dominance grew out of the problems of small private banks. While in a few cases the new industrial capitalists obtained funding directly from London bankers, in the early years of the Industrial Revolution, the source of money capital for the development of industrial capital was the small private bank. However, an increasing number of country bank failures (311 in the period 1809–1830) prompted Parliament to pass legislation that limited the note-issuing power of these banks, allowed large joint-stock banks in and around London, and permitted the Bank of England to set up branches throughout the country. The Bank Charter Act of 1844 gave exclusive note-issue rights to the Bank of England, giving London the supreme financial position it long enjoyed in the world-economy.

State intervention was critical to the expansion of the capital circuits. The British government's designation of the Bank of England as its financial source provided an important vitality to London's finance sector. London's growth as the nineteenth-century world financial center was based on the immigration of foreign capitalists who set up offices in the city. The growth of London's merchant-banker sector led to the development of many supportive structures that contributed to making London a large metropolis. Retail stores were set up in London to cater to the upper classes.

But the majority of London's growing population (just under 5 million in 1881) consisted of working-class people. These worked in a variety of service jobs critical to the functioning of the world's financial center and in industries such as flour milling, breweries, iron-muggery, and dock loading. At least 40 per cent of the city's working class lived in poverty. Many of the poor were migrants, mainly laborers from southern England, and Irish and Jewish immigrants.

COTTON SPECIALIZATION IN MANCHESTER

To a considerable extent, the rise of industrial capitalism in Britain was facilitated by London's financing of industrial and trade projects in other core countries and in peripheral areas across the world-economy. Areas that used financial capital from London also depended on British manufactured products for their economic development. Thus with the extension of capital circuits in the world-system during the late eighteenth and nineteenth centuries, British industrial cities became "the workshops of the world." Because of world demand, cotton-cloth production led the way in the development of British industrial capitalism. The cotton cities experienced the most dynamic mechanization and urban growth; of these, Manchester was the most developed. Specializing in the production segment of the circulation of capital in the world-system, Manchester underwent a rapid concentration of labor and of buildings, machines, and raw materials that reached the unprecedented scale of a massive factory system.

Manchester's industrial specialization was an important factor in Britain's development as a supplier of half of the world's cotton goods. Cotton products manufactured in Manchester were exported through Liverpool to core countries such as France and Germany, and to areas in the periphery such as Africa and Asia, and especially to Latin America and India.

From a spatial perspective, Manchester specialized as a critical manufacturing node in an urban network – the means by which Britain, as a core area, was related to other core, peripheral, and semiperipheral areas in the world-system. Without these sources of raw cotton and these consumer markets, Manchester could not have achieved its remarkable industrial specialization.

Various political, spatial, environmental, and class factors were related to Manchester's cotton specialization. Manchester's spatial situation was enhanced through the development of transport systems. Two elements in the natural environment of Manchester were resources for the area's industrialization: swift streams and coal. Before steam technology, streams powered the waterwheels of the textile industry. The use of steam engines in the late eighteenth century made coal an important fuel. By the end of the

Plate 4 Industrial Age glory in Manchester

Source: Roger Keil

Industrial Revolution in the 1840s, the geographical shape of the Manchester textile-manufacturing district was substantially determined by the location of adjacent coalfields.

Manchester's specialization involved labor migration as well. While Manchester actually exported labor for rural industrial growth in the early development of industrialization, by the second and third decades of the nineteenth century the city was a destination for many who abandoned the handlooms in small communities and headed for the factories.

The central factor in Manchester's development of textile specialization was a radical change from artisan workers to wage-earning factory workers. Whereas at the beginning of industrial capitalism many British workers owned their tools and controlled their own work, by the mid-nineteenth century many worked for wages in factories. This transformation was not peaceful. Different artisan groups revolted as their trades became deskilled through factory organization. In the new conditions of class relations, women and children were used by British factory owners to displace the skilled workers.

THE ERA OF U.S. HEGEMONY

New York's financial specialization

New York's emergence as the world financial center in the period of U.S. political-economic hegemony (1945–1967) occurred in a setting of dramatic American economic expansion into the world-system. Rapid postwar development, the opening of market outlets in reconstructed western Europe, and investment opportunities in Latin America without competition from European or Japanese investors all helped thrust U.S. capital into a dominant position in the international circuits of capital. In the United States, the centralization of industrial capital was a salient feature throughout the period of hegemony.

Historically, the financial district of New York was "the child of the port." International trade stimulated the financial growth of New York. With the construction of the Erie Canal and other canal systems in Ohio and Pennsylvania, the port of New York took a dramatic leap in international trade. With this dynamic commercial setting, and with the immigration of impoverished European labor, a large number of industries (garment, chemical, tobacco) located in the city.

By 1900 the New York financial district had evolved into a complex formation of commercial and investment banks, insurance companies, stock markets, and the central offices of large and small firms. Of the 125 insurance firms in Manhattan, foreign-owned fire and marine insurance companies outnumbered any single type of domestic insurance company. With the decline of European capital by the end of World War I, New York financial institutions emerged as one of the world's specialized banking centers. During the 1945–1967 period of U.S. hegemony, the expansion of industrial capital stimulated the growth of financial capital. New York financial capitalists increased in wealth as they provided much external funding to nonfinancial corporations.

The foreign growth rate of U.S. banks has far exceeded their U.S. growth rate. New York banks have even expanded into nonfinancial operations in peripheral countries. New York's financial specialization has been particularly important for production capital in the Latin American periphery. The city has served as headquarters for many large corporations that do business in the area.

The growth of New York's financial specialization in the world-system during and after the U.S. period of hegemony involved a restructuring of the city's labor demand. Since the 1950s, the city has lost more than 400,000 manufacturing jobs in industries such as clothing, publishing, and electronics. In the 1970 census, workers in service industries outnumbered workers in manufacturing activities for the first time. But the growth of the service sector linked to the financial and corporate headquarters complexes has involved polarized concentrations of highly paid and poorly paid jobs. High-income jobs are associated with the growth of the advanced services (e.g., financial, managerial) and of international corporate headquarters. The expansion of low-income service jobs is associated with the growth of low-skill, dead-end jobs in major service industries and with the increase of business services that cater to the top-level work forces of these industries. The city's 1.5 million Hispanics and large concentrations of Asians and West Indian immigrants – symbolic again of migration in a world-economy – have been an important supply of labor for the low-wage service jobs, as well as for the resurgent sweatshops and industrial homework.

New York's financial capitalists have played a critical role in remaking the face of the city itself. For example, they have sought state aid to provide critical

Plate 5 One World Trade Center, New York

Source: Roger Keil

infrastructure in the form of highways, bridges, and even office buildings. The two 110-story World Trade Center towers in lower Manhattan were the brainchild of the Downtown-Lower Manhattan Association, chaired by David Rockefeller, then head of Chase Manhattan Bank. Designed to house international banks, import and export companies, and world trade organizations, the twin towers involved the razing of older structures and the construction of these mega-structures at local government expense (about $1 billion). Government aid in the form of special utility and street projects has also facilitated numerous private office complexes, such as the Rockefeller Center, with its 21 office towers.

New York's financial capitalists have even become the overseers of the city government itself. The very institutions that had encouraged the city to overextend itself financially, the banks holding New York municipal securities, had used the fiscal crisis in the 1970s to force the workers of New York to accept a fiscal austerity plan. Finance capital was, with the aid of corporate executives from utility and airline companies, running the city.

Oil-industrial specialization in Houston

The beginning of the U.S. period of hegemony in the 1940s coincided with an oil-industrial boom in Houston that clearly established the city as the "oil capital of the world." Post-World War II oil speciali-zation in Houston consisted of three parts: the pro-duction and refinement of crude oil from nearby oil fields and from peripheral countries, the emergence of a major petrochemical industry, and, accelerating in the 1960s, the distribution of oil technology for the world's oil fields and industries. Since the 1910s this industrial specialization was shaped in part by New York area corporate headquarters and bank com-plexes; in turn, this specialization stimulated the growth of many support industries in the Houston area, from steel companies to downtown office devel-opers. Because of its development of oil refining, pet-rochemical, and oil tools production, the Houston area achieved an elevated standing in many respects. By the 1960s and 1970s Houston oil-industrial spe-cialization was so connected to the world-economy that it was not substantially affected by the decline of U.S. economic hegemony in the 1970s.

From the perspective of circuits of capital, Houston's oil-industrial specialization involves differ-ent forms and levels of capitalistic interests. Finance capital has been involved. East Coast finance capital periodically has been a critical source of loans for some of the companies that have petrochemical, oil technology, and real estate projects in the area. In addition, British, Canadian, German, Iranian, Mexican, and Saudi Arabian capital invested heavily in the city's real estate sector in the 1970s and 1980s. As a center of oil-industrial specialization in the world-economy, Houston has been the focal point for innovation in the world's oil industry. Specialization has, indeed, charac-terized this world city.

Consider Houston in the spatial structure of the world-system. Houston's industrial specialization has served as a means by which the United States, as a core area, developed enduring relationships with cities in outlying peripheral areas. In the post-World War II era, the growing Houston economy enhanced the incorporation of oil-rich peripheral regions into a capitalist world-system dominated by the United States. Oil-industrial specialization developed in Houston, and not in some other part of Texas because of nearby oil fields. The proximity of natural resources was very important at the start of Houston's specialization, because a pipeline infrastructure did not exist.

Yet the actions of Houston's business elite to advance the area's oil industry demonstrated that the source of Houston's specialization was shaped by political and boosterism factors as well. The local business elite promoted the economic growth of the Houston area by seeking spatial and governmental financial advantages. In spite of their widely heralded "free enterprise" philosophy, members of Houston's business elite were masters at using governmental aid to accelerate the development of local infrastructure and enterprise.

Workers from rural areas in Texas and surrounding states, from other urban areas, and from peripheral areas in the world-system have supplied Houston with labor. Migrating mainly from the peripheral countries of Mexico, El Salvador, and Guatemala, the area's undocumented Hispanic population numbers about 200,000. During Houston's economic boom in the 1970s and early 1980s, these workers played a critical labor role in construction and real estate development.

THE POLITICAL ECONOMY OF DEVELOPMENT

From the standpoint of the impact on the world-system, the development of urban specialization represents the development of a global city. This is easier to recognize in cities with financial specialization (e.g., London and New York), where multinational corporations and international banking institutions made decisions concerning economic development throughout the world. But cities with production specialization may also be considered global cities. As the cases of Leiden, Manchester, and Houston demonstrate, such specialized cities have substantial world-level impact, for example, in attracting foreign migrant labor, extending international commerce, and spatially enlarging the periphery in search of raw materials and commodity markets. Thus, for major cities with specialized production, the sources of supply and demand are global.

Though widely used in urban theory since the writing of R. D. McKenzie in 1926, the concept of "specialization" has not received the analysis it requires. Instead of exploring the dimensions and development phases of urban specialization, many analysts have been content to see specialization as an outgrowth of natural-ecological processes that proceed through an invisible-hand type of logic. The analysis presented in this article offers a different view of how specialized activity develops in cities. Far from being a "natural" process, specialization is grounded in the political economy of development. For each case of specialization the intentional business (class) and political actions that undergirded its development can be identified.

REFERENCES FROM THE READING

Braudel, F. (1984) *Civilization and Capitalism, 15th–18th Century: Vol. 3, The Perspective of the World*, New York: Williams Collins and Harper & Row.

McKenzie, R.D. (1926) The scope of human ecology, *American Journal of Sociology*, 32, July: 141–54.

6

"Accumulation and comparative urban systems"

from *Comparative Urban Research* (1977)

John Walton

EDITORS' INTRODUCTION

John Walton, Professor Emeritus at the University of California, Davis, centers his research on historical and comparative sociology, political economy of development, and urban sociology. He is the author or co-author of several books exploring the urban political economy and underdevelopment in the global South, including *Urban Latin America: Political Conditions from Above and Below* (1976, with Alejandro Portes) and *Reluctant Rebels: Comparative Studies of Revolution and Underdevelopment* (1984). This essay is excerpted from an article published in 1977, when he taught at Northwestern University. In the essay, Walton sought to establish a common theoretical framework for comparative urban studies on development in cities in the less developed global South. He proposed to do so by linking urban research to the world-system and the urban political economy perspective. Moving beyond previous scholarship that focused on socio-economic processes within nation-states, Walton drew upon a consensus that had emerged by the mid-1970s in the field of urban studies that a more integrated perspective was needed to understand development issues in the global South. According to this view, the underdevelopment in cities in the global South had as much to do with the way they were integrated in the global economy as with their domestic institutions and socio-economic processes. The inter-dependency between core and periphery countries largely shapes urban structures and processes in cities in both the developed and developing countries. From the perspective of urban political economy, cities are products of dominant modes of production and accumulation.

The world-system perspective and urban political economy enabled urban scholars in the 1970s to adopt a global approach to study urban development, but an analytical framework that could guide comparative urban analyses still was missing. Walton's essay filled the gap by proposing such a framework to examine the various roles of cities in their integration into the world economy, and the resulting impact on their urban structures and process. He identified three mechanisms in the transition of peripheral countries to capitalism: (1) the creation and extension of commodity relations, which produce hierarchies of towns; (2) foreign trade, which initiates massive urbanization through migration and alters urban economy through its orientation to external demands; and (3) foreign investment, which advances the tertiarization of the labor force and also leads to property speculation. Walton then traced how these mechanisms, in turn, altered urban systems and hierarchies, and also the spatial distribution of social classes. This essay demonstrates how comparative urban analyses can be pursued under such a framework with examples from Latin America and Africa.

It is always tempting to begin discussions of a topic as broad as urbanization and development by lamenting the lack of progress in the field or denigrating other researchers for their narrow range of geographical and substantive interest and adherence to outmoded theories. Such laments combine elements of truth and falsity with exaggerated expectations and self-interest, but they never carry us very far toward a fuller understanding. Certainly the field of urbanization and development is fragmented, reasonably chaotic. But, given its complexity and relatively recent comparative study, I do not see how it could be otherwise. What seems more impressive, however, is the amount of exciting new research and nascent theoretical integration. My discussion will attempt to demonstrate this trend and pose some of the important questions we now confront.

At the most general theoretical level there is growing recognition of the fact that the categories and assumptions of conventional developmentalist thinking must be superseded by approaches to "the modern world system" (Wallerstein 1974). Evolutionary models based upon traditional, modern-industrial, and post-industrial societies must be replaced by holistic frameworks which compare the advanced societies and analyze the environment they create for the underdeveloped.

It is now commonly accepted that the underdevelopment of the "third world" must be explained in some considerable part by the mechanisms of structural dependency, unequal exchange, and accumulation on a world scale. That is, where earlier research on development tended toward an exclusively international focus in sequential studies of nation-specific obstacles to change such as economic imbalances, institutional rigidities, or traditional practices, it is now generally agreed that peripheral countries owe much of their underdevelopment to the mechanisms by which they have been incorporated into a global economy. This is not to say that richly variegated national experiences are suppressed in favor of the vagaries of a "global approach." Rather, the point is to examine comparatively how different satellite nations variously articulate with an international political economy and how these interfaces, in turn, interact with local conditions to produce the special circumstances of underdevelopment.

Less attended to, but equally valid, is the fact that the national economy, labor force, and class structure of advanced countries like the United States cannot be adequately understood without reference to a set of international exchange processes. That is, analysis of the internal functioning of the more developed societies, no less so than the underdeveloped, must begin with a global community perspective.

Recent developments in Europe, Latin America, and the United States suggest that. . .a theory of urban political economy is emerging. Representative expressions of this approach are variously labeled as Marxist or Neo-Marxist, structuralist, political economy, and dependent urbanism (Castells 1972). Despite the various terms (which seem to reflect more and less adherence to Marxian analysis), the approaches are united by a dual effort: to develop the historical and economic foundations of urbanism and urbanization, and to account for various urban forms by reference to changing modes of production and accumulation. Although the theory has implications for urbanism under a variety of economic systems (Harvey 1973), its application to date has been confined to advanced and dependent capitalist states. The distinctive role of cities in these systems is to increase productivity and stem the tendency for rates of profit to fall by creating the general conditions of production, i.e., by concentrating spatially the means of production (the labor force, the productive unit or work place, and the means of communication) in order to reduce the time and cost of the circulation of capital.

If this conception of the role of cities in capitalist economics provides a general orientation, by no means does it "explain everything" or foreclose on theoretical explanations of differentiated urban forms. Thus Harvey (1973: 201) regards urbanism as a social form "broadly consistent with the dominant mode of production. . . .like any social form, urbanism can exhibit a considerable variety of forms within a dominant mode of production, while similar forms can be found in different modes of production." And he suggests that specific urban forms, notably in the uses of space and distribution of social classes, will depend upon the particular function of the urban center in the total pattern of the circulation of surplus, i.e., of accumulation.

Although the general approach appears fertile and powerful, most of the empirical work necessary for its evaluation, elaboration, refinement, or modification remains to be done. Certainly, it is fundamental to compare urban structures in capitalist and noncapitalist systems, but equally important is the application of the approach to third world or dependent urbanism.

For example, despite the rapid expansion of research in Latin America proceeding from "dependency" theory, little attention has been given to specific and variable roles of cities in this process. Rather, as dos Santos (1969) observes, theories of imperialism and neocolonialism tend to view the international expansion of capitalism exclusively from the standpoint of the imperial centers. Typically, the effects of this expansion are assumed to be relatively homogeneous across dependent societies leading to the systematic destruction of precapitalist forms and reducing third world cities to socially and ecologically standard mechanisms for extracting and passing along surplus value through a nested set of metropolitan-satellite relationships. While this tendency in the dependency literature has come under a good deal of criticism recently, substantive efforts to remedy it through specifically urban research on variable forms of articulation between advanced and dependent economies remain to be undertaken.

Moreover, theoretical work on the political economy of urbanization has yet to specify major forms of accumulation and the circulation of surplus, how such distinguishable patterns of economic integration variously condition urban systems and socio-spatial structure, and how institutions such as the state variously mediate the Interplay of economy and urban form. Several writers seem to recognize these gaps, but so far have addressed them in rather imprecise ways. For example, Castells (1972) proposes three historical modes of dependency: colonial, capitalist commercial, and imperialist industrial and financial. Each type of domination is hypothesized to produce certain urban "spatial effects" such as accelerated growth, urbanization without corresponding development of productive capacity, social distance between city and countryside, ecological segregation, and so forth. Although there would appear to be considerable evidence that these effects occur, Castells' analysis does not take us very far theoretically because it does not really differentiate separate causes and effects, i.e., each historical mode seems to produce most of the effects listed.

In what follows we shall advance at least one formulation with which to address these issues. We begin with the recognition that today's problems of urbanization and underdevelopment in the third world had their origin in the expansion of European colonialism. Specifically, they arose with the expansion of capitalism from its European center to the new world periphery—a process best described as the penetration of precapitalist social formations and their transition to peripheral capitalism. Emphatically, this process and its effects took many different forms as a result of the historical timing of the penetration, the particular colonial power involved, and the specific conditions encountered in the peripheral society.

With respect to the indigenous conditions that colonialism encountered, it is useful to identify types of precapitalist social formations. Amin (1974) distinguishes five types; they take into consideration means of production, class organization, authority relations (e.g., in control of the land), and the presence or absence of commodity exchange. The five are primitive community, slave-owning, feudal, "Asiatic" or tributary, and simple commodity exchange. These formations are not to be confused with dominant modes of production; none exists in a pure state. Rather, they represent structural combinations of a variety of limited modes. Although we have little comparative historical research on this topic, the presumptions are that these formations responded in different ways to the introduction of colonialism, and that the transition to peripheral capitalism took different paths as a result of the socioeconomic base on which it began.

While the development of European and peripheral capitalism came about in vastly different ways, both required the creation of a "free" labor force for wage employment (i.e., "proletarianization") and the monetization of the economy. In the transition to peripheral capitalism these requisites are met by three principal mechanisms which do not necessarily occur sequentially or in isolation. The first involves the creation or extension of commodity relations—the conversion of traditional and subsistence economies to commodity producers integrated into a world market. Generally, this integration is based on the export of agricultural products and other raw materials such as minerals. It requires that subsistence farmers, craftsmen, and merchants engaged in simple exchange be converted to wage workers producing a surplus. Typically, this is accomplished through certain incentives or forms of "primitive accumulation" such as compulsory crops, the imposition of monetary taxes, and labor contracting. The entire economy need not (perhaps cannot) be monetized and proletarianized at this juncture. Plantation economies based on slave labor and the Spanish encomienda systems are illustrative forms of primitive accumulation in which only a small segment of the peripheral society entered the money economy.

But these early systems are undermined by the second mechanism: the formation of a peripheral capitalism based on foreign trade, including the expansion of the export sector and the import of colonial manufactures. Agriculture is commercialized to a greater extent for purposes of export by foreign investors. This continues the destruction of traditional commodities and the displacement of rural populations. Local craftsmen are threatened on one side by the progressive elimination of the market represented by traditional communities and small-scale agriculture, and on the other by competition from imports. The principal beneficiaries are the large landowners and comprador classes involved in commerce and export who, in turn, consume imports and luxury goods and invest in urban properties. Wealth is transferred to the towns to support commercial activity as the economy comes to be based on export trade. The urban hinterland assumes a hierarchal structure governed by extractive purposes.

The third mechanism is the formation of a capitalism based on foreign investment. Foreign capital expands beyond the export sector to vertically integrated agricultural and mining enterprises, to the commercial sector and, notably, to industry within the peripheral countries. Obviously this does not eliminate the importance of trade and export; but it magnifies those activities as the capital goods necessary for industry are imported and significant portions of this new industry are devoted to the production of export products as well as to local demand. The domestic market expands as a result of greater numbers of wageworkers and the profits realized by landowners and classes associated with export trade. But the distribution of benefits from this market expansion is highly unequal. Foreign owned agriculture and industry tend to be capital intensive, producing rural underemployment (and migration) and a small "labor aristocracy" in the industrialized urban centers (Arrighi 1973). In turn, the urban tertiary sector expands in numbers while economic opportunity is progressively fragmented in a process of "urban involution." And this process is accelerated by the penetration of foreign capital into the commercial-services sector, reducing the scope of the "informal economy."

On the basis of this formulation we may now return to the issues raised earlier about urban political economy. That is, the framework presented specifies major forms of accumulation that condition third world development and underdevelopment. Therefore, we may now inquire into how these mechanisms variously affect urban structures. Moreover, it would appear possible to do this at two levels, one concerning urbanization patterns (i.e., urban systems and hierarchies) and the second concerning urban structure (i.e., the spatial distribution of social classes). In effect, this entails advancing a series of hypotheses the accuracy of which must await historical and comparative assessment.

In summary, several important junctures in the development of peripheral capitalism (or the "development of underdevelopment") can be identified: the monetization and proletarianization of precapitalist formations, peripheral capitalism based on export of raw materials and import of manufactures, and foreign investment and industrialization in the periphery. These cumulative processes condition, among other things, levels of urbanization and urban (class and ecological) structure. The initial mechanism restructures traditional networks and produces rudimentary hierarchies of towns. The second initiates massive urbanization through migration and distorts the urban economy through its orientation to external demands. The third advances urbanization and the tertiarization of the labor force; but most important it causes inflationary speculation which is at the root of the litany of urban woes often treated separately as the problems of housing, slums, overcrowding, land invasions, squatments, the absence of services, and the lower class politics which these effects generate.

It should be stressed again that we have developed here a highly generalized framework: less a theory than a perspective which sensitizes analysis to some of the broader features of urbanization and underdevelopment and poses key questions for comparative study. For example, concerning the sources of urbanization and urban structure, the framework would suggest researchers inquire how different precapitalist formations are affected by colonialism, what influence does the timing of incorporation have on the course of peripheral underdevelopment, what are the consequences of different colonial powers and policies, and how are these "external" influences mediated by political factors and the role of the State in new nations.

Obviously we cannot begin to answer these in the present discussion, but a few speculative comparisons of Latin American and African urbanization serve to illuminate applications of the approach. More specifically, and more modestly, I would like to share some tentative contrasts that derive from my research in the northern climes of Latin America (especially Mexico and Colombia) and very recent exposure to East Africa (especially Kenya and Tanzania).

In the first instance both regions provide instances of a combination of precapitalist social formations including primitive community, slave owning, tributary, and simple commodity exchange. By way of contrast, however, slave owning or indentured labor and tributary modes may have been more common in Latin America, whereas simple commodity exchange based on long distance trading networks was more typical in Africa. This, like most of the observations that follow, is due in large part to the discrepancy in time between the colonization of the regions. Being colonized much later, Africa was able to develop trading systems more fully—particularly through contact with Muslim caravans. Some consequences of these differences, which also relate to the motives of the colonizers, are that the Spanish in Latin America came initially to plunder the treasures of tributary systems and search out new sources of precious minerals, while the Islamic Sultans, Germans, and British came to East Africa to superimpose their trade networks on existing ones. Spanish colonization more systematically destroyed indigenous society and created new networks for the extraction of raw materials while, to some extent, in East Africa indigenous forms were transformed for other purposes. As a consequence of these different circumstances, the Spaniards became "indigenous" to the Americas by creating a new colonial society on the ruins of the old (including religious conversion of the natives), while the Europeans in Africa created colonialism on top of distorted older forms. Of course, these differences had a good deal to do with the resource endowments of the respective regions. In the mining centers of Africa something like the Latin pattern may have obtained and in regions of the Americas where minerals were absent or quickly exhausted, the conquerors turned to settled agriculture under the encomienda system which has certain parallels with African plantation systems. Another factor of importance was that the Spaniards were out to settle and possess a territorial empire extending far beyond any immediate trade advantages, while the Europeans in Africa penetrated only as far as economic advantage dictated. As a result of all this the Spaniards created in Latin America a large number of cities and towns linked in a cultural, political, and economic urban system. It is suggested that these differences in precapitalist social

formations, the timing of colonization, and the purposes of the colonizers or the form of incorporation are in large measure responsible for the higher levels of urbanization and more complex urban hierarchies found in Latin America today.

What has been said so far is suggestive of different experiences with the mechanisms of transformation to peripheral capitalism. In brief, Latin America has had a longer and more complete exposure to the forms of capital penetration and may now be entering into a new phase which some have described as "technological capitalism" whereby industry in peripheral countries is less controlled by direct investment and a wide range of imports than by the sale and licensing of high technology capital goods. By contrast, African experience with foreign investment in industry is more recent and the principal form of incorporation in many areas may still be trade and importation. Put crudely, Africa may be entering into a form of neocolonial accumulation which Latin America is moving beyond.

The implication to be drawn from our framework, of course, is that urbanization levels and urban problems are greater in Latin America. I believe there is ample evidence for this conclusion, whether we look at the cold statistics on urbanization or at the relative instance of overcrowding, slums and squatments, deficient or maldistributed urban services, and sociospatial polarization.

Finally, turning to political considerations and the role of the State in urbanization and development, the very differences we have considered here suggest alternative possible futures for African and Latin American urbanism. Political independence in Africa came at a time when the mechanisms of imperialism were less advanced than in Latin America. Moreover, whatever criticisms may be leveled at the authenticity of African independence movements of the 1960s, they were certainly more anticolonial than their Latin American counterparts in the 1820s. Thus, for Africa two factors combine to suggest that earlier experiences of urbanization may not be repeated; less extensive incorporation and policies in search of greater autonomy.

More contrasts could be developed, but we are already beyond any systematic comparative evidence. My purpose here is not to provide a comparative study of urban political economy, but to suggest that such studies can and should be undertaken as steps toward a theoretical foundation for urban sociology and a richer comparative understanding.

REFERENCES FROM THE READING

Amin, S. (1974) *Accumulation on a World Scale: A Critique of the Theory of Underdevelopment*, New York: Monthly Review.

Arrighi, G. (1973) International corporations, labor aristocracies, and economic development in tropical Africa, in G. Arrighi and J. S. Saul (eds.) *Essays on the Political Economy of Africa*, New York: Monthly Review: 105–51.

Castells, M. (1972) *Imperialismo y Urbanization en America Latina*, Barcelona: Editorial Gustavo Gill.

dos Santos, T. (1969) La crise de la theorie de developpement et les relations de la dependance en Amerique Latine, *L'Homme et la Societe*, 12, Avril-Mai-Juin.

Harvey, D. (1973) *Social Justice and the City*, Baltimore: Johns Hopkins University Press.

Portes, A. and Walton, J. (1976) *Urban Latin America: Political Conditions from Above and Below*, Austin: University of Texas Press.

Wallerstein, I. (1974) *The Modern World-System: Capitalist Agriculture and the Origins of the European World Economy in the Sixteenth Century*, New York: John Wiley.

Walton, J. (1984) *Reluctant Rebels: Comparative Studies of Revolution and Underdevelopment*, New York: Columbia University Press.

7

"The world-system perspective and urbanization"

from *Urbanization in the World-Economy* (1985)

Michael Timberlake

EDITORS' INTRODUCTION

Michael Timberlake, Professor of Sociology at the University of Utah, has published extensively on globalization and the world system of cities, global urban networks, and economic development policies in the underdeveloped rural regions of the United States. Since the early 2000s, he has been collaborating with his students to examine the structures of urban networks of Chinese cities. This essay is excerpted from the introductory chapter of a book he edited in 1985, *Urbanization in the World-System*, which examines how urbanization patterns can be studied from a world-system perspective. It aimed to theorize and empirically test the ways in which urbanization patterns articulate with the dynamics of the world-economy. The book is a seminal contribution to the scholarship of urban studies from the world-system perspective. In this introduction, Timberlake forcefully argues that macro-level social change cannot be adequately studied without accounting for world-system processes. He first reviews how the study of urbanization has evolved from Max Weber's *The City* to the Chicago School's ecological approach, and then to Marxian urban sociology in the 1970s. He points out the overlap between the Marxist approach and the world-system perspective but also highlights the manner in which their interpretations depart. For example, he suggests that conventional Marxian approaches have overlooked the role of cheap labor and materials in peripheral areas that contributed to the world-system's expansion, and also the importance of political-military rivalries among competing core powers.

Urbanization processes have typically been studied by social scientists as if they were isolated in time and explicable only in terms of other processes and structures of rather narrow scope, limited to the boundaries of such areas as nations or regions within nations. However, within the past 15 years, the study of large-scale social change has been transformed by the emergence of the world-system theoretical perspective. World-system scholars have adumbrated properties of the modern world-economy that allow us to view it as a coherent whole. Much of the research pursued from the point of view of this perspective has been historical, and it has dealt with either the system as a whole or how local social formations (e.g., class relations, social movements, states) are transformed as regions of the world are first incorporated into the structure of the system, and then become subject to processes that reproduce it. The claim is not that world-system processes determine everything. Rather, the fundamental lesson is that social scientists can no longer study macro-level social change without taking into account world-system processes. Specifically, processes such as urbanization can be more fully understood by beginning to examine the many ways in which they articulate with the broader currents of the world-economy that penetrate spatial barriers, transcend limited time boundaries, and influence social relations at many different levels.

Urbanization has been one of the most frequently studied features of the modern world. Since the dramatic growth and spread of urban agglomerations beginning in the nineteenth century, scholars have concerned themselves with documenting, for different countries and regions of the world, such aspects of urbanization as the size and growth of the largest cities, the relative size of urban populations, and changes in urban hierarchies. These phenomena have then been related to other developmental processes, such as level of economic development and differentiation or political change. With few exceptions, there have not yet been attempts to interpret patterns of urbanization in light of a world-system perspective.

THE STUDY OF URBANIZATION

Within North American social science the process of urbanization has conventionally been viewed as an evolutionary outcome of, first, the elaboration of trade relations among relatively isolated localities, and then of industrial development within regions or nations. As in other fields of sociological interpretation, the organic analogy and functionalism have been brought to bear on interpretations of the process of urbanization. The conventional understanding of the growth of ancient cities, for example, stresses as fundamental such general processes as growing specialization and the evolutionary nature of technological change that induces specialization and, thus, promotes some degree of urban concentration. The importance of trade, especially long-distance trade, in giving rise to towns and diffusing technology is emphasized. Such approaches are particularly useful in identifying limitations on urban growth (e.g., limitations placed on the food supply by the level of technical development in agriculture). This general approach has been applied intensively to analyzing the connection between urbanization and industrialization.

Written at the turn of the century, Weber's (1967) seminal statistical study of nineteenth century urbanization employed an explicitly Spencerian interpretation of city growth. The growing concentration of the population in cities is viewed as a "natural" outcome of economic growth and differentiation (Weber

1967:154–229), having to do with factors such as the application of machine power for agricultural production (in the United States) and the attendant job-displacing effects that "encourage" migration to cities, as displaced farm workers seek employment.

The approach taken by Weber turned out to be far from barren. His study itself is highly useful in documenting the rise and spread of urban agglomerations around the world in the nineteenth century. Research on urbanization spawned by both early and more recent "Chicago School" sociologists and human ecologists has also used an evolutionary-organic framework and has yielded a wealth of descriptive and theoretical material on the spread of urbanization over time and space, the relationship between urbanization and other aspects of the industrial division of labor, urbanization, and regional development, and regional development and the elaboration of city systems.

From this general perspective urbanization, or expansion of the local community, and integrated regional development result from ecological processes. Regionally, urban growth is seen as a process of centripetal movement, but from the point of view of each urban center growth is viewed as centrifugal expansion. Cities expand by growth away from centers. It is the patterns of outward movement that are described by human ecologists. The central motivating factors behind centrifugal expansion posited by scholars such as these are reviewed by Hawley (1981). They include (1) sheer population pressure; (2) increasing specialization of, and competition among, functions at the center, driving out less dominant functions; (3) obsolescence of physical structures; and (4) the revolution of short-distance transportation.

Against this approach stand Marxian interpretations of urbanization. Writers in this tradition have taken less for granted about the nature of urbanization—at least they have not taken the same things for granted. More attention is paid to the fundamental ways in which urbanization processes are embedded in specific modes of production. One can overemphasize differences between the evolutionary ecological approach to urbanization and a Marxist approach. Perhaps differences are due to divergent emphases. For example, both acknowledge the importance of many of the same factors—long-distance trade, military and naval power, and slavery—in explaining the development of the ancient cities. However, a Marxian analysis of the rise of the cities, in both classical

Greece and the Roman Empire stresses the requirements of the slave mode of production. For example, Anderson (1974: 35–39) regards the "introduction on a massive scale of chattel slavery" as the "decisive innovation" of this era.

More recently, Marxian scholars have specifically directed their efforts toward understanding the nature of urbanization. In this endeavor writers have often explicitly confronted some of the assumptions of conventional urban social science. Harvey (1973, 1982), Castells (1977), Pickvance (1978), and Gordon (1978) are prominent among those who, using Marxian concepts, have begun to build a compelling framework with which to analyze the social and physical structure of urban life.

From the Marxist perspective urbanization cannot be understood apart from the mode of production under which it exists. It is assumed that the process of urbanization is different in important respects under capitalism than under other modes of production. Under capitalism there is a unique logic that has an important bearing on urban growth and decline. Investments in the built environment are made to facilitate profit-making through commodity production and exchange, and labor exploitation. Again, behind earlier, pre-capitalist urbanization was the logic of surplus extraction through tribute, conquest, and subsequent use of slave labor. For example, in contrast to urbanization under capitalism, towns in the Roman Empire were primarily "consumer towns" in which lived the owners of large amounts of land, supported by income from these estates. "The arena for strictly urban economic development was thus very limited. There was no urban institution which rivaled tenancy as a medium of exploitation" (Hopkins 1978: 79). Capitalist cities are, on the other hand, organized more around market trade and commodity production, and less around political redistribution and warfare. Thus it is crucial to view the nature of urbanization in terms of processes endemic to capitalism.

Cities emerge, grow, become interdependent, and decline in response to the logic and contradictions of capitalism—a mode of production with identifiable space requirements. Edel argues that urban phenomena can best be interpreted in light of specific "aspects of the accumulation process" (1981: 87). These include the conditions under which labor is employed in creating value and surplus value, the way in which labor is reproduced, and the way in which surplus value is circulated and reinvested. Housing problems,

urban renewal, urban disinvestment, community political organization, and many other "urban" issues can be understood within a framework that explicitly acknowledges these factors.

Similarly, Harvey (1982: 373–412) turns our attention to some of the space requirements of capitalism that produce many changes in the physical and social structure of urban areas. The circulation of capital in different forms has implications for urban structure. Harvey points out that two ways in which capitalists seek to obtain competitive advantage are by adapting new technologies and making use of new locations for production and distribution facilities. Either of these efforts to gain advantage over competitors (and thereby to garner relative surplus value) by individual capitalists may have important urbanization consequences. For example, to the extent that labor is an object of capital—by necessity responsive to changes in the location of investments—shifts in the location of capital investments will be reflected in shifts in the location of population, the decline of some population centers, the increasing significance of others, the decline of particular neighborhoods within cities, and so on. Certainly urbanization has accompanied industrialization, but not because "industrialization" per se has dominated, but because urbanization "was the expression of the capitalist logic that lay at the base of industrialization" (Castells 1977: 14).

The recognition that urbanization must be analyzed as a component process of the capitalist mode of production has been elaborated by hypotheses connecting patterns of urban development in less developed countries with mechanisms that link, in various ways, the economies, social classes, polities, and other institutions of these peripheral countries with those of the more developed countries of the capitalist core. *Dependence* is a term used to describe, in a general way, these asymmetrical economic, political, and cultural relations that characterize interaction between the core and periphery. The acknowledgment by many scholars of and in the Third World that dependence is a crucial aspect of capitalist accumulation at the world level has led, in turn, to a fruitful body of research and theory linking dependence to variation in patterns of development across countries. These research efforts have included quantitative cross-national studies, historical studies of countries, local institutions, social movements, social classes, and in-depth case studies of "marginal" classes in Third World cities.

The impact of the Marxian and dependency critiques has been to move the study of Third World urbanization beyond the stage at which passing mention of the importance of taking into account the colonial history of the countries in question constituted sufficient discussion of world-economic forces. Proponents of these perspectives have argued persuasively that urbanization must be studied holistically— part of the logic of a larger process of socioeconomic development that encompasses it, and that entails systematic unevenness across regions of the world. The dependence relation is an important theoretical concept used to pry into the ways in which the processes embodied in the world-system produce various manifestations of this unevenness, including divergent patterns of urbanization.

The world-system perspective is in agreement with the Marxian and dependency positions on many issues. "Developmental problems" in different world regions are, in part, outcomes of processes of the capitalist world-system and must be understood in terms of the system as a whole. However, whereas dependency theory applies mainly to the postcolonial Third World, the world-system perspective allows one to focus upon any level of the system (core, semi-periphery, or periphery) since the emergence of capitalism as the dominant mode of production in Europe in the sixteenth century. The world-system is dynamic, changing as capital adapts to contradictions and crises. Hence, it is not sufficient to identify the structure of the world-economy in any simplistic way. Yet there is an internal logic to the system that is understandable in terms of its processes, among the most important of which is the "peripheralization" of regions of the world-economy that accompanies accumulation, and the subsequent expansion of the system. Further, the perspective entails the argument that there are relatively stable, identifiable, structural expressions of these processes. Thus urbanization patterns, along with other developmental patterns, are expected to be shaped, in part, by the operation of the world-economy. It is this hypothesis that guides the work presented in this volume.

While the world-system perspective is tremendously indebted to both dependency theory and Marxian theories of social change, it differs from them in several important respects. It diverges from the first in focusing on the system as a whole, rather than conceiving of dependency as an internal-external relationship from the point of view of the dependent country. In contrast to more orthodox Marxian interpretations of capitalism, the world-system perspective tends not to delineate distinct stages of capitalism, which begin with the stage of competitive capitalism in the nineteenth century. Rather, Wallerstein describes the emergence of a capitalist world-economy in Europe in the "long 16th century." Its changes are then interpreted in terms of cycles of world-system development. It differs further from orthodox Marxian theory not because accumulation and class struggle are less central to the theory, but because regional unevenness, the importance of the interstate system, cultural diversity, and the structure of the world-economy are given central importance as well. In contrast, many Marxian scholars maintain that unequal development across regions is transitional. These scholars acknowledge that, due to the genesis of capitalism in northwestern Europe, a gap in the level of development emerged between the early capitalist countries and others. But the orthodox position assumes that this gap will be closed as capitalist relations of production become evenly spread over all global regions. In fact there are a growing number of discussions by Marxists of the "new international division of labor" that they see emerging in the world today. This, it is maintained, will serve to further reduce differences among countries in the extent to which capitalist relations of production have become reproduced more or less evenly.

A WORLD-SYSTEM PERSPECTIVE ON URBANIZATION

As we have seen, orthodox Marxists have indeed interpreted urban phenomena as part of the capitalist mode of production. For example, Edel (1981) has specified several "conditions of existence" of the capitalist mode of production in relation to which urban patterns should be understood. Capitalism is by its nature expansive, and reproduction of this mode of production involves accumulation. Hence urban phenomena must be related to aspects of the accumulation process, such as the way in which labor is employed, the way in which labor power is reproduced, and the way in which surplus value materializes in the circulation of commodities (Edel 1981: 22–35).

Such analyses need, further, to account for distortions caused by class conflict. It is essential to realize that capital's requirements are not satisfied without struggle, including struggle centered around the state.

Analysis of urban housing policies, levels and rates of urbanization across regions, and urban ecological structure—such as patterns of racial and ethnic segregation, metropolitan expansion, and central city decay—all could be usefully approached in terms of these requirements of the capitalist mode of production.

Factors that the conventional Marxian approaches to urbanization rarely stress, but which the world-system perspective would emphasize, include the impetus for expansion of the system provided by the availability of cheap raw materials in the peripheral areas, more highly coerced labor in these areas, and political-military rivalry among competing core powers. These themes run through neo-Marxian theories of imperialism, but world-system theorists understand them in terms of institutional features basic to the functioning of capitalism.

According to the framework sketched by Chase-Dunn and Rubinson (1979), the capitalist mode of production is constituted as a world-system, embodying several "institutional constants" throughout each of its four historical epochs. The core-periphery division of labor and labor control and the interstate system are the two institutional constants in this schema. The interstate system fragments the class struggle, confining it within the various nation-states of the system, which are themselves unequal in terms of military, political, and economic strength. Core states are relatively stronger and labor is relatively free in contrast to peripheral labor, which is often subject to coercion over and above that typical of the wage labor system within the core. Surplus product is appropriated from the periphery to the core by a number of market and nonmarket mechanisms within the context of the interstate system. Underdevelopment is reproduced within the system, in the sense that the development gap between the core and the periphery remains even though both "develop." It is sometimes argued that labor in the core benefits from the unequal exchange between zones of the world-economy, and, to the extent that this is true, there are objective reasons for the tremendous differences among nations and across zones in the intensity of the class struggle and in the conditions under which labor is employed.

The world-system is also characterized by three types of "cycles which occur in each of the four epochs" (Chase-Dunn and Rubinson 1979: 279). First are the long waves of increasing capital accumulation and increasing velocity of exchange followed by

slow-down and stagnation. There are several studies that examine relatively local patterns of urban change in terms of cycles of accumulation and stagnation, but very little work has been done on macro-urban processes in the world-economy. At this broader level, several research questions come to mind. Do rates and levels of urbanization vary in such a way as to reflect these cycles? How is this complicated by the core-periphery division of labor? Do city systems at the world level or national level become more integrated and hierarchical during upswings in the cycle?

Another cyclical feature of the world-system involves core competition. "This refers to a cycle of uncentricity vs. multicentricity in the distribution of power and competitive advantage among core states" (Chase-Dunn and Rubinson 1979: 279).

A Third World-system cycle is "the structure of core-periphery trade and control," which exhibits cyclical fluctuations between relatively free-market exchange and exchange that is highly controlled politically; for example, trade that is organized rigidly "within colonial empires." It is quite possible that this cycle, too, would have consequences for macro-urban patterns. This dimension of the world-system is obviously affected by the other dimensions, especially by the cycle of core competition, and it is unrealistic to discuss one cycle in isolation. Nevertheless a few highly speculative hypotheses can be advanced that relate this cycle to macro-urban issues. For example it is quite possible that peripheral economies will be somewhat freer to develop more differentiated economies during periods of free-market exchange. If this is true, this general economic differentiation, in turn, should be reflected in more occupational heterogeneity in the urban labor force of the periphery as a whole, and especially in the more developed peripheral countries with stronger states that are able to exert relatively more resistance to the coercion of the interstate system.

The discussion immediately above is meant as a list of implicit hypotheses, many of which are investigated in the chapters below, relating world-system theory to the macrolevel urban phenomena introduced in the preceding section. For purposes of explicating the basic features of the system, I have considered in isolation processes and structures that ideally must be considered together. The secular trends of the world-system, for example, are operating in the context of its cyclical processes, and the manifestations of the former will undoubtedly be masked or exaggerated, depending upon which moment in which cycle is under consideration.

REFERENCES FROM THE READING

Anderson, P. (1974) *Passage from Antiquity to Feudalism*, London: Verso.

Castells, M. (1977) *The Urban Question: A Marxist Approach*, Cambridge, MA: MIT Press.

Chase-Dunn, C. and Rubinson, R. (1979) Cycles, trends and new departures in world-system development, in J. Meyer and M. Hannan (eds.), *National Development in the World-System*, Chicago: University of Chicago Press.

Edel, M. (1981) Capitalism, accumulation and the explanation of urban phenomena, in M. Dear and A. Scott (eds.), *Urbanization and Urban Planning in Capitalist Society*, New York: Methuen: 19–44.

Gordon, D. (1978) Capitalist development and the history of American cities, in W. Tabb and L. Sawers (eds.), *Marxism and the Metropolis*, New York: Oxford University Press: 25–63.

Harvey, D. (1973) *Social Justice and the City*, Baltimore: Johns Hopkins University.

Harvey, D. (1982) *The Limits to Capital*, Chicago: University of Chicago Press.

Hawley, A. (1981) *Urban Society: An Ecological Approach*, New York: Ronald.

Hopkins, T. (1978) World system analysis: methodological issues, in Kaplan (ed.), *Social Change in the Capitalist World Economy*, Beverly Hills, CA: Sage.

Pickvance, C. G. (1978) Competing paradigms in urban sociology: Some epistemological issues, *Comparative Urban Research*, 6, 2.3: 20–27.

Weber, A. (1967) *The Growth of Cities in the Nineteenth Century: A Study in Statistics*, Ithaca, NY: Cornell University Press.

8

"Global city formation in New York, Chicago and Los Angeles: an historical perspective"

from *New York, Chicago, Los Angeles: America's Global Cities* (1999)

Janet L. Abu-Lughod

EDITORS' INTRODUCTION

Janet Abu-Lughod (1928–2013) made seminal contributions to the study of urbanization in global perspective for over three decades, primarily through her studies of "Third World urbanization" (Abu-Lughod and Hay, 1977) and through a series of classic monographs on urban development in Cairo, Egypt (1971) and Rabat, Morocco (1980). In the late 1980s, Abu-Lughod (1989) completed a landmark historical study that underscored the extent and dynamism of transnational urban systems, based upon dense commercial interdependencies, well before the consolidation of mercantile capitalism in 16th-century western Europe. Abu-Lughod's methodology resonates very closely with that of Braudel (Ch. 2) due to its emphasis on the centrality of urban systems to economic life, its adoption of a macrospatial perspective for the analysis of such systems and its focus on the long-term trajectory of historical-geographical change.

The selection included below is excerpted from the introductory and concluding chapters of Abu-Lughod's book *New York, Chicago, Los Angeles: America's Global Cities* (1999), which examines the divergent patterns of restructuring that have emerged in New York, Chicago, and Los Angeles since their origins. Two key points should be kept in mind. First, Abu-Lughod insists on the importance of a long-term historical perspective to the understanding of contemporary patterns of urban development. For Abu-Lughod, therefore, the current round of globalization is not as new and unique as some scholars have claimed. Rather, it represents the latest rupture within the long history of global capitalist urbanization. Second, Abu-Lughod provides a broad geohistorical sketch of urban development in the USA, suggesting that the history of each of her three cities can only be understood adequately in relation to broader, national and transnational political-economic trends. Thus, despite her criticisms of certain strands of global cities research, Abu-Lughod's study of "America's global cities" forcefully demonstrates one of its key propositions—namely, that a global political-economic and spatial perspective is required in order to decipher local developmental outcomes.

The theme of "global cities" has recently captured the imagination of urbanists, but as I shall argue, much of this exciting literature has been remarkably ahistorical, as if contemporary trends represent a sharp break from the past, if not an entirely new phenomenon. Furthermore, both the general descriptions of "world cities" and the accompanying causal analyses that attribute their commonalities to general forces residing at the highest level of the international economy neglect variations in global cities' responses to these new forces.

Contemporary scholars, trying to define the "global city," imply that it is a relatively new phenomenon that has been generated *de novo* in the present period by the development of an all-encompassing world system – variously termed late capitalism, postindustrialism, the informational age, and so on. Among the hallmarks of this new global city are presumed to be an expansion of the market via the internationalization of commerce, a revolution in the technologies of transport and communications, the extensive transnational movement of capital and labor, a paradoxical decentralization of production to peripheral regions accompanied by a centralization in the core of control over economic activities, and hence the increased importance of business services, particularly evident in the growth of the so-called FIRE economic sector –finance, insurance, and real estate. Accompanying these changes, and often thought to result from them, is a presumed new bifurcation of the class structure within the global city and increased segregation of the poor from the rich.

The value of such insights cannot be denied, but it is questionable whether these phenomena are as recent as is claimed. For example, all of these characteristics, at least in embryo form, had already made their appearance in New York City before the last quarter of the nineteenth century, when that city was clearly recognizable as a "modern" global city. And even though the pace and scale of today's globalized economy – and thus of the global cities that serve as its "command posts" – are faster and vaster, and the mechanisms of integration more thoroughgoing and quickly executed, the seeds from which the present "global city" grew were firmly planted in Manhattan during the middle decades of the nineteenth century. Chicago and Los Angeles eventually (and sequentially) followed that model with time lags of thirty and sixty years, respectively, although they naturally did so in a changed world context and under revised regimes of production, circulation, and consumption, as well as politics.

The relatively recent appearance of these global cities in the United States alerts us to proceed cautiously in comparing global cities without taking into account variations in the depths of their historical heritages. Most megalopolitan agglomerations that today serve as "world" or "global" cities – for example, London, Paris, Amsterdam, Tokyo – developed over many centuries. They thus contain accretions of successive types of settlements that have layered, one upon the other, vastly different patterns of development and reconstruction, until the composite whole becomes difficult to grasp. Not only are their landscapes difficult to "read," but they are not easily compared with one another, because the national political and cultural contexts in which they developed are so different. Even to compare the differential impact of global forces on America's three largest metropolitan centers requires much closer attention to the specific historical and geographic contexts in which they developed over the last century. Such an approach needs to take into consideration: the changing shape of the world system that constitutes the largest context for developments within them; the history of the expansion of the United States over the course of the nineteenth and twentieth centuries, within which the national urban hierarchy developed; and the more detailed histories of these individual urbanized regions that, over time, have generated the physical and social "terrain" onto which the newer global forces are now being inscribed and with which they interact.

Cities as nodes in networks are not a new phenomenon. Indeed, the fact that cities lie at the center of complex networks constitutes their *essential* feature. Throughout world history, certain cities – some of them imperial capitals remarkably large for their times, but a few relatively tiny "city-states" – have served as key nodes through which wider circuits of production, exchange, and culture have been coordinated, at least minimally. But in these earlier manifestations of integration, the territorial reach of even the most extensive "transnational/transimperial" systems was limited to only small fractions of the globe. Entire continents were excluded or were in touch at their peripheries only with the outer fringes of core regions. Nevertheless, urbanization per se was, in fact, both a symptom and a consequence of the construction of such regional systems, whose cores exerted dominance over their agricultural hinterlands and/or via rivers or even the edges of the sea, and increased the surplus available to the cities through conquest and/or tribute or through favorable terms of trade with distant points.

The first of these mini-world-systems climaxed toward the beginning of the second millennium B.C.E. when the three river-valley cradles of urbanism – along the Nile, the Tigris-Euphrates, and the Indus Rivers – came in more intimate contact with one another by multiple networks of trade that threaded through deserts, skirted the shores of the eastern Mediterranean and the Arabian Sea, transited the Red Sea and Arabo-Persian gulfs, and sent out probes to more distant areas in Anatolia, the Iranian plateau, and the zones beyond the Indus River Valley. (Almost contemporaneously, another minisystem was developing in the Yellow River region of China, one that would eventually form linkages with regions south and west of it.)

A second surge in integration began during the Hellenic Age, when Alexander's conquests briefly unified the eastern Mediterranean and reached beyond it – as far as India. This system climaxed during Roman imperial hegemony, when the entire littoral of the Mediterranean Sea became part or a central core that eventually stretched into western Europe as far north as England and reached, via trading circuits, not only

the eastern coast of the Indian subcontinent but, indirectly, even China.

Despite a brief hiatus caused by the fragmentation of western Europe (glossed somewhat inaccurately as the "fall of Rome" but more accurately described as the devolution of the so-called Western Roman Empire), the persistence of Eastern Christianity and an Islamic expansion throughout the Mediterranean and Eastern worlds led to the emergence of a third partial world system that extended over an even larger area. This system climaxed in the thirteenth century when very large portions of Europe, Eurasia, the Middle East, and North Africa, coastal zones of east Africa, India, Malaysia, and Indonesia, and even China were becoming more interactive – in both commercial and cultural contacts and through military conflicts. The unprecedented unification of Central Asia and China under the Yuan dynasty intensified such interactions. Needless to say, northwestern Europe was then still at the periphery of this system and the New World was not yet connected to it.

Perhaps shaken by a series of pandemics that culminated in the Black Death, whose highest mortalities occurred in zones most tightly integrated into the ongoing world system, there was another hiatus.

Plate 6 Vintage skyscrapers, Chicago

Source: Roger Keil

Within this breathing space, a fourth world system began to be reorganized, admittedly on the basis of the old but expanding rapidly through the so-called Age of Discovery to encompass parts of the New World and eventually other "terra incognita" in the South Pacific and southern Africa. This was the early phase of what Immanuel Wallerstein has called "the modern world-system," and it constituted the context within which New York first developed, albeit as a subordinate node.

During this period of early modern restructuring, the "balance of power" began to shift away from the Mediterranean and Asia to the increasingly powerful Atlantic sea powers, first Portugal and Spain and then England and the Low Countries (including Holland and Spain, which were then in a common "nation"). In the process, the formerly forbidding Atlantic was added as the third central sea (albeit more treacherous) of the evolving system, joining the Mediterranean and the Indian Ocean-South China Sea, which continued to serve as major pathways of trade, commerce, conquest, and the movement of peoples. But minor European incursions were also being made into the Pacific as well. By the end of the nineteenth century, this system climaxed in classical European colonialism,

achieved through the conquest of Africa and portions of Asia. By then, most countries in the Americas had been liberated.

Because throughout these earlier eras transport by water remained considerably cheaper than transport over land, the key points of exchange were, almost without exception, river or sea ports (the exceptions, of course, were oases along desert routes). It is in the context of the modern world-system, then, that the ports along the northeastern seaboard of North America became linked to a European core, and that, eventually, New York solidified its lead in the U.S. subsystem, a lead that, although later challenged by inland and Pacific coast cities, has never really been superseded.

For much of the first centuries of its existence, then, New York remained a key American link into a world system that focused increasingly on the Atlantic. Throughout the nineteenth century and into the early years of the twentieth, American history reads as the integration and eventual consolidation of a transcontinental subsystem, spreading from east to west. Even when the midcontinent was settled up to the Mississippi, and St. Louis (soon to be overtaken by Chicago) became the hinge for the drive to Manifest

Plate 7 Bonaventure Hotel and its neighbors in downtown Los Angeles

Source: Roger Keil

Destiny, New York retained and indeed strengthened its dominance as a core in its own right. It was, almost from its start, a "global city." Chicago could never have achieved the eminence it did without its prime outlet to the sea, New York.

It is important to recall that the integration of Chicago with the nascent U.S. system to its east and south was initially by water, the historically preferred transportation pathway. In the first quarter of the nineteenth century, decades before the railroad terminals consolidated Chicago's lead as midcontinental nexus, the outlets to the Atlantic coast via the Erie Canal-Great Lakes system and to the Caribbean Sea via the internal thoroughfare of the Mississippi River were already in place. What the rails did that waterways could not do, however, was link the zones west of the Mississippi to Chicago and from there on to New York. Without these rail linkages, Los Angeles's later growth (at least in the form it took) would have been inconceivable.

It was not until the tiny Mexican settlement of Los Angeles, conquered a bare quarter of a century before, was finally connected to the U.S. network via railroads – at first indirectly through San Francisco in 1875 and then via a direct route a decade later – that its modern growth spurt began. And it was not until the twentieth century, after the formation of America's first "overseas empire" (thanks to territories ceded in the 1898 Spanish-American War), that the Pacific became a true, albeit still a secondary, focus of American geopolitics. Heightened by these strategic interests, the sea circuit from the Pacific to the Caribbean was significantly shortened a dozen or so years later by the construction of the Panama Canal. Thus New York was the point of departure for Manifest Destiny, Chicago was its midwestern switching yard, and Los Angeles ultimately became its terminus.

It would be an error, however, to think in such geographically determined ways. Although an advantageously located site is a sine qua non for urban development, the agency of "men" (and they were mostly men in those days), acting politically and economically, has always intervened to favor certain of several otherwise equiprobable locations and to mobilize private and public financing to exaggerate the potential of such favored sites. And changes in technological capacity often have served to reduce or increase the viability of any natural setting.

Thus New York's port, so favorably endowed by nature, did not expand dramatically until the commercial invention of direct port auctions gave the city's brokers a monopoly over foreign trade, and until the engineering achievement of a through waterway to the Great Lakes made New York the dominant break-in-bulk point in internationally linked trade. And it was the capital accumulation facilitated by sophisticated institutions of insurance, banking, and credit that consolidated New York's lucrative role as broker for the slave-produced cotton crop, in preference to any southern port.

Similarly, both drainage of Chicago's waterlogged site and the clever machinations of land-speculating politicians in attracting rail termini and "hub" functions were essential in consolidating that city's lead over potential competitors, just as the later engineered reversal of the flow of Chicago's river reduced the need for portage to the Mississippi. "Nature's metropolis" may have drawn upon a rich agrarian and mineral hinterland, but it was, in the last analysis, the city's skill at centralizing the processing of these raw riches by means of machines and accounting inventions that made it "the metropolis of midcontinent."

The case of Los Angeles is even clearer, because initially the region had neither a water supply sufficient to support a major city nor a natural harbor able to compete with the better-endowed ports at San Francisco to the north or San Diego to the south. Only the political clout of local businessmen, exploiting access to both local and national public funds, enticed a continental rail terminus to the area, secured distant water for the municipality's monopoly (assisted by the engineering skills of the compulsively driven genius immigrant Mulholland), and gained the enormous federal subsidies necessary to construct an expensive, artificially enhanced massive port complex.

Unhappily, wars also play their part in creating locational advantages out of potentials. Just as Los Angeles's modern history was born in the 1847 conquest, expanded in the 1898 Spanish-American War, and further consolidated with the construction of the Panama Canal just before World War I, so the city was not decisively catapulted into the first ranks of the American urban system until World War II, when the Pacific arena drew the United States into an irreversible involvement with the "East" (to its west). The Second World War also boosted the economies of New York and Chicago: the former primarily through its ports, from which lend-lease shipments were funneled to Europe, and its expanding shipbuilding and airplane manufacture directed to the

European theater; the latter through the burgeoning demand for war materiel produced by its heavy Fordist industries.

By then, the world system was moving into the culminating phase of late-modern globalization. The evidence is obvious. One has only to contrast the First World War with the Second. The first had really encompassed only a portion of the European-Atlantic "world." The second signaled that the world system had incorporated the countries of Asia and the Pacific Rim as well. To this day, the postwar period has seen the "reach" of this system extend to virtually all parts of the globe, including most of Central and South America. Only a few mountainous redoubts, some interior deserts in Africa, Asia, and Australia, and a handful of off-course islands lie temporarily beyond global reach, and their days are numbered.

Weapons of war, produced first in the United States for its own defense, have fueled the remarkable economic prosperity of the Southwest, including Los Angeles; have partially infused the economies of the Northeast and Mid-Atlantic states, including New York's extended region; and have, by their absence, further undermined the economies of the Midwest, including Chicago's. But weapons produced for export have also enhanced the hegemonic position of the United States in the world economy and, through sales to Third World countries and the deployment of forces in subregional conflicts, have reconfigured the shape of the entire world system.

History does not end with globalization. The present fates of the urbanized regions of New York, Chicago, and Los Angeles are linked to a changing geography of power, and thus, ultimately, to the shape of the larger system. Reflecting the Janus-like position of the contemporary United States as both an Atlantic and a Pacific power, and the increased integration of North America with the Caribbean and the Latin American continent, the three seacoasts of the United States have become even more important magnets for people, both through internal migration and external immigration. In recent decades the population of the United States has continued to decant toward those coasts, not only in the conventional directions of east and west but southward as well. The rapid rise of gateway Miami almost to world-class status is certainly linked to the growing importance of the Caribbean and

"our neighbors to the south," as that zone of influence is increasingly integrated with the American core, if only, it sometimes seems, by illegal traffic in drugs. Chicago's tragedy is that it is not in these growth zones.

Technological advances have continued the age-old process of disengaging decisions from actions on the ground, with the ironic effect of facilitating the dispersal of production and people while increasingly centralizing what many analysts now refer to as "command functions." We saw this at earlier moments: the substitution of the commerce in "chits" in New York for the "real" midwestern wheat that remained in place in Chicago's silos; the removal of factories to the outskirts of cities at the same time company headquarters expanded in city centers, where telephones and later computers could monitor production farther afield and even abroad; the diffusion of stock ownership at the same time professional managers concentrated their hold over important decision making.

To some extent, these processes continue, but the scales at which they now operate often disengage or camouflage any clear lines between causes and effects, between those who command and those who labor, as capital and labor move with increasing freedom beyond not only the metropolitan boundaries but national borders as well. This disengagement means that healthy growth in command functions is not incompatible with dire destitution in those parts of the system (whether highly localized, at the national level, or at the global level) that are "out of the loop." Such marginalized zones can now be found in Manchester and Sheffield, England, in downtown Detroit, in the South and West Side ghettos of Chicago, in South-Central Los Angeles, in Bangladesh, and in many parts of the African continent.

History does not end, nor do changes in the world system cease. As sites to satisfy the world's burgeoning demand for consumer goods and services relocate to Asia, the hierarchy of global cities is being reconfigured and many new centers are being added. China, the location of some of the world's oldest and largest cities, may be reclaiming her status as one of the cores of the evolving "fifth iteration" of this world system. This changing context will inevitably have an impact on the global cities of the West, but we can only guess at the role political and technological developments may play in this process.

REFERENCES FROM THE READING

Abu-Lughod, J. (1971) *Cairo: 1001 Years of the City Victorious*, Princeton, NJ: Princeton University Press.

Abu-Lughod, J. (1980) *Rabat: Urban Apartheid in Morocco*, Princeton, NJ: Princeton University Press.

Abu-Lughod, J. (1989) *Before European Hegemony: The World System, A.D. 1250–1350*, New York: Oxford University Press.

Abu-Lughod, J. and Hay, R. (eds.) (1977) *Third World Urbanization*, Chicago, IL: Maaroufa Press.

ONE

9

"Global and world cities: a view from off the map"

from *International Journal of Urban and Regional Research* (2002)

Jennifer Robinson

EDITORS' INTRODUCTION

Jennifer Robinson is a South African urbanist based in the UK, where she teaches in the Department of Geography at University College London. Robinson has written a large number of articles and book chapters on colonial and postcolonial urban development, with particular reference to South African cities. Robinson's 1996 book, *The Power of Apartheid*, explored the relationship between state power and space in South African cities, and she has also written extensively on feminist theory and politics. Most recently, Robinson has published a number of influential critiques of contemporary urban theory, including the chapter below, from the point of view of postcolonial theory.

According to Robinson, the dominant approaches to global city theory have normalized the distinctive sociospatial features and developmental trajectories of North American and Western European cities: the latter have been presented as the paradigmatic "model" in terms of which all other cities are to be interpreted, regardless of their particular locations or histories. Within such a framework, cities located "off the map"—that is, in the developing countries of the global South—are almost invariably said to be lacking the characteristics that would qualify them as genuinely "global" cities.

Much like Fulong Wu (Ch. 16), Veronique Dupont (Ch. 17), and Gavin Shaktin (Ch. 18), Robinson is concerned to analyze the contextually specific patterns of globally induced urban restructuring that have been unfolding in cities located outside the developed capitalist North. However, whereas Wu, Dupont and Shatkin suggest various ways in which extant approaches to global city theory might be modified or expanded in order to take such cities into account, Robinson aims her critique at the epistemological core of the theory itself. For Robinson, the entire conceptual apparatus of global city theory is problematic insofar as it is grounded upon basically static, decontextualized categorizations and typologies. Without denying the importance of global economic inequalities or the global urban hierarchy, Robinson suggests that a significant conceptual reorientation is required to grasp the distinctiveness of cities in the developing world. To this end, Robinson proposes a number of methodological innovations.

First, the question of a city's "relevance" to the global economy needs to be explored not only with reference to its role as a basing point for transnational corporate activities, but on the basis of a broader range of possible global-local linkages. Second, based on work by Amin and Graham (1997) among others, Robinson mobilizes the notion of the "ordinary city," which she views as a conceptual basis on which diverse forms of urbanism, and the broad range of connections that link cities to global processes, might be more adequately appreciated and theorized. Robinson concludes her critique by underscoring its policy implications. For Robinson, the goal of becoming a global city is seriously unrealistic for most urban centers in the developing world. Other models of "successful" urban development are required, she argues, so that policymakers might harness the socioeconomic potential of developing cities while also alleviating their most pressing social, infrastructural and environmental problems. Although originally articulated from a decidedly contrarian perspective, Robinson's critique of "first-generation" global city theory has now arguably acquired paradigmatic status: it has inspired a generation of subsequent research on cities in the global South.

There are a large number of cities around the world which do not register on intellectual maps that chart the rise and fall of global and world cities. They don't fall into either of these categories, and they probably never will – but many managers of these cities would like them to. Some of these cities find themselves interpreted instead through the lens of developmentalism, an approach which broadly understands these places to be lacking in the qualities of city-ness, and which is concerned to improve capacities of governance, service provision and productivity. Such an approach supports some of the more alarmist responses to mega-cities, which are more commonly identified in poorer countries. But for many smaller cities, even the category mega-city is irrelevant. My concerns in this chapter extend beyond the poor fit of these popular categories, though. I would like to suggest that these widely circulating approaches to contemporary urbanization – global and world cities, together with the persistent use of the category "third-world city" – impose substantial limitations on imagining or planning the futures of cities around the world.

One of the consequences of the unreflexive use of these categories is that understandings of city-ness have come to rest on the (usually unstated) experiences of a relatively small group of (mostly western) cities, and cities outside of the West are assessed in terms of this pre-given standard of (world) city-ness, or urban economic dynamism. This chapter explores the extent to which more recent global and world city approaches, although enthusiastic about tracking transnational processes, have nonetheless reproduced this long-standing division within urban studies.

I do this by reflecting on some fashionable approaches to cities from a position off their maps. Of course, the cities I am concerned with are most emphatically on the map of a broad range of diverse global political, economic and cultural connections, but this is frequently discounted and certainly never explored within these theoretical approaches. There is a need to construct an alternative urban theory which

reflects the experiences of a much wider range of cities. This will involve disrupting the narrow vision of a (still) somewhat imperialist approach to cities, which has been reinforced by the strident economism in accounts of global and world cities. Elements of urban theory have become transfixed with the apparent success and dynamism of certain stylish sectors of the global economy, despite (and perhaps because of) their circumscribed geographical purchase and most unappealing consequences. These studies have been valuable, and offer great insight into the limited part of the world and economy that they study. My suggestion, though, is that these insights could be incorporated in a broader and less ambitious approach to cities around the world, an approach without categories and more inclusive of the diversity of experience in ordinary cities.

GLOBAL AND WORLD CITIES

In considering the dynamics of the world economy in relation to cities, a structural analysis of a small range of economic processes with a certain "global" reach has tended to crowd out an attentiveness within urban studies to the place and effect of individual cities and the diversity of wider connections which shape them (King 1995). Although status within the world city hierarchy has traditionally been based on a range of criteria, including national standing, location of state and interstate agencies and cultural functions, the primary determination of status in this framework is economic. The world cities approach assumes that cities occupy similar placings with similar capacity to progress up or fall down the ranks. The country categorizations of core, periphery and semi-periphery in world-systems theory have been transferred to the analysis of cities, and overlain on an extant but outdated vocabulary of categorizations (such as first/third world) within the field of urban studies. On this basis, from the dizzy heights of the diagrammer, certain significant cities are identified, labelled, processed and placed in a hier-archy, with very little attentiveness to the diverse expe-riences of that city, or even to extant literature about that place. The danger here is that out of date, unsuit-able or unreliable data, and possibly a lack of familiar-ity with some of the regions being considered, can lead to the production of maps which are simply inac-curate. These images of the world (of important) cities have been used again and again to illustrate the perspective of world cities theorists.

A view of the world of cities thus emerges where millions of people and hundreds of cities are dropped off the map of much research in urban studies, to service one particular and very restricted view of sig-nificance or (ir)relevance to certain sections of the global economy. This methodology also reveals an analytical tension between assessing the characteris-tics and potential of cities on the basis of the pro-cesses which matter as viewed from within their diverse dynamic social and economic worlds, or on the basis of criteria determined by the external theo-retical construct of the world or global economy. Although aiming to emphasize connections and not attributes, a limited range of cities still end up catego-rized in boxes or in diagrammatic maps, and assigned a place in relation to *a priori* analytical hierarchies.

The discursive effectiveness of the global city hypothesis depends on the pithy identification of the "global city" – a category of cities which are claimed to be powerful in terms of the global economy. If the global city were labelled as just another example of an industrial district (perhaps it should rather be called: new industrial districts of transnational management and control), it might not have attracted the attention it did.

FILLING IN THE VOIDS: OFF THE WORLD CITIES MAP

I certainly appreciate that the focus of global and more recent world cities work is on a limited set of economic activities, which are assuming an increas-ingly transnational form, and in which relatively few cities can hope to participate. But it is the leap from this very restricted and clearly defined economic anal-ysis, to claims regarding the success and power of these few cities, their overall categorization on this restricted basis and the implied broader structural irrelevance of all other cities, which is of concern.

These theoretical claims and categorizing moves are both inaccurate and harmful to the fortunes of cities defined "off the map."

The "end of the third world" is perhaps an accurate assessment of changes over the last three to four decades in places like Hong Kong, Singapore, Taiwan, South Korea and even Malaysia, and the appearance of these major urban centers in rosters of first- and second-order global cities reflects this. But in parts of the world where global cities have not been identified – the "voids" of world and global city approaches – the experience of many countries and cities has been much more uneven. For many, the 1980s and 1990s have been long decades of little growth and growing inequality. It is, however, inaccurate to caricature even the poorest regions as excluded from the global economy or doomed to occupy a slow zone of the world economy.

It is hard to disagree that some countries and cities have lost many of the trading and investment links that characterized an earlier era of global economic relations. A country like Zambia, for example, now one of the most heavily indebted nations in the world, and certainly one of the poorest, has seen the value of its primary export, copper, plummet on the world market since the 1970s. Its position within an older international division of labor is no longer economi-cally viable, and it has yet to find a successful path for future economic growth. En route it has suffered the consequences of one of the World Bank/IMF's most ruthless Structural Adjustment Programmes. However, Zambia is also one of the most urbanized countries on the African continent, and its capital city, Lusaka, is a testimony to the modernist dreams of both the former colonial powers and the post-independence govern-ment. Today, with over 70 per cent of the population in Lusaka dependent on earnings from the informal sector, the once bright economic and social future of this city must feel itself like a dream.

Lusaka is certainly not a player in the new global economy. But copper is still exported, as are agricul-tural goods, and despite the huge lack of foreign cur-rency (and sometimes because of it) all sorts of links and connections to the global economy persist. From the World Bank, to aid agencies, international political organizations and trade in secondhand clothing and other goods and services, Lusaka is still constituted and reproduced through its relations with other parts of the country, other cities, and other parts of the region and globe. The city continues to perform its functions of national and regional centrality in relation to political and financial services, and operates as a significant market (and occasionally production site) for goods and services from across the country and the world.

It is one thing, though, to agree that global links are changing, some are being cut, and that power relations, inequalities and poverty shape the quality of those links. It is quite another to suggest that poor cities and countries are irrelevant to the global economy. When looked at from the point of view of these places which are allegedly "off the map," the global economy is of enormous significance in shaping the fortunes of cities around the world. For many poor, "structurally irrelevant" cities, the significance of flows of ideas, practices and resources beyond and into the city concerned from around the world stands in stark contrast to these claims of irrelevance.

And to pursue a more polemical line, mineral resources crucial to the global economy are drawn from some of the poorest countries of the world, where financiers and transnational firms negotiate with warlords, corrupt governments and local armies to keep profits, production and exports flowing. Widening the compass of analysis might help to encourage a more critical edge to the global and world cities literature. Moreover, it is precisely through avoiding "risky" investments (and pursuing vastly exploitative and violent forms of extraction instead) in the poorest countries and cities in the world that the western financial "mode of production" is able to aim to secure the stable shareholder returns which maintain post-Fordist finance based economies. To the extent that they are absent from this aspect of the global economy, these places may well be central to sustaining it.

Most importantly, the particular "global economy" which is being used as the ground and foundation for identifying both place in hierarchy and relevant social and economic processes is only one of many forms of global and transnational economic connection. The criteria for global significance might well look very different were the map-makers to relocate themselves and review significant transnational networks in a place like Jakarta, or Kuala Lumpur, where ties to Islamic forms of global economic and political activity might result in a very different list of powerful cities. Similarly, the transnational activities of agencies like the World Bank and the IMF who drive the circulation of knowledge (World Bank, 2000) and the disciplining power to recover old bank and continuing bi-lateral

and multi-lateral debt from the poorest countries in the world would draw another crucial graph of global financial and economic connections shaping (or devastating) city life.

ORDINARY CITIES

A diverse range of links with places around the world are a persistent feature of cities. They can work for or against cities everywhere and are constantly being negotiated and renegotiated. To aim to be a "global city" in the formulaic sense may well be the ruin of most cities. Policy-makers need to be offered alternative ways of imagining cities, their differences and their possible futures – neither seeking a global status nor simply reducing the problem of improving city life to the promotion of "development." In developmentalist perspectives cities in poor countries are often seen as non-cities, as lacking in city-ness, as objects of (western) intervention. Ordinary cities, on the other hand, are understood to be diverse, creative, modern and distinctive, with the possibility to imagine (within the not inconsiderable constraints of contestations and uneven power relations) their own futures and distinctive forms of city-ness (Amin and Graham 1997).

Categorizing cities and carving up the realm of urban studies has had substantial effects on how cities around the world are understood and has played a role in limiting the scope of imagination about possible futures for cities. This is as true for cities declared "global" as for those which have fallen off the map of urban studies. The global cities hypothesis has described cities like New York and London as "dual cities," with the global functions drawing in not only a highly professional and well-paid skilled labor force, but also relying on an unskilled, very poorly paid and often immigrant workforce to service the global companies. These two extremes by no means capture the range of employment opportunities or social circumstances in these cities. It is possible that these cities, allegedly at the top of the global hierarchy, could also benefit from being imagined as "ordinary." The multiplicity of economic, social and cultural networks which make up these cities could then be drawn on to imagine possible paths to improving living conditions and enhancing economic growth across the whole city.

A more cosmopolitan urban theory might be more accurate or helpful in understanding the world; it might also be more resourceful and creative in its output. But interrogating these categorizations of cities and theoretical divisions within urban studies matters primarily, I think, because they limit our potential to contribute to envisioning possible city futures. And given the gloomy prognoses for growth in poor cities within the context of the contemporary global economy, creative thinking is certainly needed!

From the viewpoint of global and world cities approaches, poor localities, and many cities which do not qualify for global or world city status, are caught within a very limited set of views of urban development: between finding a way to fit into globalization, emulating the apparent successes of a small range of cities; and embarking on developmentalist initiatives to redress poverty, maintain infrastructure and ensure basic service delivery. Neither the costly imperative to go global, nor developmentalist interventions which build towards a certain vision of city-ness and which focus attention on the failures of cities, are very rich resources for city planners and managers who turn to scholars for analytical insight and assessment of experiences elsewhere. It is my opinion that urban studies needs to decolonize its imagination about city-ness, and about the possibilities for and limits to what cities can become, if it is to sustain its relevance to the key urban challenges of the twenty-first century. My suggestion is that "ordinary-city" approaches offer a potentially more fertile ground for meeting these challenges.

THE POLICY IMPERATIVE: THE POLITICAL CASE FOR ORDINARY CITIES

If cities are not to remain inconsequential, marginalized and impoverished, or to trade economic growth for expansion in population, the hierarchies and categories of extant urban theory implicitly encourage them to aim for the top! Global city as a concept becomes a regulating fiction. It offers an authorized image of city success (so people can buy into it) which also establishes an end point of development for ambitious cities. There are demands, from Istanbul to Mumbai, to be global. But, calculated attempts at world or global city formation can have devastating consequences for most people in the city, especially the poorest, in terms of service provision, equality of access and redistribution (Robins and Askoy 1996; Firman 1999). Global and world city approaches

encourage an emphasis on promoting economic relations with a global reach, and prioritizing certain prominent sectors of the global economy for development and investment. Alternatively, the policy advice is for cities to assume and work towards achieving their allocated "place" within the hierarchy of world cities.

Most cities in poorer countries would find it hard to reasonably aspire to offering a home for the global economy's command and control functions. More feasible for many poorer cities is to focus on some of the other "global functions" that Sassen (1994) associates with global cities. These include promoting attractive "global" tourist environments, even though these have nothing of the locational dynamics of command and control global city functions. Disconnected from the concentration of arts and culture associated with employment of highly skilled professionals in global cities, the impulse to become global in purely tourist terms can place a city at the opposite end of power relations in the global economy, while substantially undermining provision of basic services to local people. In addition, Export Processing Zones may be "global" in the sense that they are "transnational spaces within a national territory" (Sassen 1994: 1), but they too involve placing the city concerned in a relatively powerless position within the global economy, which is unlikely to be the city's best option for future growth and development. These are not places from where the global economy is controlled: they are at quite the other end of the command and control continuum of global city functions. More than that, the reasons for co-location would not involve being able to conduct face-to-face meetings to foster trust and cooperation in an innovative environment. Rather, they are to ensure participation in the relaxation of labor and environmental laws which are on offer in that prescribed area of the city. Cities and national governments often have to pay a high price to attract these kinds of activities to their territory. Valorizing "global" economic activities as a path to city success – often the conclusion of a policy reversioning of world cities theory – can have adverse consequences for local economies.

This is a familiar story, but one which scholars are more likely to blame on others capitalists, elite urban managers, than on their own analyses, which are seldom the object of such reflection. It is when attention slips from economic process to a sloppy use of categorization that I think the most damaging effects of the world and global city hypotheses emerge.

Categorizing a group of cities as "global" on the basis of these small concentrated areas of transnational management and coordination activity within them is metonymic in that it has associated entire cities with the success and power of a small area within them. In the process a valid line of analysis has reproduced a very familiar hierarchization of cities, setting certain cities at the top of the hierarchy to become the aspiration of city managers around the world.

This has happened just as a burgeoning postcolonial literature became available to critique earlier categorizations of cities into western and third world, a categorization which had emphasized difference and deviation from the norm as bases for analysis and which had established certain (western) cities as the standard towards which all cities should aspire. Instead of pursuing the postcolonial critique, urban studies has replicated this earlier division by accepting the categories of world/global city as analytically robust and popularizing them in intellectual and policy circles. Global cities have become the aspiration of many cities around the world; sprawling and poor megacities the dangerous abyss into which they might fall should they lack the redeeming (civilizing) qualities of city-ness found elsewhere. This may not have been the intention of urban theorists, but ideas have a habit of circulating beyond our control. It is my contention that urban theory should be encouraged to search for alternative formulations of city-ness which don't rest upon these categories and which draw their inspiration from a much wider range of urban contexts.

The political need for a new generation of urban theoretical initiatives is apparent. How can the overlapping and multiple networks highlighted in the ordinary city approaches be drawn on to inspire alternative models of development, which see the connections, rather than conflict, between informal and formal economies? Approaches which explore links between the diversity of economic activities in any (ordinary) city, and which emphasize the general creative potential of cities, are crucial, rather than those which encourage policy-makers to support one (global) sector to the detriment of others. (See for example, Benjamin 2000; Simone 2001.)

CONCLUSION

The academic field of urban studies ought to be able to contribute its resources more effectively to the creative imagining of possible city futures around the world. One step in this direction would be to break free of the categorizing imperative, and to reconsider approaches which are at best irrelevant and at worst harmful to poor cities around the world. I have suggested that in place of world, global, mega-, Asian, African, former Socialist, European, third-world etc. cities, urban studies embark on a cosmopolitan project of understanding ordinary cities.

A second step must be to decolonize the field of urban studies. Theoretical reflections should at least be extremely clear about their limited purchase and, even better, extend the geographical range of empirical resources and scholarly insight for theorizing beyond the West and western-dominated forms of globalization. This has been initiated in a restricted form, through the transnational emphasis of global and world cities approaches, and the growing interest in globalization within a developmentalist frame. But a more cosmopolitan empirical basis for understanding what cities are, and how they function, is essential to the future relevance of the field of urban studies. In an age when most people now live in cities, and most of this urban population is in poor countries, irrelevance is a very real possibility for a field whose wellsprings of authorized theoretical innovation remain firmly fixated on the West and its successful satellites and partners.

This is not to insist that every study consider everywhere. But there is considerable scope for the spatial trajectories of theoretical imaginings to come closer to the spatiality of cities themselves, which are constituted on the basis of ideas, resources and practices drawn from a variety of places – not infinite, but diverse – beyond their physical borders. The conditions of incorporation though, are crucial. Firstly, simply mobilizing evidence of difference and possibly deviation within the frame of dominant theory is not enough. Consideration needs to be given to the difference the diversity of cities makes to theory (not simply noting the difference that they are). How are theoretical approaches changed by considering different cities and different contexts, by adopting a more cosmopolitan approach? And secondly, as with cities themselves, power relations and their geographies cannot be avoided. If a cosmopolitan urban theory is to emerge, scholars in privileged western environments will need to find responsible and ethical ways to engage with, learn from and promote the ideas of intellectuals in less privileged places. This is not a call to western writers to appropriate other places for continued western intellectual advantage. It is a plea to acknowledge the intellectual creativity of scholars and urban managers in a wider range of urban contexts.

This will involve a critical analysis of the field's own complicity in propagating certain limited views of cities, and thereby undermining the potential to creatively imagine a range of alternative urban futures. It will require more cosmopolitan trajectories for the sources and resources of urban theory. Much innovative work is already being undertaken by scholars and policy-makers around the world, who have had to grapple with the multiplicity, diversity and ordinariness of their cities for some time. Ordinary cities are themselves enabling new kinds of urban imaginaries to emerge – it is time urban studies caught up. More than that, I would suggest that as a community of scholars we have a responsibility to let cities be ordinary.

REFERENCES FROM THE READING

Amin, A. and Graham, S. (1997) The ordinary city, *Transactions of the Institute of British Geographers*, 22: 411–29.

Benjamin, S. (2000) Governance, economic settings and poverty in Bangalore, *Environment and Urbanisation*, 12, 1: 35–56.

Firman, T. (1999) From 'global city' to 'city of crisis': Jakarta metropolitan region under economic turmoil, *Habitat International*, 23, 4: 447–66.

King, A. (1995) Re-presenting world cities: Cultural theory/social practice, in P. Knox and P. Taylor (eds.) *World Cities in a World-System*, London: Routledge, 215–31.

Robins, K. and Askoy, A. (1996) Istanbul between civilisation and discontent, *City*, 5–6: 6–33.

Robinson, J. (1996) *The Power of Apartheid: State, Power and Space in South African Cities*, London: Butterworth-Heinemann.

Sassen, S. (1994) *Cities in a World Economy*, London: Sage.

Simone, A. (2001) Straddling the divides: Remaking associational life in the informal African city, *International Journal of Urban and Regional Research*, 25, 1: 102–17.

World Bank (2000) *Cities in Transition: World Bank Urban and Local Government Strategy*, Washington, DC: World Bank.

10
"Space in the globalizing city"

Peter Marcuse

EDITORS' INTRODUCTION

Peter Marcuse is a lawyer and Professor Emeritus of Urban Planning at Columbia University. He is one of the most influential writers on cities of the past three decades; his academic and political writings have been an inspiration to generations of planners and urban theorists. Among his major contributions are studies of housing and gentrification, planning theory, the sociospatial reordering of what he has termed the "quartered city", and sociospatial change on both sides of the Atlantic, with a particular focus on the United States and Germany. More recently, Marcuse has become one of the most recognized voices in the international debate on cities and globalization. While critical of the global cities approach in the narrow sense, Marcuse has systematically explored the specific effects of globalization on urban sociospatial structures. In two widely recognized publications with Dutch urban geographer Ronald van Kempen, Marcuse developed a sophisticated matrix for the investigation of sociospatial restructuring in globalizing cities (Marcuse and van Kempen 1999, 2002). The contribution below is based on this work and it proposes to study spatial patterns in globalizing cities.

What difference has globalization made in the space within cities? More specifically, what are the characteristic internal structures of cities today, and how, if at all, are they different from what they were before globalization? What aspects of globalization account for what changes we find? Is the result a new global city? If not, what does account for the spatial structure of cities today? These are frontier questions for anyone concerned about cities today, and suggests the research agenda this paper attempts to outline.

To start with, a definition of globalization is necessary. Then we pose in more detail the key questions that globalization raises about cities. We then go on to examine the actual patterns of space in cities, and changes in them, in the current period. We conclude by asking whether globalization has created a new form of global city, and what other forces may be involved in shaping cities today.

FORMAL DEFINITIONS

One seldom finds a formal definition of globalization in the literature. Very often, simply a listing of parallel developments is given: greater mobility of capital, greater mobility of labor, rapid development of communication technology, computerization, world-wide conveyor belt in manufacturing, shift from manufacturing to services, integration of all economies in a single market, permeability of borders, decline of the nation-state, homogenization of culture, and so on. That is not very satisfactory; it mixes causes and effects, fails to distinguish the inherent from the temporary or circumstantial, the determining from the result, the necessary from the variable. What almost all uses of the term have in common, however, is the dating: those changes, that set of events, that began about 1970. Sometimes the dating is referred to the "oil shock" of 1973, sometimes to the breakdown of the Bretton Woods arrangements, sometimes symbolically to events such as the destruction of the Pruitt-Igoe public housing development and the end of the dominance of the "modern" in architecture. So considering globalization to encompass those events that began or developed dramatically after about 1970 is a useful initial working definition.

For our purposes, however, a more substantive definition is useful. It may run as follows: Globalization, in its really existing form, is the further strengthening and internationalization of capital using substantial advances in communications and transportation technology (see also Marcuse 1997; 2002). The definition is important for two reasons. First, it highlights two separate components of globalization, the social/economic and the technological, and suggests that the former determines the latter, not the opposite, as is often implicitly assumed. Second, it notes that there may be many forms of globalization, and we are concerned with its specific form in the world of the 1970s.

Since the two-part definition of globalization is critical for the discussion that follows, it is worth detailing a bit further. Really Existing Globalization is:

1 the qualitative leap in information and transportation technology since c.1970, permitting increased internationalization of information and physical production;
2 used by dominant social-economic-political groups to produce a further increase in the concentration of private economic power since c.1970 (both within the economic sphere and vis-à-vis government), permitting increased internationalization of control over economic and political processes;
3 with consequences for cross-border integration of production and investment, cultural homogeneity, United States dominance, inequality and polarization, environmental quality, popular movements, culture, etc. (and perhaps for the spatial structure of cities, the topic of this chapter).

This is a definition of really existing globalization; it is not a definition of all possible forms of globalization. Specifically, one could imagine a form of globalization in which advances in technology, #1 above, were used, contrary to #2 above, to improve the standard of living of all, promote world-wide democracy, and reduce inequality of opportunity, with results quite the opposite of #3 above. When the World Social Forum adopts the slogan, "Another World is Possible," it does not mean a world without globalization, but a world in which globalization takes a quite different form from its really existing model.

I would hypothesize that a detailed examination of the spatial patterns in cities we witness today would show that they are in fact significantly, but only partially, linked to really existing globalization; that the patterns affect all cities, not only global cities; that, to the extent they are linked to really existing globalization, they are linked as much to its second aspect as to its first; and that thus the specific form they have taken is not an inevitable product of globalization, but rather of its specific current form. And I would finally hypothesize that the forces shaping contemporary cities are not simply the products of the current form of globalization, but that they, like really existing globalization itself, are the results of more deeply embedded forces of longer standing in our economies and political structures.

THE FOUR KEY QUESTIONS FOR RESEARCH

As to the spatial patterns we witness in cities today, we might ask the following questions:

1 Is the pattern new in the present period of globalization? Has it produced a new spatial model of a city?
2 To the extent it is new, to what aspect of globalization is it attributable? To oversimplify, are they the result of technological change or of social/economic/political change?
3 Is the pattern an aspect of all cities, all globalizing cities, or only global cities?
4 To what extent is the pattern not new, and, for what is not new, what are its underlying determinants?

This chapter only suggests, at the end, a very partial answer to some of these questions, but first suggests what the spatial patterns are that need examination, with the suggestion that the detailed answers are the appropriate agenda for further work for those interested in the production of better spaces for all people within which to live.

CONTEMPORARY SPATIAL PATTERNS

What in fact are the spatial patterns we witness in cities today, and to what extent have they changed in the present period of globalization? Let us begin with the conception of a quartered city (Marcuse 1989). It may be seen both in the spatial arrangement of residential life and in the spatial arrangement of business activities. Is the fact that cities today, at least in the advanced industrialized economies of the west, are not "dual," but more like "quartered," cities, new (Marcuse 1991)? The answer becomes clear if we recapitulate the argument about the patterns of the contemporary city briefly.

The residential cities

Within the city of today we see a set of sometimes overlapping but quite different residential cities:

- The luxury city, with luxury housing, not really part of the city but in enclaves or isolated buildings within it, occupied by the top of the economic, social, and political hierarchy;
- The gentrified city, the city of winners, occupied by the professional-managerial-technical groups, whether yuppie or muppie without children;
- The suburban city, sometimes single-family housing in the outer city, other times apartments near the center, occupied by skilled workers, mid-range professionals, upper civil servants;
- The tenement city, sometimes cheaper single-family areas, most often rentals, occupied by lower-paid workers, blue and white collar, and generally (although less in the United States) including substantial social housing;
- The abandoned city, the city of the victims, the end result of trickle-down, left for the poor, the unemployed, the excluded, where in the United States home-less housing for the homeless is most frequently located.

The economic cities

These felt divisions in the residential city are roughly paralleled by divisions in the economic city:

- The city of controlling decisions include a network of high-rise offices, brownstones or older mansions in prestigious locations, but are essentially locationally not circumscribed; it includes yachts for some, the back seats of stretch limousines for others, airplanes and scattered residences for still others;
- The city of advanced services, of professional offices tightly clustered in downtowns, with many ancillary services internalized in high-rise office towers, heavily enmeshed in a wide and technologically advanced communicative network;
- The city of direct production, including not only manufacturing but also the production of advanced services, in Saskia Sassen's phrase, government offices, the back offices of major firms, whether adjacent to their front offices or not, located in clusters and with significant agglomerations but in varied locations within a metropolitan area, sometimes, indeed, outside of the central city itself;
- The city of unskilled work and the informal economy, small-scale manufacturing, warehousing, sweatshops, technically unskilled consumer services, immigrant industries, closely intertwined with the cities of production and advanced services and thus located near them, but separately and in scattered clusters, locations often determined in part by economic relations, in part by the patterns of the residential city; (spatially, the overlap with the city of advanced services is substantial, for the service economy produces both high and low end jobs in close proximity to each other: janitors in executive offices, etc.);
- The residual city, the city of the less legal portions of the informal economy, the city of storage where otherwise undesired (NIMBY) facilities are located, generally congruent with the abandoned residential city.

These patterns are spatial, but they are not rigid, in the old sense in which Burgess and Park tried to describe city structure. And their spatial pattern varies widely from city to city, country to country; Los Angeles, for instance, has a pattern I have described as fluid, separations as of oil and water together with walled enclaves, rather than the more clearly bounded and more homogeneous quarters of New York City. But the spatial patterns are always there, if differing in intensity and sharpness. Initial work on the 1990 census in the United States has demonstrated that fact.

The built environments

Paralleling the divisions within the residential and the economic cities are different concrete forms of the built environment. Again, they may be listed in outline form:

- The citadel, as the assemblage of residential and commercial space used by the upper classes, separated from the rest of the city, often gated and always secured, increasingly in the form of mega-projects;
- The older city neighborhoods, near the core, rehabilitated, "regenerated," gentrified by the professional and managerial groups whose role in the cities is growing both quantitatively and qualitatively;
- The edge cities, in-between cities, ex-urban centers, in which the older forms of suburbia are combined with jobs and all of the cultural and recreational and commercial necessities of life formerly assumed to be found only in the urban core;
- The diluted and manipulated areas of social and working class housing, in which mixes of incomes and population groups are designed to attenuate any coalescence of "lower-class" based behaviors or mobilizations;
- The ethnic enclaves, in which immigrants and the lowest-aid workers integrated into the mainstream economy are able to find the possibilities of life supportable by their jobs towards the bottom of the economic structures of the city;
- The excluded ghettos, the areas to which those at the very bottom of the economic ladder, and not needed in the dominant economy, are confined, with reduced public services and neglected physical surroundings.

The soft locations[1]

Within these divisions there is a particular set of locations in which the processes of change are particularly striking, and which appear at first blush to be linked to the processes of globalization and post-Fordist economic change were likely to have a particular impact. Recapitulating but adding to that list:

- waterfronts
- currently centrally located manufacturing
- brown fields (formerly industrial sites)
- central city office and residential locations
- central city amusement locations and tourist sites

- concentrations of social housing
- residential locations on the fringe of central business districts
- ethnic areas of concentration
- suburbs
- historic structures
- public spaces

The new aspects

What is new about these patterns? It is, it seems to me, an under-debated question, but an important one. A preliminary answer would include at least the following:

- The growth in the size of the gentrified city and the shrinking of others: expanding gentrification.
- The growth in the size of the abandoned city: increasing ghettoization.
- The dynamic nature of the quarters, in which each grows only at the expense of the others: tensions among quarters, with displacement as the mechanism of expansion, and both walling in and walling out more common.
- The growing importance of the identity of the quarter for the lives of most residents: the defensive use of space, including the intensity of turf allegiance.
- The shift in location and structure of certain types of economic activity, such as manufacturing, finance, and consumer services.
- The role of government, not only acceding to but promoting the quartering of the city in the private interest, fortifying both the gentrified and the abandoned city: the subsuming of the public interest under the private, the dominance of the neo-liberal agenda.

But in each case (and more might be added), these changes represent trends manifest well before the period of globalization, and are represented in cities in various relationships to globalization (see Marcuse 2002; Marcuse and van Kempen 2002).

THE APPLICATION OF THE RESEARCH AGENDA AT THE CITY SCALE

As to each of these spatial patterns of the contemporary city, the four key questions as to their nature and causes raised at the beginning of this paper may be addressed. To give only a few examples:

- *Waterfronts*. The contemporary pattern is certainly the abandonment of shipping uses at old port facilities and the construction of giant facilities for much larger container-oriented shipping. That appears a result of new technologies of transportation. Is it (or is it not?) also the result of a practice in which the expense of the necessary new harbor facilities is publicly paid, but the profit from its use private?[2] Is the essential motivation for the change the replacement of labor by capital, and if so is this use of technological possibilities dictated by social priorities or the priorities of capital? Is the change in transportation technology greater than previous changes? Is it limited to a few global cities, or is it general?
- *Ghettoization, polarization, walling and the construction of citadels*. The pattern is a widening gap between the rich and the poor, linked to the gap between the well-educated and the ill-educated. Is the existence and impact of that gap a product of the greater education advances in technology demand of workers? Or is it a result of the social/political strength of the haves to skew public investment in education in their favor, and to maintain grossly unequal wage rates for different kinds of economic activities? Is the division of cities represented by these phenomena something new, or is there historical continuity in them?
- *Brownfields*. Is their abandonment a result of the technological obsolescence of the manufacturing processes that had taken place there, or of the skewed responsibility for the externalities of production, which permits abandonment without cleaning up, and zoning that follows rather than directs the economic uses of land? Are major changes in land use a new phenomenon historically? Are they limited to a handful of cities, or are they general?
- *Megaprojects and skyscrapers*. Is the use of ever more sophisticated building technology to build higher and larger simply the replacement of better technology replacing outdated technology, or is it determined by the symbolic desire to express power and the social desire to insulate, withdraw, and secure?
- *Gentrification*. Is gentrification, now at the global scale (Smith 2002), the result of increasing technologically-enabled prosperity, or of the replacement

of industrial by service activities in city centers, or is it the result of shifts in the balance of political power that have permitted the imposition of neo-liberal policies for the determination of land use? What have been the historical determinants of changes in the social uses of the built environment?

The first three of our four key questions may be specified in similar fashion to every one of the changes in city spatial structure outlined above. The fourth question is at a different scale.

THE RESEARCH AGENDA AT THE SCALE OF GLOBAL STRUCTURES: THE DETERMINANTS OF THE STRUCTURE OF GLOBALIZING CITIES

To look at the determinants of the spatial structure of cities in broader historical perspective, we need first to determine whether what we are examining is a set of phenomena confined to a few cities, or whether it is a general characteristic of cities in the contemporary period. It is a question that deserves considerable attention, and at least the broad outlines of the answer are evolving.

The difficulty arises from the misuse of the concept of the global city. Its original meaning called attention to a hierarchical pattern among cities, based on their position in a network of cities increasingly linked by newly developed modes of communication and transportation, later with a focus on financial networks among them (Friedmann and Wolff 1982). The term has however been very widely used in a related but essentially different sense, by city boosters that would like to see their city also designated "global," meaning modern, up-to-date, "in," real parts of a global network, at the cutting edge of globalization. It is in this usage then often followed by a call for a certain arrangement of space within the city, for instance extra-high skyscrapers, as appropriate to a "global" city, differentiating the "global" city from "non-global" cities.

But the patterns described in this article which are characteristic of the so-called global cities are by no means confined to them. Every spatial pattern found in New York City, London, and Tokyo can also be found in Cleveland, Vancouver, Detroit, Stuttgart, Accra, or Calcutta. Some aspects characterizing the key global cities may be further developed in cities much further down the hierarchy; even among the global cities, there are major differences in urban form (Abu-Lughod 1999). If anything, we may speak of "globalizing cities," cities – virtually every city in the world – that are impacted by the processes of globalization, that reflect tendencies that are common to all cities linked to each other in a global pattern (Marcuse and van Kempen 1999).

Is globalization, really existing globalization as here defined, then the dominant, determining factor shaping all globalizing cities, explaining all the patterns described above?

And the answer suggested here is: no. The changes described here are only in part, and ultimately only in small part, the results of the really existing globalization of the post-1970s. Every one of the changes has a long history; every one is influenced by multiple factors; every one can be linked to more fundamental causes than those newly arrived with globalization.

Beyond that, if one looks more narrowly at the impacts on space that can be traced to globalization, the patterns outlined above, another suspicion arises, but one as to which there is as yet no clear empirical evidence. It may well be that even putting all these influences of globalization together, they affect only a minority of a city's population, only the lesser part of a city's spaces. Granted that, to some degree, everything that happens in or to a city affects everyone and every place within that city – granted that no man is an island, nor no woman either, nor no spaces either except in the narrowest geographical sense – the question is how important that effect is. The impact of globalization on the outer boroughs of New York City can be easily seen: the airports, the change in job opportunities, the back offices, the immigrant quarters, the gated communities. But how many of the outer boroughs residents are significantly affected by what is new in these areas (and of course much is not new, and while internationally linked pre-existed the globalization of the period after 1970), and how many continue their lives and lifestyles much as before, if not entirely the same? The question is worth exploration.

And in any event the various aspects of city spatial structure described in this paper have all been with us for several centuries, in some cases even longer. They are aspects perhaps more striking, perhaps more extensive, perhaps even more important, today than ever before. But they are not new, and they are not special to globalizing cities.

So if globalization is not adequate to explain the shape of contemporary cities, what is? The details of

an alternative answer should be part of a further research agenda. But I would suggest, at least as a working hypothesis, that the character of contemporary cities can be traced to one crucial fundamental aspect that they share: They are all phenomena of the Capitalist City. The Capitalist City goes under many different names. It is often spoken of as the industrial city, and indeed the shift from an agricultural or mercantile form of production to an industrial one, and certainly the rise of manufacturing has had an indelible imprint on today's cities. Sometimes it goes by the name of the modern city, not to designate an architectural style but to differentiate it culturally from its predecessors, with their more traditional values and customs. And sometimes the defining characteristic is found in the rise of new economic classes and their changed relationship to each other as in the work of Karl Marx or Karl Polanyi, in which economic aspects are given priority. Perhaps, indeed, one might build on the analysis of globalization presented at the outset of this paper, suggesting it is a combination of specific advances in productive technology and social relationships, to argue that the Capitalist City is itself, and has been for centuries, always the result of changes in technology and social relations, in evolving combinations, with the globalizing city only the latest permutation of those changes, perhaps more extreme and more widespread than before, but nevertheless cut by the same tailors of the same cloth.

The general formulation of the fourth of the key research questions, given the above hypothesis, might then be twofold:

- As to any given characteristic of globalization (technological change, concentration of control, polarization, cultural homogenization, mobility, etc.), did that characteristic pre-exist the period of globalization, and, if so, when did it first appear, and does it differ today only in extent, rather than nature?
- As to any given characteristic of globalization, does it appear also as a characteristic of national systems (as opposed to global), and if so does it differ globally from its national manifestation only in extent, rather than nature?

Again, to give some examples for empirical research:

- *The shift from manufacturing to services.* Certainly the global assembly line, the new global division of labor, is a characteristic of globalization. But it is not a result of the declining importance of manu-facturing; on the contrary, both the volume of goods produced and the number of workers engaged in manufacturing have increased in the last 40 years. What has happened is a relocation of manufacturing activities from higher-wage to lower-wage countries. The driving force is one well known in industrial relations everywhere: Profits can be increased if labor costs can be reduced. The quest for higher profits exerts a basic and on-going pressure to hold wages down. That is part of capitalism. It was seen most recently in the move of textile plants from the United States north to its south, where labor legislation was less protective, unionization more limited, and wages lower. The threat, and its actualization, to move a plant from a higher to a lower wage region has been a standard part of labor negotiations for at least a century. Moving over-seas is only the latest manifestation of the underling dynamics of capitalist production, facilitated (as in earlier times) by improvements in the technology of transportation and communica-tions, but not determined by them. Whether this is so or not, and, if so, to what extent and how such moves differ today from earlier periods are matters capable of empirical investigation.

- *Brownfield sites.* Similarly the relocation of manu-facturing activities to new sites, and the abandon-ment of (polluted) old sites, has to do with the relationship of land prices to the costs of produc-tion, in the context of other costs of location, speci-fically, accessibility. New technology has increased the options for business firms, and awareness of the impact of pollution has increased, but the dynamics are centuries old. The disconnect between the private profit-determined impact of relocation and its social costs and benefits is a characteristic of capitalism, not of globalization. Its extent can be empirically reviewed.
- *The mobility of capital.* Indeed, the mobility of capital has increased, but it has been increasing steadily over the centuries and may even not be, proportion-ately, higher today than at certain earlier times in history. Whether its impact on the spatial form of cities is different today than before is an empirical question. Cursory examination of how foreign capi-tal invests in real estate development in foreign cities suggests that its patterns are no different from those of domestic capital; foreign capital uses domestic real estate appraisers, consultants, market analysts,

brokers, even architects (if in consultation with "world-class" ones). Often, in fact, foreign capital co-ventures with local. Does the product then differ (except perhaps in size) from what it would be with purely domestic capital?

Other examples, more briefly:

- *Gentrification.* Is gentrification simply a continuation of the urban restructuring, the uneven spatial development of capitalism, that has been going on for centuries, with variations on in extent and in the relative use of public vs. private means?
- *Edge Cities.* Is the search for the most efficient loca-tions to accommodate the demands of expanding profit-oriented production, and its shifting spatial needs, largely a continuation of efforts that in earlier periods resulted in the exponential growth of some towns (from Gary to Wolfsburg) and the founding of new towns, and the pattern of private market-driven suburbanization almost everywhere?
- *Centralization of control functions.* The impact of the centralization of control functions, resulting in larger and more ornate/elaborate structures in cities (e.g. downtown Manhattan, I.G. Farben in Frankfurt, Avenida in Sao Paulo, etc.), has its predecessors in the tendencies to monopolies and, internationally, cartels, over a previous century, tendencies long associated by economists with laws of capitalist societies. Is the rate of develop-ment different today than what it has been?

In general, the researchable question might be posed: Is the spatial pattern that seems today to be the product of globalization any different in its tendency than similar patterns of change on an intra-national level in previous years (e.g. locational changes, capital investment, centralization of control, etc.)?

The agenda for research – and for the action that should follow from research – is thus extensive. It can encompass, at the scale of individual cities, the initial questions raised above: To what extent have the distin-guishing characteristics of the contemporary city, as outlined in the first part of this paper, been the pro-duct of the developments and the actors involved in really existing globalization; if they are products of globalization, are they the result of technological changes or of changes in the balance of social, poli-tical, and economic power; do they taken together create a new model of a global city applicable to all, or some, or only to global cities? And then such an agenda should go further and explore what other models of the contemporary city might equally or better reflect its reality. If research is not to be merely empiricist, but to deal with the large questions of policy that confront us, it cannot shy away from addressing the large questions that detailed examination of the smaller ones raises. Research at the scale of the city and the time scale of contemporary globalization must be accompanied by research at the scale of the economic structures and the time scale of the post-feudal period. Foremost among the research questions at both scales, this paper suggests, is the problem of tracing the intertwined but separate impact of technological development and social relations, to see what changes in either, or their relationship, might lead us to the better city that we would wish to see.

NOTES

1 We use the term "soft" by analogy to its use in zoning practice, where a "soft" site is spoken of as one not developed to the limits its legal zoning permits, i.e. one viewed as ripe for change and new development.
2 I owe the example to Susan Fainstein's 2005 study of the Port Authority of New York and New Jersey.

REFERENCES FROM THE READING

Abu-Lughod, J. (1999) *New York, Chicago, Los Angeles: America's Global Cities*, Minneapolis: University of Minnesota Press.

Fainstein, S. (2005) Ground Zero's landlord: The role of the Port Authority of New York and New Jersey in the reconstruction of the World Trade Center site, in J. Mollenkopf (ed.) *Contentious City: The Politics of Recovery in New York City*, New York: Russell Sage Foundation.

Friedmann, J. and Wolff, G. (1982) World city formation: An agenda for research and action, *International Journal of Urban and Regional Research*, 6: 309–44.

Marcuse, P. (1989) "Dual city": A muddy metaphor for a quartered city, *International Journal of Urban and Regional Research*, 13, 4: 697–708.

Marcuse, P. (1991) Housing markets and labour markets in the quartered city, in J. Allen and C. Hamnett (eds.) *Housing and Labour Markets: Building the Connections*, London: Unwin Hyman: 118–35.

Marcuse, P. (1997) Glossy globalization, in P. Droege (ed.) *Intelligent Environments*, Amsterdam: Elsevier Science Publishers.

Marcuse, P. (2002) Depoliticizing globalization: From Neo-Marxism to the network society of Manuel Castells, in J. Eade and C. Mele (eds.) *Understanding the City*, Oxford: Blackwell: 131–58.

Marcuse, P. and van Kempen, R. (eds.) (1999) *Globalizing Cities: A New Spatial Order?* Oxford: Blackwell.

Marcuse, P. and van Kempen, R. (eds.) (2002) *Of States and Cities: The Partitioning of Urban Space*, Oxford: Oxford University Press.

Smith, N. (2002) New globalism, new urbanism: Gentrification as global urban strategy, *Antipode*, 427–50.

PART TWO

Pathways

Plate 8 Istanbul Financial Center under construction

Source: Roger Keil

INTRODUCTION TO PART TWO

The contributions to this section examine the applications of global city theory to investigations of urban restructuring in some of the most globally integrated metropolitan centers of western Europe (London, Amsterdam, Zurich), North America (New York, Houston, Los Angeles), Latin America (Medellín and Bogotá), and Asia (Shanghai, Delhi, Phnom Penh). While the founders of global cities research ventured a number of broad generalizations regarding urban structures and processes within global cities, others have been concerned to study the distinctive, place-specific ways in which cities are articulated to the world economy. Beyond their underlying commonalities as basing points for global capital accumulation, global cities have proven extremely diverse in their economic functions, political strategies, and institutional arrangements, as well as in various features of their built environments.

Early contributors to world city theory attempted to explain certain key developments and processes internal to a given city with reference to its external positioning in the global urban system. Subsequent research, including the contributions to this section, has reversed this line of causality, demonstrating that the history, spatiality, institutional configuration, and sociopolitical environment of a city exercise a powerful influence upon its mode of insertion into the world economy (Lees, Shin, and Lopez-Morales 2015). From this point of view, a city's position in the world economy cannot be taken for granted, but must itself be analyzed as an expression and outcome of complex, contested socioeconomic processes and strategies within that city and the broader (generally regional and national) space-economies in which it is embedded. One of the basic methodological challenges of global cities research, therefore, is to uncover, simultaneously, not only the general roles that cities have come to play in the contemporary world economy but also their distinctive histories, geographies, political economies, and developmental trajectories.

For this reason, global city theory should not be viewed as postulating a convergence of urban political-economic and spatial structures towards a single, generic model. Instead, this conceptual framework is better understood as a means to decipher the interplay between general, cross-national trends of urban restructuring and place- and territory-specific outcomes. While confronting this intellectual task presents major methodological and empirical challenges, the contributions in this section demonstrate that doing so can generate illuminating analyses of urban restructuring.

Following the pioneering work of the 1980s, case study-based research on global city formation was pursued with enhanced methodological sophistication and with reference to an increasingly broad range of cases (Brenner 2001; Robinson 2016). Global city theory thus came to serve as a key analytical reference point for the investigation of contemporary urban restructuring. At the same time, the empirical extension of global cities research opened up new debates regarding its appropriate theoretical foundations and methodological applications. Was the concept of the global city to be viewed as a description of a particular type of city (for instance, a transnational headquarters location or a global command and control center), or could it be mobilized more effectively to examine the more general trend towards globalized urban restructuring? Were analogous patterns of political-economic and sociospatial restructuring unfolding in cities that fulfilled similar roles in the world economy, or was the globalization of urbanization contributing to a further differentiation of urban institutional configurations and sociospatial forms? Were local political institutions and social movements helpless victims of transnational capital, or could they find ways to defend local interests and priorities?

To address such questions, the contributions to this section explore a representative selection of case studies and comparative analyses of global city formation. Several common threads of investigation, analysis, and interpretation emerge:

- *Remaking the built environment.* The chapters track various ways in which the urban built environment is transformed in and through the process of global city formation. New transportation and telecommunications infrastructures are established to ensure global connectivity; large-scale property developments, mega-projects, and office towers are constructed to meet the perceived requirements of transnational capital and middle-class lifestyle demands; central business districts and luxury housing development projects are extended into older working class neighborhoods; and a new "landscape of power" (see Ch. 12 by Zukin) is constructed to market the city, both to its inhabitants and to potential investors, as a global command and control center. The transformation of the built environment takes the most drastic forms in China and India. The essays by Wu (Ch. 16) and Dupont (Ch. 17) examine how local elites have "beautified" Shanghai and Delhi to prepare for international mega-events and investors, and how the poor are rendered invisible in the process (see also Ren 2013 and Ghertner 2015).
- *Growth coalitions.* Several of the chapters underscore the role of growth-oriented political alliances—generally composed of national, regional, and local state officials; transnational corporate elites; and local property owners or "rentiers"—in establishing the local preconditions for global city formation. Such alliances serve not only to mobilize the tremendous infrastructural, technological, and financial resources that are required to ensure a city's seamless global connectivity, but also to blunt local opposition to large-scale mega-projects and real estate developments that are generally associated with global city formation. For instance, Wu's essay (Ch. 16) highlights how global city formation in Shanghai is inseparable from interventions from both the central and local government, as well as domestic and transnational capital. Shanghai's growth coalitions are composed of the local state and business interests, and completely exclude urban residents.
- *State strategies.* The chapters illustrate the central role of state institutions, at once on national, regional, and municipal scales, in promoting global city formation, whether through regulatory realignments, financial subsidies or some combination thereof. From London and New York to Houston and Detroit, it is evident that states actively facilitate the reorganization of urban space to accommodate the demands of globally oriented firms, real estate developers, and political elites. This often entails the construction of new, quasi-public institutions or public-private partnerships lacking in democratic accountability, that oversee and finance the development of globalized urban spaces while protecting private investors from "excessive" risk (Ch. 12 by Zukin). In non-democratic regimes such as China, the state decides the pace, direction, and magnitude of urban restructuring with little participation from the civil society (Ch. 16 by Wu). Local social forces may also attempt, with varying degrees of success, to harness state institutions in order to block particular pathways of global city formation, leading in turn to the negotiation of "territorial compromises" that may protect certain local interests while nonetheless facilitating (a modified form of) globalized urban development (Ch. 15 by Schmid). State institutions thus become a key political terrain in which the form, pace, and geography of globalized urbanization are fought out by diverse local and supralocal social forces.
- *Uneven development and territorial inequality.* In addition to their attention to urban built environments, the chapters also map out some of the wide-ranging sociospatial transformations that have ensued in conjunction with global city formation. Whereas Zukin (Ch. 12) focuses on the reconfiguration of downtown urban cores through large-scale office developments and luxury residential developments, Hill and Feagin (Ch. 13), Soja (Ch. 14), and Schmid (Ch. 15) examine some of the broader, region-wide expressions of global city formation. Existing patterns of uneven development and sociospatial inequality are exacerbated, and new clusters of globally oriented firms may be established beyond the traditional city core, such as the case of the Millennium City of India—Gurgaon, developed on the outskirts of Delhi

(see Dupont Ch. 17). Consequently, as Soja (Ch. 14) forcefully suggests, polycentric patterns of urban development appear to be superseding traditional, monocentric models of urban spatial form. Clearly, these trends cannot be explained entirely with reference to the forces of globalization. However, the contributions to this section do suggest that they are being articulated within global city-regions around the world, with powerful consequences for urban social space and everyday life.

The chapters in this section can be also viewed as engaging in a common effort to develop comparative perspectives on the process of global city formation. Some make use of global city theory through individualizing case studies that are intended to illuminate the particularities of the places under investigation, albeit with reference to a broader geoeconomic context (Ch. 15, 16, 17). Others, by contrast, mobilize global city theory in order to explore the similarities and differences among distinct cases of globalized urban development; they thus adopt a more explicitly comparative methodology (Ch. 12, 13, 14). By highlighting the persistent differences among these cities, these authors are also able to develop theoretically informed interpretations of their quite striking commonalities. Most of the contributions to this section mobilize some kind of comparative perspective in order to concretize and extend some of the more general propositions of global city theory. In sum, this section explores the applications of global city theory to various forms of place-based research on globalized urbanization.

REFERENCES AND SUGGESTIONS FOR FURTHER READING

Binelli, M. (2012) *Detroit City is the Place to Be: The Afterlife of an American Metropolis*, New York: Picador.
Brenner, N. (2001) World city theory, globalization and the comparative-historical method: Reflections on Janet Abu-Lughod's interpretation of contemporary urban restructuring, *Urban Affairs Review*, September: 124–47.
Ghertner, A. (2015) *Rule by Aesthetics: World-class City Making in Delhi*, New York: Oxford University Press.
Huyssen, A. (ed.) (2008) *Other Cities, Other Worlds: Urban Imaginaries in the Globalizing Age*, Durham and London: Duke University Press.
Lees, L., H. B. Shin, and E. Lopez-Morales (eds.) (2015) *Global Gentrifications: Uneven Development and Displacement*, Bristol: Policy Press.
Ren, X. (2013) *Urban China*, Cambridge, UK: Polity Press.
Robinson, J. (2016) Thinking cities through elsewhere: Comparative tactics for a more global urban studies, *Progress in Human Geography* 40, 1: 3–29.
Weinstein, L. (2014) *The Durable Slum: Dharavi and the Right to Stay Put in Globalizing Mumbai*, Minneapolis: University of Minnesota Press.

11 Prologue
"Istanbul was our past, Istanbul is our future"

from *Aljazeera* (2017)

Hamid Dabashi

I have now blissfully forgotten how many times I have visited Istanbul, or why it is I feel so much at home there. Last time I was there was during the last World Cup, Brazil 2014, which as it happened, coincided with the Muslim month of Ramadan, when many were fasting. I remember sitting in a cafe/bar in the heart of Istanbul, near the Taksim Square, watching Germany destroy Brazil in the semi-finals, surrounded by Turkish, Arab, German, French, Brazilian, Iranian, and Russian football fans. It was a sheer joy of being in a Muslim city where women dressed as they wished, with or without an item of modesty, happily in possession of the streets of their homeland without anyone ever bothering them. Next to them were European visitors, shoulder to shoulder with tourists from across the Arab and Muslim world. You would hear as much Turkish as you did Arabic, Persian, English, French, German, or Russian. That was and remains the real Istanbul.

Before the horrific nightclub attack in the Reina, on the shore of the Bosporus Strait, in Istanbul on New Year's Eve 2016, is lost into yet another cycle of vicious, mind-numbing violence – which now extends from Orlando to Paris, Berlin, Damascus, Baghdad, Cairo, deep into Pakistan and beyond – we might want to pause for a moment and wonder what these heinous crimes actually mean. What do they signify, how are we to read them? Why would an innocent gathering of young people from around the Arab and Muslim world with their Turkish friends be a target of such a vicious attack? "In continuation of the blessed operations that Islamic State [of Iraq and the Levant] is conducting

against the protector of the cross, Turkey," according to reports ("Istanbul" 2017), ISIL has assumed responsibility for this cowardly act, further adding: "a heroic soldier of the caliphate struck one of the most famous nightclubs where the Christians celebrate their apostate holiday."

COSMOPOLITAN URBANISM

This is habitually inane gibberish that may or may not be an indication of ISIL having actually perpetrated this crime. But the question is: What is this inanity targeting? What is it, that it is opposing? What kind of sentiment, however crudely, does it want to provoke? The answer lies in the location and timing of this attack: A nightclub where a group of young people from around the world had gathered to celebrate the new year on the Christian calendar. Whoever was behind it, this attack is on the culture of tolerance, on the factual pluralism of Muslim countries now in many ways represented in Istanbul.

The young people in that club represent a new breed of Turks and their friends from around the (Muslim) world. The terms "secular" or "Westernised", which you keep hearing on these occasions, are terribly flawed, deeply misguided. Such clubs, cafes, markets, bookstores, movie theatres, or opera houses are all specific insignia of a living, thriving urbanity – the figurative emblem of a deeply rooted cosmopolitanism that is definitive to Istanbul.

NOTHING'S WRONG WITH CELEBRATING THE NEW YEAR

There is absolutely nothing wrong with marking and celebrating the new year on Christian calendar, or even Christmas, in any Muslim country. The birthplace of Christianity is in Palestine, where other sacrosanct sites of Islam and Judaism are also located. Christ was from historic Palestine, a Jewish rabbi born and raised in Nazareth. These subterranean creatures that call themselves ISIL, or their kindred souls in any other part of the Muslim world, both inside and outside Turkey, are not just viciously violent, they are pathetically ignorant.

Muslim countries have always been home to thriving Jewish, Christian, Zoroastrian, Hindu, Buddhist, etc. communities. Muslims have lived alongside these communities in successive empires – from the Abbasids to the Seljuks to the Ottomans, the Safavids, and the Mughals. How could any such cosmopolitan empire be limited to the myopic zealotry of any particular sect of hateful fanatics?

It is now habitual to refer to the victims of this pernicious attack in the Ortakoy neighbourhood as "foreigners." These young men and women may have come from anywhere, from India to Morocco. But they were not "foreigners" in Istanbul. They were at home in Istanbul – which is home to any human being with an urbanity of culture and demeanor to her and his character and culture.

What we see today in Istanbul is no accident, nor is it the sign of "Westernisation" or "secularisation" of Istanbul – all of them nasty Orientalist nonsense, entirely ignorant of Islamic social and intellectual history. Quite to the contrary. This is the perfectly normal post-colonial growth of Istanbul from deep roots of its Ottoman lineage, a vastly and deeply pluralistic society, welcoming artists, literati, intellectuals,

journalists, and political activists from four corners of the world. How did Istanbul accommodate all of those varied communities throughout its history and today we hear calls of intolerance from certain voices, even within the Turkish society? Because, up until its fateful encounter with European imperialism, Istanbul was the epicentre of a confident cosmopolitan culture.

TOLERANCE AND PLURALISM

Today, Muslims and non-Muslims, in and out of Islamic world, are facing a vicious battle, not of identity, but of alterity – not who they are, but who their nemesis is.

Muslims are not the enemies of Christians or Jews, nor are Christians and Jews the enemy of Muslims. What we have are, in fact, battles of sovereignty among the ruling states entirely bereft of legitimacy from their respective nations. As many states have degenerated into pure institutions of violence – very much on the model of ISIL – they inevitably pit against each other the most pernicious common denominators of divisive hatred. Against all odds, the glorious cosmopolitan urbanity of tolerance and pluralism of Istanbul will triumph against all forces of fanaticism, foreign or domestic to Turkey, and as it was a landmark of our past, it will beacon us all to our future.

REFERENCES FROM THE READING

Istanbul: ISIL claims responsibility for Reina attack. (January 2, 2017) Al Jazeera, http://www.aljazeera.com/news/2017/01/isil-claims-responsibility-turkey-nightclub-attack-170102082008171.html.

12

"The city as a landscape of power: London and New York as global financial capitals"

from L. Budd and S. Whimster (eds.),
Global Finance and Urban Living (1992)

Sharon Zukin

EDITORS' INTRODUCTION

Sharon Zukin is Broeklundian Professor of Sociology, Brooklyn College, at the CUNY Graduate Center in New York City. She has written extensively on the changing built environments and cultural landscapes of US cities undergoing major economic transformations. Her most influential publications include *Loft Living* (1988), *Landscapes of Power: From Detroit to Disney World* (1991), *The Cultures of Cities* (1995), and *Naked City* (2009). The selection below is excerpted from a chapter in an edited volume focused on the local sociospatial changes that were unfolding in global financial centers. Zukin's comparative case study analyzes the built environments of London and New York City, where a variety of state-subsidized mega-projects were constructed during the course of the 1980s in order to promote downtown redevelopment and to reinforce each city's position as a global financial center. Whereas entirely new quasi-governmental bodies were established in London to build King's Cross and Docklands, the construction of Battery Park City and the redevelopment of 42nd Street/Times Square in New York City continued a longer trend of governmental support for large real estate capital. Zukin frames her analysis of these mega-projects around the crucial distinction between *vernacular* – the spaces of everyday life – and *landscape* – the spaces of power dominated by capital and state institutions. For Zukin, the built environments of global financial centers are zones of intense contestation in which the forces of vernacular and landscape continually clash. The basic question, for Zukin, is: should the space of the global city be organized to facilitate the everyday social reproduction of working people, or should it be reshaped to serve the demands of global corporations and financial elites? As Zukin shows, the local governmental institutions of New York and London appear to favor the latter goal, but in so doing, they also expose the city to the turbulence of global markets. Thus the "landscape of power" forged in each city is never entirely secure or stable; it may be threatened both from within, through social unrest and resistance, and from without, through externally imposed geoeconomic dislocations. Zukin's chapter is a useful example of a case study that draws upon empirical material from two major global cities in order to develop a broader set of arguments regarding the politics of urban sociospatial restructuring.

The global cities that have captured attention in recent years owe much to intense competition in international financial services. Partly a matter of providing world-class commercial facilities, and partly a matter of image-creation, the effort to attract geographically mobile investment activity changes a city's perspective. The old, diversified urban center is cleaned up for new offices and cultural consumption; in the process, it becomes more expensive. Not surprisingly, governmental priorities shift from public goods to private development.

Local officials in New York and London have pursued this sort of growth with ambivalence. Forced to shed traditional allies from labor unions and the left, they have been disciplined by fiscal crisis, co-opted by property developers, taken in hand by financial institutions and quasi-public authorities, and scolded by national political leaders. For their part, national governments have eagerly anticipated the smaller world and larger cities that follow global financial markets. In the United States, the tradition of local autonomy precludes central government's taking a key role in directing New York's growth. In the UK, however, the national government has intervened to spur the City's eastward expansion. From 1980, activist conservative governments in both Britain and the United States broadened official support for the expansion of financial markets – and for an inevitable struggle between New York and London over priority of place. Competitive efforts to capture global markets in a single place would lead cities to the same general strategies.

It's all very well to describe such changes in terms of a traditional agglomeration economy. Yet the degree of conscious competition between cities that is involved, in a political context of seeking private resources for public problems, creates a new version of "cities in a race with time." Municipal officials of London, New York and even Tokyo must worry not only about the competitive ability of firms that are based within their realms – and the employment opportunities and tax revenues that might slip from within their grasp; they must also build an infrastructure that attracts and retains world-class financial actors. This infrastructure includes advanced telecommunications facilities, a computer-literate workforce and new skyscraper office buildings that lift urban identity from the modern to the spectacular. The interrelated effects of economic structure, institutional intervention and cultural reorganization are most directly perceived in change in the landscape: creating the city as a landscape of power.

LANDSCAPE AND VERNACULAR

Cities always struggle between images that express a landscape of power and those that form the local vernacular. While power in modern times is best abstracted in the skyscraper outline of a city's financial wealth, the vernacular is most intimately experienced in low-lying residential neighborhoods or *quartiers* outside the commercial center. Much of the social quality of urbanity on which a world city depends reflects both the polished landscape and the gritty vernacular as well as the tension between them. New Yorkers point with pride, for example, to their "city of neighborhoods," whose architectural and economic diversity suggests the cultural and social heterogeneity of its population. Londoners praise their city's informal domesticity that rests on relatively small houses and a mixture of social classes in each area. By contrast, each global city has at least one densely built, centrally located, high-rise district that drives both property values for the metropolitan region as a whole and office employment in financial and other business services.

The coherent vertical landscape of this center – the Wall Street area in downtown Manhattan and the City of London – is circumscribed by a segmented horizontal vernacular of working-class districts and low-income, immigrant ghettos. Yet the opposition between landscape and vernacular refers to more than mere architectural or art-historical categories. Just as *landscape* shows the imprint of powerful business and political institutions on both the built environment and its symbolic representation, so does *vernacular* express the resistance, autonomy and originality of the powerless. Their opposition, moreover, suggests an important *asymmetry* of power. Since elites are capable of imposing multiple perspectives on the surrounding vernacular, landscape implies their special contribution as well as the entire material and symbolic construction (Zukin 1991).

The juxtaposition between landscape and vernacular in New York is easily visualized by taking the subway line between Manhattan and Brooklyn across the Manhattan Bridge. On the Manhattan side of the East River, at the southern tip of the island, several generations of twentieth-century skyscrapers raise a profusion of gothic, flat-top and mansard roofs against the sky. The Woolworth Building was the tallest skyscraper in the world when it was built in 1913. Now [1992] it is dwarfed by the elongated twin cigar-box

Plate 9 Global city at street level, New York

Source: Roger Keil

towers of the World Trade Center, completed in the early 1970s, and the lower yet no less massive commercial and residential development of Battery Park City. From the train windows one also sees in miniature along the waterfront the few remaining early nineteenth-century merchants' warehouses and the wholesale fish market that make up South Street Seaport, a commercial redevelopment of trendy bars, retail stores and tourist shops in a municipally designated historic landmark zone. Directly beside the elevated tracks are the red-brick, late nineteenth century tenements and loft buildings of Chinatown. The train rushes past open windows where ceiling fans revolve and fluorescent lights shine on Chinese garment workers, who are engaged in the only expanding manufacturing sector left in Manhattan. North of the bridge, where the subway re-enters the earth, a dense array of public housing projects is aligned along the shore.

In London's East End, a more horizontal landscape is also chronologically segmented by financial capital and social vernacular, juxtaposing rich and poor in close proximity. The Docklands Light Railway connects the City's stern office buildings with the brighter, more reflective steel-and-glass or restored commercial

centers at Canary Wharf, Tobacco Dock and the Royal Albert Dock, two miles away. The railway rushes past council flats whose building walls are striped with graffiti. It also passes clusters of small single-family council houses that stand as solitary out-croppings on ground cleared for new construction. Docks that have not yet been demolished or redeveloped are stark monuments to Victorian industry: huge, empty, lacking a function in the eight-square-mile, purpose-built financial quarter around them, where Canary Wharf alone occupies 71 acres.

Yet none of this is permanent. Just as landscape has often been transformed into low-rent quarters for the poor or artisanal workshops, so has vernacular been changed into a new landscape of power.

This spatial metaphor unifies some of the disparate changes we see around us in the city, and couches our unease with an unfamiliar material reality. But change has not been so sudden as developments like Docklands and Battery Park City imply. Much of the landscape has been re-made incrementally. After all, London and New York have always had high property values, especially downtown and in the City. Further, gentrification proceeds by individual houses and streets; although it changes the character of an area, it rarely occurs on so dense a scale as greatly to raise aggregate income or educational levels. And despite the perennial tension between the developers' urge to tear down and reconstruct and the residents' desire that things remain as they are, most large redevelopment projects, although often compromised in size and style, have in fact been built. Seeing the landscape of the city is believing in its mission, for visual consumption holds the key to the old geographical center's creative destruction. The very syntax, the rules by which we perceive it, have been changed and, as the rules have changed, so has the way we use the city.

A structural precondition of these shifts has been the abandonment of most urban manufacturing. Since the 1960s, New York and London have lost half their blue-collar jobs to automation, regional decentralization and the internationalization of industrial production. Certainly jobs in processing industries (e.g. chemicals, foods) and product assembly had located years earlier on the periphery or in the suburbs. Yet low rents and easy access to markets kept the garment and printing industries in the center, along with a decreasing number of traditional crafts and, of course, the docks. The displacement of these jobs, however, freed the areas in which they were located for other uses. Property values that had been restricted by manufacturing plants and working-class housing could now reflect the inherent utility of a central location. Higher land values exerted pressure, in turn, on remaining industrial uses. The social meaning of this space was, moreover, freed for reinterpretation and new appropriation. The previously "closed" vernacular of working-class districts, immigrant ghettos and industrial areas that were always regarded as dangerous and suspicious, especially after dark, was now "open" as landscape to those who could consume it. These factors fed a wave of private-market property investment from new office construction to gentrification.

Historic building and district classifications have greatly contributed to opening up the vernacular to a broadly defined upper middle class. So have restaurants, gourmet groceries and stores that sell artisanal or artist-designed products. These amenities lure consumers with cultural capital, i.e. the experience, education and time to seek them out. And artists' districts – notably, Manhattan's loft district of SoHo – have also opened up the dingy charms of the vernacular by incorporating them into a landscape of cultural consumption (Zukin 1988).

While individuals change their use of the center, geography reflects the center's new functions. Clerkenwell and parts of Hackney lose their somewhat heterogeneous social character, emerging as enclaves of history that can best be apprehended by means of a guide. Spitalfields sheds the uneasy shadow of artisans' workshops, the wholesale market and Jack the Ripper to be presented as a prime site for commercial redevelopment. Docklands is re-created as a direct expansion of the financial center on the one hand, and an urban frontier, on the other. And if the railroad yards and canals at King's Cross are unified by Norman Foster's new urban center, they will lose their "inaccessible," somewhat desolate and inchoate quality: King's Cross will take its place between Camden and Islington in a coherent North London landscape.

A TALE OF TWO CITIES

With big projects like King's Cross and Docklands, London seems to escape traditional institutional bounds, i.e. the instrumental roles that are constituted by various territorial levels of government, social communities and public-private distinctions. In New York, on the other hand, big projects like Battery Park City or the 42nd Street Redevelopment at Times Square seem only to extend institutional precedents by which developers get to build what they want. For over 150 years New York's high land values have driven low-rent, low-class uses out of the center, including working-class neighborhoods, manufacturers and small crafts shops. But isn't each city really more complex? London's growth has been restricted by the Green Belt since the 1930s, as well as by political pressure by central government and an extraordinarily high percentage of housing stock in council housing; yet one observer of recent urban changes says that London is geographically and politically more mutable than New York. For its part, although New York has always been in thrall to business elites, a change of mayors (albeit usually within the Democratic Party) and extraordinary mobilization by citizens' groups, often in the law courts, have sometimes shifted the purpose and reduced the size of new construction, and occasionally slowed the pace of change.

Despite these differences, the landscapes of power that are emerging in New York and London are strikingly similar. This is hardly surprising, for redevelopment of the center in both cities is commissioned and designed by the same worldly superstars, including developers, architects and private-sector financial institutions. Just as skyscrapers have become the *sine qua non* of place in the global hierarchy of cities, so do US, Japanese and Canadian builders and bankers represent the basis of global market rank. A city that aims to be a world financial center makes deals with Olympia & York and Kumagai Gumi, welcomes Citibank and Dai-Ichi Kangyo and transplants Cesar Pelli as well as Skidmore Owings and Merrill and Kohn Pederson Fox.

Developers tend to use the same strategic tools. They seek to diversify their portfolios by spreading investment projects around the world, and find these investment sites according to computerized projections made by the same accounting, property and management firms, which are also organized on a worldwide basis. Wrongly or not, the progressive lowering of barriers to multinational trade in services, as in London's Big Bang of 1986 and the consolidation of Western European markets in 1992, has encouraged this foreign participation in local property markets. Yet unified markets do not require a large number of first-tier financial capitals. And the close links between trading practices in global market centers probably leads to a coordinated overexpansion of office space, like the coordinated retrenchment of personnel in New York and London following the stock exchange crash of 1987 and the increased volatility in the market in 1989. In other words – despite the common wisdom – it may no longer be a good idea to use property development in London as countercyclical hedging against loss of investment value in New York.

New York's office market should at any rate provide a cautionary tale. While foreign banks cut investments and employment less in New York than in London following the 1987 crash, and domestic financial firms either delayed massive layoffs or limited them to specific low-yield activities, office vacancy rates rose as US firms trimmed their payrolls selectively and physically consolidated operations. Even the large money center banks have added slack to New York's office market. While generally profitable and still expanding overseas, these financial institutions decided to sell some of the remarkable corporate headquarters buildings with which they were identified and move back-office operations outside Manhattan. Chase Manhattan Bank, long the linchpin of and major property investor in the Wall Street area, moved back-office operations to a new building in downtown

Plate 10 Urban redevelopment, Soho, London

Source: Roger Keil

Brooklyn. Citibank sold most of its diagonal-roof head-quarters to a Japanese insurance company, building a distinctive new skyscraper across the East River in Long Island City. Storing assets in face of Third World debt and Japanese competition may have been only one among many reasons for Citibank's action, for the bank is playing an important part in the projected commercial redevelopment of the waterfront in that area of Queens.

But the pattern of playing musical chairs with office relocations has long been typical of New York's commercial property markets. City government sub-sidies and zoning changes have recently lured major developers from the overbuilt East Side to the West Side of midtown Manhattan, and office building owners attract large corporate tenants by offering special facilities, rent reductions, build-out allowances for tenants' improvements, and even a share of equity. These conditions allowed developer William Zeckendorf Jr to sign up the Madison Avenue advertising firms Ogilvy & Mather and N.W. Ayer and the Wall Street law firm Cravath Swain & Moore as tenants in Worldwide Plaza, his new mixed-use complex on Eighth Avenue north of Times Square.

Governmental connections also aid developers' efforts to create a locational advantage. During the decades that David Rockefeller directed Chase Manhattan Bank, he formed a business group, the Downtown-Lower Manhattan Association, in 1957. The association's desiderata for protecting Chase's property investment in the area began with relocating the wholesale food markets to the Outer Boroughs, and included an ambitious program to rebuild the municipal administrative center, develop middle-income housing, attract university facilities and establish new bellwether skyscrapers that could compete with the more modern offices in midtown. The city government implemented most of this wish list during the 1960s. Moreover, when David's brother Nelson was elected governor of New York State – a post he held from the 1960s to the mid-1970s – powerful state institutions were created and utilized for large-scale urban redevelopment. A new public authority, the New York State Urban Development Corporation, superseded local powers to condemn property, design projects and finance their construction by issuing bonds. The Battery Park City Authority controlled new waterfront construction. Meanwhile, the older Port Authority of New York and New Jersey, which manages most of the mega-scale transportation systems in the

metropolitan region, used its jurisdiction over the port to commission its redevelopment. Mainly due to these institutions, the World Trade Center was conceived and built on waterfront landfill. To aid the process of marketing this mammoth space, some say, a local law requiring new fire protection devices was passed, hastening tenant relocation from older office buildings throughout lower Manhattan into the World Trade Center. The Rockefellers plied a winning strategy in the downtown office market with their influence over government, ability to move both public and private investment capital into new construction and fine-grained understanding of the value of architecture as symbolic capital. This was in any event a period of manufacturing decline and business service boom, accentuated in property markets by the levers of public and private power.

Until recently, in fact, commercial property development in New York was seldom constrained by popular protest or political opposition. In contrast to housing and highway construction, where community groups made their pressure felt, the business district in the center was both materially and symbolically regarded as the province of business elites. Industrial decentralization toward the Outer Boroughs, New Jersey and Long Island had reduced manufacturers' and workers' influence over land use in housing over commercial development. By the sixties, large-scale removal of Manhattan residents occurred on only two, noncommercial project sites: the cultural complex at Lincoln Center and the primarily residential Upper West Side urban renewal area, both of which were intended to complement rather than expand the midtown business center. Thus by 1970 no residential community remained to voice complaints about a "change of character" in the center.

During the 1980s, however, the enormous expan-sion of New York's landscape of power was seen as having taken a toll on both "quality of life" and "afford-able housing," two shibboleths of liberal opposition to business strategies. Yet without a tenacious working-class presence and pressure groups, as in Tower Hamlets and nearby boroughs of London, institutional bargaining focused mainly on trade-offs. On the one hand, the city government fully acknowledged the right to create private value by permitting taller buildings with more rentable space in congested parts of the center. On the other hand, city officials tried to impose an obligation on developers to restore or renew public value, by adding plaza-like spaces open to public use, incorporating theatres or historic landmark structures instead of tearing them down, or refurbishing parks and subway stations near their building sites.

IN THE LONG RUN

The material and symbolic reconstruction of the center is a long-term historical process. The sudden appearance of a Docklands or Battery Park City masks the gradual effects of structural change, notably, a shift from organizing the city as an assortment of concrete production spaces toward visualizing it as a coherent space of abstract financial processes and consumption. By the same token, the seeming abruptness with which urban redevelopment schemes are adopted and abandoned reflects temporary booms and busts in property markets and changing political alliances, as well as financial market advances and declines. Yet, in addition to redeveloping capitalist space, the landscape of power reflects both the incremental institutional interventions of the national and local state and the conscious cultural reorganization of business, political and artistic elites – those who wield *all* the levers of social power.

Battery Park City and Docklands offer concise examples of the morphological, geographical and aesthetic processes that are involved. They are primarily sites that government has had to market many times: first to private investors and developers, then to banks and public and private bodies with the authority to approve large-scale financing (e.g. bonds), also to a restive citizenry that presses for both quality-of-life improvements (e.g. low-rise building, low-density zoning) and affordable housing, and finally to a broad segment of the public that has both financial and cultural capital to consume the space the city has produced. These processes lead, on the one hand, to an aestheticisation of the project at the expense of a focus on critical social needs such as housing and jobs. On the other hand, they lead to image-making that obscures the removal or incorporation of the segmented vernacular by the landscape of power of a world financial center.

Despite great differences between the political systems of New York and London, and different relations between their financial communities and the national economy, both major projects of contemporary new construction – Battery Park City and Docklands – show the same trajectory of a landscape of power.

Significant to globalization, the redevelopment of the center in New York and London has proceeded almost in tandem. While Docklands required, and still requires, the building of more transportation infrastructure, Battery Park City has merely spawned new streets and parks, essentially consumption amenities on the water (plazas, marinas, fountains). Each project has been delayed by the continued belief among investors and developers that the project might be unmarketable, although the offices of the World Financial Center are over 90 per cent filled. Each crisis of investors' confidence was overcome by expanding the project's commercial facilities and reducing or removing the proportion of low-income housing. But the major obstacle to construction in both cases was eased by maximizing the *private* role in development. Battery Park City Authority relinquished most of its instrumental role to private developers in 1979, and London Docklands Development Corporation proceeded the same way when it replaced the earlier, borough-dominated London Docklands Joint Commission in 1981. A belated interest in community development on the part of LDDC in 1989–90 led the corporation to lobby central government for more social services without, however, denying the priority of private-sector demands.

A multi-centered city that would somehow dilute the arbitrary expansion of the center negates the cheek-by-jowl juxtaposition of neighborliness and imperial power that characterized New York and London at their commercial height in the not-too-distant past. The old contrast between a singular landscape of power at the center and a segmented vernacular has deliberately been destroyed.

REFERENCES FROM THE READING

Zukin, S. (1988) *Loft Living: Culture and Capital in Urban Change*, 2nd ed., London: Radius/Century Hutchinson.

Zukin, S. (1991) *Landscapes of Power: From Detroit to Disney World*, Berkeley and Los Angeles: University of California Press.

Zukin, S. (1995) *The Cultures of Cities*, Cambridge, MA: Blackwell.

Zukin, S. (2009) *Naked City: The Death and Life of Authentic Urban Places*, New York: Oxford University Press.

TWO

13

"Detroit and Houston: two cities in global perspective"

from M. P. Smith and J. Feagin (eds.),
The Capitalist City (1989)

Richard Child Hill and Joe Feagin

EDITORS' INTRODUCTION

Joe Feagin's work was already introduced in the previous section of this volume (see Ch. 5, by Rodriguez and Feagin). In this chapter, originally published in an edited volume on cities and economic restructuring under global capitalism, Feagin collaborates with Richard Child Hill (Professor Emeritus of Sociology at Michigan State University) to compare the process of urban restructuring in two major US cities, Detroit and Houston. The authors seek to decipher the developmental dynamics behind each of the two cities by examining the specialized economic niche of each within world-scale divisions of labor. By studying a combination of geoeconomic pressures and internal political-economic struggles, they try to chart the long-term evolution of each city. Hill and Feagin's analysis illustrates Friedmann's hypothesis (see Ch. 3) that a city's position in the world economy strongly influences its internal sociospatial development. As they note at the outset of their analysis, by the 1970s, both Detroit and Houston had adopted diverging approaches to political-economic regulation. While Detroit had embraced a welfarist model in which state institutions sought to correct market failures and organized labor wielded considerable political influence, Houston pursued a neoliberal model of "free market" capitalism that privileged big business and imposed only minimal regulatory constraints upon capital accumulation. Notwithstanding these different approaches, the authors argue, both cities suffered similar structural crises, albeit at different moments, when faced with economic fluctuations within their niche industry—automobiles in Detroit, and oil and petrochemicals in Houston. Following the global economic recession of the 1970s and the oil shocks of the 1980s, this sectoral specialization rendered both local economies particularly vulnerable to shocks and fluctuations within the international division of labor.

Since the late 1980s, however, Houston and Detroit have embarked on very different trajectories of growth and decline. Houston diversified its economy towards high-tech industries and professional services, and its population increased to 2.2 million as of 2010. By contrast, Detroit's population shrank to 681,000 in 2013, when the city declared bankruptcy (Maraniss 2015). The very different developmental pathways of the two cities can be attributed to a variety of historical, institutional, and policy factors, but Hill and Feagin pinpoint one important factor—Houston's comparative advantage in harnessing federal government support. The authors also emphasize a number of common features of North American urban development that emerged during the postwar period in both cities—including extensive residential suburbanization, industrial decentralization, and institutionalized racism, all of which are readily visible today.

INTRODUCTION

Detroit and Houston became urban archetypes in the United States in the 1970s. Detroit was the snowbelt city in decline; Houston, the booming sunbelt metropolis. Each city was held up as an object lesson for the other. It was commonly argued, for example, that if Detroit wanted to revitalize, the city should embrace something like Houston's freewheeling, boomtown philosophy with all that implied: aggressive business promotion, weak labor unions, social inequality, low taxes, few social services and unplanned urban sprawl. Others thought that as Houston matured, the Motor City's brand of welfare state capitalism would take hold and chart the Oil City's future. If so, Houston would experience growth in worker organization, an institutional partnership between Big Business and Big Labor, more political power among minority groups, tax increases, expanded social services and eventually, business disinvestment and fiscal crisis. But the 1980s held something different in store from what is implied in these laissez-faire versus welfare state contrasts.

Houston, in fact, was to experience an economic crisis in the 1980s not unlike the one which began to confront Detroit a decade earlier. Ironically, by the mid-1980s, it was the two cities' similarities, not their differences, that were most obvious. This suggests that answers to questions about a city's political-economic future are to be found as much beyond as within its local boundaries. Cities are spatial locations in a globally interdependent system of production and exchange. That global system is in crisis and transition. So the path a city follows in the future will depend upon the niche it comes to occupy in a changing international division of labor. It seems fruitful, therefore, to conceptualize how the city as a "localization of social forces" (Zukin 1980) is articulated with the city as a nodal point in the world capitalist system (Friedmann and Wolff 1982).

Here we take a holistic methodological approach. We assume that cities are not discrete and independent entities, but rather are interconnected parts of a world system of cities. Explanation, in this scheme of things, comes from locating parts within a larger whole in such a way as to render their complementary and contradictory relationships meaningful.

DETROIT: CRISIS IN THE MOTOR CITY

Specialization and growth

Detroit grew with the automobile industry. By importing raw materials and semi-finished products and converting them into durable finished goods to be exported throughout the world, Detroit's factories formed the heart of a vibrant international production system. Apart from auto-parts suppliers, three complementary industries have played a particularly salient role in Detroit's "metal-bending" economy: non-electrical machinery, fabricated metals and primary metals.

The Motor City's "Big Three" – General Motors, Ford and Chrysler – came to number among the world's largest corporations. The United Auto Workers (UAW) became the nation's biggest industrial union. Confrontations between the "Big Three" and the UAW ushered in a postwar era of collective bargaining. Productivity bargaining, cost-of-living adjustments and group insurance plans brought Detroit's industrial workers the highest standard of living to be found in any major North American metropolis. But it was also in Detroit that ethnic minority unemployment, inner city poverty, decaying neighborhoods and violent street unrest came to symbolize the urban crisis of the 1960s.

Crisis and reorganization

Suburbanization, institutionalized racism and uneven urban growth set the terms of political discourse in Detroit during the 1960s. By the early 1970s, however, it was capital flight to other regions and abroad that most focused public attention. The Great Lakes manufacturing empire was crumbling in the face of regional and international shifts in business investment and employment growth. In 1972, Detroit's civic elite commissioned a study of the exodus of capital from the Motor City. The study's conclusion was alarmingly simple: with outdated production facilities, a public infrastructure in poor condition, high taxes and a strongly unionized and aggressive labor force, Detroit would be hard-pressed to retain business activity, let alone attract new investment.

Plate 11 Model T in former Ford Piquette Avenue Plant, Detroit

Source: Xuefei Ren

In 1980, the US auto industry experienced an economic slide unparalleled since the Great Depression. Economic recession, rising energy prices, a saturated market for energy-inefficient cars and increased foreign competition sent the Big Three's profits plummeting. The way the auto production system is laid out in the United States meant hardship piled upon hardship for people in Detroit and the industrial cities of the Great Lakes.

With their survival at stake, the auto giants introduced changes in product design; more global concentration and centralization of capital; redesign of the labor process in relation to new technologies; and transformation in the industry's international division of labor. The auto transnationals are also working out a new international division of labor designed to maximize global profits by minimizing production costs through global resourcing: that is, by locating different segments of the production process in different regional and national locations according to the most favorable wage rates and government subsidies.

Decentralization and uneven development

Viewed along the dimensions of time and space, the trajectory of economic development in metropolitan Detroit during this century can usefully be divided into three periods: (1) the era of city building, 1910–49; (2) the era of suburbanization, 1950–78; and (3) the era of regional competition, 1979 to the present. One era does not give way to the next so much as each new period forms a layer upon the ones that came before.

The era of city building in Detroit coincided with the creation of the assembly line and mass production. When Henry Ford built his Crystal Palace in Highland Park in 1913, the modern factory system was born. From then on the auto industry expanded according to a well-defined spatial logic: a factory, then complementary plants and residential development clustered along industrial corridors following railroad lines.

The era of suburbanization can be dated from 1951, the year the central city's population peaked at 1.85 million residents. The United States experienced unparalleled economic growth during the early postwar years and the logic of industrial expansion stayed much the same, but now it extended beyond the city's limits. The Big Three built 20 new auto plants in the Detroit area during the decade following World War II, all beyond the boundaries of the central city. Complementary industries, commercial development and residential enclaves followed, like metal shavings drawn to a magnet, but this time it was the suburbs that boomed, not the central city. As industrial growth extended to the suburbs, and as commercial capital concentrated on the urban periphery, the principal axis of uneven development shifted from cities within the city to the line that divided the city from its suburbs. In Detroit that line became a racial barrier as the division between central city and suburb came to coincide all too closely with that between black and white.

The Detroit metropolitan area was a thriving economy, stimulated by high levels of capital investment. It contained nearly half of Michigan's population. Residents of this Detroit were mostly white; they lived in single-family houses located in the suburbs; and they earned an income above the state average. But the other Detroit, the central city, had become more and more like a segregated urban enclave during the decades following World War II. Home to hundreds of thousands of poor and unemployed people, the city was now pitted against the suburbs in a dual pattern of uneven urban development (Taylor and Peppard 1976). Deindustrialization now spread out from the central city, and down river, into white suburban Detroit. Now residents in Detroit's industrial, working-class white suburbs came to share many problems with their central city neighbors. But even as capital flight and automation were dealing a hard blow to Detroit's industrial suburbs, a new type of regional growth pole was emerging. Epitomized by Silicon Valley outside of San Francisco, and Route 128 outside of Boston, it is the science city, or the technopolis, as the Japanese like to call it. At the core of the technopolis are universities with strong science and engineering faculties, government-subsidized

research parks and closely linked high-tech companies specializing in high-value production. Oakland County's Technology Park in suburban Detroit is billed as the "workplace of the 21st Century," and fits this blueprint precisely. The era of regional competition is dominated neither by industrial nor by commercial capital. Rather, the driving force seems to be the creation of new information technologies and their application to all sectors of the economy.

For their part, Detroit officials have tried to revitalize the Motor City by following a corporate center redevelopment strategy. The linchpin in Detroit's redevelopment effort is the Renaissance Center (RenCen), a towering riverfront office, hotel and commercial complex meant to compete with outlying office centers and symbolize the city's corporate future. But this attempt to revitalize the central city ran into serious trouble from the beginning. The heart of the matter was Detroit's depressed downtown real estate market – the most salient indicator of the central city's weak position in the regional economy. A region of independent yet relatively autonomous cities is emerging. Knit together by a regional division of labor, these cities remain deeply divided along lines of race, class and municipal boundary. The principal fault line of uneven development no longer runs among areas within the city, nor between the city and its suburbs; rather it travels among competing cities within a region of cities. In the era of regional competition there is no longer one Detroit, nor two Detroits; there are many Detroits.

HOUSTON: THE CAPITAL OF THE SUNBELT

Specialization and growth

[Since the 1920s] Houston has become the center of a world oil and petrochemical production system: 34 of the nation's 35 largest oil companies have located major administrative, research and production facilities in the metropolitan area. In addition to these corporate giants, there are 400 other major oil and gas companies there. Thousands of smaller oil-related companies have attached themselves to these major petroleum companies.

The expanded flow of profits to the oil – petrochemical sector has provided the direct capital and borrowing capacity for other capital which lies

Plate 12 Packard Factory, Detroit

Source: Xuefei Ren

behind much of Houston's industrial and real estate (spatial) growth. At the heart of investment decision-making by these companies is business leader concern for a "good business climate," a code word for an area with lower wages, weaker unions, lower taxes and conservative politics. Companies that function as locators, such as the Fantus Company, have advertised Houston as having one of the best business climates in the US. Houston has grown because of cheaper production costs (e.g. weaker unions, lower wages), weaker physical and structural barriers to new development (e.g. no ageing industrial foundation) and tremendous federal expenditures on infrastructure facilities (e.g. highways) and high-technology defense industries.

In 1973 the OPEC countries gained control over their oil, and once-dominant US companies became primarily suppliers of technology and marketing agents for OPEC oil. US company profits on Middle Eastern oil fell, but the sharp rise in world prices brought great increases in profits on oil controlled by US companies elsewhere. In the 1973–1975 recession employment in goods-producing industries dropped 6 per cent in cities such as Dallas, but grew by 18 per cent in Houston, because its manufacturing firms produce for the oil world's industry. The rise in OPEC oil prices in 1973–1974 gave a boost to oil exploration and drilling, thus stimulating the Houston economy in a time of national recession. Between 1968 and 1980 the percentage of Houston employment in oil exploration, drilling and machinery expanded.

Prior to the 1973–1974 price-rise an economic diversification trend was underway, with growing investment in non-oil projects. With the sharp rise in the oil price, oil companies and allied bankers moved away from diversification to a heavier emphasis on investments in oil projects. In the late 1970s there was yet another rise in the OPEC oil price, which further stimulated companies to invest in oil.

Crisis and reorganization

Yet, in 1982–1987, Houston was looking a lot like Detroit. Job announcements brought long lines; tax revenues had plummeted; public sector workers were laid off; bond ratings had slipped; firms were going bankrupt in increasing numbers; and corporations were closing plants and shifting work overseas. World-wide recession had led to an oil glut. The downturn rippled its way through drilling pipe and oil rig production; through construction and trucking; and eventually through retail stores and real estate. Industrial production declined in Houston more rapidly than the national average. The unemployment rate grew more rapidly in Houston than in the nation. In 1984, the number of bankruptcies continued to escalate.

Decentralization and uneven development

The oil industry has brought periods of rapid growth to Houston. Coupled with the commitment of the local elite to auto-centered transit and private enterprise in housing – and a fierce opposition to mass transit and public housing – this rapid growth created a decentralized city. Commercial and industrial corporations have commissioned or leased a vast array of megastructures (industrial parks, shopping malls, multiple-use projects and office towers built in business centers). Scattered between and beyond these business centers are residential areas, including condominium apartment buildings and sprawling suburban subdivisions.

Houston's developers have been pioneers in multiple-use developments (called MXDs in developers' publications). These megastructure projects illustrate the central role of oil and gas companies in Houston's physical development. The large office and MXD complexes scattered in the seven "downtowns" in Houston are populated primarily by oil and gas companies and by the legal, accountancy and other business service firms serving the oil industry.

Yet, not all Houston residents profit from growth. Houston's low-income and minority homeowners and tenants have suffered greatly from market-oriented growth. The central city houses large numbers of black and Mexican-American, low- and middle-income families. Many areas of the central city have suffered from gentrification, the replacement of poorer families with better-off professional, technical and managerial families. Gentrification has displaced residents of ethnic areas as well as elderly whites. The

Fourth Ward is one of Houston's oldest black communities, with the misfortune of being in the path of expansion of the central business district. The area is populated by tenants living in single-family dwellings and in a major public housing project. Because of its proximity to downtown, developers have their eye on the area. A number of prominent consultant reports have suggested that the area should be redeveloped. The major public housing estate there is scheduled for demolition, significantly reducing the amount of housing for middle-income families.

Houston is also facing a major infrastructure crisis. The hidden side of its "good business climate," its low taxes and scaled-down government, has been a neglect of sewerage, water, flood prevention and other infrastructure facilities. This neglect has simply postponed the cost of paying for decaying or seriously inadequate facilities. Hundreds of billions of dollars will be required to meet Houston's escalating infrastructure costs. And that does not include the human costs of this neglect. Paying for infrastructure repair will require massive tax increases, which are even less likely in an era of slow economic decline. The "free enterprise" city has cost, and will continue to cost, its citizenry heavily in monetary and social expenses not normally enumerated in promotional and news media accounts of Houston's growth.

CONCLUSION

Because Detroit and Houston are spatial locations in a global system of production and exchange, the forces shaping their convergent and divergent paths have not been bounded by municipal, regional or even national lines. Three themes have ordered our tale of these two cities: (1) specialization and growth; (2) crisis and reorganization; and (3) decentralization and uneven development. It is to these reference points that we turn to draw our concluding comparisons.

Specialization and growth

Detroit and Houston evolved as specialized nodes in internationally organized production systems, one centered on the auto industry, the other on oil and petrochemicals. The Motor City grew and prospered, expanded horizontally and vertically, all in time with the beat of plant, warehouse and office investment for the production of cars. Detroit developed as a one-industry town located near supplies of labor and raw materials. The mass production of automobiles spur-red a vast and growing demand for fuel, and at the other end of an auto–oil "pipeline," more than 1,000 miles away, Houston began to prosper not long after Detroit's emergence as the Motor City. Experiencing a series of growth spurts, particularly in the 1920s, the 1940s and the 1960s, Houston became the oil capital of the world. It too expanded horizontally and verti-cally, with its seven business centers and hundreds of major plants, office towers and shopping centers. The Oil City too developed as a one-industry town located near crucial raw materials and labor pools.

Detroit's auto specialization generated a leapfrog logic of spatial development. Industry, commerce and residences decentralized into suburban rings, assisted by federal highway and housing finance programs. But even in the midst of postwar prosperity, large numbers of blacks were confined to blighted areas in the inner city. Houston's own distinctive specialization created a complex geographical network of oil-gas centers encircled by suburban belts. Houston's spatial sprawl was assisted by the same federal highway and housing programs that conditioned suburbanization in the Motor City. The auto and housing industries played the most aggressive role in the postwar "Highway Lobby" that pressured the federal government to support new highway and housing programs, while the oil companies remained more in the background. Yet Houston's oil interests were just as firmly committed to a low-density, decentralized urban environment. Black and Mexican-American workers and their families were a large island of poverty in a central city bordered by huge megastructures, like Houston Center.

Crisis and reorganization

The era of prosperity for the city of Detroit was tem-porary. Detroit's economic hegemony in world auto markets and its expanding employment in auto firms proved transitory. Global crisis and reorganization in the auto industry brought massive economic decline to the central city. As its industry left town, its human problems grew.

Yet even the "shining buckle of the Sunbelt," once thought to be immune to urban crisis, experienced the fundamental contradictions of a capitalist world system. The 1982–1987 recession was Houston's worst ever; the Oil City had even boomed through most of the Great Depression. The 10 per cent unemployment,

over 1,000 bankruptcies, oil refinery and other oil com-pany layoffs and cutbacks in the petrochemical industry indicate that Houston too is showing the effects of crisis and reorganization in the world oil-gas industry.

Capital moves on a world market stage, and with modern modes of transportation and communica-tion, big businesses can accelerate investment at a high velocity (Bluestone and Harrison 1983). Working people and their families, on the other hand, move in locally bounded communities. They cannot chart a new course with capital's velocity. Corporation's invest-ment space outdistances people's living space and that is the fundamental urban contradiction in the world capitalist system.

Decentralization and uneven development

Capitalists themselves are caught up in the investment-living space contradiction, as indicated by Henry Ford II's failed attempt to revitalize Detroit's central city with a multi-million dollar investment in the Renaissance Center. The same contradiction is revealed in Houston's oversupply of office towers and residential blocks which now have record-setting vacancies. During the next two decades Houston may come to bask in the same light that is now refracted through the "Rust Belt." And given Houston's starved social services sectors, the human and monetary costs of Houston's decline could well surpass Detroit's own experience.

So, ironically, the declining center of a crumbling Great Lakes manufacturing empire and the booming buckle of the Sunbelt both turn out to be one-industry towns whose export industries are going through a global reorganization that bodes well for neither's economic future. Officials in Detroit and Houston recognize the need to restructure and diversify their local economies. And both are emphasizing hi-tech complexes and suburban office-commercial parks; but that development policy presages further inequa-lity and uneven development along lines of race, class and territory.

The extent to which each city can muster new comparative advantages and revitalize its economic base will continue to be affected by the kind of rela-tionship each establishes with the federal government. And here another irony comes into play. For one big advantage Houston, the self-proclaimed bastion of free enterprise, retains over Detroit, a fading outpost of the welfare state, is Houston's greater ability to garner largesse from the federal government; not the kind distributed by the Department of Health, Education and Welfare, but the sort passed out by the Pentagon. Today's massive defense spending on the Pentagon's production system, the military-industrial complex, is spawning new industries, setting the direction for future economic development, and enriching or impoverishing regions. Even so, the critical issues always seem to remain the same. New urban development continues to be targeted to the privileged few. Power over urban development continues to be concentrated among a handful of individuals and corporations whose reach spans far beyond the metropolis. And urban development continues to be uneven, unpredictable and precarious, since the most important development decisions are made in private offices, not in publicly accountable places.

REFERENCES FROM THE READING

Bluestone, B. and Harrison, B. (1983) *The Deindustrialization of America*, New York: Basic Books.

Friedmann, J. and Wolff, G. (1982) World city formation: an agenda for research and action, *International Journal of Urban and Regional Research*, 6, 3: 309–44.

Maraniss, D. (2015) *Once in a Great City: A Detroit Story*, New York: Simon & Schuster.

Taylor, M. and Peppard, D. (1976) *Jobs for the Jobless: Detroit's Unresolved Dilemma*, East Lansing, MI: Institute for Community Development.

Zukin, S. (1980) A decade of the new urban sociology, *Theory and Society*, 9: 575–601.

T W O

14

"The stimulus of a little confusion: a contemporary comparison of Amsterdam and Los Angeles"

from L. Deben (ed.), *Understanding Amsterdam: Essays on Economic Vitality, City Life and Urban Form* (2000)

Edward W. Soja

EDITORS' INTRODUCTION

Edward Soja (1940–2015) was one of the most creative and influential contemporary urbanists. From the early 1970s, Soja worked in the Department of Urban Planning at the University of California, Los Angeles (UCLA), where he developed his own perspective on the globalization of urbanization amidst ongoing discussions of such matters within the so-called "LA School" of urban studies (see Dear 2002; Scott and Soja 1996). Soja's major books were focused extensively on Los Angeles and Southern California (Soja 1989; 1996; 2000), which he used as a geographical focal point for the development of broader theoretical arguments regarding the contemporary global urban condition. Much of Soja's work advanced what he termed a "postmodern" viewpoint, but in so doing, he also drew upon the tools of urban political economy, including neo-Marxian urban theory and global city theory. Soja's brilliantly energetic writing style, his creative theoretical eclecticism and his astute powers of observation have given his work a broad, interdisciplinary appeal both within and beyond the field of urban studies.

In the contribution below, Soja embarks upon a sharp comparative analysis of Los Angeles and Amsterdam, two cities that would appear to be incommensurably different. Indeed, Soja begins the chapter by briefly reviewing these differences at once on structural and experiential levels with reference to the downtown centers of each city. However, as Soja indicates, once one shifts to the regional or metropolitan scale, and examines the sprawling mega-urban galaxies in which both of these cities are embedded, any number of similarities between them are suddenly brought into focus. Soja examines, in particular, (a) the restructuring of the urban form; (b) the internationalization of the regional economy; (c) the consolidation of post-Fordist forms of industrial organization; and (d) the intensification of socioeconomic polarization within each of these urban regions. Soja's claim is not that Amsterdam and Los Angeles have become identical, or that they are converging towards a single model of global city formation. Rather, Soja is suggesting that these cities are both undergoing broadly analogous forms of restructuring due to their embeddedness within an emergent global urban system, and for this reason, their analytical juxtaposition can illuminate important aspects of urban life under contemporary capitalism.

At first glance, a comparison of Los Angeles and Amsterdam seems as impossible as comparing oranges and potatoes. These two extraordinary cities virtually beg to be described as unique, incomparable, and of course to a great extent they are. But they are also linkable as opposite and apposite extremes of late twentieth century urbanization, informatively positioned antipodes that are almost inversions of one another yet are united in a common and immediate urban experience. First I will annotate the more obvious oppositions.

Los Angeles epitomizes the sprawling, decentered, polymorphic, and centrifugal metropolis, a nebulous galaxy of suburbs in search of a city, a place where history is repeatedly spun off and ephemeralized in aggressively contemporary forms. In contrast, Amsterdam may be the most self-consciously centered and historically centripetal city in Europe, carefully preserving every one of its golden ages in a repeatedly modernized Centrum that makes other remnant mercantile capitalist "Old Towns" pale by comparison. Both have downtowns of roughly comparable area, but only one of 100 Angelenos live in the city's center, whereas more than 10 per cent of Amsterdammers are Centrum dwellers.

Many residents of the City of Los Angeles have never been downtown and experience it only vicariously, on television and film. Very few now visit it to shop; and surprisingly few tourists take in its local attractions, at least in comparison to more peripheral sites. Amsterdam's Centrum receives nearly 8 million tourists a year and is packed daily with many thousands of shoppers. Amsterdammers may not be aware of the rest of the city, but they certainly know where the center can be found.

It has been claimed that nearly three-quarters of the surface space of downtown Los Angeles is devoted to the automobile and to the average Angeleno freedom and freeway are symbolically and often politically intertwined. Here the opposition to Amsterdam's Centrum, second only to floating Venice in auto-prohibition, is almost unparalleled. It is not the car but the bicycle that assumes, for the Amsterdammer, a similarly obsessive symbolic and political role, but it is an obsession filled not with individualistic expression and automaniacal freedom as much as with a collective urban and environmental consciousness and commitment. This makes all the contrasts even more stark.

Amsterdam's center feels like an open public forum, a daily festival of spontaneous political and cultural ideas played at a low key, but all the more effective for its lack of pretense and frenzy. Its often erogenously-zoned geography is attuned to many different age groups and civically dedicated to the playful conquest of boredom and despair in ways that most other cities have forgotten or never thought possible. Downtown Los Angeles, on the other hand, is almost pure spectacle, of business and commerce, of extreme wealth and poverty, of clashing cultures and rigidly contained ethnicities. Boredom is assuaged by overindulgence and the bombardment of artificial stimulation, while despair is controlled and contained by the omnipresence of authority and spatial surveillance. Young householders are virtually nonexistent. In their place are the homeless, who are coming close to being half the central city's resident population despite vigorous attempts at gentrification and dispersal.

In compact Amsterdam, the whole urban fabric is clearly readable and explicit. From its prime axis of the Damrak and Rokin, the city unfolds in layers like a halved cross-section of an onion, first in the "old side" and "new side," then in the neat crescents of the ringing canals from the inner to the outer Singel girdles, and finally in segments and wedges of inner and outer suburbs. This morphological regularity binds Amsterdammers to traditional concepts of urban form and function.

In comparison, Los Angeles seems to break every rule of urban readability and regularity, challenging all traditional models of what is urban and what is not. One of America's classic suburbias, the San Fernando Valley, is almost wholly within the jig-sawed boundaries of the monstro-City of Los Angeles, while many inner city barrios and ghettoes float outside on unincorporated county land. There is a City of Industry, a City of Commerce, and even a Universal City, but these are not cities at all. Moreover, in an era of what many have called post-industrial urbanization, with cities being emptied of their manufacturing employment, the Los Angeles region has continued its century-long boom in industrial growth in both its core and periphery. It is no surprise, then, that Southern California has become a center for innovative and non-traditional urban theory, for there seems little from conventional, established schools of urban analysis that any longer makes sense.

And then there is that most basic of urban functions, housing. One of the most interesting features of the success of the squatter movement in Amsterdam was the absence of a significant housing shortage. Although much of the Centrum is privately owned, the rest of the city is a vast checkerboard of public, or social housing. Even what the Dutch planners consider the worst of these projects, such as the huge Bijlmermeer high-rise garden suburb, served effectively to accommodate the thousands of migrants from Surinam and other former colonies during the 1970s.

The squatter movement was more than just an occupation of abandoned offices, factories, warehouses, and some residences. It was a fight for the rights to the city itself, especially for the young and for the poor. Nowhere has this struggle been more successful than in Amsterdam. Nowhere has it been less successful than in Los Angeles. In the immediate post-war period, Los Angeles was poised to become the largest recipient of public housing investment in the country, with much of this scheduled to be constructed in or around downtown. In no other American city did plans for public housing experience such a resounding defeat by so ferociously antisocialist campaigners. The explosion of ethnic insurrections in the 1960s and early 1970s cruelly accelerated the commercial renewal of downtown at the expense of its poor residential inhabitants. On the central city's "new side" grew a commercial, financial, governmental, and high cultural fortress, while the "old side," beyond the skyscraper walls, was left to be filled with more residual land uses, from the tiny remnants of El Pueblo de Nuestra Senora de Los Angeles to the fulsome Skid Row of cardboard tenements and streetscapes of despair, to the Dickensian sweatshops and discount marts of the expansive Garment District.

The core of my oppositional comparison is thus amply clear. But what of the periphery? Are there comparative dimensions that are missed when we focus on the antipodal centralities of Amsterdam and Los Angeles? For the remainder of this chapter, I will set the two cities in a larger, more generalizable context that focuses on contemporary processes of urban restructuring. Here, the cities follow more similar paths than might initially seem possible. These similarities are not meant to contradict or erase the profound differences that have already been described, but to supplement and expand upon their emphatic and extreme particularity.

My own research and writing on the urban restructuring of Los Angeles has identified a series of intertwined trends that have become increasingly apparent not only in Los Angeles but in most of the world's major urban regions. Each trend takes on different intensities and forms in different cities, reflecting both the normality of geographically uneven development and the social and ecological particularity of place. More important than their individual trajectories, however, is their correlative interconnectedness and the tendency for their collective impact to define an emerging new mode of urbanization, significantly different from the urbanization processes that shaped the industrial capitalist city during the long post-war boom period.

It is appropriate to begin with the *geographical recomposition of urban form*. As with the other trends, there is a certain continuity with the past, lending credence to the argument that restructuring is more of an acceleration of existing urban trajectories than a complete break and redirection. The current geographical recomposition, for example, is in large part a continuation on a larger scale of the decentralization and polynucleation of the industrial capitalist city that was begun in the last half of the nineteenth century.

There are, however, several features of the recent round of polynucleated decentralization that suggest a more profound qualitative shift. First, the size and scale of cities, or more appropriately of urban regions, has been reaching unprecedented levels. The older notion of "megalopolis" seems increasingly inadequate to describe a Mexico City of 30 million inhabitants or a "Mega-York" of nearly 25, stretching from Connecticut to Pennsylvania. Never before has the focus on the politically defined "central" city become so insubstantial and misleading. Complicating the older form still further has been the emergence of "Outer Cities," amorphous agglomerations of industrial parks, financial service centers, and office buildings, massive new residential developments, giant shopping malls, and spectacular entertainment facilities in what was formerly open farmland or a sprinkling of small dormitory suburbs. Neither city nor suburb, at least in the older senses of the terms, of these reconcentrated poles of peripheral urban and (typically "high-tech") industrial growth have stimulated a new descriptive vocabulary.

The growth of Outer Cities is part of the recentralization of the still decentralizing urban region, a paradoxical twist that reflects the ability of certain areas within the *regional metropolis* to compete within an increasingly globalized economy. Over the past

Plate 13 Amsterdam

Source: Roger Keil

25 years, the decentralization of manufacturing and related activities from the core of the older industrial capitalist cities broke out from its national containment. Jobs and factories continued to move to suburban sites or non-metropolitan areas within the national economy, but also, much more than ever before, to hitherto non-industrialized regions of the old Third World, creating a new geographical dynamic of growth and decline that not only has been changing the long-established international division of labor but also the spatial division of labor within urban regions.

The geographical recomposition is paradigmatically clear in Greater Los Angeles. Within a radius of 60 miles (100 kilometers) from the booming Central Business District of the misshapen City of Los Angeles there is a radically restructured regional metropolis of nearly 15 million people with an economic output roughly equivalent to that of the Netherlands. At this scale, a comparison with Amsterdam seems totally inappropriate. But if we shift scales, a different picture

emerges. A 100-kilometer circle from Amsterdam's Centrum cuts through Zeeland, touches the Belgian border near Tilburg, curves past Eindhoven to touch the German border not far from Nijmegen, and then arcs through the heart of Friesland to the North Sea. Most of the nearly 15 million Dutch live within this densely urbanized region and its scale and productivity come remarkably close to matching its Southern California counterpart.

The southwest quadrant of this "Greater Amsterdam" coincides rather neatly with the Randstad, which can be seen as a kind of Outer City in itself, but with the defining central core being not an old urban zone but the determinedly preserved rural and agricultural "Green Heart". Around the Green Heart are the largest cities of the Netherlands: Amsterdam, Rotterdam, The Hague, and Utrecht, each experiencing a selective redistribution of economic activities between central city, suburban fringe, and more freestanding peripheral centers. As a whole, the Randstad contains the world's largest port (Los Angeles-Long Beach is now probably second), Europe's fourth largest international financial center (after London, Zurich, and Frankfurt) and fourth largest international airport (Schiphol, surpassed in traffic only by Frankfurt, Paris, and London).

Like the Greater Los Angeles regional metropolis. Greater Amsterdam has been experiencing a complex decentralization and recentralization over the past 25 years. How useful this larger scale regional comparison of geographical recomposition might be to a further understanding of urban restructuring I will leave to others to determine. For present purposes, however, it at least forms a useful antidote to the tendency of urban observers to persist in seeing the contemporary period of restructuring too narrowly due to an excessive focusing on the long-established central city, thereby ignoring a dimension of an entirely different order from the one they traditionally know.

The recomposition of urban form is intricately connected to other sets of restructuring processes. Already alluded to, for example, has been the *increasing internationalization of the regional metropolis,* leading to the formation of a new kind of world city. Amsterdam in its Golden Age was the prototypical model of the world city of mercantile capitalism and it has survived various phases of formation and reformation to remain among the higher ranks of contemporary world cities, whether combined in the Randstad or not. What distinguishes the global cities of today from those of the past is the *scope* of internationalization, in terms of both capital and labor. To the control of world trade (the primary basis of mercantile world cities) and international financial investment by the national state (the foundation of imperial world cities) has been added the financial management of industrial production and producer services, allowing the contemporary world city to function at a global scale across all circuits of capital. First, Second, and Third World economies have become increasingly integrated into a global system of production, exchange, and consumption that is sustained by an information-intensive hierarchy of world cities, topped today by the triumvirate of Tokyo, New York, and London.

Los Angeles and Amsterdam are in the second tier of the restructured world city hierarchy, but the former is growing much more rapidly and some predict it will join the top three by the end of the century. Amsterdam is more stable, maintaining its specialized position in Europe on the basis of its concentration of Japanese and American banks, the large number of foreign listings on its Stock Exchange, the strong and long-established export-orientation of Dutch companies, and its control over Dutch pension funds. The banking and financial services sector remains a key actor in Amsterdam's Centrum, feeding its upscale gentrification and drawing strength from the information-rich clustering of government offices, university departments, cultural facilities, and specialized activities in advertising and publishing.

A characteristic feature of increasing internationalization everywhere has been an erosion of local control over the planning process, as the powerful exogenous demands of world city formation penetrate deeply into local decision-making. Without a significant tradition of progressive urban planning, Los Angeles has welcomed foreign investment with few constraints. Its downtown "renaissance" was built on foreign capital to such an extent that today almost three-quarters of the prime properties in the Central Business District are foreign-owned or at least partially controlled by overseas firms. The internationalization of Amsterdam has been more controlled, but the continued expansion of the city as a global financial management center is likely to pose a major threat to many of the special qualities of the Centrum.

The other side of internationalization has been the attraction of large numbers of foreign workers into almost every segment of the local labor market, but especially at lower wage and skill levels. Los Angeles today has perhaps the largest and most culturally diverse immigrant labor force of any major world city, an enriching resource not only for its corporate entrepreneurs but also for the cultural life of the urban region. Amsterdam too is fast-approaching becoming a "majority minority" city, a true cosmopolis of all the world's populations. With its long tradition of effectively absorbing diverse immigrant groups, Amsterdam appears to have been more successful than Los Angeles in integrating its immigrant populations into the urban fabric. One achievement is certain: they are better housed in Amsterdam, for Los Angeles is currently experiencing one of the worst housing crises in the developed world. As many as 600,000 people, predominantly the Latino working poor, now live in seriously overcrowded conditions in dilapidated apartments, backyard shacks, tiny hotel rooms, and on the streets.

Intertwined with the geographical recomposition and internationalization of Los Angeles and

Amsterdam has been a pervasive *industrial restructuring* that has come to be described as a trend toward a post-Fordist regime of flexible accumulation in cities and regions throughout the world. A complex mix of both deindustrialization (especially the decline of large-scale, vertically integrated, often assembly-line, mass production industries) and reindustrialization (particularly the rise of small and middle-size firms flexibly specializing in craft-based and/or high technologically facilitated production of diverse goods and services), this restructuring of the organization of production and the labor process has been associated with a repatterned urbanization, a new dynamic of geographically uneven development.

A quick picture of the changing post-Fordist industrial geography would consist of several characteristic spaces: older industrial areas either in severe decline or partially revived through adaptation of more flexible production and management techniques; new science-based industrial districts or technopoles typically located in metropolitan peripheries; craft-based manufacturing clusters or networks drawing upon both the formal and informal economies; concentrated and communications-rich producer services

districts, especially relating to finance and banking but also extending into the entertainment, fashion, and culture industries; and some residual areas, where little has changed. It would be easy to transpose this typology to Greater Los Angeles, for much of the research behind it has been conducted there. Although the post-Fordist restructuring has not gone nearly as far in Amsterdam, the transposition is also quite revealing.

The Centrum has been almost entirely leached of its older, heavier industries and 25 per cent of its former office stock has been lost, primarily to an impressive array of new subcenters to the southeast, south, and west and to the growing airport node at Schiphol. One might argue that this dispersal represents a sign of major decline in the inner city, due in part to a shift from a concentric to a more grid-like pattern of office and industrial development. Just as convincing, however, is a restructuring hypothesis that identifies the Centrum as a flexibly specialized services district organized around international finance and banking, university education, and diverse aspects of the culture and entertainment industries (fashion, especially for the twenty-somethings, television and

Plate 14 Street scene, downtown Los Angeles

Source: Roger Keil

film, advertising and publishing, soft drugs and sex, and, of course, tourism).

A fourth trend needs to be added, however, before one goes too far in tracing the impact of post-Fordist industrial restructuring. This is the tendency toward *increasing social and economic polarization* that seems to accompany the new urbanization processes. Recent studies have shown that the economic expansion and restructuring of Los Angeles has dramatically increased poverty levels and hollowed out the middle ranks of the labor market, squeezing job growth upward, to a growing executive-professional-managerial "technocracy" (stocked by the largest urban concentrations in the world of scientists, engineers, and mathematicians), and downward, in much larger numbers, to an explosive mix of the "working poor" (primarily Latino and other immigrants, and women, giving rise to an increasing "feminization of poverty") and a domestic (white, African-American, and Mexican-American, or Chicano) "urban underclass" surviving on public welfare, part-time employment, and the often illegal opportunities provided by the growing informal, or underground, economy. This vertical and sectorial polarization of the division of labor is reflected in an increasing horizontal and spatial polarization in the residential geography of Los Angeles. Old and new wealth is increasingly concentrated in protected communities with armed guards, walled boundaries, "neighborhood watches," and explicit signs that announce bluntly: "Trespassers will be shot;" while the old and new poor either crowd into the expanding immigrant enclaves of the Third World City or remain trapped in murderous landscapes of despair. In this bifurcating urban geography, all the edges and turf boundaries become potentially violent battlefronts in the continuing struggle for the rights to the city.

Here again, the Amsterdam comparison is both informative and ambiguously encouraging, for it too has been experiencing a process of social and economic polarization over the past two decades, and yet, it has managed to keep the multiplying sources of friction under relatively successful social control. The Dutch "Job Machine," for example, shows a similar hollowing out of the labor market, with the greatest growth occurring in the low-paid services sector. Official unemployment rates have been much higher than in the U.S., but this difference is made meaningless by the contrasts in welfare systems and methods of calculating the rate itself. Overall job growth has been much lower than in the U.S. and, except for the producer services sector, there has been a decline in high-wage employment thus limiting the size of the executive-professional-managerial "bulge." Increasing flexibility in the labor market, however, is clearly evident in the growth of "temporary" and "part-time" employment, with the Netherlands having the largest proportion of part-time workers in the EEC and perhaps the highest rate (more than 50 per cent) in the Western world for women.

With its exceptional concentration of young, educated, often student households, high official levels of unemployment, still solid social security system, and distinctive patterns of gentrification, an unusual synergy has developed around the personal services sector and between various age and income groups. Income polarization has been producing a growing complementarity between the higher and lower income groups with respect to the flexible use of time and place, especially in the specialized provision of such personal services as domestic help and baby-sitting, late-night shopping, entertainment, and catering, household maintenance and repair, educational courses and therapies, fitness centers, body-care activities, etc. Such activities in Amsterdam take place primarily in the underground economy and are not captured very well in official statistics. But they nonetheless provide a legitimate and socially valuable "survival strategy" for the poor and unemployed that has worked effectively to constrain the extreme effects of social polarization that one finds in Los Angeles or New York City. Moreover, it is a strategy that draws from the peculiar urban genius of Amsterdam, its long tradition of grass-roots communalism, its sensitive adaptation to locality, its continuing commitment to libertarian and participatory social and spatial democracy, and its unusual contemporary attention to the needs of the twenty-something generation.

I had originally intended to conclude by addressing *postmodernism and post-modernization* as a fifth restructuring theme and to explore the extent to which this restructuring of the "cultural logic" of contemporary capitalism can be traced into the comparison of Amsterdam and Los Angeles. In my own recent research and writings, I have argued that a neoconservative form of postmodernism, in which "image" replaces reality and the simulated and "spin-doctored" representations assume increasing political and economic power, is significantly reshaping popular ideologies and everyday life all over the world and is fastly becoming the keystone for a new mode of social regulation designed to sustain the development of (and control the resistance to) the new post-Fordist regimes of "flexible" and "global" capitalist accumulation and the accompanying "new urbanization processes" discussed on the preceding pages. After experiencing Amsterdam, where resistance to the imposition of this neoconservative restructuring seems exceptionally strong, it is tempting just to add another polar opposition to the comparison with Los Angeles where this process is probably more advanced than almost anywhere else on earth. But I will leave the issue open for future research and reflection.

REFERENCES FROM THE READING

Dear, M. (ed.) (2002) *From Chicago to LA: Making Sense of Urban Theory*, Thousand Oaks, CA: Sage.

Scott, A.J. and Soja, E. (eds.) (1996) *The City: Los Angeles and Urban Theory at the End of the Twentieth Century*, Berkeley and Los Angeles: University of California Press.

Soja, E. (1989) *Postmodern Geographies*, Cambridge, MA: Blackwell.

Soja, E. (1996) *Thirdspace*, Cambridge, MA: Blackwell.

Soja, E. (2000) *Postmetropolis*, Cambridge, MA: Blackwell.

TWO

15
"Global city Zurich: paradigms of urban development"

Christian Schmid

EDITORS' INTRODUCTION

Christian Schmid is a geographer, sociologist, and urban researcher. He is Titular Professor of Sociology at the Department of Architecture, ETH Zurich, and researcher at ETH Studio Basel/Contemporary City Institute. Since 1980, Schmid has been active as a video activist and an organizer of cultural events. He was a founding member of the International Network for Urban Research and Action (INURA) in 1991 and of the Ssenter for Applied Urbanism (SAU) in Zurich. Schmid has conducted much of his research on Zurich and, along with several German and Swiss collaborators, has played a key role in introducing global city theory to a German-language readership (Hitz et al. 1995). Schmid has authored, co-authored, and co-edited numerous publications on urban theory, sociospatial theory, Henri Lefebvre, and comparative urban development. He is currently collaborating with Neil Brenner in research on planetary urbanization. Through his work at the ETH Future Cities Lab Singapore, he is also directing a comparative project on urban restructuring in Tokyo, Hong Kong/Shenzhen/Dongguan, Kolkata, Istanbul, Lagos, Paris, Mexico City, and Los Angeles (see Ch. 65).

In this contribution, which summarizes some of the results of his studies in the 1990s and early 2000s, Schmid situates urban development in Zurich in relation to its role as a "headquarter economy" within European and global circuits of capital. Schmid traces the evolution of Zurich into a global city during the 1970s, emphasizing the clash between a modernizing growth coalition and various locally rooted oppositional forces that opposed the expansion of the Central Business District (CBD) into surrounding residential neighborhoods. As Schmid indicates, a "territorial compromise" was established in the wake of these struggles that slowed down inner city restructuring while pushing many global city functions outwards into the city's peripheries, such as Zurich North. Schmid's case study provides a nuanced perspective on a number of aspects of global city formation in Zurich—including conflicts between parochial and cosmopolitan forms of urban culture; the transformation of urban form through the decentralization of global city functions; the degradation of the built environment in the suburban periphery; and the formation of new cultural milieux in revitalized inner city neighborhoods such as Zurich West. Schmid also devotes considerable attention to the contested politics of global city formation, emphasizing both the periodic reworking of municipal and regional growth coalitions and the everyday contestation of urban restructuring through diverse, neighborhood-based social movements. Schmid concludes by situating global city formation in Zurich in global perspective. For Schmid, the process of urban restructuring in Zurich is necessarily contextually specific, but it is also indicative of a number of more general trends and conflicts that can be witnessed in global cities throughout the world.

Zurich today is a global city, one of a group of global control centers of the world economy. In international comparisons, Zurich has been routinely placed at the second or third rank in global city hierarchy, together with cities like San Francisco, Sydney or Toronto (Friedmann 1995).

Fifty years ago, Zurich was an industrial town with a strong position in the machine-building and armament industries. The transformation of Zurich into a global city began in the 1970s, with the increasing deregulation and globalization of financial markets. Zurich became the undisputed center of Switzerland as a location for finance, and a headquarter economy established itself, which specialized in the organization and control of global financial flows. In 2001 only around 7 per cent of all jobs were still in the manufacturing sector (not counting construction), while 36 per cent were in the core sectors of the global city economy (financial industries, insurance and business services). Yet, even though it directly depended on global lines of development, the transformation of Zurich into a global city was still a contradictory process, which was also strongly marked by local conflicts.

This radical economic transformation has caused fundamental changes in the urban development of Zurich. Two differing historic models of urbanization can be distinguished. The first model developed in the 1970s and was characterized by the process of global city formation. It was growth oriented, but it was also grounded upon a strong regulation of urban development and by the conservation of inner-city areas. In the 1990s, with the process of metropolitanization and the expansion of the global city into the region, a second model of urbanization established itself that has been characterized by a neoliberal policy of urban development, the emergence of new urban spatial configurations and a new definition of the urban.

GLOBAL CITY FORMATION: TERRITORIAL COMPROMISE AND URBAN REVOLT

In the decades after World War II, urban development in Zurich was defined by an encompassing growth coalition consisting of right-wing and left-wing forces, following a relatively moderate strategy of modernization. The conditions of this policy changed in the early 1970s: the protest movement of 1968 led to a radical questioning of functionalistic approaches to

urban development; meanwhile, the global economic crisis ended the "golden age" of Fordism. Subsequently, in the mid-1970s, the transformation of Zurich into a global city began.

Through the process of global city formation, globally defined strategies to establish a "headquarter economy" collided with the locally defined everyday concerns of many residents. The growth coalition fell apart, and the city was subsequently divided into two camps quarreling about the appropriate model of urban development. On the one hand, a new "modernizing coalition" took shape, consisting of right-wing parties and the growth-oriented sections of the trade unions, which promoted the development of Zurich as a financial center, the extension of the Central Business District (CBD) and the construction of new traffic infrastructure. On the other hand, in the wake of the social movements of 1968, an alternative position emerged which was critical of urban growth. Left-wing parties and various action groups and neighborhood organizations united in a heterogeneous and fragile "stabilization alliance"; their goal was to fight for a livable city, low rents and the preservation of residential neighbourhoods in the inner-city. Occasionally, this alliance was also supported by conservative forces. Through the Swiss system of direct democracy, in which many questions and projects must be decided by referendum, these opposing positions were transferred directly to the level of institutional politics. Both parties had their victories and defeats in this conflict, but neither side was ultimately able to win decisively. Thus, for two decades, from the mid-1970s to the mid-1990s, urban development in Zurich was in fact determined by a precarious political stalemate that generated a specific type of "territorial compromise." This territorial compromise entailed a rejection of large-scale modernization strategies and it considerably slowed down the transformation of inner-city residential neighbourhoods. Yet, global city formation and the dynamics of urbanization were not fundamentally challenged.

However, this territorial compromise also encompassed a second line of conflict, which did not immediately reveal itself. At the level of everyday life, the demands of cosmopolitan open-mindedness and urban culture created by global city formation clashed with the highly localized forms of social regulation which had been inherited from the Fordist period, and which were oriented towards social control and conformism. As of the 1970s, public life in Zurich was still characterized by a crushing parochialism that left very little space for the development of new lifestyles or alternative forms of cultural expression. This situation eventually caused a social explosion: on May 30, 1980, an urban revolt began. With riots, happenings and actions of all kinds, a new cosmopolitan urban generation demanded what Henri Lefebvre (1968) once called "the right to the city." Although the urban revolt collapsed after two years, its consequences subsequently became evident; the urban movement had changed the city's everyday life, its cultural sphere and its public spaces. A cosmopolitan ambience emerged. The city government began to promote diverse types of cultural projects, and a cultural and artistic milieu established itself, radiating far beyond Zurich. This created the basis for a successful economic sector of "cultural production," including design, image production, events, etc. This economic sector today plays a key role in international competition between global cities. The urban revolt thus became an important catalyst for the process of global city formation in Zurich.

As a result of these two lines of conflict, a new model of urbanization was established which combined the goals of modernization, stabilization and both economic and cultural globalization. This model contained a concept of the city that was simultaneously metropolitan and exclusive, derived from the classical European image of the city as a coherent, dense and innovative whole. This concept ultimately reduced the focus of urbanity to a narrow fraction of urban reality, to downtown Zurich. Seen from the urban center, all areas outside this restricted district were considered to be elements of a drab and uninteresting urban periphery.

While the inner-city evolved into a culturally and socially pulsating urban center, the opportunities for the construction of new offices and the expansion of the central business district were massively restricted. Service and financial enterprises were compelled to establish their additional offices at other locations. In various places outside Zurich City, the new strategic centers of the global city headquarter economy were developed. This process can be seen as an "explosion of the center" – global city functions were increasingly spread over an extended region, which was now structured as a center (Sassen 1994). Thus, a new urban configuration evolved, characterized by the regionalization of economy and society.

EXOPOLIS: THE CASE OF ZURICH NORTH

As in many other cases, the demarcation of this new global city region is quite difficult, since it is not formed as a coherent unit. A growing number of towns and villages in the densely populated lowlands of Switzerland have come under the influence of Zurich's headquarter economy and have become metropolitan in character. Therefore, depending on the criteria selected, a great variety of regions can be delineated around Zurich. Whereas the municipality of Zurich (Zurich City) has a population of 360,000, the global city region can be estimated at approximately 2 million inhabitants (Figure 1).

Analysis and deconstruction of this urban universe has only just begun. It is an amoebae-like urban space, which is characterized by floating centralities and by the constant emergence of ever new and surprising urban configurations. A showcase example for these new urban configurations is the Glatt Valley, north of Zurich, where the airport is also located. A series of "edge cities" (Garreau 1991) have developed here, forming a kind of fragmented twin city of Zurich. This "new" city is called simply "Zürich Nord" (Zurich North). It is one of these amorphous implosions of archaic suburbia that Soja (1996: 238) named "'Exopolis' – 'the city without' – to stress their oxymoronic ambiguity, their city-full non-cityness. These are not only exo-cities, orbiting outside; they are ex-cities as well, no longer what the city used to be" (Figure 2).

In a narrow sense, Zurich North includes 8 municipalities and two districts of Zurich City. With 147,000 inhabitants and 117,000 jobs, this area is today the fourth largest city in Switzerland – it is even bigger than Berne, the capital. The emphasis is on activities of the headquarter economy, predominantly producer services, banking and information and telecommunications (IT) industries. The concentration of these activities in this area is based less on the effect of the immediate (physical) vicinity than on the opportunities for flexible interconnections in a broader logistical space that stretches from the airport to the national highway system and electronic networks.

Some 20 years ago, Zurich North was composed almost entirely of classical middle-class suburbs. Planning was in the hands of the individual municipalities, which as a rule followed a simple planning concept: they tried to preserve the historic core of the

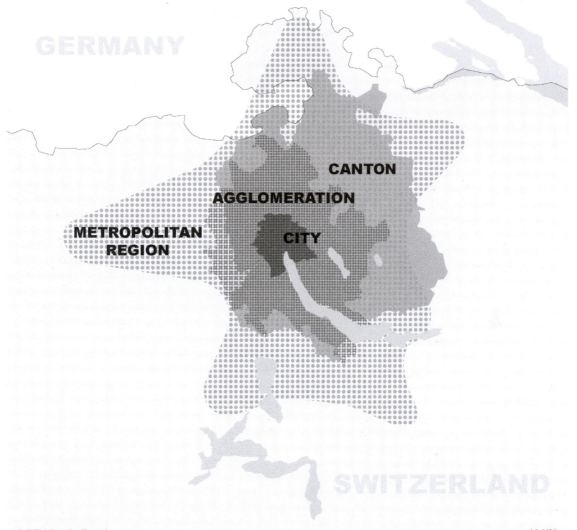

© ETH Studio Basel   10 KM

Figure 1 Zurich region

Source: Christian Schmid

settlement, expanding the housing zone concentrically around the core and placing industrial zones at the outskirts of the municipal territory. But, in the wake of deindustrialization and global city formation, it was not industrial operations, but the headquarter functions of global corporations, that were located within these formerly industrial zones. Consequently, satellites of the headquarter economy developed, consolidating in an odd kind of belt located at the periphery of the old cores of the settlement. Here,

high-quality business intermingles closely with highways or even waste incineration plants. The geographic center of this belt is a forest, and so a kind of circular town emerged with an "empty" center. This shape corresponds exactly to the "doughnut model" that Soja (1996) developed to describe Orange County.

Fascinating as this urban patchwork may appear, it nevertheless produces severe problems. Since the new centralities are dispersed over a wide area, this fragmented non-city is largely dependent on private

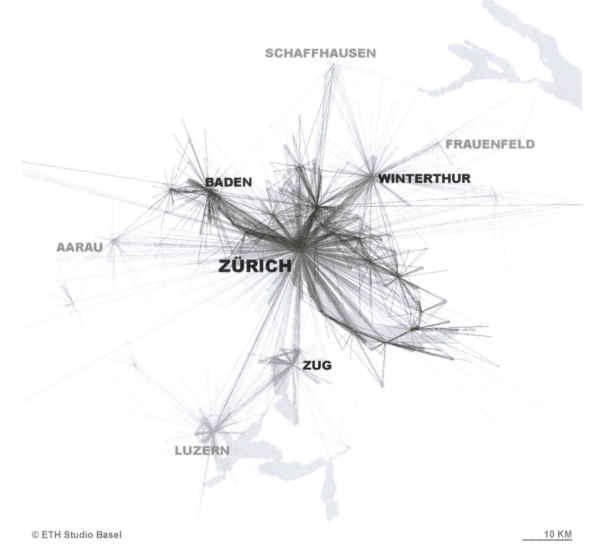

© ETH Studio Basel   10 KM

Figure 2 Commuters to Zurich

Source: Christian Schmid

cars, a situation that produces traffic jams and air pollution. In addition, the environment is often not very attractive. In most of the new centers, there is a lack of urban infrastructure, restaurants, meeting places and cultural establishments, and there are few places that create a sense of identification or an urban atmosphere. Accordingly, many residents and employees are not at all happy with the quality of everyday life in this urban patchwork, which still has yet to overcome its peripheral status.

In contrast to many other urban peripheries, in Zurich North this lack of urban character has increasingly been perceived as a deficiency. Attitudes have gradually been changing, and there is now an explicit agenda of creating a "real city" from this patchwork by achieving a certain architectural and social coherence. The first initiatives for coordinated planning among municipalities already appeared in the beginning of the 1990s, and in 2001 an association of eight municipalities was created. The new label for Zurich North was "Glattstadt" (City of Glatt Valley). This name is meant to stand for a new region with its own identity, while also underscoring its separation from Zurich City. Accordingly, Zurich City is not included in this new organization, even though its northern neighborhoods formally belong to Zurich North.

The most important project in this new cooperation is the construction of a tram line, which was approved in a referendum in spring 2003. This line is not only meant to open up and connect the various new centers of Zurich North, but is also the symbol of the newly discovered self-confidence of this new "city of the future." This is why it is officially called "Stadtbahn" (city train) instead of tram, the traditional term for streetcars in Zurich.

Thus the "model of exopolis" remains delicately balanced: on the one hand there are attempts to make Zurich North into a "true city." On the other hand, however, the underlying historically developed patch-work structure of the area is continually creating new difficulties and surprises (Figure 3).

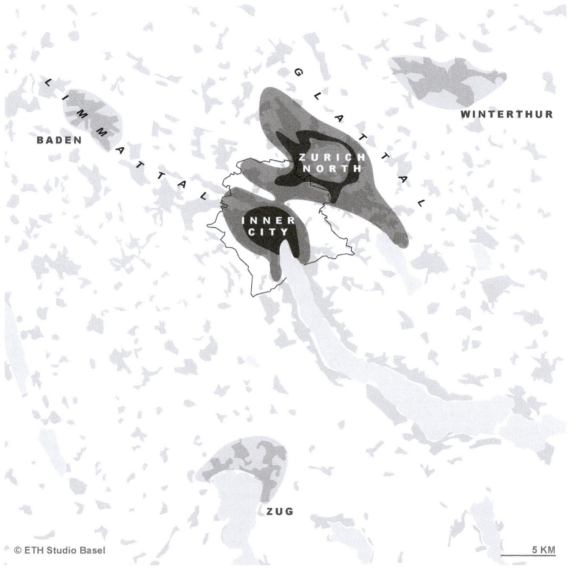

© ETH Studio Basel   5 KM

Figure 3 Zurich's spatial configurations

Source: Christian Schmid

CONTRADICTIONS IN THE GLOBAL CITY: THE METROPOLIS ALLIANCE

While the global city has been expanding into the region, the situation in Zurich City has also undergone fundamental changes. During the course of the 1990s, a major paradigm shift in urban development has become apparent.

This paradigm shift originated in a double crisis. On the one hand, in the beginning of the 1990s Switzerland – like most west European countries – was in the grip of a long-term economic crisis, and growing deficits forced public authorities to impose new budgetary restrictions. On the other hand, the social consequences of globalization became visible: like many other global cities, extensive economic and social polarization and fragmentation became apparent. These developments were accompanied by fundamental shifts in the political landscape. In reaction to globalization and urbanization, an aggressive right-wing populism emerged for the first time in Switzerland. It grew first at the margins of the unravelling metropolis, in the suburban and periurban areas, but soon it began to take hold in Zurich City as well. Right-wing populist forces mobilized against the social and cultural open-mindedness of the 1980s. Through an aggressive campaign, these forces succeeded in transforming drug policy and the asylum question into central political issues. In subsequent years, the political and social climate clearly deteriorated.

Against these right-wing populist political activities, the positions of moderate right-wing and social-democratic forces drew closer. The antagonisms between the modernizing coalition and the stabilization alliance softened, and the moderate forces began to seek out pragmatic solutions across ideological and partisan boundaries. Thus the territorial compromise, which had existed for two decades, was broken and a new hegemonic political alliance emerged – the metropolis alliance. For a number of the main issues, trail-blazing compromises were reached, first over the drug question and then over traffic policy. This political shift also resulted in a fundamental change of urban development policy. While urban planning in Zurich City up to the 1990s had attempted to protect the quality of everyday life, to preserve the historical built environment and to defend the "city of residents" against the headquarter economy, now the focus was on competition: international investors, global capital and affluent residents were to be attracted to Zurich.

Viewed in historical context, this development entailed a disavowal of the basic principles of urban planning that had dominated the development of Zurich City for about a century, which had aimed to establish a clearly defined, coherent urban structure for the entire city. At the level of urban planning, the city was increasingly assimilated into the region.

THE RECONSTITUTION OF THE CITY: THE CASE OF ZURICH WEST

While in the course of the 1990s the processes of urbanization in the city and the region were interlinked and the disintegration of the city into the region advanced, a remarkable reversal occurred – the reconstitution of the city and the reproduction of the old center/periphery dichotomy.

Based on the analysis of national referenda in the last 20 years, this tendency can be illustrated in detail at the political level (Hermann and Leuthold 2003). While in the region there was a strong tendency towards right-wing populist positions, Zurich City, and in particular the inner-city neighborhoods, showed an increasingly left-liberal orientation. This tendency was evident in the entire German speaking part of Switzerland, but most distinctly in Zurich. The pattern of political polarization does correspond to socioeconomic disparities; rather, it reflects different preferences in everyday life. Within the Zurich region, different lifestyles have evolved. While suburban life still has a great attraction for many, others seek a distinctly urban lifestyle. The proximity of cultural facilities, a cosmopolitan milieu and a trendy image have become important location factors, not just for the lifestyle-conscious urban professionals, but increasingly for companies as well. These factors are still to a large extent concentrated in the center (Figure 4).

The new urban feeling has been manifested most visibly in the trendy new neighbourhood "Zurich West." As late as in the 1980s, this inner-city area was still one of the main centers of Swiss engineering industry. Due to the process of deindustrialization, a growing number of industrial activities relocated. At that time, the area was earmarked for the expansion of the financial sector. Because of the stalemate in urban planning, and the consequences of the economic and real estate crisis, development projects remained frozen for years. Eventually, the impressive

T W O

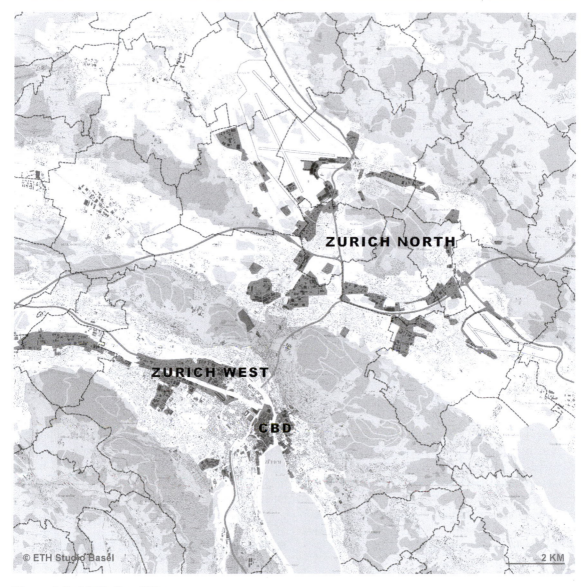

Figure 4 Zurich North and West

Source: Christian Schmid

industrial landscape, with its imposing warehouses and its austere charm, became a utopian place, a projection zone for fantasies, and a promise of opportunity. Small businesses, illegal or semi-legal bars and discos, theaters, hangouts, artist's studios and projects of all sorts began to appear throughout the area.

From the beginning, this transformation was also linked to market-based developments. One of the first projects was a condominium with luxury lofts and a multiplex movie theater. The result was a highly urban blend of both the commercial and the ephemeral,

something extraordinary, and not only for Switzerland. The new combination of working, living and entertainment as well as the unconventional atmosphere of the new neighbourhood attracted a wide range of additional business activities, from hotels to international consulting firms. A veritable cultural zone developed, housing several renowned institutions of arts and culture. The brownfield was thus transmuted into an elegant urban neighbourhood, which was presented to astonished visitors as a "Swiss Greenwich Village." Many pioneer projects from the early days have more recently been displaced, but a number of remarkable alternative projects succeeded in securing land while real estate prices were still low.

So Zurich West today represents a new inner-city development model. Nevertheless, the new neighbourhood differs radically from the existing downtown area. It presents an amazingly high density and diversity of varying utilizations and social groups. These are, however, hardly interrelated, but rather live side by side in an overlay of social and economic networks extending over the entire metropolitan region. The area basically consists of individual islands belonging together less on the basis of interactive processes than by sharing an urban milieu and a metropolitan image. This is not only an effect of the large-scale structure of the built environment, originating in the area's industrial history, but is also the result of the changed everyday routines of the metropolitan population.

PARADIGMS OF URBAN DEVELOPMENT

More than a decade ago, urban researchers discovered fundamental transformations and postulated a new urban era in urban development. They stated that contemporary cities did not resemble the traditional cities of the past (Garreau 1991; Soja 1996). Los Angeles, with its polycentric and excentric development, was declared the "paradigmatic industrial metropolis of the modern world" (Soja and Scott 1986). In the meantime, things have settled down at Exopolis. What is it that's new? What does the paradigm shift consist of? The example of Zurich reveals some reference points for confronting these questions.

1 The polycentric development of cities has become a general phenomenon. Even smaller cities follow this developmental trend. At the same time, the example of Zurich also illustrates that specific local traditions, contradictions and fields of conflict may have a decisive influence on urban development. The specific form of urbanization is determined not only by economic development, but is also a result of debates on the concept of the "city," of both struggles and compromises.
2 The process of metropolitanization breaks up the unity of the city. It is no longer possible to define the urban clearly. It is composed of overlaying configurations and unexpected constellations. In Zurich, two ideal-typical configurations can be distinguished,

giving an impressive illustration of the change—on the one hand the "exopolis model," as exemplified in Zurich North, on the other hand the "inner-city model," as manifested in Zurich West. Both areas stand for differing urban forms that are developing simultaneously. Yet the two models do not differ as much as might initially seem to be the case. In Zurich North, there is an attempt to reintroduce a classical conception of urbanity into the excentric urban chaos and to create new, coherent urban structures. On the other hand, the new "inner-city model" does not correspond to the traditional image of a downtown neighborhood, with its dense network of social interaction. Indeed, it represents a conjunction of regional networks that are barely interlinked on an everyday level.
3 Despite the trend towards polycentricity, the relationship between center and periphery remains highly contradictory. In the case of Zurich, the dichotomy between center and periphery did not weaken in this process, but was actually strengthened. Politically and culturally, center and periphery have drifted further apart. While the center exploded and disintegrated into the region, the city was meanwhile reproducing itself at the level of everyday life.

These are only some aspects of the paradigm shift in urban development that are currently transforming living conditions in Zurich. The new model of urbanization has proved to be contradictory and indeterminate.

REFERENCES FROM THE READING

Friedmann, J. (1995) Where we stand: A decade of world city research, in P. Knox and P. Taylor (eds.), *World Cities in a World System*, Cambridge: Cambridge University Press: 21–47.

Garreau, J. (1991) *Edge City: Life on the New Frontier*, New York: Doubleday.

Hermann, M. and Leuthold, H. (2003) *Atlas der Politischen Landschaften. Ein Weltanschauliches Porträt der Schweiz*, Zürich: vdf-Verlag.

Hitz, H., Lehrer, U., Keil, R. Ronneberger, K., Schmid, C. and Wolff, R. (eds.) (1995) *Capitales Fatales: Urbanisierung und Politik in den Finanzmetropolen Frankfurt und Zürich*, Zürich: Rotpunktverlag.

Lefebvre, H. (1968) *Le Droit à la Ville*, Paris: Anthropos.

Sassen, S. (1994) *Cities in a World Economy*, Thousand Oaks, CA: Pine Forge Press.

Soja, E. W. (1996) *Thirdspace*, Cambridge, MA and Oxford, UK: Blackwell.

Soja, E. W. and Scott, A. J. (1986) Los Angeles: Capital of the late twentieth century, *Environment and Planning D: Society and Space*, 4: 249–54.

16

"From 'state-owned' to 'City Inc.':
the re-territorialization of the state
in Shanghai"

from *The Making of Global City Regions* (2007)

Fulong Wu

EDITORS' INTRODUCTION

Fulong Wu, Bartlett Professor of Planning at University College London, is one of the most prolific scholars in urban China studies. Since the 1990s, he has authored and co-authored more than a hundred journal articles and book chapters. He has also edited and co-edited many books on China's urban transformation, including *Rural Migrants in Urban China* (2013), *China's Emerging Cities* (2007), and *Globalization and the Chinese City* (2006). His most recent book is *Planning for Growth: Urban and Regional Planning in China* (2015, Routledge). Drawing upon both qualitative and quantitative methods, Wu has systematically examined how Western concepts such as the global city, urban entrepreneurialism, neoliberalism, and gentrification can be usefully applied to interpret Chinese urbanism. His writings suggest that, despite the significant degree to which Chinese cities vary from their Western counterparts, these borrowed concepts can shed light on urban restructuring in post-reform China. However, like other China experts, Wu emphasizes the major role of the local state—especially municipal governments—in carrying out urban policy and programs. Perhaps in no other country in the world do local state initiatives so deeply shape urban growth.

The essay included here is excerpted from a book chapter in *The Making of Global City Regions: Johannesburg, Mumbai, São Paulo, and Shanghai* (2007). Here, Wu zooms into Shanghai's urban transformation and explains how its dramatic rise as China's global city has been possible. He identifies a number of crucial factors at work: the central government's approval to build Pudong district, the decentralization of power and resources to the Shanghai city government, and the flow of investment from international and domestic sources due to favorable investment policies since the 1990s. However, as Wu's chapter also shows—in close parallel to the other studies included in this section—the global city-making project is costly in social terms. The devastating social consequences of becoming East Asia's financial center have been experienced in every corner in Shanghai. As property prices have spiked, the local state has opted to displace and dispossess the poor. Wu discusses the massive demolition drives in Shanghai that led to the eviction of millions of urban households. Quoting Saskia Sassen, Wu asks: who is making claims on our city?

The global city region is the best place to observe how Castells's (1996) "spaces of flows" are being "reterritorialised" into local politics. For Scott (2001) global city regions represent a new economic and political order. Economically, the global city regions see new concentrations of post-Fordist industries. This is because these city regions "are invariably important centers of resourcefulness and invention for all sectors of production, but especially for post-Fordist industries where the basic conditions leading to these outcomes are so abundantly concentrated" (Scott 2001: 820–21). Politically, the neoliberal governmentality is becoming dominant and traditional citizenship is "vigorously and increasingly in question" (Scott 2001: 821). Globalization has introduced new actors into local politics. It has also transformed existing actors in the city. In the case of European cities, business interests are being restructured at this scale along with the extension and the deepening of market logics (LeGales 2002). Organized interests are becoming more territorialized: for example, chambers of commerce and industry, as powerful local actors, have successfully invoked the principle of economic competition as the organizing principle for political actors. The dominance of powerful business leaders is a major feature of entrepreneurial urban politics (Peck and Tickell 1995).

Saskia Sassen views this changing governance as a process of "making claims on the city": global capital has made claims on nation-states, and in consequence the changing power relationship often contributes to strengthen the advantages of certain types of economic actors and to weaken those of others (Sassen 1996). While the American version of an entrepreneurial city shows a strong role of chambers of commerce and industry and business elites, the European model suggests greater involvement of the local government and its strategic coalition with business in reoriented local politics. The making of new claims on the city does not always mean the dominance of multinationals and the weakening of local politics. Instead, the transformation of power structures could be more driven by proactive local governance as an institutional response to the challenge brought about by economic globalization. It must be reminded that the state is not necessarily a static entity. Rather, it can actively transform itself into a new structure, often through localization and territorialization, and thus survive or even consolidate its presence.

China, as a previous state socialist country, represents a totally different context. Globalization, decentralization, and marketization are major processes that capture profound social and economic changes since the market reform initiated in 1979. The remaking of global city regions has been seen in several megalopolises such as the Hong Kong-Shenzhen-Guangzhou and the rest of the Pearl River Delta, Shanghai-Nanjing-Hangzhou and the lower Yangtze River Delta, and Beijing-Tianjin-Tangshan and the cities along the Bohai Bay. The emergence of global city regions indicates the spatial reorganization of the Chinese economy and changing governance in the transition toward a more market-oriented economy and greater integration with the global economy. In the post-reform period, the Chinese city is witnessing the formation of territorially based entrepreneurialism, which is associated with the re-globalization strategy, that is, using globalization as an opportunity to transform the role of the state and, in turn, to legitimize its presence in the economic sphere. In this process, the spatiality of the globalizing city becomes the important medium and vehicle to achieve the aspiration of the state. Overall, on the surface, Shanghai departs significantly from the classical neoliberal project, which emphasizes the triumph of the market and minimization of state intervention. However, in essence, the case of Shanghai only suggests some variation of this worldwide spread of neoliberal governance under globalization, because we have to situate the state in the new structure of power so as to understand the novelty of state functionality.

This essay will begin an investigation of changing power structures—both formal and informal arrangements—in Shanghai since the early 1990s. The changing power relationships are understood through the changing strategy of accumulation, which in turn is a local response to globalization and the global city in the making. Rather than returning to the much-heated debate on globalization and the waning nation-state, I attempt to understand how the national state is rescaled and territorialized into the global city region and how the local state is reconfigured to suit the changes in governance. It is argued that although globalization imposes a similar threat and imperative to all cities, the local response is historically and geographically contextualized. They are globalizing cities in developing countries. Their responses vary according to the interaction between historical legacies and new visions of change.

The Shanghai case suggests that situated in this periphery of globalization, the state's capacity to take the lead in the building of both physical and institutional environments is critical in the formation of the global city region. While globalization is the critical factor in triggering off a series of changes in power structures, the context of "postsocialism" and hence of strong state intervention is also a co-determinant in the trajectory of urban changes in the Chinese case. Changes in the global periphery are now being incorporated in a bigger picture of unfolding neoliberalization in the contemporary world, but the exact process that occurs at locales at the periphery varies. In the case of Shanghai, we see active decentralization of administrative and fiscal duties and continuing utilization of state policy instruments to enforce economic restructuring in Shanghai.

RESCALING THE STATE AND CHANGING FORMAL POWER STRUCTURES

The power structure has changed since the adoption of economic reform and open-door policies in China. The market-oriented reform is characterized by reengineering power relationships, first, between the central and local states and, second, between the public and private sectors. First, the change in fiscal arrangement has led to greater autonomy of local governments. Economic decentralization is said to be the recipe for fast growth of the Chinese economy, and the rearrangement of central-local relations has been widely documented.

Decentralization of the fiscal system went through three stages in Shanghai. During the first, from 1978 to 1984, expenditure was linked to revenue and the total proportion of revenue submission to the central government was revised on a yearly basis. During the second, from 1985 to 1993, the central government relaxed control over local expenditure and the baseline of expenditure was raised from 2 to 3.5 billion yuan. Meanwhile, revenue submission was fixed to 10.5 billion yuan in 1988, and Shanghai was allowed to use its surplus revenue. During the third stage, from 1994 onward, the tax-sharing system was operated and the Shanghai municipal government had an independent tax base and autonomous expenditure. The fiscal reform created progressive institutional innovations through which local government agencies have been mobilized. Similar to other places in China, local governments are now becoming ambitious actors who are subject to harder budget constraints.

Consequently, their behavior is transformed into that of "industrial firms" (Walder 1995). The operation mechanism is very different from the one characterized by the "soft budget" under state socialism. What is different is Shanghai's strategic importance as China's "dragon-head" and gateway to connect to the global economy. It has thus regained its favorable position in fiscal deals with the central government.

Behind the increase in local revenue and expenditure is the strong determination of the central government to create an international Chinese economic, finance, and trade center in the Yangtze River Delta. This is known as the "dragonhead" strategy. The expansion of local autonomy can be seen as the willingness of the central government to reformulate its economic strategy rather than as the result of hard negotiations by local government to gain more freedom in economic decision making. For example, an often-neglected fact is that some working teams composed of senior officials were sent by the central government to Shanghai to help it prepare a better deal. The increasing presence of Shanghai officials in the central government in the 1990s further strengthened the voice of Shanghai in its request for more progressive arrangements.

Corresponding to fiscal decentralization, Shanghai has launched a series of institutional reforms to further mobilize lower-level agents. The urban districts, formerly with limited functions, gained a whole array of administrative powers, including planning, public works maintenance, the approval of local foreign trade deals, and commercial administration. Since the mid-1990s Shanghai has begun to adopt a new structure of local government, namely "two levels of government and three tiers of management." The two levels of government refer to the municipal and district governments, and three tiers of management refer to municipal, district, and subdistrict (or street office) management. The street office is not a level of government, but rather the agency of district government. The rise of the private sector, the inflow of rural migrants, and increasingly mobile individuals require the establishment of governance beyond the formal work-unit system. In order to manage those who are not formally affiliated to a state work unit, the street office has been given a new role of comprehensive management. This level of government has thus become an active participant in governance and has begun to manage its own economic base by attracting and registering enterprises. The result of economic

Plate 15 Shanghai selfie in front of Pudong skyline across Huangpu River

Source: Roger Keil

decentralization is that multiple actors have been created at the levels of local government.

Community construction, as defined in the government policy, is a discursive strategy to consolidate the state's penetration into the grass-roots level and to maintain a governable urban society in order to support the strategy of global city remaking. In practice, this policy emphasizes a shift of many functionalities previously taken by the state work units—pension distribution, unemployment support and poverty relief, property management, and organized consumption—to local street-level organizations. While the policy sometimes does promote self-organization of mass and local democracy, in the practice of service privatization, retrenchment of the welfare function, and consumption-oriented urban development, the shift in fact favors those who have purchasing power and thus form the consumer groups, such as the homeowners' association. This policy is in part a response to the impact of globalization, which has exacerbated the mobility of resources and population and economic restructuring and, in turn, created challenges for the state to manage a changing society. It is increasingly impossible to accomplish these functions through the top-down approach. As a result, many responsibilities are transferred to the communities. These processes are strengthening local-level governance.

In summary, globalization together with a more market-oriented development approach has forced the state to adopt a more decentralized power structure. This process is not very different from those seen in other parts of the world, although different routes and forms are taken. If there is something unique in the post-reform context, the process is actively driven by the state itself. In this sense, it is more a process of re-territorialization of the state itself rather than just re-territorialization of mobile capital.

MAKING COALITIONS WITH CAPITAL: INFORMAL POWER STRUCTURE

While the formal structure does not give a prominent status to external investors, in the competition for investment overseas investors have begun to show increasing leverage by forging relationships with local political elites. It is observed that the making of "relationships" (guanxi) blends the traditional

practices of gift exchange with the new context of entrepreneurialism. For example, Hsing (1998) attributes the success of Taiwanese investors in mainland China to their cultural and linguistic affinity with the local elites. This allows "footloose" capital to be embedded into local politics through informal and often fragile coalitions.

The informality of the relationship between local elites and business actors should not, however, be romanticized, and cultural practices play a role of lubrication rather than determination. In questioning China's guanxi capitalism, Castells (2000: 311) argues that the reason why overseas Chinese business networks play such an important role is "not that they and their southern China partners both like steamed cod. It is because China's multiple link to the global economy is local, that is, it is performed through the connection between overseas Chinese business and local and provincial governments in China." This is accompanied with a simultaneous transition in China's local governance toward decentralization, as discussed earlier. Changing central-local politics have now been extensively documented with reference to decentralization of property rights and changing budgetary and fiscal systems (Walder 1995; Oi and Walder 1999). But these changes had been so far mainly attributed to China's economic reform and internal changes in the central and local relationships. Shanghai's case suggests that interplay between globalization and localized networks.

The interplay can be observed from two angles. First, the interplay reflects the change in professional services under globalization—in this case, the prominent role of so-called global intelligence corps (GIC, i.e., elite architectural firms)—in the production of mega-projects or the citadels of global cities. These GIC firms are solicited by the Shanghai municipal government in its international planning consultation and thus play a key role in Shanghai's global publicity. Second, the interplay is made possible by key politicians who overcome institutional hurdles and build pro-growth coalitions on a project-by-project basis. The development projects reveal the informal power structure built upon coalitions between the state and capital. The transformed local politics are now more oriented toward undertaking the "central task of economic construction," without "questioning whether such a road leads to capitalism or socialism," as urged by Deng Xiaoping, the architect of China's economic reform.

THE CHANGING RELATIONSHIP BETWEEN THE PUBLIC AND PRIVATE SECTOR

Marketization has changed the relationship between the public and private sector. The omnipotent state and state-owned enterprises have retreated from comprehensive welfare provision to households. The market mechanism has been introduced into housing consumption and urban services provision. After the abolition of in-kind work-unit housing, commodity and low-profit economic housing become major housing sources. In services provisions such as water, gas, and electricity, service and utility companies have gradually replaced government-funded public works and diminished their monopoly position through increasing user charges. For example, as early as the 1980s, Shanghai adopted an installation fee for gas provision. The initial installation fee amounted to 1,500 yuan plus a compulsory 500 yuan construction bond, which was equivalent to 50% of annual household income by that time (Lu 2002: 20). This innovative practice has greatly eased the capital bottleneck and increased investment in infrastructure. Since the 1990s the user-charge practice has been widely used in bus fares, gas supplies, wastewater discharge, and municipal sanitation services. Utility companies have been set up to take over responsibilities of service provision. The Shanghai Urban Construction Investment and Development Company, established in 1992, used a variety of financing mechanisms, such as construction bonds, the stock market, and service concession, to raise capital, and its funds accounted for about 90% of Shanghai's total urban infrastructure revenue in 1996 (Yusuf and Wu 2002: 1230).

Recently, the *entrepreneurial city* (literally translated from Chinese as the "city of business management") has become a popular term. The concept emphasizes the use of market instruments to manage urban assets—both physical and symbolic—so as to fund urban development. The government can cooperate with the private sector by transferring its physical assets (e.g., land) or institutional assets (e.g., the right of naming and franchising). Private funds are encouraged to purchase and operate public works and services. For example, the Shanghai Urban Construction Investment and Development Company recently sold the operation rights of the Nanjing-Shanghai highway, recovered its original investment, and further invested in new projects.

CONCLUSION

The development of Shanghai suggests a strong dimension of local politics in the formation of global city regions. Because of the weak community, the state is able to set the agenda without much resistance from below. Residential relocation has been carried out swiftly, making space for global capital. Industrial restructuring has generated redundant workers in traditional manufacturing industries such as textiles, iron and steel production, and traditional light industries. The property right of land under state socialism enables a monopoly position of the municipality, which has been used as an instrument to solicit development projects.

However, the case of Shanghai does not simply follow the development state model. Seemingly, the dominance of the state in urban development contradicts the global trend toward neoliberalism. But when the role of the state is examined in the transformed power structure, formally and informally, it is clear that the city is a variant to the neoliberal city. This does not necessarily mean a convergence to the universal global, neoliberal city. Rather, Shanghai perhaps represents a more advanced stage of neoliberalization, where the state plays an active role by using its governing function to support the market. The reterritorialized state can better support the market and pave the way to new methods of capital accumulation.

Although the state retains or has even consolidated its role in governance, the transformation of the power structure shows that this is no longer the model of state-led growth. More specifically, the state no longer owns the means of production and does not seek to treat the city as a state-owned asset. Instead, the city is now becoming an entrepreneurial, gigantic "City Inc." Or, in other words, the city itself is becoming the site of new governance So far the transformation of the power relationship is asymmetric, in the sense that the local community is weak and that capital is better accommodated to the project of global city building. Institutional and market forces reinforce marginalization and have led to a poverty of transition. However, the power structure is not fixed. The relationships forged in the entrepreneurial state project will be continuously contested along with the changing external and internal political economic environments.

REFERENCES FROM THE READING

Castells, M. (1996) *The Rise of the Network Society*, Vol. 1, Oxford: Blackwell.

Castells, M. (2000) *The End of Millennium*, Vol. 3, Oxford: Blackwell.

Hsing, Y. T. (1998) *Making Capitalism in China: The Taiwan Connection*, Oxford: Oxford University Press.

LeGales, P. (2002) *European Cities: Social Conflicts and Governance*, Oxford: Oxford University Press.

Lu, H. L. (2002) Governance of urban public life and urban residents, in Z. Yi and H. L. Lu (eds.), *Urban Governance and Quality of Citizen*, Shanghai: Shanghai Social Science Academy: 1–37.

Oi, J. C. and Walder, A.G. (1999) *Property Rights and Economic Reform in China*, Stanford, CA: Stanford University Press.

Peck, J. and Tickell, A. (1995) Business goes local: Dissecting the "business agenda" in Manchester, *International Journal of Urban and Regional Research*, 19: 55–78.

Sassen, S. (1996) *Losing Control? Sovereignty in an Age of Globalization*, New York: Columbia University Press.

Scott, A. J. (2001) Globalization and the rise of city-regions, *European Planning Studies*, 9: 813–26.

Segbers, K. (ed.) (2007) *The Making of Global City Regions: Johannesburg, Mumbai, Sao Paulo, and Shanghai*, Baltimore, MD: Johns Hopkins University Press.

Walder, A. (1995) Local governments as industrial firms: An organizational analysis of China's transitional economy, *American Journal of Sociology*, 101: 263–301.

Wu, F. (ed.) (2006) *Globalization and the Chinese City*, London: Routledge.

Wu, F. (ed.) (2007) *China's Emerging Cities: The Making of New Urbanism*, London: Routledge.

Wu, F. (2015) *Planning for Growth: Urban and Regional Planning in China*, London: Routledge.

Wu, F., Zhang, F., and Webster, C. (eds.) (2014) *Rural Migrants in Urban China: Enclaves and Transient Urbanism*, London: Routledge.

Yusuf, S. and Wu, W. (2002) Pathways to a world city: Shanghai rising in an era of globalization, *Urban Studies*, 39: 1213–40.

17
"The dream of Delhi as a global city"

from *International Journal of Urban and Regional Research* (2011)

Veronique Dupont

EDITORS' INTRODUCTION

Veronique Dupont, a senior research fellow at the Institute of Research for Development in France, has published widely on the sociospatial dynamics of large Indian metropolises, with a focus on Delhi. Delhi, like other large Indian cities, has been undergoing a rapid makeover since the Indian economy began to liberalize in the early 1990s. Migrants flock to the capital by the tens of thousands from all across the country, and generally settle in the city's many slum quarters. As for the middle class, new residential colonies offering modern amenities have sprung up across the capital. And on the outskirts of Delhi, in the state of Haryana, a previously little-known place called Gurgaon has become a booming new suburb. Labelled the "Millennium City of India," Gurgaon boasts sparkling new office buildings, upscale shopping malls, and privately financed infrastructure, all of which cater to the new middle class and businesses. As Dupont explains, Delhi is not a global city of high finance, but it does interact with other global cities by offering investment opportunities and outsourcing services. She highlights the increasing housing shortage, which was signifcantly worsened through the slum clearance campaigns connected to the 2010 Commonwealth Games. She also discusses the controversial role of the High Court of Delhi, which in recent years has often ruled in favor of the middle class in their effort to "beautify" the capital by displacing the poor. This article examines the underlying socioeconomic currents behind the project of global city making and its sociospatial consequences.

Delhi has a long history of growth and change. Today's rulers have new aspirations for the capital city—those of transforming it into a "global city". Is Delhi a strategic site for the new global economy? A city with decisive functions on an international scale? If the Indian capital is yet far from meeting John Friedmann's (1986) and Saskia Sassen's (1991) criteria, the ambition to attain this status is clearly expressed in the Master Plan for Delhi 2021: Vision-2021 is to make Delhi a global metropolis and a world-class city. This is also a recurring slogan in the Chief Minister's speeches, especially deployed at a time preparations were underway for the Commonwealth Games, hosted in Delhi in October 2010.

What facts and realities form the basis of this ambition? And what are the consequences on the restructuring of the urban landscape, especially for those who do not fit into this vision? Delhi's recent development confirms a trend first evidenced by Harvey (1989) in advanced capitalist countries, and now observable in emerging economies such as India: the shift to entrepreneurialism in urban governance resting on public-private partnership and involving the creation of an attractive urban imagery as part of the speculative construction of place in the context of competitiveness among cities. The international sporting event in Delhi provides a privileged moment to examine the transformations and processes at work in the Indian capital. In the same way as the Olympics elsewhere, the Commonwealth Games are used by the city's authorities as a "catalyst of urban change" and an "international showcase" to enhance the city's global recognition and image; the Games, along with other urban projects, including shopping malls and the Delhi Metro rail system, epitomize the "concentration on spectacle and image rather than on the substance of economic and social problems" (Harvey 1989: 16). As the Delhi study will also show, the assertive "new" Indian middle class, character-ized by its "attitudes, lifestyles and consumption practices associated with commodities made available in India's liberalising economy" (Fernandes 2004: 2415) has concurred with the restructuring of the urban space, as well as its concomitant polarization and underlying exclusion process—what Fernandes (2004) refers to as a "politics of forgetting" marginalized social groups.

INDIA'S OPENING UP TO THE GLOBAL ECONOMY

The ground for the agenda of transforming major Indian metropolises into global cities was prepared by the opening up of the Indian economy to the interna-tional market. A first series of reforms, dismantling most protectionist measures applicable to foreign firms, was introduced gradually from 1985 onwards and confirmed in 1991 by an unambiguous policy of economic liberalization. The politico-administrative decentralization reforms of the 1990s, which trans-ferred more important responsibilities to the munici-palities, have also changed the context of development in big cities.

The vital contribution of towns and cities to the national economy (although home to less than one-third of the country's population, they generate three-quarters of gross domestic product) and the driving role of metropolises were officially recognized only in the late 1980s. The national urban strategy that was then adopted hinged on the concepts of decen-tralization, deregulation and privatization. It was aimed at enhancing the economic efficiency of cities, promoting infrastructural development and better response to housing demands. Public-private partner-ships, strengthening of the private sector's role and the cost-recovery principle were also promoted. Several deregulatory measures have altered the urban context of development and management, in parti-cular: deregulation of the cement sector, which boosted the construction sector, abolition of the "license Raj" or mandatory licensing for the majority of industrial plants (previously, licensing could be a means of keeping industries away from metropolises), the repeal in 1999 of the Urban Land Ceiling and Regulation Act (which imposed a ceiling on the size of urban property owned) in order to stimulate the land market. In urban governance, the cornerstone of reforms was the 74th Amendment to the Constitution (1992) that decentralized strategic governance and promoted participatory democracy. In particular, this amendment transferred to urban local bodies the res-ponsibility for urban development, inclusive of pro-viding infrastructure and services as well as mobilizing the required financial resources—through taxes, levying users' costs and by attracting private invest-ment, both domestic and foreign. The new urban context was thus marked by a reduction of government subsidies and the state's partial withdrawal from several traditional areas.

The economic liberalization and political decen-tralization reforms are also the product of the pres-sures exerted by international donor agencies (World Bank and International Monetary Fund) on India. Therefore, there is a dialectic relationship between neoliberalization and globalization in Indian cities; the global has strongly influenced the 1990s reforms and the setting of the new urban scene, and, in return, the implementation of a neoliberal agenda in Indian cities favoured the incorporation of the biggest metropolises into the larger global movement.

Globalization of India's big cities is further notice-able in the influence of international organizations on the formulation of urban programmes. In Delhi, the *bhagidari* scheme, the standard bearer of new urban governance, which was launched in 2000 as a citizen-government partnership programme, reflects notably the focus of the World Bank on participation and good governance, while the *bhagidari* workshops are inspired by global theories of corporate governance.

Plate 16 Billboard in 2010 during the Commonwealth Games, Delhi

Source: Xuefei Ren

City development strategies promoted by the World Bank and the United Nations Centre for Human Settlements also echo the city development plans at the core of the Jawaharlal Nehru National Urban Renewal Mission (JNNURM), a major programme launched in December 2005 and funded by the central government. Compared to previous national urban programmes, the JNNURM is distinctive in that it makes access to central subsidies dependent on the introduction of a number of reforms and the production of city development plans, encouraging municipalities to project themselves into the future and improve the productivity and efficiencies of cities, while simultaneously ensuring that they are equitable and inclusive. Beyond this growth-focused strategy and stated concern for equity and inclusion, one should read the culmination of a neoliberal urban vision; thus critics have already established strong scepticism about the achievement of the sub-mission

aimed at integrating the urban poor and fear, by contrast, further exclusion through more slum demolitions and displacement. In conclusion, the urban reforms and programmes highlight the longevity of global pressure applied to Indian cities.

DELHI SHINING

Delhi's population has grown at a remarkable rate for an urban agglomeration of its size, reaching 18.6 million in 2016. Migration has played a major role in this demographic evolution: after the massive influx of refugees from Pakistan following the country's partition, migration continued to contribute significantly to urban growth in the following decades.

Delhi has for long been a premium market for India's Northwest region. It is endowed with multiple economic functions, standing out not only in trade and commerce but also banking, finance, insurance, hotel and tourism, as well as manufacturing. It has greatly surpassed its original function, namely public administration, to which its national-capital status is linked. Yet, within the organized sector (accounting for 15% of the total employment in 2006), the public sector (central and state governments, quasi-government and local bodies) remains the largest employer (72% of the workforce in 2008).

Delhi and its ring towns have capitalized well on the new economic opportunities that arose after 1990. The availability of a large skilled English-speaking manpower has attracted many multinational companies in the information and communication technologies sector (Alcatel-Lucent, Niksun, IBM, Dell, Nokia and others) as well as in the banking and hotel sectors. Gurgaon, in Delhi's southern suburbs, is a prime example of the emergence of a prominent outsourcing and offshoring hub over the last decade, with numerous business process outsourcing (BPO) firms. Foreign multinationals have also invested in industry, especially the automobile and two-wheeler sector, for example, Maruti Suzuki (a subsidiary of the Japanese Suzuki Motor Company) and Hero Honda (a joint venture of India's Hero group and the Japanese Honda Motor company), which both set up their headquarters in Gurgaon.

Sassen identified export processing zones as "transnational spaces for economic activity" that "symbolize along with offshore banking centres, high-tech districts and global cities the new forms of economic globalization" (2006: xiii, 31). In India, this qualification seems even more appropriate for second generation export processing zones (EPZ), the Special Economic Zones (SEZ) that replaced the former in 2000, and whose rules and regulations were formalized in 2005 in the SEZ Act as notified in 2006. The SEZ policy's main purpose is to provide an internationally competitive and hassle-free environment for exports, and to attract private investment through highly attractive fiscal incentive packages. While EPZs were just industrial enclaves, SEZs are integrated townships with fully developed infrastructures.

Delhi is not a hub of international finance. Yet, since the 1990s, it has displayed its ability to interact with other global cities. Like other large Indian metropolises, it provides the global market with some direct investment opportunities and outsourced services. Hence, some scholars argue that Delhi could be considered a new type of global city, fitted into a network of complex flows, mobilizing information and communication technologies, and increasingly using the internet.

TRANSFORMATION OF THE URBAN LANDSCAPE

Another aspect of Delhi's "globalization" concerns the urban landscape and its rapid changes, which is following an international model of modernization that tends to lead to a certain repetition and standardization of urban forms, for example, the proliferation of high-rises, shopping malls and business centres, gated housing complexes and the multiplication of freeway flyovers, as observed in other aspiring global cities.

The most spectacular changes affect the outlying districts, especially in Gurgaon and Noida, where new neighbourhoods have emerged, often projected as genuine urban entities (that is, as new townships) constructed and marketed by private property developers. Large high-end residential complexes are found there, equipped with modern infrastructure and amenities, and integrated with other urban services and functions—commercial, educational and recreational. References to a Western model and style through publicity hype, including names of residential complexes (Malibu Towne, Riverdale, Manhattan Apartmansions, Beverly Park, etc.), or references to other large Asian metropolises such as Hong Kong and Singapore ("Sentosa City, A mini-Singapore in India"), are especially intended to attract the wide diaspora of Non-Resident Indians (NRI) who want to invest in their home country or plan to return home one day. References to NRI architects as planners and designers of housing projects further illustrate the global diffusion of architectural standards that is supported by international migration. These new suburban developments, not unlike those seen in Bangalore and Mumbai, shorn of a distinct Indian identity, characterize the making of what King (2005) calls "spaces of global cultures," which are also "new spaces of exclusion" (Broudehoux 2007). High-standing residential estates and their logic of exclusiveness reflect a phenomenon of greater socio-spatial fragmentation, reinforcing a tendency towards the ghettoization by choice of the rich.

The Commonwealth Games, hosted in the Indian capital in October 2010, provided a good opportunity for the government to legitimize its action and the deadline for town planners and builders to put the finishing touches to the capital's 'face-lift' and remodel it to meet its globalization-in-the-making requirements. Although the Commonwealth Games do not have the same international scale, stature and prestige as the Olympics, the city authorities used this event in a similar way: as a key instrument of urban policy, to galvanize urban restructuring, enhance the city's status and advertise its position on the global stage. The media further contributed to the aestheticization of city space, as symbolized by the slogan "From Walled City to World City" popularized by the *Times of India*, one of the leading newspapers. This concentration on spectacle is an integral part of the city marketing strategy.

In Delhi, as in other Indian metropolises, the emulation of the global-city model entails an asymmetric interrelation between the global and the local. Delhi's role in the new global economy may not be strategic at the global scale, but the impact of the Indian economy's globalization, efforts to attract foreign capital and to project the image of a 'world-class city' are momentous when analysed at the metropolitan scale. Furthermore, the urban landscape's reshaping and changes in consumption patterns cannot be perceived only as an exercise of international image-building and superficial modernization. As shown in the next section, urban regeneration and beautification have dramatic consequences for those who do not fit into the model.

THE DARK SIDE OF THE GLOBALIZING CITY

The dark side of the "Delhi Shining" and globalizing city will be highlighted in this last section: from Delhi's poor performance in terms of providing access to adequate housing and informalization of employment and the exclusion of supposedly undesirable citizens thought to mar the image of a global city-in-the-making.

Shortage of adequate housing was estimated at about 300,000 units in 2006. This deficiency has led to overcrowding and lack of comfort in dwelling units. According to the 2001 census, 39% of urban households in Delhi had only one room for an average household size of five persons. Furthermore, 24% of dwelling units had no drinking water and 21% no

toilet. The large number of people living in slums further illustrates the failure of the public policy of urban planning and housing, especially the Delhi Development Authority's inability to provide affordable housing for the poor despite its initial objectives. In 1998, the population living in squatter settlements was estimated at around 3 million, scattered in around 1,100 clusters of varied sizes throughout the urban area and accounting for about 27% of the total urban population—compared to 5% in 1951 and 18% in 1991—but occupying less than 6% of the city land, essentially public land. Large-scale slum demolitions since 2000 have altered this situation—without providing any satisfactory alternative. Another disquieting trend is the increasing informalization of Delhi's labour force: the proportion of unorganized sector workers in total employment has risen from 74% in 1992 to 85% in 2006.

As a capital city, Delhi has always received particular attention from governments and town planners. As the country's showcase, its image is to be enhanced. As a projected global city, its urban landscape must be attractive, modern and neat. The Indian economy's increasingly significant integration into the global movement and the efforts expended to attract foreign capital have impacted on the capital's socio-spatial reorganization, especially through a slum-clearance policy. The aim is to 'clean up' and 'beautify' the capital's landscape, to rid it of disturbing elements such as, for example, the view of a slum from the window of a deluxe hotel welcoming potential foreign investors.

From 1990, when the Delhi Government adopted a new slum policy, until 2007, around 65,000 squatter families were relocated in resettlement colonies on the rural fringes, relegated to the city's far reaches, up to 30 kilometres from the city centre. Although slum demolitions occurred in the entire urban area, the larger operations affected the central and southern zones of the urban agglomeration and the airport vicinity too, that is, zones characterized by the presence of commercial and business districts and a concentration of residential colonies for higher income groups, and where the capital's reconstruction has been more conspicuous.

Between 2004 and 2006, such operations also dramatically affected the Yamuna River's embankments (leading to the eviction of over 300,000 people), to make way to the redevelopment of the riverfront

Plate 17 Gurgaon, Delhi

Source: Xuefei Ren

and the construction of the Games Village, despite vehement protests from environmentalists, since the river's immediate vicinity is a floodplain area and groundwater recharge zone. The economic rationale for the demolition of slums and their relocation in peripheral zones is that the value of the land on which slum clusters are situated in the city is much higher than that in the relocation sites. In the making of Delhi into a world-class city, even unclaimed spaces on which the poor squatted became prime land, "ripe for development" (Baviskar 2006).

There are still other unwanted citizens in the global city of the Games. Some 300,000 stalls selling freshly cooked meals and snacks may also face evacuation before the event. Beggars are also seen as a menace by the Delhi government, especially in view of the 2010 international sporting event, as their appearance did not fit in with the vision of a world-class city. To put the finishing touches to its clean-up operations, the

Delhi government also attempted to capture stray animals, including monkeys, dogs and cows, and remove them from the capital's roads before visitors arrived for the international Games.

The government's and town planners' agenda of transforming Delhi into a global city and the beautification and cleaning drive conducted to that end received support from another increasingly important actor on the scene of urban governance in India: the courts. Their role is even more critical in the capital city, which is also the seat of the Supreme Court of India. In many cases the intervention of the courts was a response to petitioners representing the interests of industrialists or middle-class resident welfare associations. The judicial endorsement confirms the increasing intolerance towards the informal and further transforms street vendors into illegitimate workers, and slum dwellers and beggars into illegal denizens.

CONCLUSION

Delhi's experience in emulating the global-city model is not unique. Other metropolises in India and different emerging countries have undergone similar processes, both in terms of promoting export-oriented activities and image-building. Like elsewhere, this drive for global competitiveness has had negative consequences, especially for the poor, through 'cleansing' the city of slums and other alleged undesirable elements, and exacerbating socio-economic inequalities. The ascent of urban entrepreneurialism in Delhi and its translation into a "revanchist city" suggest familiar trends of neoliberal urbanism, widespread in Western cities (Brenner and Theodore 2002), combined in India's big cities with an outward-looking global bias.

Delhi provides a fascinating laboratory for observing the multidimensional effects of the ambition to transform a metropolis according to a perceived global-city model, and the contradictions inherent in the adoption of such an agenda. Given, on the one hand, the magnitude of the informal in this megacity (both in the housing and employment spheres) and, on the other hand, the increasing assertiveness of social groups who have benefited most from the economic reforms, the imposition of a new economic model and urban aesthetics generates tensions which are reflected in a magnifying mirror. What also makes the case of the Indian capital remarkable is the conjunction of structural factors (the overriding power of the state and the decisive intervention of the courts in urban affairs) with an international event (the Commonwealth Games). These specific dynamics colour the shift to entrepreneurialism in urban governance with an original Indian shade.

REFERENCES FROM THE READING

Baviskar, A. (2006) Demolishing Delhi: World class city in the making, *Mute Magazine*, 5 September. Available online: www.metamute.org.

Brenner, N. and Theodore, N. (eds.) (2002) *Spaces of Neoliberalism: Urban Restructuring in North America and Western Europe*, Oxford: Blackwell Publishers.

Broudehoux, A-M. (2007) Spectacular Beijing: The conspicuous construction of an Olympic metropolis, *Journal of Urban Affairs*, 29, 4: 383–99.

Fernandes, L. (2004) The politics of forgetting: Class politics, state power and the restructuring of urban space in India, *Urban Studies*, 41, 12: 2415–2430.

Friedmann, J. (1986) The world city hypothesis, *Development and Change*, 17: 69–83.

Harvey, D. (1989) From managerialism to entrepreneurialism: The transformation in urban governance in late capitalism, *Goegrafisca Annaler B*, 71, 1: 3–17.

King, A. D. (2005) *Spaces of Global Cultures: Architectures, Urbanism, Identity*, London and New York: Routledge.

Sassen, S. (1991) *The Global City: New York, London, Tokyo*, Princeton, NJ: Princeton University Press.

Sassen, S. (2006) *Cities in a World Economy*, 3rd Edition, Thousand Oaks, CA: Sage.

TWO

18

"'Fourth world' cities in the global economy: the case of Phnom Penh, Cambodia"

from *International Journal of Urban and Regional Research* (1998)

Gavin Shatkin

EDITORS' INTRODUCTION

Gavin Shatkin teaches urban planning at the School of Public Policy and Urban Affairs and the School of Architecture at Northeastern University. Shatkin's work focuses on urban poverty and globalization in Southeast Asian cities. In particular, he has explored the consequences of globalization for issues such as urban infrastructure provision, the role of non-governmental and community-based organizations, the restructuring of urban governance, and patterns of urban inequality. In this chapter, Shatkin develops a powerful critique of the assertion that sub-Saharan African, Latin American, and Asian city-regions have become structurally irrelevant to the world economy under contemporary conditions. This claim, which has been articulated frequently by global cities researchers, implies that contemporary globalization has produced a "Fourth World" characterized by economic stagnation, marginalization and social upheaval, and which contributes only minimally to world-scale processes of capital accumulation. Against such arguments, Shatkin demonstrates various ways in which cities in the less developed zones of the world economy contribute to, and are in turn being reshaped by, global economic processes. Even in the absence of large-scale flows of foreign direct investment, Shatkin argues, there are significant technological, economic, organizational, and sociocultural linkages between cities in the less developed world and the cities of the global North. Shatkin illustrates this argument with reference to the case of Phnom Penh, Cambodia, which had been insulated from global economic forces during the era of socialist rule from the mid-1970s to the late 1980s. As Shaktin demonstrates, the 1990s witnessed the city's rapid reintegration into the world economy, leading in turn to a variety of local sociospatial transformations. State spending on public infrastructure was reduced, a private real estate market was established, foreign capital flowed into the built environment, the presence of large numbers of United Nations peacekeeping workers created new markets for various consumer amenities, and new technological and trade linkages to other Southeast Asian cities were established. Foreign investment in the real estate sector in Phnom Penh has intensified after 2000, with capital flowing in from Singapore, Korea, Malaysia, and China, and property speculation has made real estate in Phnom Penh among the most expensive in Asia. In addition, new sociospatial inequalities emerged as both squatter settlements and various types of luxury facilities (hotels, casinos, restaurants) proliferated throughout the city. Shatkin uses this case study as the basis for a more general claim about cities and globalization in the less developed world: global economic forces have shaped and reshaped such cities, but in contextually specific ways that cannot be adequately understood by positing their exclusion from the world economy.

The development of innovations in transportation, telecommunications and information technologies, and the consequent emergence of a global economy, has presented opportunities for nations of the developing world. However, many countries, particularly much of sub-Saharan Africa, and parts of Latin America and Asia, have largely not participated in the globalization of the economy. As a consequence, Castells (1996) contends that globalization and consequent economic restructuring has resulted in the disappearance of the third world and the emergence of a "fourth world" of regions that are increasingly excluded and "structurally irrelevant" to the current process of global capital accumulation. Cities in such excluded countries are often characterized as being distinguished by their economic stagnation, the increasing marginality of their populations and their potential for social upheaval.

While it is true that participation in the global economy is unequal among nations, this chapter argues that the rather simplistic depiction of the process of urbanization in the so-called "fourth world" cities in much of the global cities literature is inaccurate. First, although the relative share of many least developed countries (LDCs) in global trade and investment is decreasing, these countries nevertheless remain integrated into the global system in important ways. The diffusion of new technologies, global and regional economic change, the increased regulatory power of international aid and lending institutions and changes in the flows of information, goods and people affect LDCs in ways that have major social and spatial consequences for cities. Second, the idea of a "fourth world" of excluded nations is prone to many of the same criticisms as the concept of a "third world" which it is meant to replace. LDCs experience quite different patterns of urbanization based on place-specific factors such as local histories (particularly colonial histories), social, political and cultural systems, and modes of integration into the global economy. Thus the depiction of LDCs as places of uniform marginalization and despair is misleading, and the relevance of globalization for cities in LDCs extends beyond the dynamic of exclusion. Rather than treating the experience of cities in LDCs as an unfortunate footnote to the phenomenon of globalization and economic restructuring, the role of such cities in the process of capital accumulation, and the impact of globalization on their development, should be a topic for research and debate.

This chapter examines the case of Phnom Penh, Cambodia, a city that has undergone dramatic spatial and social restructuring in recent years despite the relatively low levels of foreign direct investment and industrial growth in the country's economy. These changes will be reviewed in the context of recent historical developments in Cambodia as well as the country's interaction with the global economy.

GLOBALIZATION AND CITIES IN LEAST DEVELOPED COUNTRIES

Discussions of the impacts of globalization on developing countries generally focus on the relocation of industry from the developed to the developing nations and the subsequent industrialization of parts of the developing world based on a "new international division of labor." One consequence of this phenomenon has been the rapid increase in the amount of foreign direct investment (FDI) flowing to developing countries, especially since the late 1980s. While the amount of FDI flowing to developing countries continues to constitute a small percentage of total flows, this nonetheless has led to a significant increase in industrial growth in many parts of the world. In particular, the emergence of the Asian newly industrialized countries (NICs) and the rapid industrial growth of China and the countries of the Association of Southeast Asian Nations (ASEAN) have heralded an era in which Asia has become a major industrial region.

Analysis in studies on the impact of globalization on cities in developing countries has focused overwhelmingly on such rapidly industrializing areas despite the fact that FDI has flowed disproportionately to a handful of geographic regions. Between 1989 and 1992, 72 per cent of FDI flowing to developing countries went to only 10 countries, while the 48 "least developed" countries received only 2 per cent of global FDI (Broad and Landi 1996). In particular, Asia accounted for almost half of FDI to developing countries from 1983–91, and nearly two-thirds of such investment in the late 1980s (Halfani 1996). Meanwhile, parts of Asia and Latin America, as well as much of sub-Saharan Africa, have not experienced rapid export-led industrialization, and have seen their share of world trade decline.

It is this relative exclusion of certain parts of the world from participation in global trade and industry-led economic development that has prompted some commentators to write of the phenomenon of "marginalization" resulting from globalization, and to speculate about the emergence of a "fourth world" (Friedmann 1995; Castells 1996). Cities in such contexts are often discussed primarily in relation to the extremes of poverty they contain. However, the idea of exclusion is misleading for two reasons. First, a focus on declining shares in global trade and FDI in least developed countries disregards the many important ways in which LDCs interact with and remain integrated in the global economy, for example through the export of raw materials, the tourism industry and the underground economy. Second, to ascribe observed social, political and cultural changes to the phenomenon of economic marginalization is too economistic. In order to arrive at a proper understanding of the process of urbanization in LDCs, it is necessary to examine the ways in which countries interface with the global economy, as well as the social, cultural and historical legacies that each country carries into the era of globalization, including their colonial heritage and geopolitical situation. It is also necessary to broaden the view of globalization beyond the process of relocation of industry from the developed to the developing countries to include the various social, political and economic forces at work in integrating the world.

Thus, while LDCs are not major targets for FDI, and have not become important industrial producers, they nevertheless play a role in globalization and are heavily impacted by it. In order to understand the implications globalization has had for cities in LDCs, this chapter will now turn to a case study of urbanization in a city in an LDC – Phnom Penh, Cambodia.

THE CASE OF PHNOM PENH

Since the beginning of the French colonial era, and even before, Phnom Penh's development has reflected Cambodia's role in the world economy. This role has changed over time – from provider of raw materials during the colonial and immediate post-colonial era, to "sideshow" in one of the major hotspots in the Cold War, to pariah state during the era of socialist rule from 1975 to 1989. Since 1989, however, the country has reintegrated into the global economy, and this has had dramatic implications for the city's social and spatial development.

In the modern era, Phnom Penh has been the predominant urban center in the country. Yet, in the late 1980s, Cambodia was still in the process of reconstruction and rehabilitation following the utter devastation of the country during the years of the Khmer Rouge regime, which had lasted from 1975 to early 1979. The government of the People's Republic of Cambodia (PRC), which replaced the Khmer Rouge regime, was nominally socialist and depended heavily on Vietnam and the Soviet Union. In the interests of facilitating the reconstruction of the country in the short term, the government initially limited attempts at communalization of agriculture, allowed markets to function largely unregulated and condoned a lively cross-border trade with neighboring Thailand and Vietnam. By 1989 the agricultural sector had reached production levels approximately equal to the prewar levels of the late 1960s, and basic goods and services were available on the market (FitzGerald 1993). Until 1989, however, the country remained largely cut off from the global economy, in part due to a United Nations embargo imposed on the country from 1982–87.

During this period of isolation, Phnom Penh continued to play its traditional role as the political and economic center of the country. Yet the singular form of administration chosen by the government, dictated by the imperative of rehabilitation and reconstruction, gave the city a unique dynamic of development. The government owned rights to all property in the city, although residents retained relatively unrestricted usership rights. There was no real estate market, and land development was largely confined to state construction or renovation of state-owned properties. Markets functioned relatively unrestricted, yet the lack of large sources of capital or a substantial market for consumer goods inhibited large-scale commercial development. It was in this context that a series of sweeping economic and political changes transformed the urban landscape.

Today, Cambodia's low gross national product, its tiny share of world trade and its near irrelevance as a site of manufacturing production, place the country firmly within the "fourth world." Yet the tremendous social and spatial transformation Phnom Penh is experiencing is very much a consequence of the particular attributes of the global economy – the advent of new technology and resulting hypermobility of capital, the increasing difficulty governments face in regulating capital flows, the increasing pressure on states to conform to international norms of economic practice and the rapid economic development of

Cambodia's neighboring countries. Globalization has changed Phnom Penh's meaning for its inhabitants, creating opportunities for some while immiserating others.

ECONOMIC REFORM AND REINTEGRATION INTO THE GLOBAL ECONOMY

Cambodia's reintegration into the global capitalist economy began with a number of dramatic changes that took place in the late 1980s (FitzGerald 1993). In 1989, Soviet bloc aid to Cambodia declined significantly in the wake of the collapse of the Soviet Union. In the same year, Vietnam, which had been providing crucial military assistance to the country in its continuing civil war with the Khmer Rouge, withdrew all of its troops. In April of that year the National Assembly embarked on the road to liberalization of the economy. The government began to develop more laissez-faire economic policies, while simultaneously undertaking a peace initiative.

Three aspects of the reform had a particularly dramatic impact on Phnom Penh's development. The first was the decollectivization of agriculture in 1989. Under the agricultural reforms the socialist agricultural policy of collective labor under solidarity groups was disbanded, and land title was given to peasant families. The continuing lack of adequate infrastructure in rural areas and the increase in employment opportunities in cities has led to increasing rural-urban migration. Due to the high degree of primacy in the urban system, this migration has overwhelmingly been to Phnom Penh. Secondly, by decreasing the role of the state in the national economy, the reforms also had an impact on urban planning in Phnom Penh. The role of the state in planning for urban growth was weakened due to decreases in sources of state revenue, such as state ownership of industry and the state monopoly of foreign trade. The move towards austerity in government fiscal policy was backed by the International Monetary Fund (IMF). Later, the United Nations Transitional Authority in Cambodia (UNTAC) pursued a strategy of reducing government expenditure. The cumulative result of these policies was decreased government expenditure on infrastructure and social services. Finally, the reform which perhaps had the most far reaching implications for urban development was a sub-decree that overnight changed urban

residences from state property to private property. A booming real estate market soon developed as many people sought to supplement their incomes with the sale of their property. As many single-family dwelling contained multiple families, this resulted in considerable displacement.

In October of 1991, the Agreements on a Comprehensive Political Settlement of the Cambodian Conflict were signed in Paris, initiating one of the largest peacekeeping operations ever undertaken by the United Nations. For 21 months in 1992 and 1993 the United Nations implemented the United Nations Transitional Authority in Cambodia (UNTAC). The arrival of UNTAC in 1992 had several major impacts on Phnom Penh's development. First, it created a new market in the country for luxury goods and amenities. Imports increased six times in real terms during 1992 and 1993, and there was increasing investment from both domestic and foreign sources in leisure facilities. Second, it contributed to local employment both through direct employment and indirectly through the purchase of services in the country. Third, the influx of foreign capital in the form of investment and aid during the UNTAC period accelerated the process of commodification of land in Phnom Penh. Thus at the end of the UNTAC era in 1993, Cambodia had developed contacts with the global economy through trade, through the presence of international aid and lending institutions, through vastly increased transport and telecommunications links and through investments (particularly in real estate and tourism facilities). In addition, the opening of the economy set the stage for Cambodia to be a location for FDI and tourism.

CAMBODIA IN THE EMERGING ASIAN REGIONAL SYSTEM: INVESTMENT AND TRADE

The changes that have taken place in Cambodia come at a time when Southeast Asia is one of the most rapidly industrializing regions in the world. In particular, the countries of the Association of Southeast Asian Nations (ASEAN) have experienced rapid economic growth as they have become a major target for FDI in manufacturing and services. Cambodia's dependence on trade with ASEAN states is particularly marked – ASEAN countries, particularly Thailand, Singapore and Malaysia, have increasingly viewed Cambodia as a potential market, as a source of raw materials, and as a target for investment in tourism-related services and, to a lesser extent, industry. Between 1993 and 1995, trade with ASEAN accounted for 59 per cent of total exports and 69 per cent of imports (IMF 1996). These numbers would likely rise if illegal exports (primarily of natural resources) and imports (of manufactured and luxury goods) were accounted for – smuggled goods were estimated to account for 40 per cent of consumer goods imports in 1995 (*Cambodia Daily* 7 February 1995). Cambodia's major export is lumber, and the major imports are cigarettes, alcohol and petroleum.

Cambodia has seen a significant increase in foreign investment initiatives in recent years, the most important of which have involved ASEAN neighbors. These investments initially focused almost exclusively on resource extraction and hotels and tourism. The country has seen an increase in the number of international arrivals, from 17,000 in 1990 to 220,000 in 1995 (EIU 1996; Mullins 1999). The increase in the number of international arrivals has led to a major increase in investment in airport facilities and the formation of a national airline, which is a joint venture with a Malaysian company. Beginning largely in 1995 and 1996, an embryonic textile industry began to develop, centered for the most part in Phnom Penh. While textiles accounted for only $3.5 million in exports in 1994, by 1996 there were 38 factories employing 16,000 workers, and exports totaled $33 million in the first six months of the year (EIU 1997). Another major area of development has been in the banking sector. In 1995 there were 29 registered banks in the country, including joint ventures with Thai, Malaysian and French banks. There has been increasing concern over the dubious practices of some of the domestic banks, many of which are suspected of being fronts for money laundering operations.

The prevalence of money laundering reveals another major economic and social development – Cambodia's increasing links with the global criminal economy. While improvements in transport and communications infrastructure have made it possible for international criminal organizations to move goods and coordinate activities in the country, Cambodia's technically deficient and often corrupt criminal justice system has had difficulty controlling their activities. While these criminal links are difficult to document, there is evidence that Cambodia is an increasingly important site in the global flows of narcotics, smuggled goods, prostitutes and illegal immigrants.

In sum, while Phnom Penh is not an important location of industrial production, the country has come to play a particular role in the Southeast Asian regional economy – as a source of increasingly scarce raw materials, as a playground for the region's wealthy elite and as a major point in the region's criminal economy. The city's medium-term economic future is likely to be shaped by this role, although there is a possibility of continued industrial growth in the textile sector.

SPATIAL AND SOCIAL TRANSFORMATION IN PHNOM PENH

The combination of rapid urban growth, economic adjustment and the influx of cash from abroad has brought about a rapid growth in income disparity within Phnom Penh. Many of the new migrants to the city have taken up low paying jobs in the construction industry, as pedicab and motorcycle taxi drivers, as petty sellers and in other low paying service occupations. Another factor in the high rates of urban poverty is the situation of civil servants – the government has resisted pressure from the IMF to cut the number of civil servants, yet has been unable to increase wages, which currently represent only a fraction of the cost of living for a family in the city. Many civil servants have taken second and third jobs. Finally, there exists a considerable gender imbalance in Phnom Penh due to the years of war, and 29.4 per cent of households in Phnom Penh are female-headed (Royal Government of Cambodia 1994). Single mothers are at extreme risk of poverty due to their responsibilities in the household, their generally lower level of education than men and the undervalorization of female labor. These women are often forced to take on multiple sources of employment in order to subsist. The visibility of women working as petty sellers on the streets and in markets is testimony to this. In sum, the service sector has become the core of the urban economy. Service employment accounted for an estimated 74 per cent of employment in the city in 1994, and 84 per cent of all non-agricultural employment (ibid).

Another dramatic and highly visible impact of the changes that have taken place in Phnom Penh in recent years is the proliferation of squatter settlements throughout the city. In 1989, squatter settlements were virtually unknown in the city, as the government had followed a policy of allowing newcomers to the city to settle relatively freely in unoccupied buildings and on vacant land left behind by the Khmer Rouge. Yet by 1994, an estimated 120,000 people, or approximately 12–15 per cent of the city's population, were living in squatter encampments (Urban Sector Group 1994). A particularly notable feature of Phnom Penh's squatter settlements is the high proportion of residents who have moved to the settlements from elsewhere in Phnom Penh. A survey conducted by the author in 1995 indicated that some one-third of the squatters in the city lived in Phnom Penh immediately before moving to the squatter settlement. This indicates that dislocation due to the rising cost of land and housing, caused by the demand for land as a source of investment and the increased cost of housing due to the booming market for wealthy Khmer and expatriate business people and aid workers, is a major cause of squatting. The continued housing crisis and the likelihood of rapid urban growth in the future will most likely mean growth of the illegal housing submarket into the foreseeable future.

The growth of squatter settlements has occurred in contrast to an obvious increase in wealth in the city. A remarkable construction boom is underway, as opulent nightclubs, restaurants, bars, hotels and casinos catering to a new Khmer elite, foreign investors, tourists and aid workers have appeared throughout the city. This has transformed the streetscape, as the main boulevards that once were quiet thoroughfares are now lined with flashing lights and thronging with people. Major boulevards are also beginning to experience congestion, as the number of privately owned cars and motorcycles has increased dramatically. Likewise, many of the city's residential buildings are newly constructed or have recently been renovated.

Thus the social and political changes that have taken place in the last several years are manifest in the spatial changes in the city form. The city has taken on a more cosmopolitan feel while emerging social problems related to the growing income gap are increasingly visible.

CONCLUSION

This chapter has attempted to use the case of Phnom Penh, Cambodia, to demonstrate some ways in which globalization has impacted on cities in least developed countries. The intent is not to assert that cities in other LDCs can be expected to experience identical forms of urban development. The intent, rather, is to provide an example of how the particular attributes of the global economy impact on cities in LDCs through certain points of contact – FDI, improved telecommunications and transport links, links with regional economies and the increased influence of international aid and lending organizations.

In the particular case of Cambodia, globalization has had political, social and economic impacts that have affected Phnom Penh. Politically, increased links with the global economy, and the increased influence of foreign and domestic capital in policy decision-making, have meant that access to political power is increasingly a function of access to sources of wealth. In addition, increased links with market economies in the region have exposed the country to political influence from these countries. Socially and economically, links to the global economy have meant different things to different people. For those who have been able to take advantage of the opportunities presented by the changes that have occurred – mostly people with access to resources and education – links to the global economy have meant a chance to join Phnom Penh's emerging middle and upper classes. For many others – particularly low level public sector workers, those with little education and female-headed households – these changes have often had neutral or negative effects on quality of life. The emergence of large squatter settlements in the city, and the growth of the low-income service sector indicates to some degree the types of social issues the country is likely to confront in the new socio-economic order. Finally, globalization has also had cultural impacts, as the country is increasingly marketed as a tourist destination.

The degree of the impact of globalization on a city, and the form this impact will take, will differ significantly based on the regional context and historical circumstances. This complexity and variation in experience is not adequately represented by analyses that focus on the exclusion of least developed countries in the process of globalization. Theorists would do well to abandon their assumptions regarding the exclusion of these countries, and begin to look at ways in which these countries are integrated into and impacted by global economic forces. Such an approach holds greater promise in the effort to identify the potential problems and prospects faced by least developed countries in the global era.

TWO

REFERENCES FROM THE READING

Broad, R. and Landi, C. (1996) Whither the North-South gap, *Third World Quarterly*, 17, 1: 7–17.

Castells, M. (1996) *The Rise of the Network Society*, Cambridge, MA: Basil Blackwell.

EIU (Economist Intelligence Unit) (1996) *Economist Intelligence Unit: Cambodia Country Report*, 4th Quarter.

EIU (Economist Intelligence Unit) (1997) *Economist Intelligence Unit: Cambodia Country Report*, 1st Quarter.

FitzGerald, E. V. K. (1993) The economic dimension of social development and the peace process in Cambodia, in P. Utting (ed.) *Between Hope and Insecurity: The Social Consequences of the Cambodian Peace Process*, Geneva: UNRISD.

Friedmann, J. (1995) Where we stand: A decade of world city research, in P. Knox and P. Taylor (eds.) *World Cities in a World-System*, Cambridge, MA: Cambridge University Press: 21–47.

Halfani, M. (1996) Marginality and dynamism: Prospects for the Sub-Saharan city, in M. Cohen and B. Ruble (eds.) *Preparing for the Urban Future*, Washington, DC: Woodrow Wilson Center Press.

International Monetary Fund (IMF) (1996) *Direction of Trade Statistics Yearbook*, Washington, DC: International Monetary Fund.

Mullins, P. (1999) International tourism and the cities of Southeast Asia, in D. Judd and S. Fainstein (eds.) *The Tourist City*, New Haven, CT: Yale University Press: 245–60.

Royal Government of Cambodia (1994) *Report on the Socioeconomic Survey–1993/1994 Cambodia (First Round)*, National Institute of Statistics, Ministry of Planning, August.

Urban Sector Group (1994) *Twelve Month Program Proposal*, Phnom Penh, Cambodia.

19

"Medellín and Bogotá: the global cities of the other globalization"

from *City* (2011)

Eduardo Mendieta

EDITORS' INTRODUCTION

Eduardo Mendieta is Professor of Philosophy at Pennsylvania State University. In this chapter, he examines the dual role of colonialism and globalization in shaping the fortunes of Medellín and Bogotá, the two leading cities in Colombia. Since the early 2000s, both Medellín and Bogotá have experienced a rapid makeover, as local governments redirected investment into inner-city areas (Brand and Dávila 2011). Cable cars have been installed in hill-top informal settlements in Medellín; and more than a hundred kilometers of Bus Rapid Transit (BRT) were constructed in Bogotá. Both projects aimed to improve the mobility of the poor and to integrate informal neighborhoods into the urban fabric. The two cities have thus become the new public face of urban Latin America; their former mayors currently tour the globe to offer putative "best practices" to other aspiring cities across the global South.

This article adopts an historical approach to such developments, connecting the rebirth of Medellín and Bogotá to the recent history of drug trafficking in the 1980s, and also to earlier developments in the colonial era. In the mid-16th century, the major Latin American urban centers were colonized by Spain; the anti-rural biases of Spanish colonizers helped spur the growth of dynamic urban centers, often to the neglect of rural communities. After independence, rural areas continued to be ignored by the Bogotá elite, while the major cities underwent rapid industrialization and population growth. In addition, as Mendieta shows, the large rural population was displaced during the bloody ten-year civil war (1948–1958) between Colombia's two major national political parties. Peasants headed to the cities and took whatever jobs they could find. Years later, in the 1980s, Colombia's major urban industry, textiles, suffered when production shifted to Southeast Asia. Ironically, it was the drug trade that offered a new source of jobs for skilled workers in cocaine processing and distribution. Drug profits quickly seeped into the built environment, fueling the construction of parks and urban beautification in the 1980s. Through this discussion, then, Mendieta traces the rapid transformation in contemporary urban Colombia to the long history of internal violence, mass displacement, and declining industrial employment in that country.

Globalization, like Modernity, would not have been possible without colonialism. Nothing like the study of global cities reveals the imbrications between colonialism and globalization, for most contemporary global cities were at one point or another colonial outposts. Indeed, while most cities are caught in the dynamics of global processes, they also remain chained to colonial histories that have shaped not just their geographical locations and urban landmarks, but also their political and demographic fates. The cities of globalization are built on layers of history that reach beneath to pre-Columbian eras. But these layers are not inert brick and mortar, crushed, buried and paved over by cement, steel, glass and asphalt. These layers remain determining and guiding. In the following I will look at two cities from Colombia, Medellín and Bogotá, to explore, illustrate and gain some insight into the ways in which globalization and colonialism have shaped and continue to shape the cartographies of global mega-urbanization.

CONQUEST AND COLONIZATION: COLONIAL URBANIZATION

From the year of the first landing on the New World, in 1492, to the establishment of the first Spanish cities, there transpired about three decades. These three, and in some cases four decades, were extremely important in the formation of the Spanish *Conquistador* mentality. The Spanish conquistadores and the Crown came to see their empire in Latin America as a network of cities, cities that were both political acts and nodes in a network through which alone power circulated. The city was a dispositif of domination not only because it was the exemplification of a sovereign's fiat, always accompanied by all the accouterments and pomp of the court with its displays of sword and cross, but also because these cities were established as military forts and seats of political and legal power. When the Spanish did encounter major Amerindian urban centers, the Spanish established their cities on the ruins of the destroyed cities. When a conquered territory had to be controlled and exploited, cities as stopovers or byways were established that connected the network of urban centers. Within the first half of the 16th century, Spain claimed an immense territorial expanse between 40 degrees north, or what is today the southwest of the USA, and 40 degrees south, what is today central Chile and Argentina.

By the middle of the 16th century all the major urban centers in the Spanish colonies had been established: Havana, 1519; Mexico City, 1521; Quito, 1534; Lima, 1535; Buenos Aires, 1536 (re-founded in 1580); Bogotá, 1538; Santiago de Chile, 1541. These cities were tools of military and administrative control that also served as fortifications against racial mixing. The Spanish culture, with its religion and civic practices, was to be re-enacted and transplanted to the New World via the new cities, which more often than not were named after a city on the Iberian Peninsula. Thus, the Spanish conquest was marked by a hyper-urbanism and an anti-ruralism that was to mark the *criollo* (Creole) and later Latin American mentality to such an extent that some of the major literary productions in the 20th century still reflect the anti-countryside, anti-rural, anti-jungle sentiment (Rama 1996). By the middle of the 18th century the Latin American continent had been divided into administrative centers whose primary role was to facilitate the transfer of resources from the New World to Spain.

The South American subcontinent is cut in half vertically by the Andes that sit on top of it like a dorsal fin. The long distances, fractured territory and lack of viable ways for crossing the continent along a horizontal axis made commerce and political power flow vertically and through the coasts. The interior of the continent was left unexplored, essentially. The colony and imperial administrative center clung to the forts on the coast or fortified cities built close to or along major rivers that empty into either of the oceans. This pattern of colonial administration determined the way in which future independent nations would be relatively ineffective at imposing political sovereignty over territorially unified geographies. The Spanish colonies were characterized by a form of sovereignty by proxy—the Viceroys were representatives of the Crown and they had to be Spanish born—that never matched or extended over any territory in a noticeable or permanent way. Territory and sovereignty remained uncoupled or incommensurate.

POLITICAL INDEPENDENCE WITHOUT TERRITORIAL INTEGRATION

If during the colonial period sovereignty was spectral, that is, a mere semblance of the trans-Atlantic court, then during the post-independence period political sovereignty remained elusive. Colombia occupies a distinctly strategic place on the South American subcontinent. It sits on the northeastern corner of the continent, jutting out northwards with a narrow land strip that connects Central America and South America. This land mass would prove a major source of contention and eventually a destabilizing role. The country thus has access to the Atlantic and the Pacific. The country is actually divided into five major regions, each with its own climates and patterns of colonization. The Pacific coast is covered with thick, luscious jungles that suffer from torrential rains. Its impenetrability and distance from Bogotá made it a destination for escapee slaves and indigenous populations displaced by the Spanish Conquest. The Cauca River irrigates the Cauca region, which sits between the Western and Central mountain ranges. This same river empties into the Magdalena River, which transverses between the Central and Eastern mountain ranges. The Cauca links the southwest with the northwest of the country, providing a most advantageous waterway to the Atlantic. This region has three of the largest urban centers in Colombia, and two of the wealthiest cities in the country: Calí, Medellín and Barranquilla, which is a major port city on the Atlantic, close to Santa Marta, a port city that had been a fort built to defend the gold shipping industry to Spain.

The Cauca valley has been one of the wealthiest regions of the country. During the colonial period, it was the source of the gold that was transported to Spain. In fact, gold was the primary export from Colombia to Spain during the colonial period. Most of the gold was collected from run-off rivers that empty into the Cauca, which in turn was used to link the south to the Atlantic coast (Hylton 2008). In the 19th century, gold was replaced by coffee as the primary and major export. In fact, as the region made the transition from mining to agriculture, processes of land distribution gave rise to patterns of violence and clientelism that have continued to shape the region. While most of the large land holdings in the country were distributed close to the capital, north and west of it, Antioquia by contrast was marked by small and medium-sized land holdings. It was this pattern of small land holding that allowed the region to become rapidly the major coffee producing area in the country. The small landowner, with his family, toiled the land producing enough food to subsist but also enough coffee to sell on the market. Coffee growing and processing are labor intensive, and given the absence of large masses of slaves, the self-sustaining farm became the ideal agricultural unit. In this way, then, the region was able to develop and retain primitive capital that allowed it to invest directly in the development of the regional infrastructure bypassing the capital city. Eventually, these financial reserves would be used in the middle of the 20th century to launch Medellín into the second industrial center in the nation, surpassing the capital in textiles and heavy machinery (Hylton 2008).

The Cauca region, headed by Antioquia, developed a distinct non-Metropolitan, anti-Bogotá identity,

namely, the *paisa* identity. *Paisas* identify as unpretentious, non-cosmopolitan, fervent Catholics, who are more mestizo and Creole than mulatto or indigenous. *Paisa* is a distinctly racial identity, which has affected the way in which new urban immigrants have been differentially integrated in the city (Wade 1995). They also identify by their work ethic and entrepreneurial spirit. The local elites emerged connected to the colonial past in a way that made them independent from the political elites from Colombia. In other words, the Cauca region has developed in relative autonomy from the seat of political power in the capital city, which at times has put them in tension and in direct contradiction with the metropolitan elites of the capital city.

THE HOT WARS OF THE COLD WAR

One of the most important 20th century events in Colombian history was the *Bogotazo* and the unleashing of what is called *La Violencia*. As was noted, political power was begrudgingly and contentiously shared by the two parties, whose ideologies were almost diametrically opposed. Any other political position or party was from the outset excluded or delegitimated. This bipolar and tense confrontation would periodically erupt into local violence, in which local political chiefs would or would not consent with the political decisions reached in the capital. Every election some violence would erupt in the provinces when political decisions were reached that challenged or displaced the local political bosses.

What is distinctive about *La Violencia* is its rural character and the relative absence of its violence in the major metropolitan centers. While the cities enjoyed peace and growth, especially Medellín and Bogotá, the rest of the country suffered a violent process of re-structuring. Yet, this rural violence would not have been possible without the already historically established absence of the state from the rural areas of the country. The liberal auto-defense military outfits not only provided security, they also became units that provided a minimum of civil services: they helped build schools, roads and hospitals. After the 1960s, when the FARC was established, the guerrillas were the face of a para-government that provided those services that the capital city had not brought to the provinces in 150 years. It can be said that the central government left those regions to fend for themselves. This default semi-autonomy was further enhanced with the political compromise that emerged from the solution that was worked out to end the civil war.

In 1958, the leaders of the two parties agreed to a set of rules that became known as the National Front, which specified explicitly the compulsory half and half sharing of the elected and appointed positions, and that each party would alternate heading the presidency. The country would be run by political elites from both Medellín and Bogotá, without consultation, inclusion or participation of other political sectors of the country. As Forrest Hylton has called this period, which lasted from 1957 until 1982, it was a time in which 'political lockout' was institutionalized. The National Front effectively defused the electoral process, depoliticized the public and turned over the government to oligarchies that ruled the country for most of the 20th century. Every presidential candidate and actual president has had a father, or close family member, in the executive branch over the last 80 years. As citizens were locked out of the government, the rural regions of the country were left to their own means. Meanwhile, the country was undergoing a massive urbanization, partly due to the displacement of farmers from the violence-rent areas, and an unprecedented economic growth, even as most of Latin America languished or regressed economically (Bushnell 1993). Economic growth has been matched by a lack of distribution of wealth and the absence of the distribution of social benefits through a state even marginally attentive to the needs of its population. The 10 years of *La Violencia* and three decades of neutralizing political apportionment resulted in a polity with one of the lowest voter turnouts in the continent, as well as the impression that electoral or civil society politics are useless in the processes of political change. Politics had been reduced to cronyism, clientelism and nepotism by the ruling elites.

THE JEWEL CITIES OF INFORMAL GLOBALIZATION: MEDELLÍN AND BOGOTÁ

Medellín, which sits sprawling on the slopes of the Western and Central *cordilleras* has been a hub of commerce since the colonial period. It has been a stopover city on the way from the south to the Atlantic coast. Its local wealth, recycled in the local economy,

allowed the local elites to develop its own industrial base independently of the capital's allocations or sanctions. During the second half of the 20th century Medellín underwent a rapid and traumatic growth. Some of this growth was precipitated by the violence that was displacing peasants at the head of the Cauca valley and the adjoining mountains. The rapid industrialization of the city, however, was also a major magnet. Accelerated urbanization brought with it severe challenges. Most of the new arrivals settled on colonias on the slopes of the mountains surrounding the Antioquian capital (the name of the state). Most of these new arrivals tended to be more mestizo and darker than the older urban dwellers. Medellín grew fivefold between 1951, at the height of *La Violencia*, and 1985, when the National Front came to an end, from 358,189 to 2,095,147 inhabitants. While the city was the national seat for the largest textile industries, with Coltejer leading the textile industry sector, the city was rocked by persistent unemployment and visible poverty. Many of the recently arrived displaced farmers had to make ends meet by engaging in whatever work could be had. Many of these youths would find profitable careers as smugglers of cigarettes and electronic contraband that would flow up and down the Cauca River from and through Calí through Medellín to Barranquilla and Santa Marta. Recently urbanized peasants re-energized a line of commerce and contraband that had its political–geographical roots in the colonial period. Gold and coffee trade routes became marijuana, cigarettes and coca trade routes. Many of the future drug lords and capos, such as Pablo Escobar, had learned their ways and established their own transport networks during these years of urban expansion in Medellín (Hylton 2008).

An important factor that over-determined Medellín becoming the capital of coca production was that as textile production and maquiladora assemblage shifted from Latin America to South East Asia in the early 1980s, the industrial basis of the city was decimated. The textile workers, technical personnel and middle management bosses were left unemployed. Many of these unemployed technicians would be employed by the drug cartels, which needed skilled workers and fairly sophisticated chemical facilities. Thus, a convergence of factors has led Medellín to become the largest producer of coca for the world. It is approximated that about 80% of the coca consumed in the world is produced and processed in Colombia. Most significantly, most of the coca is grown now in Colombia, as opposed to being imported from Ecuador and Bolivia, where it used to be grown. The city itself has benefited greatly from the immense amounts of drug dollars that have flowed into the economy. Some of the techniques of laundering the drug money have provided for urban renewal projects, the building of parks, public facilities and the general beautification of the city (Hylton 2007).

Meanwhile, Bogotá lives off of the national capital that it refuses and is unable to distribute and allocate to the starved provinces. The capital city has benefited from a sustained industrialization and the largesse of a state controlled by racially discriminating elites who have failed or been unwilling to take up the causes of the rural poor. For most of the 20th century, Bogotá has lived as if isolated and embattled on the heights of the Andes. The violence of *La Violencia* was short lived and most of the guerrilla violence of the last three decades has not touched it.

Today, Colombia has one of the largest displaced populations in the world, second only to that of Sudan (Hylton 2006). Most of these displaced peoples come from the regions where it is believed that there are reservoirs of oil. In addition, continuing the practices of *La Violencia*, paramilitary terror is used to dispossess and appropriate land. Colombia is undergoing a major regressive land reform in which the small farmer is being forcefully and violently absorbed by the large landowner. Meanwhile, the FARC, after almost half a century, remains the only rural source of medicine, schools, security and the possibility of a job that may provide for a future education. The USA's continued militarization of its Colombian foreign policy exacerbated the evisceration of the political process in the country since the inception of the Cold War. Today, this very same policy has entered a complex dynamic in which US military aid buttresses a disengaged central government, while most of the country is left to be ruled by violence, terror and extortion. Not unlike the violence of the conquistadores that was sanctioned by Papal bulls, the contemporary violence of paramilitaries and narco-traffickers that is sanctioned by neoliberalism and economic profiteers of US elites is a tool of dispossession and means to neutralize the work of political deliberation. Again, not unlike the middle of the 20th century shortly after *La Violencia*, the contemporary violence of the paramilitaries and the militaries in search of so-called narco-guerrillas,

has become a means to displace farmers, indigenous populations and Afro-Colombians from lands that may have some of the largest reserves of oil in South America. Thus, the global war on terror is synergizing in Colombia into a war for oil and coca.

CONCLUSION

I have sought to illustrate how many of the global processes that insert cities into global networks capitalize on patterns established during the colonial period. I have looked at how the political geography of Colombia has determined the divergent but parallel fates of two of its major cities, which unwittingly have become players in what we can call the informal or black market globalization. Medellín has become infamous because of its drug lords, and the fact that it produces most of the coca in the world market. The city also has become globally known because it lived through one of the most violent periods in the history of any world city. We cannot understand how Medellín became the global capital of coca if we do not understand the ways in which political sovereignty has been uncoupled from territorial unification since the colonial period, which was due precisely to the specific dynamics of European colonization. Half a century of civil war in Colombia is not just the result of colonial inheritances, it is also the consequences of Cold War and neoliberal policies that have dismantled the country's abilities to determine its political fate and to use its resources for the health of its citizenry. The Cold War metamorphosed into the War on Drugs, which has now metastasized into the Global War on Terror. Colombia has been the one Latin American country where all these three global wars have been fought and it is the one in which they have become one. Medellín is the urban face of these global wars become one.

REFERENCES FROM THE READING

Brand, P. and Dávila, J. D. (2011) Mobility innovation at the urban margins: Medellín's Metrocables, *City*, 15, 6: 647–61.

Bushnell, D. (1993) *The Making of Modern Colombia: A Nation in Spite of Itself*, Berkeley: University of California Press.

Hylton, F. (2006) *Evil Hour in Colombia*, London: Verso Books.

Hylton, F. (2007) Remaking Medellín, *New Left Review*, 44, March–April: 71–90.

Hylton, F. (2008) Medellín: The peace of the pacifiers, *NACLA Report on the Americas*, 41, 1: 35–44.

Rama, A. (1996) *The Lettered City*, trans. J. C. Chasteen. Durham, NC: Duke University Press.

Wade, P. (1995) *Blackness and Race Mixture: The Dynamics of Racial Identity in Colombia*, Baltimore: Johns Hopkins University Press.

PART THREE

Relations

Plate 18 Toronto aerotropolis

Source: Roger Keil

INTRODUCTION TO PART THREE

This section explores relational properties of cities and the conceptual, methodological, and empirical challenges of grasping global urban connectivity across scales, contexts, and historical periods. In some sense, all approaches to globalized urbanization involve some sustained consideration of the connections among cities in a worldwide context. Indeed, as previous chapters illustrate, the relationships between global cities have been imagined in diverse ways, with reference to hierarchical or horizontal forms of connectivity, governed variously by empires and nation states, and established through diverse modes of political and economic power, across tangled historical geographies of world systems in the *longue durée*. The internal conditions of individual urban regions are shaped by their roles in the global hierarchy of economic activities. In turn, the external relationships of cities have been conditioned by their capacities to articulate wider networks of global flows. It is these local capacities, meanwhile, that are often a central target for municipal or regional economic development strategies. As Taylor and Derudder (2016: 41) note, such strategies involve "an urban process of making spaces of flows: these are myriad exchanges between cities in a field of interactions." For Taylor and Derudder (2016: 98), global cities involve a new politics of the "hinterworld"—the attempt to enhance "the pattern of a city's connections across the world."

The focus on global urban connectivity has not only inspired research on inter-urban networks, but has also stimulated a reimagination of the global city region. With its spatially differentiated geography of center, periphery, and in-between-spaces, the global city region functions "as key spaces where relational interconnectivities take hold to redefine metropolitan place and globalized space in equal measure" (Keil and Addie 2015: 892). The chapters included below explore these issues and the broad question of inter-urban relations in four major ways.

First, inter-urban relations are studied by examining locational strategies of firms (see Prologue by Taylor and also Ch. 22 by Beaverstock et al.). In this case, cities and firms (or their branch offices) form two layers of nodes in the global urban network. As large producer service firms often locate their branch offices in globalizing cities, these cities are woven together into a global urban network, with business transactions, travels of employees, and flows of information and expertise among branch offices. Network analysis is a frequently used method to analyze the structure of global urban networks with data of branch office locations of global firms. The idea of an interlocking urban network brought by locational strategies of firms is the basis for a thorough empirical analysis of the centrality of Frankfurt and London in the global network of finance (Beaverstock, Hoyler, Pain and Taylor, Ch. 22).

Second, there is an emphasis on physical and material infrastructures as the skeleton of the global urban network. In many instances, these infrastructures are quite literally the preconditions of the kind of economic firm-to-firm relationships emphasized by Taylor and his colleagues. Whether it is through the mobility provided through air travel or through the communication enabled by "grids of glass," (Ch.23) the global urban network in this view rests on material relationships that are vehicles of economic activity. However, as the contributions to this section illustrate, the viruses (as elaborated on in Ch. 24) and packages (as shown in Ch. 26) that are variably distributed through the network are not just inactive props of a global society and economy, but are themselves agents, or actants, whose agency (and that of those who are governing them) reflects back onto the constitution of the network itself. This means that there is a wide variety of objects moving through the "space of flows" regulated by various institutional environments. Such thinking on the materiality of networks has, of course, been de rigueur in urban studies for some time now.

Third, relational properties of cities are envisioned through social, institutional, and functional connectivity among places. Manuel Castells (1996) expressed this type of relationality most concisely in his contradistinction of a space of flows and a space of places. For Castells, in the relationships between the space of *flows* and the space of *places*, a new type of society, *network* society, emerges. This society is both urban and global. Contributions to this section investigate the interactions of these spaces through their functioning and failing connectivity: technical, material, social, and ecological.

Fourth, we can distinguish "mobilities" as another way to study relational properties of cities. The notion of mobility explicitly positions itself vis-à-vis territoriality in the urban network. The mobilities approach conceptualizes "the social content or movements or people as objects from place to place at various scales and the immobilities and moorings that underpin and challenge these dynamics" (McCann and Ward 2011: xx). In their chapter, McCann and Ward (Ch. 29) discuss accordingly policy mobilities by examining how policy templates circulate among urban decision makers and practitioners.

In later sections of the book, we will see other examples of urban connectivity. Global urban connectivity is central to cultural production and representation in globalizing cities (Part 6), and also to the ongoing reimagination of the frontiers of global urban research (Part 7).

REFERENCES AND SUGGESTED FURTHER READING

Castells, M. (1996) *The Rise of the Network Society*, Oxford: Blackwell.

Graham, S. (ed.) (2010) *Disrupted Cities: When Infrastructure Fails*, London: Routledge.

Keil, R. (2011) Transnational urban political ecology: Health and infrastructure in the unbounded City, in G. Bridge and S. Watson (eds.) *The New Blackwell Companion to the City*, Oxford: Wiley-Blackwell: 713–25.

Keil, R. and Addie, J. P. (2015) It's not going to be suburban, it's going to be all urban': Assembling postsuburbia in the Toronto and Chicago regions, *International Journal of Urban and Regional Research*, 39, 5: 892–911.

McCann, E. and Ward, K. (2011) Introduction. Urban assemblages: Territories, relations, practices and power, in E. McCann and K. Ward (eds.) *Mobile Urbanism: Cities and Policymaking in the Global Age*, Minneapolis and London: University of Minnesota Press: xiii–xxxv.

Taylor, P. J. and Derudder, B. (2016) *World City Network: A Global Urban Analysis*, 2nd Edition, London and New York: Routledge.

20 Prologue "Specification of the world city network"

from *Geographical Analysis* (2001)

Peter J. Taylor

The world city network is an unusual form of network with three levels of structure: cities as the nodes, the world economy as the supranodal network level, and advanced producer service firms forming a critical subnodal level. The latter create an interlocking network through their global location strategies for placing offices. Hence, it is the advanced producer service firms operating through cities who are the prime actors in world city network formation. This process is formally specified in terms of four intercity relational matrices—elemental, proportional, distance, and asymmetric. Through this specification it becomes possible to apply standard techniques of network analysis to world cities for the first time.

Topologically, a network consists of nodes and links that display a pattern of connections. The world city network is an unusual social network. This is not just because of its large scale. The problem arises with the idea of city as actor. In other social networks, individuals in roles or nation-states through governments can be reasonably interpreted as the key agents in their network production and reproduction. Cities are different: they do have decision-making administrations and competition between cities is part of world city network formation, but it constitutes just a small component of the overall process. In the original conception of world cities as control and command centers it is the multinational corporations who are the key actors, the loci of the decisions behind the control and command. Thus, as well as the world economy as the supranodal level, there exists a critical subnodal level within the network: it is the behavior of firms

within and across cities that creates world cities as the nodes of the network. World city network formation is more an outcome of global corporate decisions than the collective works of urban policymakers. This is critical for theory and practice: this subnodal process makes the world city network an unusual and difficult-to-specify social network.

The starting point is that it is large global service firms, not cities per se, that are the key actors in world city network formation. In order to carry out their business they seek out knowledge-rich environments, world cities, in which they can prosper. The success of each firm is dependent upon their location strategies of having offices in selected world cities. These are the office networks of firms through which they provide their global service. The ideal is to be able to produce a seamless service for every client whatever the locational scale and complexity of a given project. Each world city, therefore, is constituted as a particular mix of advanced producer service offices. In short, the world city network is a complex amalgam of multifarious office networks of corporate service firms.

The world city network is an interlocking network but, as indicated, it is not an exact analogue of other such networks. In particular, this relates to the cities' being nodes but not being primary actors. This specific feature of the world city network has important implications for both theory and practice: in the world city network the nodes may operate a weak "adjunctive competition" but the network has been able to develop because of the cooperative behavior within the prime agents, the firms, through their world city offices.

21
"Local and global: cities in network society"

from *Tijdschrift voor economische en sociale geografie* (2002)

Manuel Castells

EDITORS' INTRODUCTION

Manuel Castells has been one of the most influential voices in urban studies since the 1960s. He currently holds the Wallis Annenberg Chair in Communication, Technology, and Society at the University of Southern California in Los Angeles. Among other things, he revolutionized the ways in which scholars have seen the relationships of cities and urbanization with capitalism and the state, has laid the groundwork for the study of urban social movements, and has drawn our attention to the growing significance of information in what he calls the "network society." The present chapter is about an important aspect of that network society, the relationship between the spaces of places and spaces of flows. As he outlines below, the ways cities are constituted today are changed by technological development, globalization, and the rise of networking as a mode of social organization.

INTRODUCTION: THE NETWORK SOCIETY

I explore two sets of relationships – that between the local and the global, and that between certain dimensions of identity and functionality as they impinge on spatial forms – and will try to show how they interact in the spatial transformation of the information society. Some people call this the network society. We are indeed living in a period of historical transformation. In my analysis, this process involves the interaction of three features that, though distinct, are related to each other. The first is the revolution in information technology that started in the 1970s and then expanded all over the world. The second is the process of globalisation, which incidentally is not only economic. The third feature is the emergence of a new form of organisation that I call networking.

This is the specific kind of power networking that works through information technology. This power networking is changing the way we perceive, organise, manage, produce, consume, fight and counter-fight – embracing practically all dimensions of social life. The interaction between the revolution in information technology, the process of globalisation and the emergence of networking as the predominant social form of organisation constitutes a new social structure: the network society. The network society has many different manifestations, depending on country, culture, history and institutions. However, some basic commonalities emerge when we consider specific features of this network society. This chapter focuses on the spatial transformation. We need a theory of spatial forms and processes, which can be adapted to the new social, technological and spatial context in which we live.

KEY SPATIAL PROCESSES

First, I think we are rapidly moving in the direction of an urbanised world. By mid century, between two-thirds and three-quarters of the total population will be living in 'some kind' of urban agglomeration. Certainly we will not be living in the countryside as we now know it. We may be living in urbanised villages, though. This process of urbanisation is concentrated disproportionately – and increasingly so – in metro-politan areas of a new kind. These urban constellations are scattered throughout huge territorial expanses. Today, we have not only metropolitan areas but also big 'metropolitan regions', and these are very special indeed. They are a mix of cities, countryside, centre and periphery – they are not necessarily part of one urban continuity. Some people call them edge cities, others call them conurbations. The novelty of this process lies in the ability to connect functionally a huge number of people and activities throughout a large expanse of space. That space is constantly being remodelled and reconstructed by the transformation of the communication, transportation and telecom-munications systems.

An empirical definition of what a real conurbation is has changed for the USA, at least. It used to be the telephone network, but now, with the internet, it has become global so you do not have any specificity. Now, it is the television market. What the television station considers as their market, that is the city. But it is not a city, of course; it is a market link to a connection between residential and working places. In really big cities – not little European conurbations – in Hong Kong, Shenzhen, Macau, Zhuhai and the other cities of the Pearl River delta, all the way to Canton, about 65 million people work and live in a highly interrelated functional area. Certainly not everybody does every-thing in that area, of course. Take Japan – the conurbation of Tokyo, Yokohama and Nagoya (now functionally extended to include Kobe and Osaka, and Kyoto to a large extent) is another huge urban constellation.

This is the kind of phenomenon I do not call a city; my current term is metropolitan region (which can-not be more than a provisional one until we find a serious empirical interpretation of what is going on). It is a new kind of urbanised agglomeration that we are generating. These magnets of economic, cultural, political and urban growth are absorbing more and more of their population and activities in their hinter-land. In fact, they become nodes in global networks

of cities. Indeed, advanced telecommunications, the internet and fast computerised transportation systems (I remind you that planes, trains and ships are all com-puterised transportation systems) allow for a simul-taneous spatial concentration in huge areas and thus for decentralisation. Therefore these systems are introducing a new geography of networks and urban nodes throughout the world, throughout countries, between metropolitan areas and within metropolitan areas. This is the new urban geography.

CURRENT SOCIAL PROCESSES

On the social side, there is a trend for social rela-tionships to be characterised simultaneously by two processes: individuation (not individualisation but the building of meaning vis-à-vis the individual project) and communalism. Both processes use spatial pat-terning and online communication. Individuation is both spatial and virtual: physical proximity and online connectivity. The same applies to communalism: vir-tual communities and physical communities develop in close interaction. We now have enough empirical research to go beyond these fantasies about virtual communities being different from physical commu-nities in a world in which the internet has become a key communication mode. We have both online and off-line social interaction, creating a hybrid pattern of sociability. Something else that should be emphasised – though not a spatial phenomenon, it does have extraordinary consequences for spatial structure and dynamics – is the crisis in the patriarchal family. This has different manifestations depending on the culture and the level of economic development. This crisis gradually shifts sociability from family units, in the traditional sense, to networks of individualised units. Most often, these are made up of women and their children in relationship to other women with their children, but these units may also consist of all kinds of individualised cohabitation partnerships. This has extraordinary consequences for the uses and forms of housing, neighbourhoods, public space and trans-portation systems.

This crisis coincides with changes in the business world. Here, we see the emergence of the network enterprise as a new form of economic activity, which is a highly decentralised yet coordinated form of net-work. At the same time, we see the emergence of decentralised and coordinated management patterns.

This network enterprise is not a network of enterprises. Rather, it consists of enterprises that are internally organised as networks and then connected with other networks of other enterprises. The network enterprise has very substantial spatial consequences. The most important is a return to the work-living arrangements of the pre-industrial age or of the period of industrial craft work. Interestingly, these arrangements for working and living in the same place often take over the old industrial spaces, transforming them into informational production sites. For instance, in San Francisco's multi-media gulch, the city's last remaining industrial buildings were transformed into spaces for multi-media production sites. What is multi-media? Manufacturing or services? It is both! It is the production of dreams, which is the most powerful form of manufacturing in our world. It is a very material production in many ways, but it is software, so it is informational production. It is a different kind of manufacturing. It is a production organised in terms of the people living there, working there and socialising there. And whether we refer to experiences in London, Tokyo, Beijing, Taipei, Barcelona or Helsinki, we find exactly the same kind of work-living arrangements in the advanced software-based industries.

Urban areas around the world, another key trend, are becoming increasingly multi-ethnic and multi-cultural. Another trend is that the global criminal economy is local at the same time. It is solidly rooted in the local urban fabric. The cities are being taken over in many ways by this global criminal economy. In other words, the global criminal economy does not start from localities and depressed areas. It is a global business that penetrates the urban areas in different ways. It reaches into the poor ghettos but at the same time also links up to money-laundering and other activities. The breakdown of communication patterns between individuals and between cultures is another major trend. This leads to the emergence of defensive spaces, which are in fact at the root of the formation of sharply segregated areas: gated communities for the rich, territorial turfs for the poor.

THE CHANGING NATURE OF PUBLIC SPACE

At the same time, in reaction to the trends of suburban sprawl, major metropolisation and individualisation of residential patterns, urban centres and public space become critical expressions of local life. This is in fact a reaction and an interaction. In other words, public space is really critical. In most planning projects everywhere in the world, the revitalisation of urban life and of the city as a communicative space has become paramount. In fact, it is becoming the most salient selling device for private residential development. In principle, support for the vitality of public space is still a major trend. I say only in principle, because the commercial pressures and the globalisation of tourism and business travel are mimicking urban life in many cities rather than actually rebuilding urban space. Many public spaces around the world – and thus in your cities too – are also being transformed into theme parks, where symbols rather than experience create a life-size urban virtual reality. Ultimately it is the next best thing to being projected in the media and then selling the city. In that sense, the Las Vegas phenomenon – building all the greatest cities in the world in Las Vegas – can also be reproduced, whereby the greatest cities of the world become Las Vegas themselves. It is a consequence of the commercialisation of public space, of the massive diffusion, and of the suburban and exurban sprawl.

On the other hand, it is a consequence of the increasing individualisation, whereby consumption items become individually appropriated. Thus, you have individualisation of the residential and work experience, on the one hand, and individualisation of the consumption of the city, on the other. All in all, the new urban world seems to be dominated by a double movement: inclusion in trans-territorial networks and exclusion by the spatial separation of places. The higher the value of people and places, the more they are connected in interactive networks; the lower their value, the lower their connectivity. In extreme cases, some of the places are by-passed by the new geography of networks. This is indeed what happens in depressed rural areas around the world, in declining regions, or in urban shantytowns. Then the infrastructure of these networks – not only of communication networks, but also of water, electricity, roads or advanced communication systems – reinforces this segregation. The work on splintering urbanism clearly shows how these spatial and social trends towards splintering spaces are in fact materially articulated and reproduced in the design of telecommunication infrastructures [see Graham in Ch. 23]. In this way, the world is not socially segregated simply by the market or by people moving or not moving. It is also segregated by the spatial layering of major communication infrastructures – for example, where you have fast internet access or not, where you have fibre-optic cable or not, where you have advanced transportation systems or not. In Europe, the localities by-passed by the high-speed trains are being segregated.

THE EMERGENCE OF THE NETWORK-STATE

The constitution of these mega-metropolitan regions without a name, without a culture, and without institutions weakens the mechanism of political accountability, of citizen participation, and of effective management. In other words, there is increasing contradiction between the actual spatial unit and the institutions of political representation and metropolitan management. On the other hand, however, local governments in the age of globalisation emerge as flexible institutional factors that are able to react, to adapt more quickly to global trends. Whatever is left of political legitimacy, which is not much, is left mainly at the local level. So, a new form of state emerges. In this particular network configuration, the network-state becomes the actual institution that is managing cities and regions in our context. In that sense, local governments become a node in the chain of institutional representation and management. The local authorities are able to input the overall process, but with added value because of their capacity to represent citizens at a closer range.

What we observe is an increasing gap between the actual unit of working and living in the metropolitan region on the one hand, and the mechanism of political representation and public administration on the other. In this context, urban social movements have not disappeared by any means; they have merely mutated, essentially around two main lines. The first is the defence of the local community affirming the right to live in a particular place and to benefit from adequate housing and urban services in that place. The second, and I would say probably the most proactive, is the environmental movement. It acts on the quality of cities within the broader goal of achieving a better quality of life. The environmental movement is not simply a movement for a better life but for a different life. In that sense, it is as much a cultural movement as it is a traditional urban economy-oriented movement.

In my view, these are the main spatial trends. They are based on pure observation and certainly can be challenged by different observations. But this is what I would distil from my observation of current changes worldwide in terms of the spatial transformation. Let me try to make sense of what is going on with the help of some concepts that bring the discussion to a more analytical level.

AN ANALYTICAL FRAMEWORK

I think that the transformation of cities in the network society can be organised – in terms of the building blocks of a new theory – around three bipolar axes. The first relates to function, the second to meaning, and the third to form. Functionally speaking, the network society is organised around the opposition between the global and the local. Dominant processes in the economy, in technology, the media and authority are organised largely in global networks. But day-to-day work, private life, cultural identity and political participation are essentially local and territorial. Now, cities as communication systems that work throughout history are supposed to link up the global and the local. But that is exactly where the problem starts. Cities are in fact being torn by these two conflicting logics that destroy the city as a sociospatial communication system when they try to simultaneously respond to logic. So, while we have to be in the networks, we have to be at the same time rooted in locality and identity. Communalism refers to the enclosure of meaning in a sheer identity. That enclosure is based on a system of values and beliefs to which all other sources of identity are subordinated. Society, of course, only exists at the interface of individuals and identities mediated by institutions. This interface and this mediation are at the source of the network-state.

The network-state relates to the grassroots, to the people themselves. Civil society was always seen in relationship to the state. Social integration is now problematic because of the urban transformation represented by a third and major axis of opposing trends, this one concerning spatial forms.

FLOWS AND PLACES

We have dealt with function, and we have dealt with meaning. Let us now look at the issue as a question of form. In terms of form, the major bipolar opposition is between what I call the space of flows and the space of places. In the space of flows, separate locations are linked up electronically in an interactive network that connects people and activities in different geographical contexts. Now, the spatial flows – let us say, the financial networks, the international production networks, and the media networks – are not a-territorial. They consist of territories which are distant, which are linked to different geographic hinterlands. But they are electronically connected; their function and their meaning come from their connections. Thus, they do not exist separately. In that sense, they are not purely electronic networks. The electronic networks link up the specific places, and it is this hybrid space that is the space of flows. The space of places organises experience and activity around the confines of locality. What is critical in our society is that cities are structured and restructured simultaneously by the competing logics of the space of flows and the space of places. Cities do not disappear into the virtual networks. Rather, they are transformed in the interface between electronic communication and physical interaction. They are transformed by the combination in practice of cities, networks and places but without fully integrating them.

TYPES OF URBAN INTERACTION

The key challenge for the new urban civilisation is to restore communication. To restore communication means the building and development of communication protocols. This is not a metaphor; it is a concept from information technology theory. Which kind of communication protocols? Let me present three of them. The first kind is the physical protocol of communication. How do you restore communication in a fragmented sprawl? Well, you have to introduce new forms of symbolic nodality that will identify places in this endless sprawl. Symbolic nodality reconstructs spatial meaning in the city. That is why architecture again becomes very important. Architecture always had been about the marking of places. Urban design has always been about the marking of urban forms in relation to culture and meaning.

The second level of urban interaction refers to social communication patterns. That is, it concerns how people can start being together, sharing cities without being able to speak to each other and without

going through the public institutions. So, in the practice of the city, the answer lies in public spaces, including what I call the social exchangers or communication nodes. These are the stations, airports, all those places where people have to bump into each other because they have to change trains or planes or buses. And these are the squares, which have some kind of social activity. These spaces are in fact the devices with which to reconstruct sharing communication and therefore city life. I call this level of urban interaction the sociability of public spaces in the individualised metropolis.

The third level of urban interaction refers to the new combination of electronic communication and physical face-to-face communication as new forms of sociability. Virtual communities as networks of individuals connect to face-to-face sociability, thereby recreating some form of sociability. The analysis of code-sharing in the new urban world requires the study of the interface between physical layout, social organisation and electronic networks of communication. In this sense, the analysis of the new network of spatial mobility in the mobile phone era is a critical frontier for the new theory of urbanism. The mobile phone also transcends the boundaries of space. The places of the space of flows – that is, the corridors and the halls that connect places around the world – will have to be understood as exchangers and social refuges, as homes on the run as much as offices on the run.

Under these conditions, a dominant trend emerges towards the disintegration of cities as communicative devices. The beginning of general urbanisation could be at the same time the end of urban civilisation, which is based on communication and sharing, even sharing in a conflictive manner. This is the current situation, but at the same time there is a counter-offensive.

CONCLUSION: THE INTEGRATION OF URBAN LIFE

Urban life should be seen as a world of social interaction and meaning operating on the basis of the appropriation of a space by sociability and by the society that goes beyond the functionality of integration in the global networks. In that sense, the process of reconstructing urban life is the process of reconstruction of the city as a communication system in its multi-dimensional sense. Restoring functional communication through metropolitan planning, providing spatial meaning through a new symbolic nodality created by innovative spatial projects, and reinstating the city in its urban form through the practice of urban design, focused on the preservation, restoration and construction of public space – these are the critical issues in the new type of urbanism. There are a number of very good examples that always combine an emphasis on public space, competitiveness in the global networks, a strong emphasis on local governments and citizen participation, and the ability to reintegrate symbolic nodality, symbolic representation in the reconstruction of space. Ultimately, the meaning of cities depends on the governance of cities. It is a political problem in the traditional sense in the sense of the polis, the challenge is to reconstruct society through the practice of living together, in the process of communicating with each other in the urban civilization.

22

"Comparing London and Frankfurt as world cities: a relational study of contemporary urban change"

from a report for the Anglo-American Foundation for the Study of Industrial Society (2001)

Jonathan V. Beaverstock, Michael Hoyler, Kathryn Pain, and Peter J. Taylor

EDITORS' INTRODUCTION

In this contribution, Jonathan Beaverstock, Michael Hoyler, Kathryn Pain, and Peter J. Taylor put the conceptual and methodological thinking introduced by Peter Taylor in this section's Prologue to work in a study of London and Frankfurt around the time of the projected move of the European Central Bank to the German financial center (see also Ch. 47). This topic has gained renewed relevance through the more recent Brexit vote in 2016 in Britain. This very brief excerpt from a much longer study done for the Anglo-German Foundation for the Study of Industrial Society at the time is an exemplary utilization of the comparative methodology using the "network model of inter-city relations." This methodology was deployed widely and in variable ways by a large group of researchers at the Global and World Cities (GaWC) research center at the University of Loughborough to which all authors have had a relationship at one time. Presently, only Michael Hoyler remains at Loughborough while Jonathan Beaverstock is now the International Director (Associate Dean) for the Faculty of Social Sciences and Law, and Professor of International Management in the School of Economics, Finance and Management at the University of Bristol; Kathy Pain is Professor and Research Director of Research of Real Estate and Planning at the University of Reading; and Peter Taylor is Professor of Human Geography in the Faculty of Engineering and Environment at Northumbria University in Newcastle upon Tyne, UK. Taylor is considered the "inventor" of the world city network approach. His book *World City Network: A Global Urban Analysis* was recently published in second edition in collaboration with Ben Derudder (Routledge, 2016). Derudder, Hoyler, and Taylor are also, with Witlox, the editors of the influential *International Handbook of Globalization and World Cities* (Edward Elgar, 2011). Kathy Pain is the editor, with Peter Hall, of *The Polycentric Metropolis: Learning from Mega-City Regions in Europe* (Routledge, 2006) and, with John Harrison, of *Global Cities, Volume IV: Planning and Governance of Cities in Globalization* (Routledge, 2012).

This project has a quite straightforward purpose: to investigate relations between London and Frankfurt with the coming of a single European currency. Generally speaking, we have confirmed a network model of inter-city relations at the expense of the simplistic competition model that dominates so much public discussion of inter-city relations. A simple competitive model is wholly inadequate for understanding London–Frankfurt relations. This is not to say there has not been relative movement between the status of the cities – there has, but this is not indicative of any 'fierce competition' as it has sometimes been portrayed. We discuss the complexity of London–Frankfurt relations using the idea of tensions that have to be overcome in day-to-day operations in a world city network. This is where we deal with the 'competition/co-operation' conundrum as co-operation between cities operating alongside competition between firms. Our results consider the flows within networks that constitute London–Frankfurt relations. As boundaries give way to increasing interdependencies in a network society, co-operation between cities is found to be a priority for firms and institutions. We have introduced network thinking into consideration of London–Frankfurt relations. London and Frankfurt prosper by being within both similar and different webs of connections that have conditioned contemporary relations and whose dynamics will determine future relations. Finally, we hope that this research has shown that, under conditions of contemporary globalisation, cities are at least as important as countries in trying to understand fundamental social change in the world today. Relations between cities cannot be 'enframed' as small versions of simple international relations. With globalisation, there has been the creation of a world city interlocking network that relates to, but exists separate from, traditional international relations and the comparative study of countries.

LONDON AND FRANKFURT AS WORLD CITIES

This is a study of two cities under conditions of contemporary globalisation. Financially, globalisation operates through three time zones, and here we are concerned with two leading cities in the 'middle' time zone, Europe. The dominant city in this region is London. However, Frankfurt is generally perceived to be the up-and-coming financial centre in Europe, so

much so that there have been suggestions that it might be 'catching up' with London or even poised to take over its leadership position. The location of the new European Central Bank (ECB) in Frankfurt and the launch of the euro with the UK, and therefore London, outside 'euroland' have fuelled such speculation. Here we use these events as a lever to investigate the relations between London and Frankfurt, to see whether these relations are changing, and to assess whether Frankfurt is indeed a serious rival to London.

Cities do not operate as separate 'islands' of activity – their raison d'être is their connections with other cities. London and Frankfurt are thus both part of the same network of cities that straddle the world providing financial and other services. It is through studying their different roles in this network, and how they relate to each other within the network, that we can make intelligent assessments of the changing relations between the cities and whether Frankfurt is becoming a threat to London's pre-eminence in Europe. This is the approach we take in this study: we investigate flows, linkages and connections between London and Frankfurt.

First, we set London-Frankfurt relations within the context of recent studies of 'world cities'. We interpret world cities as the nodes in a world city network defined by the office networks of global service firms. Thus this particular 'Anglo-German' research is not the usual comparison between the two countries, but focuses instead on the relations between the most important nodes in the world city network within the UK and Germany, London and Frankfurt. From the existing data we get a picture of London as a supreme 'all-rounder' world city with a global presence, and Frankfurt more as an international financial service centre with a largely European remit. But what the data does not show is whether there are signs that Frankfurt is becoming a more encompassing world city with a global presence like London.

Second, we review the speculation on Frankfurt challenging London that built up a head of steam in the period up to the launch of the euro. Just by asking whether Frankfurt is catching up, the reporting of relations between the two cities is enframed as a competition. There is a mindset that has become embedded in the story of recent London–Frankfurt relations that depicts the cities as rivals. It seems that the metaphor for describing these inter-city relations is the world of international relations. But this is not the only way of looking at relations between cities. Relationships between members of a network are complementary in nature. Therefore, we need to consider the mutuality between London and Frankfurt in the world economy at least as much as their famous rivalry.

Third, we describe the methodology we use to test this speculation and to take the argument further into the nature of London–Frankfurt relations. The quantitative analysis reported earlier provides background material for considering the competition between London and Frankfurt but it cannot begin to uncover the processes involved. These mechanisms can only be explored through careful study of the actual behaviours and attitudes of the key persons within the institutions and advanced producer service firms doing business in London and Frankfurt. This requires a completely different methodology, a qualitative approach involving key players involved in the inter-city relations. Thus in-depth interviews were carried out with important personnel in financial and business service firms in London and Frankfurt and with some institutional representatives in both cities.

In London 10 banks and 15 other non-banking producer service firms were interviewed. In Frankfurt nine banks and 14 other non-banking producer services were interviewed. Most firms interviewed ranked in the European top 10 of their respective service sectors. Of the non-banking producer services interviewed in both cities, the research collected data from those sectors allied closely to the financial industry: law, accountancy, management consultancy and advertising. The interviews took place over two specific time census periods. Census one, using face-to-face taped one-hour interviews, evaluated changing Frankfurt–London relations one year after the launch of the euro. Census two, using 30-minute in-depth telephone interviews, evaluated changing Frankfurt–London relations up to two years after the launch of the euro. Each face-to-face and telephone interview took place with a senior member of the firm, at the grade of partner and/or vice-president and above, who had responsibility for the firm's adjustment to the euro with respect to Frankfurt–London intercity relations. The questions raised in both these interview census periods focused on three major themes pre- and post-euro: Frankfurt catching up with London; changing business relations in Frankfurt

Plate 19 Frankfurt skyline from south of the Main River

Source: Roger Keil

and London; and sectoral (e.g. banking, law) responses in London and Frankfurt. The interviews with regulators, professional bodies and state agencies were interactive in nature. Summaries of findings from the first interview census period were provided as the basis of face-to-face interviews with a selection of key state, financial and sectoral bodies which represented both Frankfurt and London's interests in the euro. Eighteen institutions were interviewed in London and eight in Frankfurt, and questions focused on sectoral responses to the euro in a London–Frankfurt context.

THE FINDINGS

The findings from this research can be summarized as such: Three major findings stand out from the parts of the interviews that focused on the effects of the euro on London–Frankfurt business relations:

- The new currency has had no wholesale effects on changing business relations between the two cities.
- Frankfurt's position in Europe is strengthening, but not at the expense of London.
- London remains the main European financial centre.

Despite the fact that the euro has only recently been launched, these findings are so strongly evident in our interviews that we can dismiss the idea that the euro will have an appreciable influence on London–Frankfurt relations in the near future. Research findings suggested that the introduction of the euro and the location of the ECB have not had a significant influence on business relations between service firms in London and Frankfurt. Frankfurt has not grown at the expense of London. London has not experienced leakage of financial services or market share to Frankfurt. But, the research findings have highlighted the overwhelming complexity of London–Frankfurt changing relations. It is no surprise to find that the introduction of the euro has not resulted in wholesale financial service reconfiguration in the EU. The political decision to locate the ECB in Frankfurt, in tandem with the introduction of the euro, quite simply will not undermine London's global financial status. London's financial centre has a deeply embedded matrix of firms, players, institutions and cultural practices which are tightly spatialised in well-established knowledge networks and discourses. As we shall see in the following chapter, firms play an important role in the reproduction of the complexity of London's changing business relations with Frankfurt, pre- and post-euro.

THE COMPLEXITY OF LONDON-FRANKFURT RELATIONS

The reason why the euro and ECB have had minimal effects on business relations between London and Frankfurt is that inter-city relations do not operate as a simple cause–effect mechanism. Relations between major cities are highly complex affairs and should not be viewed as a primitive competition with one city attempting to knock the other off top spot. As one of our respondents put it: 'The trouble is it's a huge jigsaw'. We have to understand the London–Frankfurt link within that larger whole.

We interpret the world city network as an interlocking type of network created by financial and business service firms through their location policies for offices. In their global strategies, firms have to decide where they need to have offices in order to provide the seamless service clients demand for cross-border business. In this way certain key cities have become the 'places to be' in order to service particular sub-markets, national and regional, within an integrated world market. It is these cities that have become world cities, nodes in the world city network. Thus world cities are 'interlocked' one into another through the office networks of service firms. The result is a world city network as the amalgam of offices each receiving and dispensing flows of information and knowledge within and between cities.

The key point that arises from viewing world cities in this way is that, as part of an interlocking network, the key agents in the production and reproduction of the network are the firms and not the cities themselves. Hence the increasing importance of cities under conditions of contemporary globalisation is not due primarily to the promotional activities of city governments. To understand how a particular city is faring, its links to other cities through business firms are the crucial determinant of success. It follows that city competition may be less important than city synergy or mutual reinforcement of cities by cities in the ongoing reproduction of the network by firms. This is consistent with the argument that it is only firms that compete in the world market and not nations or, by extension, cities.

We can summarize the major findings on the complexity of London–Frankfurt business relations as such:

- The fundamental tension facing firms in both London and Frankfurt is negotiating trans-border (global) reach against local sensibilities.
- There are other critical tensions in organisation, knowledge production, operationalisation and locational issues.
- The primary inter-city outcome of this inter-firm competition is a co-operative relation between London and Frankfurt within a wider network.

RESOLUTIONS AND OUTCOMES IN CITIES

Attempts to resolve the many tensions that face service firms in their provision of cross-border and global financial and business services take many forms, as we have just shown. In a dynamic globalising world these tensions are never 'solved'; rather they are managed in ways to accommodate, as competitively as possible, to contemporary conditions. Managing tensions may involve either intra-firm actions or inter-firm relations and, as we have seen, in both cases cities are directly implicated as business service centres. It is through this myriad of processes that service firms create the interlocks that produce the world city network. In our model, firms compete in the world financial and business services market and the outcome is a world city network. If this is the case then we would expect that evidence on inter-city relations should be marked more by co-operation than by competition. We focus upon the relation between London and Frankfurt, and each city's relations with selected other cities. In a further complication to the jigsaw, we find that the balance in evidence for co-operation or competition varies by scale of operation.

COMPETITION BETWEEN CITIES: GLOBAL AND DOMESTIC SCALES

Evidence for a London–Frankfurt conflict is all but missing from this study. This is the underlying reason why the launch of the euro and the Frankfurt location of the ECB are deemed to be unimportant in the relations between the two cities. Quite simply, the 'huge gap' or 'big difference' between the two cities

means that London and Frankfurt are not really rivals. There may have been some small relative improvements in Frankfurt's position, especially in finance, but this has no immediate impact on London's role as 'the European metropolis'.

Where inter-city competition is mentioned, it is of a different nature. According to one institutional interview in London, 'London is not competing against Frankfurt, it's competing against New York'. London is seen as being at the very top of the world city hierarchy where it competes only with other such 'global cities' and not with Frankfurt, which is 'a sort of regional German city'.

The problem for Frankfurt is that 'there's a lot of competition between German cities'. As one London institution put it: 'if I would be Frankfurt I would be more worried about Berlin than I would be about London'. In fact the existence of a 'German system of cities' of which Frankfurt is just one member is a point frequently made. Whereas 'Great Britain is London', our respondents saw Frankfurt in relation to other German cities – Berlin is mentioned 21 times, Munich and Düsseldorf 16 times, Hamburg 15 times, Cologne and Stuttgart three times each, and Leipzig twice. Germany's more decentralised urban system means that although Frankfurt may lead in financial services, it is not the leading city for other services – for instance, Frankfurt is ranked by respondents behind Hamburg and Düsseldorf in advertising, behind Düsseldorf and Munich for management consultancy, and Berlin is identified as a future rival in legal services.

Clearly London and Frankfurt are seen to be engaged at two very different levels of competition. Despite the remit of this study, respondents talk very little about regional competition at the European scale. The idea of a 'triumvirate' of London, Paris and Frankfurt is mooted at one point, but competition between them is hardly discussed at all. In Europe it is complementarity and not competition that is emphasised.

CO-OPERATION BETWEEN CITIES: EUROPEAN AND GLOBAL SCALES

'There is no rivalry between offices' is the common refrain of our respondents, and therefore it is 'wrong to see it in terms of office versus office'. All firms operate through their network of offices to provide clients with the best international product they can. It is clear that market pressures on firms are creating co-operation across cities: 'we cooperate very strongly in Europe, that's something our clients demand'. The only example where there was competition between offices in the same organisation was where offices operated as autonomous profit centres; the norm, however, was to use network teams to eliminate intra-firm rivalry and 'pull together . . . that's much more important than geographical issues'.

Contacts and flows between offices vary by sector and firm, depending on the mix of projects at any one time. But the interaction between cities is continuous: 'we are integrated, we talk every day with almost all our offices'. It is usual in one firm to have a Frankfurt person in London and vice versa every day. For another, 'co-operation between our London and Frankfurt offices . . . has grown considerably in the last five years'. And the reason for this is the mutual benefits of being in both cities, as witness the use of London as a platform for global business by German firms and Frankfurt as a centre for German M&A business by London-based firms.

And, of course, this is not just a feature of London–Frankfurt relations and Europe. Many firms mentioned other equivalent inter-city relations (Paris–London, New York–Frankfurt, Beijing–Frankfurt, for example). The response to globalisation has been to develop 'global management' because 'we are serving global clients and therefore we have to put together multi-disciplinary teams from across the world'. The key operating units in all this are the office networks, and this synergy within firms has produced synergy between cities as the world city network.

The situation is best summed up by two London-based institutions. According to one, 'London is the European interface' and therefore the 'increasing strength of Frankfurt is feeding into London not drain-ing away from it'. For the other, London is the 'bridge-head in Europe' for non-European inward investors and 'the two cities who benefit from this mainly are, at the moment, Frankfurt and Paris'. Mutuality, comple-mentarity and co-operation are the hallmarks of a network. They are clearly present in the world city interlocking network within which London–Frankfurt relations are a prime example.

LONDON–FRANKFURT RELATIONS IN A NETWORK SOCIETY

Castells (Ch. 21) has argued that the rise of a network society has resulted in spaces of flows coming to dominate spaces of places. In globalisation this is represented by the challenge to the mosaic space of nation-states by 'borderless', communication/computer-based processes, for example in inter-national financial markets. Castells recognizes the world city network as a prime result of the creation of such new global spaces of flows. Within this inter-locking network there are many flows beyond the office linkages discussed in the previous chapter. Four important networks that constitute London–Frankfurt relations are the following: knowledge networks, cultural networks, power networks and governance networks.

Here we consider how our respondents report these different networks of flows to inform London–Frankfurt relations. We examined each of the networks in turn and finally the space of inter-city flows in interweaving networks. The key findings are as follows:

- Knowledge networks have the potential to revo-lutionize spaces of flows and redefine shifting relations between London and Frankfurt.
- Cultural networks are a key determinant of, and stimulus to, inter-city business relations.
- Power networks exhibit shifting relationships and strong mutual dependencies that reflect established patterns of investment.
- Governance networks reach out between the cities regulating and shaping London–Frankfurt business flows through a web of public–private and local–global relationships.
- Interweaving networks are characterized by prox-imity, connections and complex interdependen-cies making co-operation integral to inter-city relations.

COMPETITION AND CO-OPERATION REVISITED

As place and role boundaries fade and complexity increases in a networked society, traditional distinc-tions between city competition and co-operation are likely to have less relevance to the reality of inter-city relations. The need for proximity and connectivity in service business and increasing interdependencies between economy, society and governance make co-operation integral to contemporary and future inter-city relations.

Throughout our examination of inter-city networks in this chapter, co-operative relations have been shown to take precedence over competition as a stimulus to business flows that benefit both London and Frankfurt. Inter-city co-operation is clearly essential to firms and, inter-city co-operation within the institutional sphere could also be beneficial to both cities.

Finally, although not a core topic within our investigation, serious concerns about London's transport, congestion and environmental problems, skills shortages, housing costs and the sustainability of further economic development were expressed almost universally in the London interviews. In many cases, respondents indicated that these were the only reasons they could imagine for moving from a London location. Again, these are compelling reasons for inter-city institutional co-operation in addressing the policy,

investment and governance challenges associated with successful global financial centres.

REFERENCES FROM THE READING

Derudder, B., Hoyler, M., Taylor, P., and Witlox, F. (2011) *International Handbook of Globalization and World Cities*, Cheltenham: Edward Elgar.

Hall, P. and Pain, K. (eds.) (2006) *The Polycentric Metropolis: Learning from Mega-City Regions in Europe*, London: Routledge.

Harrison, J. and Pain, K. (eds.) (2012) *Global Cities, Volume IV: Planning and Governance of Cities in Globalization*, London: Routledge.

Taylor, P. and Derudder, B. (2016) *World City Network: A Global Urban Analysis*, London: Routledge.

23

"Global grids of glass: on global cities, telecommunications and planetary urban networks"

from *Urban Studies* (1999)

Stephen Graham

EDITORS' INTRODUCTION

Stephen Graham is a pioneering scholar of late 20th- and early 21st-century urbanism. Throughout much of the 1990s, Graham worked in the Urban Planning Department at the University of Newcastle, where he co-founded the Center for Urban Technology. After holding a post as Professor of Human Geography at the University of Durham, Graham has now returned to Newcastle where he is Professor of Cities and Society at the Global Urban Research Unit and is based in Newcastle University's School of Architecture, Planning and Landscape. A major strand of Graham's work has focused on the historical, political, economic, and cultural geographies of urban infrastructural systems. Rather than treating these urban infrastructural arrangements—from water, heating, sewage, and electrical systems to transportation and communications grids—in technocratic terms, as simple instruments for the provision of public and private goods, this new scholarship has attempted to decipher their complex social, political, and cultural dimensions and their role as expressions of historically specific forms of urban power. With his long-time collaborator, Simon Marvin, Graham has been at the forefront of this new research on urban infrastructural networks. Their two major books, *Telecommunications and the City* (1996, Routledge) and *Splintering Urbanism* (2001, Routledge) have explored at length (a) the worldwide reorganization of urban infrastructural arrangements under contemporary globalizing capitalism; and (b) the role of such transformations in a broader rearticulation (or "splintering") of urban space. More recently, Graham has also occupied himself with the study of the relationships of urbanization and war, and the intersection of violence, technology, and urbanization. In 2016, he published the critically acclaimed volume *Vertical: Looking at the City from Above and Below* (Verso).

Graham's chapter below introduces his approach to the study of urban infrastructure and applies it to the restructuring of telecommunications systems in global cities (see also Graham 2004). The global urban system is linked together not only through the activities of transnational corporations, but also through dedicated, high-performance informational networks based upon transplanetary optic-fiber connections. Global cities contain the highest concentrations of advanced telecommunications infrastructures within their respective national urban systems. This trend is being further accelerated and intensified through the deregulation and privatization of telecommunications systems, a process that has triggered an intensive reorganization of telecom, media, and cable firms across Europe and North America, as ever-larger, global conglomerates are formed in order to compete on a global scale. Finally, Graham considers the intra- and inter-urban spatial transformations resulting from the construction of such transplanetary technological networks. On the one hand, the effective operation of planetary telecommunications grids depends upon concrete physical installations within the dense built environments of global cities: optic-fibers must be "looped" under roads and sidewalks, into and out of buildings. On the other hand, "global grids of glass" must be constructed that link major metropolitan centers across oceans and continents: this entails creating transoceanic optic-fiber networks and transterrestrial satellite linkages that wire distant cities together to maintain instantaneous connectivity. Graham concludes by underscoring the persistent problem of the "digital divide." While the new informational and telecommunications technologies may link global cities more closely together, they have to date failed to include significant zones of the world economy within their web. At present, therefore, such technologies appear to serve primarily the needs of hegemonic economic actors and organizations.

The growing centrality of key large urban regions, or global cities, to the economic, social, political and cultural dynamics of the world presents a particularly potent example of the reconfiguration of space through telecommunications. In such cities, the most sophisticated electronic infrastructures ever seen are being mobilised to reconfigure space and time barriers in a veritable frenzy of network construction. Such processes seem likely to maintain the electronic competitive advantages of the largest global cities for some time to come. But the wiring of cities with the latest optic-fibre networks is also extremely uneven. It is characterised by a dynamic of dualisation. On the one hand, seamless and powerful global-local connections are being constructed within and between highly valued spaces, based on the physical construction of tailored networks to the doorsteps of institutions. On the other hand, intervening spaces – even those which may geographically be cheek-by-jowl with the favoured zones within the same city – seem, at the same time, to be largely ignored by investment plans for the most sophisticated telecommunications networks. Such spaces threaten to emerge as "network ghettos," places of low telecommunications access and concentrated social disadvantage. As with many contemporary urban trends, then, uneven global interconnection via advanced telecommunications becomes subtly combined with local disconnection in the production of urban space.

Global cities research, in particular, has detailed at length how an interconnected network of such cities has recently grown to attain extraordinary status. Such cities bring together the greatest concentrations of control, finance, service, cultural, institutional, social, informational and infrastructural industries in the world. All aspects of the functioning of global cities are increasingly reliant on advanced telecommunications networks and services; such cities concentrate the most communications-intensive elements of all economic sectors and transnational activities within small portions of geographical space. It is no surprise, therefore, that there is growing evidence that such city-regions heavily dominate investment in, and use of, these technologies.

The diversifying electronic infrastructures that girdle the planet have very specific geographies and spatialities. These counter the prevailing "information age" rhetoric suggesting that advances in telecommunications somehow prefigure some simple "end of geography." Indeed, the current international shift towards liberalised, privatised and internationalised telecommunications regimes seems to be accentuating the centrality of global cities within telecommunications investment patterns.

Many research challenges remain to be faced before we can satisfactorily understand the complex interlinkages between telecommunications grids, global cities and planetary urban networks. The focus of this chapter is to explore the linkages between the growth of a planetary network of global financial, corporate and media capitals, and the emerging global and urban information infrastructures that interlink, and underpin, such centers. Very little is known about how the global wiring of the planet with a new generation of optic-fibre grids interconnects with the development of intense concentrations of new communications infrastructures within global cities. This chapter aims to develop such an understanding by attempting to address intra-urban, inter-urban and transplanetary optic-fibre connections (and disconnections) in parallel. Such an approach is necessary given the logics inherent within the "network society," which force us to collapse conventional hierarchical notions of scale – building, district, city, nation, continent or planet. As a result, it is difficult to be a specialist on urban landscapes, intra-urban shifts or urban systems in separation. Discussions of restructuring *within* cities increasingly must address the changing relations *between* them, whilst also being cognisant of the importance of these changing relations within broader dynamics of geopolitics and geoeconomics.

GLOBAL CITIES: THE SOCIAL PRODUCTION OF LANDSCAPES OF MULTIPLE RELATIONAL CENTRALITY

It is now clear that global cities grow by cumulatively concentrating the key assets which corporate headquarters, high-level service industries, global financial service industries, national and supranational governance institutions and international cultural industries rely on, within a volatile, globalising, operating environment. The growing extent of globally stretched corporate, financial and media webs, mediated by telecommunications and transport networks, seems to support a parallel need for the social production and management of places of intense centrality. This is especially so when one adds the volatilities thrown up by global shifts toward financial globalisation, economic liberalisation and the opening up of regional blocs to "free" trade. The complex mediation of economic activity in extremely volatile contexts necessitates high levels of face-to-face interaction within the high-level managerial and control functions that concentrate in global cities. Place-based social relations become central to the economic survival of high-level corporate, media and financial organisations.

The economic, social and cultural dynamics of global cities rely in essence on the control, co-ordination, processing and movement of information, knowledge and symbolic goods and services (advertising, marketing, design, consultancy, finance, media, music, etc.). Within a post-Fordist context of vertical disintegration, niche marketing, precise logistical co-ordination, internationalisation, pervasive computerisation and the powerful growth of symbolic exchange, all such activities are generating booming demands for voice, computer and image communications of all types. Above all, such conditions are supporting an enormous growth in economic *reflexivity* placing a premium on both the specific socioeconomies offered by particular global cities and intensifying electronically mediated connections between them. The reflexive nature of global city functions thus demands ongoing social relations based on trust and reciprocity. These are supported by both intense mobility and sophisticated telemediated exchange. In a nutshell, such dynamics help explain why perhaps the two most dominant social trends in the world today are an unprecedented urbanisation and a growing reliance on telecommunications-based relations.

As mediators of all aspects of the reflexive functioning and development of global cities, convergent media, telecommunications and computing grids (known collectively as telematics) are basic integrating infrastructures underpinning the shift towards intensely interconnected planetary urban networks. Inter-urban telecommunications networks (both transoceanic optic fibres and satellites) comprise a vital set of hubs, spokes and "tunnel effects" linking urban economies together into real or near real time systems of interaction which substantially reconfigure the production of both space and time barriers within and between them. Such technologies help to integrate distant financial markets, service industries, corporate locations and media industries with virtual instantaneity and rapidly increasing sophistication. But they underpin the enormously complex communications demands *within* global cities, generated by the intense clustering of reflexive practices in space. Such dynamics mean that the very small geographical areas of the main global finance, corporate and media capitals dominate the emerging global political economy of telecommunications.

Two sources of data can help to give an indication of this dominance. First, we can see how the economic sectors that are overwhelmingly located in global cities tend to dominate international telecommunications flows as a whole. For example, over 80 per cent of international data flows are taken up by the communications, information flows and transactions in the financial services sector. Over 50 per cent of all long-distance telephone calls in the US are taken up by only 5 per cent of phone customers, largely transnational corporations whose control functions still cluster in the global metropolitan areas of the nation. Secondly, there is a small amount of available data on the dominance of national telecommunications patterns by particular global cities. A recent survey (Finnie 1998) found that around 55 per cent of all international private telecommunications circuits that terminate in the UK do so within London. And about three-quarters of all advanced data traffic generated in France come from within the Paris region.

TELECOMMUNICATIONS LIBERALISATION, GLOBAL CITIES AND URBAN COMPETITIVENESS

Central business districts (CBDs) within global cities play a predominant role within fast-moving communications landscapes. They provide leading foci of rapid technological change and concentrated patterns of investment in new telecommunications infrastructures, from multiple competing providers. Global financial service industries, in particular, are especially important in driving telecommunications liberalisation. With telecom costs taking up around 8 per cent of expenditure on goods and services in global financial firms, a world-class telecommunications infrastructure and a business friendly, fully liberalised regulatory environment are thus becoming key assets in the competitive race between global cities to lure in financial and corporate operations and their telecommunications hubs.

For transnational companies (TNCs) of all kinds, liberalised telecommunications markets allow the benefits of competition to be maximised. The spread of liberalisation also minimises the transaction, negotiation and interconnection costs that stem from constructing global networks within the diverse regulatory and cultural contexts of multiple national post, telegraph and telephone systems (PTTs) across the globe. Global telecoms liberalisation is thus critical given the strategic centrality of private telecommunications networks to the functioning of all TNCs, financial and media firms. The emerging global, private regime allows lucrative corporate and financial market segments to benefit from intense, customised and often very localised competition in high-level telecommunications infrastructure and services, when this was impossible through old-style PTTs.

In short, for leading finance, transnational and media firms, sophisticated telecommunications are becoming central to business success or failure. Such firms demand a seamless package of broadband connections and services, within and between the global cities where they operate. Such demands include leased lines, private optic-fibre connections, dedicated satellite circuits, video conferencing and, increasingly, highly capable mobile and wireless services.

Plate 20 London metabolisms

Source: Roger Keil

In addition, such powerful corporate users are pressing hard to enjoy the fruits of competition between multiple network providers in global cities. Such competition tends to increase discounts, improve efficiency and innovation and, above all, reduce risks of network failure by increasing network resilience. Meanwhile, peripheral regions and marginalised social groups often actually lose out and fail to reap the benefits of competition, as they are left with the rump of old PTT infrastructures, relatively high prices and poor levels of reliability and innovation. Highly uneven microgeographies of "splintered" telecommunications development thus replace the relatively integrated and homogeneous networks developed by national telephone monopolies.

GLOBAL CITIES AND INTERNATIONAL TELECOMMUNICATIONS LIBERALISATION

As a result of such liberalising pressures, a global wave of liberalisation and/or privatisation is transforming national telecommunications regimes. The key lobbying pressures driving this shift derive from TNCs, the World Trade Organisation (WTO), the G7 countries, telecommunication industries and, especially, the corporate, financial and media service industries that concentrate in global cities. In the context of the recent WTO agreement to move towards a global liberalised telecommunications market and with most regions of the globe now instigating regional trading bloc agreements (EU, NAFTA, ASEAN, Mercosur, etc.) based on removing national barriers to telecoms competition, a profound reorganisation of the global telecommunications industry is taking place.

The initial example of metropolitan network competition was set by New York in the 1980s. Then, a whole new competitive urban telecommunications infrastructure was developed in the city by the teleport company and port authority. As the pressures of the global neo-liberal orthodoxy in telecommunications have grown, even the most resistant nation-states (such as France, Germany, Singapore and Malaysia) are now succumbing to calls for national PTT monopolies to be withdrawn, to be replaced by uneven, multiple, telecoms infrastructures which inevitably articulate centrally around large metropolitan markets.

In the shift to a seamlessly interconnected global telecoms industry, national postal, telegraph and telephone monopolies are thus being rapidly privatised and/or liberalised right across the globe. Private telecom firms are aggressively "uprooting" to make acquisitions, strategic alliances and mergers with other firms, in a global struggle to build truly planetary telecommunications service firms, geared towards meeting the precise needs of TNCs and financial firms within the whole planetary network of global cities, on a "one stop, one contract" basis.

For example, in an effort to position themselves for this "one-stop" market, BT and AT&T, both trying to reposition themselves from national to global players, have recently announced a global alliance. AT&T has also merged with the huge US cable firm Tele-Communications to improve their position within home US markets. The Swedish, Swiss, Spanish and Dutch PTTs, meanwhile, have their Global One umbrella. France Telecom and Deutsch Telekom have an alliance with the US international operator Sprint. A myriad of other firms are offering services based on reselling network capacity leased in bulk from other firms' telecommunications networks. Finally, there is a process of alliance formation between telecoms, media and cable firms. All are jostling to position themselves favourably for corporate and domestic markets, especially in the information-rich global city-regions, for the future where many envisage a globally liberalised telecommunications environment with perhaps four or five giant, truly global, multimedia conglomerates.

The position of global cities is therefore being substantially reinforced by the shift from national telecommunications monopolies to a globalised, liberalised communications marketplace. As cross-subsidies between rich and poor and core and periphery, spaces within nations are removed, infrastructure developments now reflect unevenness in communications demand in a much more potent way than has been witnessed for the past half century. Prices and tariffs are being unbundled at the national level, revealing stark new geographies which compound the advantages of valued spaces within global cities as attractive and highly profitable telecommunications markets, dominating telecommunications investment patterns within nation-states.

THE MICROGEOGRAPHICAL IMPERATIVES OF "LOCAL LOOP" CONNECTION

It is paradoxical, however, that an industry which endlessly proclaims the "death of distance" actually remains driven by the old-fashioned geographical imperative of using networks to drive physical market access. The greatest challenge of these multiplying telecommunications firms in global cities is what is termed the problem of the "last mile": getting satellite installations, optic-fibre drops and whole networks through the expensive "local loop," under the roads and pavements of the urban fabric, to the buildings and sites of target users. Without the expensive laying of hardware, it is not possible to enter the market and gain lucrative contracts. Fully 80 per cent of the costs of a network are associated with this traditional, messy business of getting it into the ground in highly congested, and contested, urban areas.

This is why precise infrastructure planning (through the use of sophisticated geographical information systems) is increasingly being used to ensure that the minimum investment brings the highest market potential. It also explains why any opportunity is explored to string optic fibres through the older networks of ducts and leeways that are literally sunk deep within the archaeological "root systems" of old urban cores. Mercury's fibre grid in London's financial district, for example, uses the ducts of a long-forgotten, 19th-century, hydraulic power network. Other operators thread optic fibres along rail, road, canal and other leeways and conduits. Whilst satellite infrastructures are obviously more flexible, they are nowhere near as capable or secure as physical optic-fibre "drops" to a building. And they, too, ultimately rely on having the hardware in place to deliver services to the right geo-economic "foot print."

GLOBAL GRIDS OF GLASS: INTERCONNECTING URBAN FIBRE NETWORKS

We now come to our final scale of analysis: that of planetary interconnection and the geopolitics of infrastructure construction across oceans and continents. Of course, dedicated optic-fibre grids *within* the business cores of global cities are of little use without interconnections that allow seamless corporate and financial networks to piece together to match directly the hub and spoke geographies of international urban systems themselves. To this end, WorldCom and the many other internationalising telecommunications companies are currently expending huge resources within massive consortia, laying the satellite and optic-fibre infrastructures necessary to string the planet's urban regions into a single, highly interconnected communications landscape. Such efforts are especially focusing on the high-demand corridors linking the geostrategic, metropolitan zones of the three dominant global economic blocs: North America, Europe and East and South-eastern Asia. Wiring North America and the North Atlantic is proceeding at a particularly rapid pace, with AT&T, Sprint, WorldCom and US Regional "Baby Bells" fighting it out with newcomers like Quest to provide the "pipes" that will keep up with demand, especially from the Internet. Other lesser infrastructures are also being laid, designed to link southern metropolitan regions in Australia, South Africa and Brazil into the global constellation of inter-urban information infrastructures.

Currently, transoceanic and transterrestrial optic-fibre and satellite capacities concentrate overwhelmingly on linking North America across the North Atlantic to Europe and across the North Pacific to Japan, reflecting the geopolitical hegemony of these three regions in the late 20th century. The first transoceanic fibre networks, developed since 1988 across the Atlantic, Pacific and Indian oceans, with AT&T playing a leading role, tended to stop at the shorelines, leaving terrestrial networks to connect to each nation's markets. Increasingly, however, transoceanic fibre networks are being built explicitly to link metropolitan cores, in response to their centrality as generators of traffic and centers of investment.

Once again, WorldCom, in its efforts to connect together its metropolitan infrastructures, is leading this process. As well as constructing a transatlantic fibre network known as Gemini between the centers of New York and London, WorldCom is building its own pan-European Ulysses network linking its city grids in Paris, London, Amsterdam, Brussels and major UK business cities beyond London. Elsewhere in the world, too, it is exploring the construction of transnational and transoceanic fibre networks to connect its globalising archipelago of dedicated city networks.

At the strategic global scale, globalisation and the rapid growth of newly industrial countries means that much effort is being spent filling in gaps in the global patterns of optic-fibre interconnection, particularly between Europe and Asia. One project, for example, known as FLAG (fibre optic link around the globe) will provide a new ultra high-capacity (120,000 simultaneous phone calls) telecoms grid over 28,000 km from London to Japan, via many previously poorly connected metropolitan regions and nation-states. The route first goes by sea from England via the Bay of Biscay and the Mediterranean to Alexandria and Cairo. Then it crosses Suez and Saudi Arabia overland to Dubai; crosses the Indian Ocean to Bombay and Penang and Kuala Lumpur; traverses the Malay peninsular to Bangkok; and finally goes by sea again to Hong Kong, Shanghai and Japan. Nynex, the New York phone company, is organising the project with support from a huge range of private investors. FLAG will also connect with the 12,000 km Pacific Cable Network linking Japan, South Korea, Taiwan, Hong Kong, the Philippines, Thailand, Vietnam and Indonesia. Other similar projects are proliferating across the globe. A 30,000 km China-US cable network, the first direct linkage between the two nations, was announced by a 14-firm consortium in December 1997. Other transoceanic cables are also being constructed: in the Caribbean (the Eastern Caribbean Fibre System); from Florida, through the strait of Gibraltar, to the Mediterranean urban system; by AT&T around the African continent (offering a high level of security whilst providing fibre drops to all the main cities on the African coast); and by Telefonica of Spain from the US and Europe to the main metropolitan regions of Latin America (Fortaleza, Rio, Sao Paulo, Buenos Aires), again with the involvement of WorldCom. But the largest project of them all, Project Oxygen, was announced in 1998: a $15 billion global network, built by 73 multinational telecommunications firms, linking the largest urban regions in 171 countries by 2003, via over 300,000 km of optic-fibre cable.

CONCLUSION

This chapter has sought to extend our empirical and conceptual understanding of the ways in which global cities and global city networks are related to rapid advances in both intra-urban and inter-urban information infrastructures. It has emerged that current advances in telecommunications are a set of phenomena which tend overwhelmingly to be driven by

large, internationally oriented, global metropolitan regions. The activities, functions and urban dynamics which become concentrated in global city-regions rely intensely on the facilitating attributes of advanced telecommunications for supporting relational complexity, distance links and snowballing interactions, both within and between cities. Such telereliance is particularly high in internationally oriented industries whose products and services are little more than telemediated flows of exchange, information, communication and transaction, backed up also by intense face-to-face contact and supporting electronic coordination and transportation.

It should be no surprise, then, that, in an increasingly demand-driven media and communications landscape, global cities should dominate investment in and use of advanced telecommunications infrastructures and services. Global cities are the main focus of the highly uneven emerging geographies of network competition. They dominate all aspects of telecommunications innovation. As we have seen, such cities are now developing their own superimposed and customised fibre networks, seamlessly linked together across the planet into a global cities communicational fabric, whilst often (at least in the initial stages of development) separating them off from traditional notions of hinterland, urban and regional interdependence and national infrastructural sovereignty. The WorldCom city networks explored above, for example, demonstrate that the emerging urban communicational landscape is rapidly becoming dominated by tailored, customised infrastructures. Through these, dominant corporate, media and financial players can maintain and extend their powers over space, time and people. Such infrastructures are carefully localised physically to include only the users and territory necessary to drive profits and connect together global corporate, financial and media clusters.

But, in so doing, such networks seem likely to further urban socio-spatial polarisation by carefully by-passing non-lucrative spaces within the city. They are, in other words, physical, infrastructural embodiments of the splintering and fracturing of urban space. The combination of intra- and inter-urban fibre networks thus materially supports the dynamic and highly uneven production of space-times of intense global connectivity, made up of linked assemblies of high-level corporate, financial and media clusters and their associated socio-economic elites, within the spaces of global cities.

Finally, it seems inevitable that the customised combinations of intra- and inter-urban fibre networks analysed in this chapter will also drive uneven development at the national level. They will be excluding non-dominant parts of national urban systems from the competitive advantages that stem from tailored, customised urban information infrastructures. Such infrastructures lay to rest any prevailing assumptions that there necessarily remains any meaningful connection between the sovereignty of nation-states as territorial "containers," and patterns of infrastructural development, especially for telecommunications infrastructures. The modernist notion that nationhood is partly defined by the ability to roll-out universally accessible infrastructural grids to bind the national space is completely destroyed by the new infrastructural logic. Instead, the logic of the emerging electronic infrastructure is to follow directly the global city networks, through direct global-local interconnection. Consequently, relatively cohesive, homogeneous and equalising infrastructure grids at the national level are now being "splintered" into tailored, customised and global-local grids, designed to meet the needs of hegemonic economic and social actors (Graham and Marvin 2001).

REFERENCES FROM THE READING

Finnie, F. G. (1998) Wired cities, *Communications Week International*, May 18: 19–22.

Graham, S. (ed.) (2004) *The Cybercities Reader*, London and New York: Routledge.

Graham, S. (2016) *Vertical: Looking at the City from Above and Below*, New York: Verso.

Graham, S. and Marvin, S. (1996) *Telecommunications and the City*, London: Routledge.

Graham, S. and Marvin, S. (2001) *Splintering Urbanism*, London: Routledge.

24

"Global cities and the spread of infectious disease: the case of Severe Acute Respiratory Syndrome (SARS) in Toronto, Canada"

from *Urban Studies* (2006)

S. Harris Ali and Roger Keil

EDITORS' INTRODUCTION

In the spring of 2003, a mysterious new pathogen, later identified as a corona virus that causes Severe Acute Respiratory Syndrome (SARS), ravaged populations in various large, internationalized urban centers between Singapore and Toronto. In that year, S. Harris Ali, a leading sociologist of environment and health, now Professor and the Chair of Sociology at York University, and Roger Keil, one of the co-editors of this volume, embarked on a groundbreaking study of the relationships of the global city network and the spread of emerging infectious disease. Ali and Keil mobilized their intellectual backgrounds in urban studies, urban political ecology, and sociology of health and environment to literally "connect the dots" between the spread of the disease and the network of some of the world's most dynamic global city regions. The excerpt below is an early conceptual piece that attempted to make a connection between the movement of international migrants and global urban networks in relation to epidemics of infectious disease. In subsequent empirical work, Ali, Keil and colleagues studied many aspects of the emerging relationships of "urban political pathology" in the global cities network (see Ali and Keil 2008). While the SARS epidemic burnt out quickly after taking more than 800 lives worldwide, the study from which the below excerpt has been taken alerted students of global cities to a more sinister side of global urban connectivity (see Keil 2011).

Toronto articulates the Canadian economy with the global economy. The regional economy of Canada's global city – largely based on high value added automotive manufacturing, finance, culture production, business services, education, health care, biomedical industries, etc. – is an important piece of a continental-izing and globalizing Canada. It is Canada's prime destination for new immigrants and tourists; its airport is the biggest in the country. In the study of Toronto as a 'global city' lies an important key to understanding the social, cultural, and geographic dimensions of an outbreak of Severe Acute Respiratory Syndrome (SARS) in 2003. SARS constituted a public health threat that involved 438 probable and suspected SARS cases, including 44 deaths in the Toronto area. SARS is characterized by a fast and violent onset of symptoms of very severe respiratory ailment; patients need to be isolated and usually rely on highly technological hospital environments for treatment (Naylor 2003; Walker 2004). Furthermore, SARS outbreaks in various Asian global cities such as Beijing, Hong Kong, Singapore, and Taipei (WHO 2004) bring to the forefront the question of how the interrelation of Toronto with other global cities influenced the transmission of the SARS virus. The relationship between Toronto and other Asian global cities is noteworthy because of the increased prospects of networked trade between Toronto and China's emerging global cities. Bolstered by the large Chinese-Canadian community in Toronto increased trade and interaction between Toronto's Asian diaspora community and the newly opened Chinese cities would not be unexpected.

In bringing together two as of yet mostly unconnected literatures – on global cities and on infectious disease – we argue that the emergence of the global city network may have notable implications for the nature of the threats posed by infectious disease in the contemporary world in relation to both disease transmission and outbreak response. By considering the case of the spatial diffusion of SARS, we examine how the nature of the threat of infectious disease outbreaks has changed through the alteration of patterns in microbial traffic, vis-à-vis the global city network. The chapter aims at breaking new ground in creating a potentially productive perspective from which the world city network can be seen in an entirely new light: instead of concentrating on global flows of capital, information, and power or global immigration, we view world city formation from the point of view of emerging and spreading infectious disease. We believe such a perspective can provide, in several ways, a new entry point for the already lively debate on connectedness in the global city universe – first, as one way of integrating issues related to the diffusion of disease into existing understandings of globalization processes; second, by taking into account the important insight of political ecologists of scale that "global flows are necessarily embedded in local processes [and this] prompts a consideration of place not merely as an isolatable physical space but as a dimension of historical and contemporary connections" (Gezon and Paulson 2005: 9). And third, by considering the relationship between the global city network and the diffusion of disease we may be able to take the first steps towards the development of what Gould refers to as "human disease topographies" that would map

"disease spaces" (1999: 196). Disease spaces refers to those ecological niches that pathogens need to exist and reproduce, and such niches include human beings. Thus, disease spaces are: "enormously complex spaces that control the movement of viruses in much the same way that the hills and valleys of the familiar topographical map control the flow of water" (1999: 196). In turn, it is hoped that such human disease topographies would eventually lead to the development of more effective ways to halt the diffusion of pathogens.

Of course, relationships between the governance of urban areas and the control of infectious disease are not new. Attempts to deal with pandemics such as the bubonic plague in 14th century Europe, as well as other major epidemics of smallpox, typhus, yellow fever, and cholera in the 18th and 19th century, served as the impetus for some of the first formally organized efforts to develop international health strategies (Banta 2001). Historically, the favoured method to contain the spread of contagious diseases involved the quarantine of ships arriving at European port cities. Such measures, however, seriously disrupted the flow of commerce and financial transactions of the day and led to tensions in the international shipping trade. Spurred on by the need to address the economic and political consequences of quarantined merchants and sequestered cargo, diplomatic dialogue ensued. This international dialogue was not limited to simply matters of economic security, but also involved issues related to public health protection. As the analysis of the SARS outbreak in Toronto demonstrates, the interrelationship between public health and economics is not only of historical significance, but it remains as a concern at the forefront of current responses to disease outbreaks in our globalized world.

Although there are some commonalities with the past, the time-space compression resulting from today's globalizing forces implies that modern disease epidemics have the potential to spread across the world much more quickly than in the past. This increased potential may be accounted for by several reasons. First, the medium involved in the transport of microbes and humans has changed quite dramatically – for example, from slow moving ships to rapidly moving airplanes. Second, the number of people involved in movements between nations and cities of the world has increased dramatically in the contemporary era. Third, as the nature of the economic and political interconnections between various parts of the world have evolved over time, the potential pathways for the travel of pathogens into populations have also changed. Today, the situation is distinctive for two reasons. First, in the modern age of air travel those infected could very likely be asymptomatic during their trip and upon arrival to their destination. Second, the routes through which pathogens can enter the populations of cities have multiplied partly because of the emergence of a dynamic interconnected network of what are known as "global cities." As air travel amongst these global cities has increased and the number of flights between them continue to multiply (Smith and Timberlake 2002), the possibilities for the spatial diffusion of a pathogen also multiply, making it more difficult to track the spread of a pathogen. This increased mobility of both individuals and microbes implies that the boundaries and structures of the global "network society" (Castells 1996) are more fluid than those of traditional societies such as those of the port cities of an earlier era.

CONCEPTUALIZING THE GLOBAL CITY NETWORK

Held and McGrew (2002: 1–4) identify several dimensions associated with globalization. First, globalization involves a widening reach of networks of social activity and power. That is, the extensivity of global networks increases. Second, the intensity of global interconnections increases as patterns and flows begin to transcend the constituent societies and states of the world order. Third, the velocity of global flows speeds up with the development of worldwide systems of transport and communication of ideas, goods, information, capital, and people. Fourth, as the extensivity, intensity, and velocity of global interactions increase, the enmeshment of the local and global becomes deepened, such that the impact of distant events is magnified, while at the same time, local events may have global consequences. In other words, the impact propensity of global interconnectedness increases. It is against this backdrop of globalization that changes in the qualitative and quantitative nature of urban centres must now be understood and applied to the analysis of disease diffusion in the contemporary world because diseases spread within a specific historical and geographic context (Haggett 2000; Gould 1999).

Debates on urbanization and globalization have, perhaps, most systematically been linked in a growing body of conceptual and empirical work which deals with the emergence of a network of world cities or global cities. Global cities may be organized in a hierarchical fashion or in "cliques," in which certain distinct clusters or strata of cities are ordered according to the economic, political, and cultural power they yield on the international scene. Hierarchically organized clusters of cities have been found to be the skeleton of the new world economy and the relatively higher position of a particular city in this hierarchy results in a competitive economic advantage and status for this city in the international context (Taylor 2004).

The global city network scheme has come under criticism for portraying a rather static view in which global cities, conceptualized as distinct entities, are locked into a network of linear relationships with each other, at a fixed scale of interaction (Smith 2003). According to this perspective on global cities, it is argued that cities are treated as self-evident scalar entities in which specific social relations, localization, and territorial concentration serve as the basic and unchanging precondition for global economic transactions. As such, it is suggested that the conventional conceptualization of the global cities network does not appear to take into account the ways in which global cities are always interactive and constantly in a process of change. Nor does that perspective give adequate analytical consideration to how the global city is polyrhythmic – "a liquid theatre alive with the unruly times of urban practices" (Smith 2003). Notably, these "flux characteristics" of global cities are important to consider because once global financial networks reach a certain level of complexity, their nonlinear interconnections generate rapid feedback loops that may lead to many types of unanticipated and emergent phenomena (Smith 2003: 566). For Smith therefore, the forces of globalization and the development of global cities are too "intermingled through scattered lines of humans and non-humans to be delimited in any meaningful sense" (2003: 570).

It is our contention that the effects of local-global interactions are not limited to purely economic and financial functions of the global cities network, but that such effects can also be discerned in relation to matters of public health within and amongst the global cities comprising the network. Notably, such unanticipated, nonlinear effects should be considered

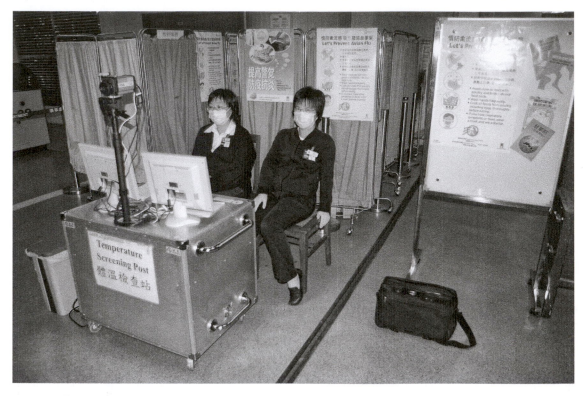

Plate 21 Temperature screening post, Hong Kong-Macau border

Source: Roger Keil

in assessing the spread of SARS within the global cities network. Thus, for example, an editorial in a national Canadian newspaper noted that:

> Globalization means that if someone in China sneezes, someone in Toronto may one day catch a cold. Or something worse – if, in Guangdong province, 80 million people live cheek by jowl with chickens, pigs and ducks, so, in effect, do we all. Global village, indeed.
>
> (Editorial Comment, *The Globe and Mail*, 28 March 2003)

The spread of SARS, therefore, to a large extent can be viewed as a "borderless" problem; one which reminds us that in the contemporary globalized context, infectious diseases cannot simply be considered as a public health issue that is exclusively confined to the developing world or pegged to a particular level of scale (such as the local or national).

In the present analysis, we will build on both the scalar, networked hierarchy model of the original

world city literature, and on the newer, topological view espoused by Smith and others. As such, we contend that the global cities network is characterized by both types of arrangements – that is, global cities are relatively fixed network nodes, bounded, historical, path-dependent, and rooted in national, regional levels, while, at the same time, they are also global diasporic historical geographies that are fluid, autonomous, self-producing, and "interactive and constantly in process" (Thrift, quoted in Smith 2003: 562). This double character is a necessary condition for the existence and functioning of the global city under the contemporary social and economic conditions. The global city can thus be characterized as requiring spatial and temporal fixes for functioning in the global economic system, but at the same time, these fixed qualities melt and reform daily, hourly, or even by the minute to conform to the rapidly changing demands of globalized capital. It is the dialectic of mobility and fixity that is truly characteristic of the urban condition under globalized circumstances. In global cities, one state cannot exist without the other. And it is precisely this dialectic that needs to be the focus of attention to understand the spread of pathogens in the contemporary era.

HUMAN MOBILITY AND MICROBIAL TRAFFIC

The classic disease triad of agent, host, and environment is often schematically portrayed as a static triangle, but Wilson (2001) notes that such a perspective inadvertently directs attention away from the dynamic interactions involved amongst the three components. In particular, since there must be appropriate contact with an infected person, animal or object for the transmission of infectious disease, spatial proximity and movement necessarily play important roles in the spread of disease. In the contemporary globalized world, spatial proximity and the dynamism of interactions takes on even greater significance than in the past. The dynamism of globalization necessitates a rethinking of how issues related to spatial proximity, microbial movement and the feasibility of geographic sequestration are interrelated – particularly, in light of how "relations among all geographical scales are continuously rearranged and reterritorialized" (Brenner 2000: 361). Under these conditions, one strategy in modifying the classic triad model to better understand disease diffusion under the modern globalizing circumstances is to incorporate concepts such as the network of global cities and mechanisms for pathogenic travel.

As part of the attempt to analytically capture the dynamic nature of globalization in the spread of disease, the concept of "microbial traffic" (Mayer 2000) is useful. Microbial traffic refers to the movement of pathogens in populations as well as the mechanisms involved in such movement. This includes a consideration of: cross-species transfer; spatial diffusion; pathogenic evolution (or change in the structure and immunogenicity of earlier pathogens); and changes in the human-environment relationship. The notion of microbial traffic can be used to re-situate the static triad model in a more dynamic context – a context that takes into account the phenomenon of the urban-global dialectic by emphasizing the movement of pathogens and humans, as well as by recognizing the changing features of the environment – all of which occur within the larger context of globalization. The first two features of microbial traffic

are especially relevant to the case of the global spread of SARS.

Cross-species transfer (from animals to humans) appears to be a mechanism common to the spread of many emerging and resurgent diseases. Contact with "exotic" species in live animal markets was thought to have facilitated the zoonotic transfer in the case of SARS, referring to the transfer of a disease from an animal host to humans (although it appears that bats may have been more likely to have been the original reservoir of the disease). The significance of the rural/urban interface becomes even more critical in the situation of rapidly emerging urban areas as an expanding urban population base increasingly extends into the hinterland as is the case in the Pearl River Delta of Southern China (where the SARS virus is said to have originated). The SARS coronavirus began its international spread as an infected senior professor of medicine from a university in Guangzhou (where he was treating SARS-infected patients) traveled to Hong Kong to attend a nephew's wedding on February 21, 2003. During his stay in the Metropole Hotel, the ailing professor infected at least 12 other guests who were visiting from other countries. Thus, the disease spread to other cities such as Toronto, Hanoi, Hong Kong, and Singapore.

THE MICROBIAL TRAFFIC OF THE SARS VIRUS AND THE GLOBAL CITY

The dynamics of the international diffusion of the SARS virus also highlights the importance of air travel for the diffusion of the pathogens in the contemporary era (Dodge and Kitchin 2004). The overall population travel flux and the international movements of passengers have accelerated over the last few decades and global cities have served as nodes in this network of travel (Smith and Timberlake 2002). Thus, not only do global cities serve as nodal points for the flow of information, commodities, financial transactions, and cultural goods that are commonly associated with global city functions, they also serve as nodal points for the transmission of disease with the airports in the global city serving as interchanges.

The movement of the SARS virus illustrated the hierarchical diffusion of the disease between global cities. Gould (1999: 199) similarly found that the international spread of HIV was controlled by hierarchical diffusion but was followed by a slow spread to smaller places from the major regional epicenters through spatially contagious diffusion. The case of SARS does seem similar to HIV with respect to hierarchical diffusion within the international context, however, the viral outbreak was contained within the City of Toronto and did not spread to smaller spaces. For this reason, in order to complete the picture with regard to the spread of SARS, we must consider not only the network configuration of global cities (i.e. hierarchical diffusion) but the nature of global cities themselves because the latter will influence local microbial traffic patterns. The relationship between Toronto and other global cities in the network, for example, is not based purely on economic, communication and resource flows, but one involving cultural and familial linkages between Toronto's diaspora communities and their respective ancestral communities. Ethno-cultural linkages provided by occasioned encounters also have implications for the microbial traffic of the SARS virus – both in relation to the microbial traffic between global cities, as well as to the lateral spread of the disease within particular local diaspora communities within global cities. Examples of both types of spread were found with reference to Toronto. Thus, urban processes such as the establishment/destruction of neighbourhoods, the in and out migration of people, and changes in the racial/ethnic composition of certain areas of the city may influence the spread of disease at the local scale of interaction. Furthermore, since many in the diaspora communities retain links with family and friends in the country of their origin, the interactions and developments unfolding at a local scale of one particular city are more likely to have some connection in some (unanticipated) way with interactions occurring in another city. It is precisely these types of unexpected, multiplying, and rapid interactional effects that contribute to the unpredictable and often surprising effects of local-global interactions.

In sum, the advantage of utilizing a global cities approach in the study of disease outbreaks in the contemporary world is that such a perspective is well-suited to the analysis of an increasingly complex, interdependent world, characterized by the flux and the movement of individuals and microbes in seemingly unanticipated ways. Furthermore, since it is projected that within the next 30 years, two-thirds of the world population will live in urban areas, a perspective that takes into account the nature of cities and the

relationships amongst cities – such as the global cities network approach – will likely become increasingly important for future analyses and insight into the future diffusion of "new and emerging diseases."

REFERENCES FROM THE READING

Ali, S. H. and Keil, R. (eds.) (2008) *Networked Disease: Emerging Infections in the Global City*, Oxford: Wiley-Blackwell.

Banta, J. E. (2001) Commentary: From international health to global health, *Journal of Community Health*, 26, 2: 73–7.

Brenner, N. (2000) The urban question as a scale question: Reflections on Henri Lefebvre, urban theory and the politics of scale, *International Journal of Urban and Regional Research*, 24, 2: 361–78.

Castells, M. (1996) *The Rise of the Network Society*, Oxford: Blackwell.

Dodge, M. and Kitchin, R. (2004) Flying through code/space: The real virtuality of air travel, *Environment and Planning A*, 36: 195–211.

Gezon, L. and Paulson, S. (2005) Place, power, difference: Multiscale research at the dawn of the twenty-first century, in S. Paulson and L. Gezon (eds.) *Political Ecology across Spaces, Scales, and Social Group*, New Brunswick, NJ: Rutgers University Press: 1–16.

Gould, P. (1999) *Becoming a Geographer*, Syracuse: Syracuse University Press.

Haggett, P. (2000) *The Geographical Structure of Epidemics*, Oxford: Clarendon Press.

Held, D. and McGrew, A. (2002) The great globalization debate: An introduction, in D. Held and A. McGrew (eds.) *The Global Transformations Reader: An Introduction to the Globalization Debate* (2nd edition), Oxford: Blackwell.

Keil, R. (2011) Transnational urban political ecology: Health, environment and infrastructure in the unbounded city, in G. Bridge and S. Watson (eds.) *The New Blackwell Companion to the City* (2nd edition), Oxford: Wiley-Blackwell.

Mayer, J. D. (2000) Geography, ecology and emerging infectious diseases, *Social Science and Medicine*, 50: 937–52.

Naylor, D. (2003) *Learning from SARS, Renewal of Public Health in Canada: A Report of the National*

Advisory Committee on SARS and Public Health, October, Ottawa: Health Canada.

Smith, R. G. (2003) World city topologies, *Progress in Human Geography*, 27, 59: 561–82.

Smith, D. and Timberlake, M. (2002) Hierarchies of dominance among world cities: A network approach, in S. Sassen (ed.) *Global Networks, Linked Cities*, New York: Routledge: 93–116.

Taylor, P. (2004) *World City Network: A Global Analysis*, London and New York: Routledge.

The Globe and Mail. (2003) Editorial: Confronting the perils of a shrinking world, March 28, www.theglobe andmail.com/news/national/editorialconfronting-the-perils-of-a-shrinking-world/article20448152/.

Walker, D. (2004) *For the Public's Health: Initial Report of the Ontario Expert Panel on SARS and Infectious Disease Control*, December, Toronto: Ministry of Health and Long-Term Care.

WHO (World Health Organization) (2004) Update 95 – SARS: Chronology of a serial killer, www.who.int/csr.

Wilson, M. L. (2001) Ecology and infectious disease, in J. L. Aron and J. A. Patz (eds.) *Ecosystem Change and Public Health: A Global Perspective*, Baltimore: John Hopkins University Press: 283–324.

THREE

25

"Flying high (in the competitive sky): conceptualizing the role of airports in global city-regions through 'aero-regionalism'"

from *Geoforum* (2014)

Jean-Paul Addie

EDITORS' INTRODUCTION

Jean-Paul Addie is an Assistant Professor at the Urban Studies Institute at Georgia State University and was previously a Research Fellow in the Department of Geography at University College London. As a critical urban geographer, his work interrogates the politics and production of urban infrastructure to address questions of mobility, governance, and social justice in an era of globalized urbanization. Lately, his research has paid particular attention to how transportation and higher education sectors shape the territorial and relational construction of urban space. Addie has especially focused on the political economy of transportation and the production of the metropolis in Chicago and Toronto and has published widely on topics such as neoliberal urban policy, suburbanization, comparative metropolitan governance, and regional transportation. In this chapter, Addie introduces the concept of aero-regionalism to explore the relationality/ territoriality dialectic and mechanisms of state territorialization at the nexus of globalization, air transport, and city-regionalism. He provides a relational geographic analysis of the impact of local institutional arrangements on the political and infrastructural integration of Chicago O'Hare Airport.

AIRPORTS AND THE CHALLENGE OF GLOBAL CITY-REGIONAL URBANIZATION

Air transportation holds a privileged position in studies of global city formation, development, and connectivity. Airports function as key interfaces through which global networks are moored in place and access to advanced air transportation plays a fundamental role in determining the relative centrality of cities within multiscalar urban systems. Efficient, cost-effective transportation connections are a prerequisite for companies to employ advanced logistical techniques and for localities to market themselves as global business hubs. A generalized shift towards the deregulation and privatization of commercial air transportation over the past four decades has only served to deepen the centrality of air infrastructure within a competitive, increasingly urbanized, global economy.

Against this backdrop, an influential policy consensus (employed varying degrees of effectiveness from Amsterdam and Phoenix to Singapore and Belo Horizonte) has crystallized around the growth potential of the "airport city" or "aerotropolis" (Güller and Güller 2003). City leaders around the world, under mounting pressure to expand air capacity to ensure their position in the world city network, are advised to embrace their airports. Developing advanced, modally integrated facilities, so the argument goes, can maximize locational advantages for New Economy industries "with the ultimate aim of bolstering the city's competitiveness, job creation, and quality of life" (Kasarda and Lindsay 2011: 174).

Investment in airport-enabling urban development, though, is neither a simple nor sustainable panacea for the challenges of economic globalization (Charles et al. 2007). Globalizing airport facilities require extensive and extended capital investment alongside the place-based accumulation of technological knowledge and organizational and geopolitical power. Aviation connectivity exposes urban centers to the threats of terrorism (Graham 2006) and enhances vulnerability to global pandemics (see Ali and Keil Ch. 24). Uneven economic development, in addition to localized environment impacts, invoke a complex and contested politics of scale surrounding airport infrastructure. As territorial gateways, airports are spaces of regulation and securitization as much as interfaces expediting global flows and economic activity.

Despite a resurgence of interest in the mobilities, territoriality, and cultural economy of air transport, relations between air infrastructure, globalization and local economic development remain undertheorized. Much global cities and global city networks literature tends to treat the politics and economics of air transport uncritically and often deploys weak scalar theorizing. Normative assumptions regarding airports' economic, social, and political impacts are substantially codified through quantitative metrics. Air traffic data are regularly used to demonstrate the relational connectivity of global city networks (Mahutga et al. 2010). However, the use of airline flows as a gauge for globalization presents significant methodological and analytical limitations, for instance the need to distinguish between destination and stopover airports (Derudder et al. 2009). Moreover, unquestioned tropes of global competitiveness underpin much political discourse on airport-oriented urban development with little consideration given to how global flows are actually grounded in, and conditioned by, local sociospatial structures (Adey et al. 2007).

Neither airports, nor their developmental logics or urban settings, can be considered as singular, homogenous, or uncontested entities in isolation (McNeill 2014). The spatiality of airports emerge as "the product of numerous interlocking geopolitical, economic, environmental, social, technical and commercial

practices that operate at a variety of spatial scales and manifest themselves in different ways in different places through time" (Budd 2009: 132). Global airports foster distinct governance modalities that interrelate with the production of city-regional space in contingent ways. Global airport infrastructures have therefore emerged as a key tool for strategic state interventions, but one that reflects the complexity of urban territoriality and political power in an era of competitive city-regionalism. The airport extends beyond its built form by drawing together a myriad of socio-technical systems and relations "as an organism of the urban economy" (McNeill 2010: 2861). They are an interface between the territorial (as immobile built environments embodying vast sums of fixed capital) and topological (channeling and regulating the movement of people and commodities) factors supporting the development of global urban centers. Consequently, the challenges of infrastructural integration and scaling governance place airports at the heart of global city formation and city-regional politics.

I refer to the processes of urban territorialization unfurling at the nexus of globalization, city-regionalism, and air transport as 'aero-regionalism'. This concept acknowledges airports as contested and contradictory urban spaces by paying close attention to how global air hubs': (1) evolving material infrastructure (i.e. terminal development, runway and flight path alignments, physical footprint, and local transport connections); (2) dynamic governance regimes (e.g. national air regulations, airport and airline management, and transportation and land use planning); and (3) multifaceted political and symbolic functions mediate the territorial and relational production of global city-regions. As I demonstrate in the case of Chicago, aero-regionalism discloses both the context for strategic state selectivity and key mechanisms for scaling state action global city-regions through a mosaic of sociotechnical and political networks.

AERO-REGIONALISM, CHICAGO-STYLE

The Chicago Department of Aviation (CDA) administers all aspects of airport operations at Chicago O'Hare International Airport and Chicago Midway International Airport. As a municipal cabinet position, CDA's Commissioner reports directly to the Mayor's Office, rendering the Chicago's chief executive the region's de facto airport manager. Municipal ownership has two central advantages for the City of Chicago. First, the City controls the economy of the region's major aviation infrastructure from everyday operations to economic development strategies and long range planning. Given the airports' importance as economic drivers, this arrangement places significant power over the regional economy in the City's hands. Second, by overseeing airports as a municipal concern, the operation and development of O'Hare and Midway are buffered from the conflicting interests of other regional actors. Projects can be developed with limited intra-governmental disruption. As a result, the planning and governance of Chicago's major globalizing infrastructure have been significantly distanced from other bodies shaping regional development, including the Chicago Metropolitan Agency for Planning (CMAP): the Metropolitan Planning Organization coordinating regional transportation and land-use planning across seven counties in northeastern Illinois. Planning expediency limits direct democratic involvement by many (mainly suburban) actors in the region while elevating politically mobilized technical knowledge of airport space over that of communities impacted by the negative externalities of airport operations.

The regional impact of Chicago's airports and persistent purported need to expand air capacity, however, suggest the City's airport governance regime is increasingly under pressure. As Chicago's air infrastructure undergoes its latest round of restructuring, suburban actors are looking to enhance their influence at O'Hare through regionalizing (or suburbanizing) the airport's physical infrastructure and economic flows.

THE O'HARE MODERNIZATION PROGRAM

The 1978 deregulation of domestic air transport in the United States exacerbated calls to expand Chicago's air capacity. Despite several terminal improvements, congestion remained an issue at O'Hare into the 1990s. Moreover, rapid growth at Atlanta's Hartsfield-Jackson International Airport threatened Chicago's established position as America's preeminent global air hub. Contrary to regionalizing strategies being pursued elsewhere in the United States that sought to "spread the burden" of metropolitan air traffic across multiple airports (Cidell 2006), Mayor Richard M. Daley refused to allow funds generated at O'Hare and

Midway to finance the development of a third Chicago airport beyond the city limits in south suburban Peotone. Rather, he turned his attention to constructing new runways at O'Hare (with the purported capacity to reduce overall delays by 79%) and opening a new passenger terminal on the western side of the airport (City of Chicago 2001). The O'Hare Modernization Program (OMP) presented Daley with an opportunity to cement his legacy on Chicago's landscape, as his father, Mayor Richard J. Daley, had done in opening O'Hare nearly five decades earlier.

While the 9/11 attacks shook the airline industry, both traffic levels and economic activity at Chicago's airports were showing signs of recovery by 2003. Popular and political opinion too was moving in favor of O'Hare expansion. Although the City maintained its tendency to operate unilaterally and guarded the details of airport development, key members of Chicago's growth machine, including the Commercial Club of Chicago, Business Leaders for Transportation, the Chicagoland Chamber of Commerce, and Global Chicago backed OMP. In the State capital, Springfield, the downfall of Governor George Ryan (who had backed Peotone Airport) and the 2003 election of Rod Blagojevich, a Democratic with strong Chicago connections, brought State backing to Daley's agenda. Blagojevich signed the O'Hare Modernization Act on August 6, 2003, removing State interference from OMP and granting the City of Chicago eminent domain powers beyond its borders for the project. Springfield thus acknowledged O'Hare's "essential role" in the national air transportation system and OMP's capacity to "enhance the economic welfare of the State" (Illinois General Assembly 2003). CDA (2012), too, adopted a regional rhetoric when estimating estimated OMP would create 195,000 jobs and $18 billion in regional economic activity while realizing c.$370 million in savings for air carriers and c.$380 million for passengers.

NEGOTIATION LOCAL POLITICS IN THE GLOBAL CITY-REGION

Chicago's aviation governance regime proved extremely effective for the City in marshaling the demands for the collective provision of regional air infrastructure. CDA developed OMP in house and away from potentially prolonged debates on regional air capacity. OMP moved ahead as a project of regional, statewide,

Plate 22 Chicago aerotropolis

Source: Roger Keil

and national economic significance, but under the guidance, and chiefly serving the interests, of the City of Chicago. After receiving State backing in 2003 and federal approval in 2005, the City commenced work on OMP and attempted to foster the view that the project was a fait accompli.

OMP, however, faced considerable suburban opposition from two coalitions with political and geographically distinct interests in, and knowledge of, Chicago's air space. The first was constituted by south suburban interests including Congressman Jesse Jackson Jr., Will County, and the South Suburban Mayors and Managers Association. The mainly lower-income, Democratic and African-American base of the inner south suburbs favored developing Peotone Airport as a means to re-center the depressed industrial south within Chicago's regional economy. The second coalition, the Suburban O'Hare Commission (SOC), represented a shifting network of predominantly wealthy, white, Republican northwestern suburbs that had protested expansion at O'Hare since the 1980s. At its height, SOC brought

together 17 municipalities concerned with the impacts of congestion, noise and air pollution on their communities. While the south suburban coalition's political connections and desire to bring an airport to Peotone necessitated the maintenance of cordial relations with the City of Chicago, SOC had no such interest and adopted a bunker mentality. SOC's most vocal members were those threatened by annexation of 433 acres for runway extensions: Bensenville and Elk Grove Village to the west, and Des Plaines in the northeast.

As OMP took off and SOC ramped up its campaign, the City of Chicago effectively undermined opposition by exploiting political fractures within the anti-OMP movement. The highly fragmented nature of the region's political geography enabled the City to target the benefits of OMP to appeal to the parochial interests of individual communities. Having been hit by the downturn in the aviation economy following 9/11, Des Plaines broke from SOC and welcomed the potential benefits of OMP for its industrial base. The restructuring of prospective OMP flight paths proved instrumental in shifting Des Plaines's political allegiances. The Village embraced the freight and cargo development vision proffered for the O'Hare area, adding c. 1.5 million feet^2 of logistics space between 2003 and 2008. Other municipalities, including Itasca, Schaumburg, and Wood Dale, welcomed the potential to mirror the established conference and business center development in suburbs to the east of O'Hare. Faced with declining support and the apparent inevitability of OMP, SOC's last stalwarts, Elk Grove Village and Bensenville, dropped their resistance as the old guard of municipal leadership was swept from office in 2008–2009 elections.

OPENING O'HARE TO THE REGION?

The issue of Western Access was a key factor in swaying suburban opinion on OMP, most pointedly in the case of DuPage County. Fearing the displacement of jobs and residences, DuPage spent millions of dollars fighting the City of Chicago in court to oppose O'Hare expansion and contributed $14,000 annually to SOC through the 1990s. The County reversed its stance on O'Hare following a change in leadership on the County Board in 2002. Incoming Board Chairman, Bob Schillerstrom, withdrew DuPage's opposition to OMP in January 2003, placing the burden of resistance on the dwindling number of municipalities in SOC.

The County's shift was, in part, a reaction to the election of the pro-OMP Blagojevich; as DuPage Board member Brien Sheahan argued, "[OMP] is a project that is going to occur. . . We're either going to have it imposed on us, or we can pull up a chair. . . and be a part of the shaping of the final plan" (cf. Meyer and Hilkevitch 2003: 1). Yet as Sheahan indicated, it also reflected a change in philosophy as the County began to think in terms of competitive regionalism. The prospect of reorienting the physical space of O'Hare westward presented the opportunity to deepen DuPage's integration within an aviation-based economy (Figure 1). Western Access catalyzed a spatial reimagining of the airport's position in the region that served to resolve the contradiction between local communities' interests and regional demands for globalizing infrastructure; "the old leadership saw [O'Hare] as an economic engine for Chicago and Cook County. The new leadership sees it as an economic engine for the greater region" (DuPage

County, interview 2008). After signing on to OMP, the County conducted a $370,000 economic development study – jointly financed with the City of Chicago – projecting the program would add $3 billion and 12,000 jobs to DuPage County's economy by 2015, increasing to $10 billion and 40,000 by 2030.

DuPage leadership's strategic shift hinged on the post-political reasoning that Western Access made OMP a universal benefit to the County. Yet Western Access continues to pose a challenge for communities close to O'Hare who fear the imposition of a regional vision (particularly the concentration of globally integrated cargo distribution facilities around O'Hare) will lock airport-adjacent municipalities into an overwhelming industrial development trajectory. Former SOC members have asserted their voices in an attempt to shape the form and function of development around O'Hare. This move has been most evident in the Illinois Department of Transportation's (IDOT) planning process for the extension of the Western Bypass and Elgin-O'Hare expressway, and the development of new transit facilities proposed in CMAP's 2010 regional plan, Go to 2040.

Western Access would enable the westward flow of economic activity from the airport, and with this, both DuPage County and municipal leaders have a renewed interest in gaining political influence over the future development of O'Hare. Regionalizing the orientation of O'Hare away from Chicago challenges the existing territorially defined basis of the City's global airport governance regime as other communities and organizations stake their claims to a seat at the table. Yet while IDOT has effectively brought together key interest groups and moved ahead with highway planning surrounding the airport, the City of Chicago continues to set the agenda at O'Hare and resists the suburbanization of its global infrastructure. This position partially supports the view that the City was always primarily concerned with maintaining political and economic control over its airports, but it also draws attention to the volatile nature of the relationships through which the City governs regional aviation. Despite the CDA and State of Illinois both vocally asserting the importance of Western Access to the regional economic effectiveness of OMP, the City of Chicago had not guaranteed the construction of a new western terminal, nor opening western access to O'Hare. Through the terms of American and United Airlines' 1985 lease agreement at O'Hare (set to expire in 2018), the airlines are obligated to finance

Figure 1 Opening O'Hare to the region

Source: Jean-Paul Addie

capital improvements at the airport in return for veto power over CDA plans. Although a tentative agreement had been reached in 2005, the impact of the 2008–2009 Financial Crisis led American and United to file a contract dispute with the City of Chicago in an attempt to scale back OMP. In late 2010, facing a global decline in air travel, CDA and the airlines agreed to postpone the development of a new western terminal until demand recovered. The particular dynamics of aero-regionalism in Chicago therefore deprived suburban communities the central benefits promised in return for backing OMP. Instead, the City of Chicago appears likely to direct the political and economic orientation of the airport towards the global core of the city-region for the foreseeable future.

CONCLUSION

Globalization may take off and land on the runways of major airports, but the complex dynamics of present-day state spatiality and global networked flows internalized within global airports resonate well beyond the taxiways and terminals of these territorial gateways. Global airports are generative of extra-local relations and flows and, as such, are often characterized as sites

of deterritorialization and points of assemblage within a relational space of flows. Yet, as they territorialize globalization in place, airports necessarily invoke "the production and continual reproduction of fixed socio-territorial infrastructures" (Brenner 2004: 56). Propelled by industry deregulation and tropes of global competitiveness, Chicago's international air hub has become a locus for substantial expansion plans at the same time as its regional governance regimes and territoriality become markedly more intricate. The 'global', so often the abstract spatial imaginary appealed to in airport discourses, is now clearly embedded within a complex, on-going, locally-contingent negotiation of scale.

Aero-regionalism contributes to our understanding of the relationship between global urbanization and air transport by demonstrating how international air hubs are not only generative nodes of economic activity located in metropolitan areas, but are fundamentally conditioned by their regional context. As the case of Chicago exemplifies, the technical and social infrastructures of global airports regionalize in important material, political, and symbolic ways, but the numerous multiscalar relations and vast fixed capital brought together by air infrastructure make them difficult for territorially defined actors to operate, plan, and govern. By embedding the territoriality and modalities of urban politics engendered by airports within broader processes of city-regionalism, aero-regionalism not only reveals the complexity of unstable and evolving regional governance processes, but discloses how the political and morphological dimensions of global city-regions are structured through a contested politics of mobility.

REFERENCES FROM THE READING

Adey, P., Budd, L., and Hubbard, P. (2007) Flying lessons: Exploring the social and cultural geographies of global air travel, *Progress in Human Geography*, 31, 6: 773–91.

Brenner, N. (2004) *New State Spaces: Urban Governance and the Rescaling of Statehood*, Oxford: Oxford University Press.

Budd, L. (2009) Air craft: Producing UK airspace, in S. Cwerner, S. Kesselring, and J. Urry (eds.) *Aeromobilities*, New York: Routledge: 115–34.

Charles, M. B., Barnes, P., Ryan, N., and Clayton, J. (2007) Airport futures: Towards a critique of the aerotropolis model, *Futures*, 39, 9: 1009–28.

Chicago Department of Aviation (CDA) (2012) About the O'Hare modernization program, http://www.cityofchicago.org/city/en/depts/doa/provdrs/omp/svcs/about_the_omp.html, accessed February 4, 2012.

Cidell, J. (2006) Air transportation, airports and the discourses and practices of globalization, *Urban Geography*, 27, 7: 651–63.

City of Chicago (2001) A proposal for the future of O'Hare, Office of the Mayor, Chicago.

Derudder, B., van Nueffel, N., and Witlox, F. (2009) Connecting the world: Analyzing global city networks through airline flows, in S. Cwerner, S. Kesselring, and J. Urry (eds.) *Aeromobilities*, New York: Routledge: 76–95.

DuPage County (2008) Personal interview, December 16.

DuPage County (2009) Personal interview, January 21.

Graham, S. (2006) Cities and the 'war on terror', *International Journal of Urban and Regional Research*, 30, 2: 255–76.

Güller, M. and Güller, M. (2003) *From Airport to Airport City*, Barcelona: Gustavo Gill.

Illinois General Assembly (2003) O'Hare Modernization Act, 620 ILCS 65/1, www.ilga.gov/legislation/ilcs/ilcs3.asp?ActID=2488&ChapterID=48, accessed on February 4, 2012.

Kasarda, J. D. and Lindsay, G. (2011) *Aerotropolis: The Way We'll Live Next*, New York: Farrar, Straus and Giroux.

Mahutga, M. C., Ma, X., Smith, D. A., and Timberlake, M. (2010) Economic globalization and the structure of the world city system: The case of airline passenger data, *Urban Studies*, 47, 9: 1925–47.

McNeill, D. (2010) Behind the 'Heathrow hassle': A political and cultural economy of the privatized airport, *Environment and Planning A*, 42, 12: 2859–73.

McNeill, D. (2014) Airports and territorial restructuring: The case of Hong Kong, *Urban Studies*, 51, 14: 2996–3010.

Meyer, G. and Hilkevitch, J. (2003) DuPage in flip-flop backs O'Hare plan: County shifts sides in expansion fight, *Chicago Tribune*, January 15: 1.

26
"One package at a time: the distributive world city"

from *International Journal of Urban and Regional Research* (2011)

Cynthia Negrey, Jeffery L. Osgood, and Frank Goetzke

EDITORS' INTRODUCTION

Cynthia Negrey is Professor of Sociology at the University of Louisville, Kentucky. She has published widely in her specialty areas of gender, political economy, and urban labor markets. Jeffery Osgood serves as the Principal Deputy to the Provost at Westchester University and as Dean of the School of Interdisciplinary and Graduate Studies. His areas of expertise include local economic development and program evaluation. Frank Goetzke is Associate Professor of Urban and Public Affairs at the University of Louisville. His research interests are in urban and regional economics, applied microeconomics and econometrics, transportation planning, and environmental policy. In this chapter, Negrey, Osgood, and Goetzke shift the discourse on globalization and world cities from the conventional emphasis on finance and advanced producer services to the global distribution networks, which they consider as "handmaidens of world trade." Focusing on UPS Airlines and its headquarter city, Louisville, Kentucky, this case study speaks at once to the networked infrastructures that sustain the global economy and to the need to do grounded empirical research in "ordinary cities." Due to space restrictions, the excerpt below does not contain the detailed documentation of growth machine politics and contested development. It focuses instead on the networked aspects of producing a distinctive "place" in a "space of flows." This chapter shows how local economic development efforts, in the context of air-cargo delivery services expansion, have transformed Louisville into a distributive world city.

This article takes off from the global cities literature but shifts the analysis away from the commonplace focus on finance and producer services to a different industry and global function, namely distribution. By doing so, a different representation of a world-city hierarchy and network emerges. At the top of the hierarchy are cities that ship the most 'stuff' via air in a network of global distribution nodes, in this case cargo airports. These nodes are close to major sites of production and global markets. The article focuses on one company, UPS Airlines, and one city, Louisville, Kentucky, in a case study that shows how local

economic-development efforts combined with expansion of air-cargo delivery services in the context of globalization to transform Louisville into a distributive world city and the flying-lobster capital of the United States.

Louisville is, at least by reputation, among the 'diverse, but ordinary' (Robinson 2003) places in the modern global economy. Florida's (2002) ranking of the 50 largest metropolitan statistical areas (MSAs) in the US placed Louisville near the bottom in creativity, based in part on its relatively low percentage of professional, managerial, technical and artistic

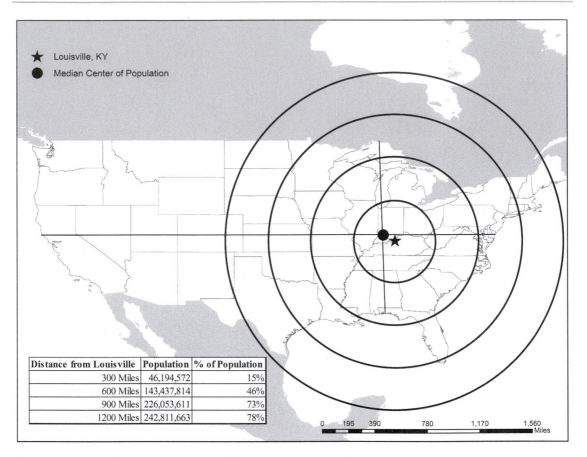

Distance from Louisville	Population	% of Population
300 Miles	46,194,572	15%
600 Miles	143,437,814	46%
900 Miles	226,053,611	73%
1200 Miles	242,811,663	78%

Figure 1 Louisville, Kentucky's proximity to US median center of population

Source: Cynthia Negrey, Jeffery L. Osgood, and Frank Goetzke

occupations; underdeveloped technology sector; and relatively few patents. Louisville is most definitely not a command-and-control center of global finance, nor is it among the alpha, beta or gamma cities in the world network of advanced producer services firms. Louisville's location is advantageous for distribution companies, however, because the winter climate is mild and it is situated within a day's drive of two-thirds of the country and about 70 miles from the median population center of the United States (see Figure 1).

Globalization is not only about capital flows and advanced producer services, of course. It is also about world trade of manufactured goods and agricultural commodities. The products of offshore manufacturing often come to the US and elsewhere as imported consumer goods, and global distribution chains are an essential part of the process. Most of the products arrive at ports on cargo ships, but much also arrives at airports on jumbo jets. Distribution warehouses store the products until they are distributed to retailers or directly to consumers (i.e. online shopping). In the other direction, jets and ships carry machine parts and components for final assembly, which are inputs in offshore manufacturing. According to Dicken (2007: 410), the circulation processes that connect together all the different components of the production network are 'absolutely fundamental', and logistics industries themselves are huge, worth around US $4 billion annually.

The essential function of the distribution services is to intermediate between buyers and sellers at all stages of the production circuit. This involves not only the physical movement of materials and goods, but also the transmission and manipulation of information relating to such movements. It involves above all the organization and coordination of complex flows across increasingly extended geographical distances. In that respect, these services have been revolutionized by technological developments in transportation and communication. The major types of organization involved in logistics and distribution include transportation companies, logistics service providers, wholesalers, trading companies, retailers and e-tailers. The growth of the market for logistics and distribution services is closely related to growth in the economy as a whole (Dicken 2007: 411–414).

Global distribution occurs in the context of a larger post-industrial service economy in the developed countries. This chapter reconstructs Louisville's development as a distributive world city coterminous with the expansion of UPS's package-sorting hub and UPS Airlines' headquarters in the city. It has been a 25-year process involving the regional airport authority positioning the airport for growth; land acquisition on the part of urban government; neighborhood opposition to noise and displacement; and a distribution company that has literally become a handmaiden of globalization.

WORLD CITIES AND TRANSPORT

World cities are commanding nodes in the world economy, exhibiting dense patterns of interaction between people, goods and information facilitated by a rapidly expanding and sophisticated global network of transport services and infrastructure (Keeling 1995: 115). Hierarchies of transport networks and services connect world cities, important components of a city's aspiration to world-city status. Thus, world cities and transport are symbiotic (Keeling 1995: 118, 129). However, globalization depends (ironically) on fixed infrastructure; as demand for this infrastructure grows, there is pressure on existing infrastructure to provide more capacity (Smith 1997: 188; Cidell 2006: 651). Airports, for example, are at the intersection of global and local, but individual characteristics of places—history, governance, politics and geography—determine whether and how demands for capacity are met. Discourses from debates over airport expansion reveal how localities position themselves. As such, globalization is inherently localized, interacting in different ways with the existing political and economic structures of places. World-class port and airport facilities can confer substantial regional advantage by lowering costs in a just-in-time economy, but these mega-projects, designed to enhance regional

competitiveness in the global economy, become objects of intense local debate.

UPS IN LOUISVILLE

UPS has been described as the 'oil that makes the gears go' in the global economy (Weber 2007). By 2008 UPS had become the ninth-largest airline in the world, with 3,000 pilots, flights around the globe, 268 planes and more aircraft on order. In addition to US airports, UPS serves airports in most of the Canadian provinces and select cities in Mexico, the Caribbean, Central and South America. UPS has an extensive presence in countries in Europe—notably in the UK and Germany—as well as the Middle East, Asia and Australasia. Table 1 shows a regional breakdown of airports served outside the US.

Although all flights do not necessarily originate in or pass through Louisville, the location of UPS Airlines' corporate headquarters in Louisville gives the city command-and-control functions in distribution comparable, at least in concept, to the command-and-control functions in finance of world cities like New York, London and Tokyo. Recently Sassen (2010: 155–6), responding to Jacobs et al. (2010), has argued that global cities arising from handling the complex functions of ports are different from cities like London (which arises not only from its past port functions, but also from finance and commerce). Air hubs such as Louisville (which did serve as a river port in its early history) are not ports, but they provide distributive functions ancillary to ports.

40	Continental Europe
21	United Kingdom and Ireland
13	Caribbean, Central and South America
11	Middle East
9	Canada and Iceland
9	'Other' Asia
6	China
5	Mexico
3	Japan
2	Australia
1	India

Table 1 Number of international airports served by UPS by region

Source: Cynthia Negrey, Jeffery L. Osgood, and Frank Goetzke

Louisville International Airport, previously known as Standiford Field, grew rapidly in recent years. In 2006 and 2007 it was the third-busiest cargo airport in North America (compared to fifth in 2000) and the ninth-busiest in the world (twelfth in 2000), according to Airports Council International's final traffic reports. The airport handled a record 1,983,032 metric tons of cargo in 2006 (a 30% increase over 6 years); most of that volume came from UPS, which operates Worldport, its all-points air-package hub and airfreight hub, at Louisville International. UPS set a record for package volume in 2006, delivering nearly 4 billion parcels.

The air group operates as a separate UPS division with employment of nearly 19,000. About half work at Worldport, primarily in part-time package-sorting jobs on overnight shifts. The others work in support positions for the airline, including some highly paid white-collar positions (e.g. pilots). About 55% of the pilots live in the Louisville area and can earn US $300,000 annually as captains. Table 2 shows Louisville's rank as a 'newly emerging' airfreight hub (Hesse 2010: 83) among the 30 leading world cargo airports, as measured by tonnage in 2007, when the worst effects of the recent recession were yet to be felt. The city's airport moved up to ninth from tenth in 2005. Eleven of the 30 top airports are in the US with the others in Asian and European countries. Memphis is the world's largest cargo airport thanks to its role as FedEx's main hub. Anchorage, at number three, is a link to Asia for both

FedEx and UPS. By 2007, Los Angeles and New York, which had in 2000 ranked at numbers three and six respectively, dropped out of the top 10 world cargo airports. Certain cities, specifically Hong Kong, Singapore, Shanghai, Tokyo, Los Angeles, New York and Dubai, have developed as major container ports as well as leading air-cargo hubs. They are key geographical nodes on the global logistics map.

UPS's first presence in Louisville was in 1954, when the company opened a package delivery facility to serve major population centers in Kentucky. In 1981, UPS opened a hub for its second-day air service in Louisville, which it offered through a number of contract carriers. Next-day air service was initiated in 1982, which followed a similar business operation, hiring other airlines to fly planes owned by UPS. By the late 1980s, UPS owned 99 planes flown by pilots from four carriers. UPS announced its intention to form its own airline in 1987 as it became increasingly difficult to coordinate different airlines into one air operation. The first UPS Airlines flights were scheduled in early 1988, when two DC-8s flew from Louisville to Chicago and Milwaukee.

Coterminous with the expansion of its air service and eventual establishment of UPS Airlines, the Louisville Regional Airport Authority voted to sell 127 acres near Standiford Field to UPS, and gave UPS the right of first refusal on land the airport might acquire adjacent to the site in the future. State and local leaders believed UPS's acquisition of the land would entice

1	Memphis, Tennessee (MEM)	16	Taipei, Taiwan (TPE)
2	Hong Kong, China (HKG)	17	Chicago, Illinois (ORD)
3	Anchorage, Alaska (ANC)	18	London, England (LHR)
4	Shanghai, China (PVG)	19	Bangkok, Thailand (BKK)
5	Incheon, Korea (ICN)	20	Beijing, China (PEK)
6	Paris, France (CDG)	21	Indianapolis, Indiana (IND)
7	Tokyo, Japan (NRT)b	22	Newark, New Jersey (EWR)
8	Frankfurt, Germany (FRA)	23	Luxembourg (LUX)
9	Louisville, Kentucky (SDF)	24	Tokyo, Japan (HNI)b
10	Miami, Florida (MIA)	25	Osaka, Japan (KIX)
11	Singapore (SIN)	26	Brussels, Belgium (BRU)
12	Los Angeles, California (LAX)	27	Dallas-Ft. Worth, Texas (DFW)
13	Dubai, UAE (DXB)	28	Atlanta, Georgia (ATL)
14	Amsterdam, Netherlands (AMS)	29	Cologne, Germany (CGN)
15	New York, New York (JFK)	30	Guangzhou, China (CAN)

Table 2 The 30 leading world cargo airports in 2007

Source: Cynthia Negrey, Jeffery L. Osgood, and Frank Goetzke

UPS to commit to build a superhub in Louisville. At the time, UPS workers sorted 150,000 parcels a day on two shifts at a complex located just south of Standiford Field. UPS signed an agreement to buy the land, anticipating the beginning of major construction in 1985 or later. When the first phase of construction of a new terminal began, the airport authority sold an additional 7.4 acres to UPS that connected the sites purchased earlier. UPS anticipated increasing its Louisville-based air service by about 40% in the short term, to 90 daily departures. Total air service employment had reached 600, including about a dozen transferred from the company's headquarters (then in Connecticut) to help manage its National Air Center in Louisville, and the company planned to recruit 500 more employees, mostly part-time package handlers.

Air-cargo delivery was booming due to competition among providers of the service, deregulation of the air-cargo industry and changes in the way America conducted business. Regulatory changes in 1977 allowed airfreight companies to own and operate their own cargo planes; increasing fuel prices and changes in plane design made it more difficult for big shippers to rely on passenger airlines; and FedEx in the 1970s had set a new standard for reliable express delivery of documents and high-priority cargo. UPS anticipated overseas growth opportunities; in the mid-1980s it operated a ground delivery service in West Germany and had only one daily overseas flight, from JFK in New York to Cologne, West Germany. The overnight-letter market was the fastest-growing but most competitive segment of the parcel-delivery business in the 1980s; UPS was credited with taking fourth-class postal away from the US Postal Service. The air division headquarters had been moved from Greenwich, Connecticut, to Louisville, bringing headquarters total employment in Louisville to about 70. By this time, the air hub employed about 2,000 workers, mostly part-time package handlers, sorting almost 300,000 parcels daily.

Expansion plans began in earnest in 1988, after UPS announced that it would invest US $20 million in the Louisville air express hub and could acquire up to 200 additional acres. The plans to expand the air hub coincided with the launch of UPS Airlines, which gave the company control of its own planes, pilots and support crews. UPS bought a major package delivery business in Italy to add to its existing base in West Germany, and it was moving to expand express service in Japan, with a goal of serving 'a good part of the

world' within two years. At the time, UPS claimed a 15% share of the air express market compared to FedEx's 58%. Coincidentally, Wal-Mart was planning a regional distribution center located in southern Indiana, within about an hour's drive of Louisville's airport. At the time of UPS's proposed expansion in the late 1980s, Standiford Field was estimated to add US $220 million to the local economy, and it had become Jefferson County's third-largest employment center (with more than 4,100 full-time workers and accounting for an estimated 3,700 additional spin-off jobs). Each dollar generated on site was estimated to create US $1.47 in total business activity in the community, and the airport contributed US $6.8 million annually to corporate, sales, property and personal income taxes. UPS had become a major player in the local economy and was projected to become the area's largest private employer by 1995 if its expansion plans went through, displacing Louisville's largest employer, General Electric, with 13,000 workers at the time of UPS's announced expansion. Local officials also hoped to attract new commercial airlines with the expansion of Standiford Field.

In 1988, UPS added 19 countries, mostly in Asia and Western Europe plus Australia and New Zealand, to its overseas operations. This was in addition to previously existing service in the US, Puerto Rico, Canada, Japan and 21 countries in Western Europe, the latter served by the company's sorting hub in Cologne. Expanded Asian services would use common carriers to deliver from the US west coast to two new UPS hubs in Hong Kong and Singapore. Small charter aircraft would deliver from these hubs to cities throughout the Pacific Rim. FedEx, by comparison, delivered to 110 countries. In 1989 plans to construct a US $18 million aircraft maintenance hangar in Louisville were announced. In 1990 UPS won approval in a competition with Dayton, Ohio-based Emery Worldwide for a new Japan route authorized under a recent trade agreement between the US and Japan. FedEx and Northwest Airlines operated the existing cargo routes. To service the new route, UPS would ship from any commercial airport in the US, some cargo going directly to UPS's west coast hub in Anchorage, Alaska, then on to Tokyo, but consolidating most cargo in Louisville for shipping to Anchorage and on to Tokyo.

By 1998 UPS had become number one in package shipping overall (well ahead of the US Postal Service and FedEx); number two in domestic air express (behind FedEx); and number three in international express (behind DHL and FedEx). A long list of companies is believed to have located in the area partly because of UPS, including better-known names like Acer America, Ann Taylor, Natural Wonders and Stride Rite; and the top 20 of these accounted for an estimated 3,500 jobs. An air hub expansion launched in 1998, the company's largest capital construction project ever, was named by Site Selection Magazine as one of the top 10 economic development deals in the world that year. Investigating further for purposes of ranking the top ten (which Site Selection does not do), the local newspaper placed the expansion at numbers one and two, based on projected employment growth (6,000) and size of capital investment (US $970 million) respectively. Subsequently, UPS's first China Express service to Beijing was initiated in 2001; the US $1 billion Louisville hub expansion begun in 1998 was completed in 2002, resulting in naming the facility Worldport; and another US $1 billion expansion begun in 2006 was completed recently.

Acquiring a complex of warehouses to the south of the airport, UPS appeared on Standiford Field's radar for the first time in 1981. UPS's first structural changes to the airport included the addition of a taxiway and apron (parking area for aircraft). Unknown at the time, this would be the first of many projects forever changing Standiford Field and surrounding communities. These infrastructure developments by UPS at Standiford Field were from the outset touted as indicators of economic growth for the city of Louisville. What was initially characterized as a 'modest ripple' would soon become a huge wave sustained by the local growth machine. In 1998, after court, social and political battles between affected and concerned residents and communities on one hand and representatives of the local growth machine on the other, and UPS found the package of amenities offered by Louisville and the state of Kentucky to be sufficient to designate Louisville as the site of its planned expansion. Louisville would now be UPS's superhub in a project named UPS Hub 2000. In 1999 the economic impact of the expansion of the airport was found to have created 29,500 jobs, which beat forecasts of 27,000 jobs for the year 2010. The total payroll was calculated to be approximately US $1.13 billion dollars. From a mere US $337 million in 1986 to US $3.8 billion in 1999, the airport now produced massive amounts of business-related expenditures in Louisville. Figure 2 illustrates the scope and extent of

the areas affected by UPS's growth at Standiford Field. The neighborhoods in question — Highland Park, Ashton–Adair, Prestonia, Standiford and Edgewood — were solidly working class and did not exhibit the characteristics commonly associated with blighted or depressed areas.

The growth of UPS at Louisville, Kentucky, the expansion of Standiford Field and its subsequent transformation into Louisville International Airport in 1995 represent the classic 'growth' versus 'anti-growth' conflict. A group of political and landed elites interested in the potential for increased revenues generated by land surrounding the airport won out, while residents lost property and communities were dismantled. Nearly 20 years later, all that remains of these neighborhoods is 77 families; 39 continue to fight against their removal. Following Louisville's airport expansion to UPS's Worldport, both the city's economy (especially the logistics sector) and suburban land use have changed rapidly. Greater Louisville, Inc. identified 96 establishments that located in the metropolitan area between 1992 and June 2007 to be near UPS's distribution facilities, accounting for 8,724 jobs, about US $500 million in new investment and US $233 million total payroll. UPS itself accounted for 8,599 new jobs from 1998 to June 2007, US $2.1 billion in new investment and US $386 million in new payroll.

In the tradition of urban economics and according to the monocentric city model, land use is determined by the trade-off between accessibility and land rent. When the airport, located at the urban fringe, becomes the business center, well connected by interstate highways it is expected that logistics companies will locate nearby in the suburban areas, especially because their facilities are also land-intensive. In the suburbs, transportation and warehousing companies can take advantage of both good transportation infrastructure as well as lower land rents. This leads to what Hesse (2008) describes as the geography of distribution, exhibiting traits of both centralization and decentralization (Keeling 1995: 116). Following Hesse's (2008) analysis, the Louisville airport with UPS represents the main transportation hub, located about five miles south of the central business district and surrounded by a ring of logistics companies located even further beyond the urban core.

The geography of distribution receives empirical support by looking at the location quotients for the transportation and warehousing industrial sector

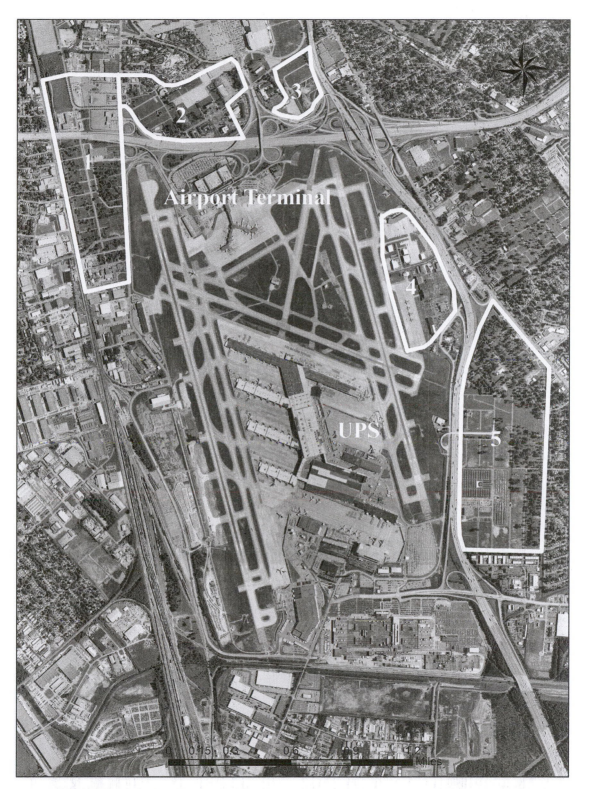

Figure 2 Neighborhoods displaced by airport expansion

Source: Cynthia Negrey, Jeffery L. Osgood, and Frank Goetzke

	1998	2006	2007
Louisville MSA			2.06
Jefferson County*	1.68	1.88	
All suburban metro counties			2.64
Bullitt County, KY	0.54	1.71	1.99

Table 3 Location quotient for the logistics sector in the Louisville, KY–IN MSA (metropolitan statistical area)

*In 2003 the Louisville City and Jefferson County governments consolidated, creating Louisville Metro, thus enlarging the legal boundaries and geographic scale of the city.

Source: Cynthia Negrey, Jeffery L. Osgood, and Frank Goetzke

(combined as the logistics industry). As shown in Table 3, throughout the Louisville metropolitan area the logistics location quotient is greater than the value 1, and also increased significantly between 1998 and 2006–07. A location quotient value larger than 1 indicates that this industry has in relative terms seen more employment in the region than nationally, representing some kind of locational comparative advantage.

CONCLUSION

As for the applicability of the world-city concept to the Louisville case, Louisville's occupational structure is unlike that of major world cities. Louisville has a relatively small percentage of the professional, managerial and technical occupations that comprise Florida's (2002) creative class, but has a relatively high percentage of working-class occupations as defined by Florida. This occupational structure does not give Louisville the bifurcated 'dual city' character of, for example, New York City. Louisville's occupational structure is 'bottom heavy' by comparison. This suggests that Louisville's social structure as a distributive world city remains similar to that of some 50 years ago when the local economy was dominated by manufacturing branch plants. Nor does distribution generate the vast wealth of finance, so Louisville's income and wealth structures would be unlike those of major financial world cities. Yet, arguably, the distribution of goods is equal in importance to the availability of

investment capital in the global economy. Without a doubt, however, the growth of UPS Airlines, both as a company and a worldwide network of air routes and urban nodes, over the past 25 years, is a defining feature of the local Louisville economy and its status as a distributive world city among the 'diverse, but ordinary' places.

REFERENCES FROM THE READING

Cidell, J. (2006) Air transportation, airports, and the discourses and practices of globalization, Urban Geography, 27, 7: 651–63.

Dicken, P. (2007) Global Shift: Mapping the Changing Contours of the World Economy (5th Edition), New York: Guilford Press.

Florida, R. (2002) The Rise of the Creative Class, New York: Basic Books.

Hesse, M. (2008) The City as a Terminal: The Urban Context of Logistics and Freight Transport, Burlington, VT: Ashgate.

Hesse, M. (2010) Cities, material flows and the geography of spatial interaction: Urban places in the system of chains, Global Networks, 10, 1: 75–91.

Jacobs, W., Ducruet, C., and de Langen, P. (2010) Integrating world cities into production networks: The case of port cities, Global Networks, 10, 1: 92–113.

Keeling, D. J. (1995) Transport and the world city paradigm, in P. L. Knox and P. J. Taylor (eds.) World Cities in a World-System, Cambridge, MA: Cambridge University Press: 115–31.

Robinson, J. (2003) Johannesburg's futures: Beyond developmentalism and global success, in R. Tomlinson, R. A. Beauregard, L. Bremner and X. Mangcu (eds.) Emerging Johannesburg: Perspectives on the Postapartheid City, New York and London: Routledge: 259–80.

Sassen, S. (2010) Global inter-city networks and commodity chains: Any intersections? Global Networks, 10, 1: 150–63.

Smith, N. (1997) The satanic geographies of globalization, Public Culture 10, 1: 169–89.

Weber, H. R. (2007) UPS looks to future on its centennial, The Courier–Journal, August 5: D1.

27

"Global cities between biopolitics and necropolitics: (in)security and circuits of knowledge in the global city network"

from Michele Acuto and Wendy Steele, eds.,
Global City Challenges: Debating a Concept,
Improving the Practice (2013)

David Murakami-Wood

EDITORS' INTRODUCTION

David Murakami-Wood holds a Canada Research Chair (Tier II) in Surveillance Studies at Queen's University in Canada. His current research focuses on smart cities and he is the author of *The Watched World: Globalization and Surveillance* (2015, Rowman & Littlefield). Together with Mark Graham's contribution in Chapter 28, this chapter has been excerpted from an important volume–*Global City Challenges*, edited by Michele Acuto and Wendy Steele in 2013. The book highlights new issues and insights concerning both the intellectual project of global city studies and the real developments in today's global cities. In this contribution below, Murakami-Wood examines the proliferation of urban security policies and measures in the network of global cities with a case study of Rio de Janeiro. Using a Foucaultian approach, he views cities and the urban networks as biopolitical entities, and he highlights the interplay of large-scale security measures and micro-scale technologies of power that regulate urban (in)securities. Similar to the essay by McCann and Ward (Ch. 29), Murakami-Wood underscores the role of knowledge agents, such as the travelling technocrats, who introduce and disseminate security applications in the global network of cities.

GLOBAL URBAN (IN)SECURITY

For Urban Studies and Geography, the city is both referent and at the same time the terrain of study, and thus, some analyses or urban (in)security draw more on top-down assumptions that consider the (in) security of the city and others from the bottom-up, that consider (in)security in the city. However, it is also the case that the referents have been shifting: on the one hand, "security is coming home" – there has been a long-term towards the (military) securitization of the urban and domestic, accelerated by 9/11 and the subsequent War on Terror; and on the other hand, critical approaches to International Relations, Security Studies and Geopolitics have meant that the formally top-down approaches have become increasingly aware of social, cultural and psycho-logical concerns. This has meant that the referent object of security in most contexts is almost always multiple and contested.

Michel Foucault (2007) argued that security was the primary object of government. This he referred to as biopolitics, the management of life itself. 'Biopower', which consists of biopolitics plus specific disciplinary mechanisms (Foucault 1977), is applied to constantly mobile and changing populations. The concept is therefore also mobile and dependent on the conception of 'life'. Equally, neither insecurity and security are simply material states of existence, but also processes, either movements towards greater security and the "capacity for taking action to secure a better future" (Ericson 2007: 4) or the risk of an increasingly insecure existence. Insecurity is a condition approaching death, or its nearest equivalent living state. Biopower (and therefore also 'necropower,' the management of death or those in the bare-life subjectivity of the 'living dead', bodies marked for death, see Mbembé 2003) operates through territory as well as on population and neither security-life nor insecurity-death are equally shared across societies or distributed evenly across space, but are continuous produced and reproduced outcomes of processes of sociospatial construction, whether resulting indirectly from spatial fixes to crises of capitalism or directly from actions with security as their specific objective.

Foucault was writing primarily with reference to two spatial scales: firstly, the body in the institution and secondly, the nation-state: cities do not figure very strongly in his analysis. However, it is clear from a historical analysis of urban formation that the city is a biopolitical mechanism, and not for the popularly understood defensive function of city walls but for the fact that cities concentrate populations in space and allow first sovereign power but then to be effected over people in a way that could not be carried out so easily over a more dispersed population, indeed in many ways a city is as much the designation of a kind of population as it is a space. The specific forms of power included violence, human surveillance through agents and spatial control through internal and external walls and gates (Coaffee et al. 2008).

The city as a biopolitical mechanism extends to the urban network. Urban networks have a long history; economic and sociospatial transformations made the global urban network of the 20th century very different from the far less intensely connected cities of previous centuries. By the early 1900s, the networks of economic relationships linking cities in the mediaeval

and early modern period had moved from geographically limited mercantilist world economies into a single more integrated world system of liberal capitalism (Abu-Lughod 1991). Thus cities in the modern industrial nation-state were not the same as cities in the pre-modern period, and cities in an era of post-industrialism and globalization are different again, however the seductions of creating a purely linear pattern should be avoided. First of all, as Foucault observed, although the governmentality (the art of government) has evolved and new ideas have emerged over time it is not that one form of governmentality entirely replaces another, but that all continue to be available and can be deployed on different territories and populations in a variety of configurations. This is partly because there is a never a singular 'population'. The first lesson from Abu-Lughod, Mbembé and other writers of (post)colonial theory and history is that we should throw out the imperial universalism that would have a western/northern development pathway as what constitutes 'human' history, and relatedly, understand that in practice colonial nations and hegemonic classes produced (and continue to produce) such discourses at least partly with the aim of constructing certain kinds of bodies and territories. In this context we can see the emergence of an urban necropolitics, where some are marked for inclusion and life, and some for exclusion and death (or bare life).

In the contemporary period, we have seen a more clearly emerging division between kinds of (in)security that affect different cities and also different groups and social classes within cities, as a rapidly neoliberalizing capitalism took hold in the second half of the 20th century. The end of Fordism and the advent of neoliberal globalization has led to the partial reconstruction of class divisions on a global scale, generated huge inequalities within cities and in particular led to the vast expansion of global south cities. Combined with the continued legacy of the colonial city (see King Ch. 49) this had led to internal divisions, with a privileged overseer class living alongside but separate from a vast pool of insecure migrant labour.

This can be macro-exclusionary, as in the new city-state models of governmentality which entirely exclude both city builders and/or the mass of the regular urban working class from the city, either being kept in external camps, as with the builders of Dubai, or having to cross national border controls to work and return home each day, as in the case of Singapore.

Plate 23 Complexo do Alemão, Rio de Janeiro

Source: Xuefei Ren

This can also happen against design intentions. In Brasilia, the 'utopian' modernist capital of Brazil, created by avowed socialist architects, Lucio Costa and Oscar Niemeyer, has become exclusionary in its evolution. Only the middle class of government workers are able to afford to live in the clean modernist housing surrounding the government core, and the working class of cleaners, shop workers and labourers are bussed in, in the morning, from less salubrious satellite cities via the underground bus station and bussed out again, in the evening.

However, for most global cities, these macro-exclusionary modes are only part of the story. Biopower in the urban context is not exercised in strict territorial divisions by sovereign power. A complex geography of micro-exclusion has emerged, produced by a shifting assemblage of state and private power: thus terms like 'splintered' (Graham and Marvin 2001) or 'fractured' (Koonings and Kruijt 2007) have come to describe the new physical and economic landscapes of global cities. In Brazil, this has been well described by Teresa Caldeira (2001) as a "city of walls", in which it is not the state's sovereign power alone that produces territorial division but the retreat of the state – or the fact that it was never quite as extensive as was imagined in the first place – and its supplementing or replacement by overlapping private security companies and economically-differentiated self-protection.

Within this fractured urban network, societies are now characterized by 'a generalised insecurity deriving from the precariousness of social and economic relations' (Garland 2001: 133) creating a situation of 'advanced marginality' (Wacquant 2008) which although its particular sociospatiality differs from place to place, has common features; a new 'precariat' whose conditions of life and existence are constantly in question not because they are entirely outside of global circuits of urban knowledge but because they are the immediate 'Other', a new 'dangerous class' (Standing 2011). The security knowledge generated and applied in the global urban network is directed at solving 'them' and the existential security problem that they represent to the wealthier classes, in other words at securitizing their existence for the benefit of the mainstream society. This can take the form of either biopolitical projects for inclusion, or necropolitical projects to exclude or remove them from the existential sphere of consideration of the wealthy, either through mass imprisonment – whether that be through a prison-industrial complex as in the USA or the transformation of whole cities into open-air prisons, as in Gaza in Palestine – continued repression or mass movement elsewhere. The global urban circuits through which such knowledge flows are not direct, and have multiple sources and connections. Such policy learning takes place not just between states or urban authorities or between elite politicians and policy-makers, but often between mid-level bureaucrats of various specific organisations, state or private, united by their technical expertise and specialisms – a group Larner and Laurie (2010) in their work on the simultaneous globalization and privatization of water policy, term 'travelling technocrats'.

(IN)SECURITY IN RIO DE JANEIRO

A particular example of this complex and differentiated reproduction of (in)security in the global city network is Rio de Janeiro, and its specific 'precariat': the residents of the informal settlements (favelas). Rio de Janeiro has a complex governmental and security landscape. The city has its own government under the Prefeito (Mayor), responsible for some aspects of security, particularly emergency services, disaster preparedness, and low-level crime through the Guarda Municipal (City Police) (GM). The city sits within the Estado (state) of Rio de Janeiro, under the Governador (Governor), responsible for the Polícia Militar (Military Police, PM), the enforcers, and Polícia Civil (Civil Police, PC), the detectives. The situation is complicated by the existence of the notorious Batalhão de Operações Policiais Especiais (Police Special Operation Battalion, BOPE), a semi-independent force within the PM. The Federal Government of Brazil is responsible for most social policy and also has its own Polícia Federal (Federal Police, PF), responsible for anti-corruption and cross-border crime.

This multilevel, multi-agency structure results in constantly changing and potentially contradictory government directions. Yet, despite nominally universal citizenship, the most marginal have remained often outside official recognition (Holston 2008). Historically, such invisibility meant no protection against arbitrary detention, torture and death, as was the case during military rule. Militias (Autodefesas Comunitárias - Community Self-Defence organisations, ADCs) remain at large in Rio de Janeiro, and as a report to the state legislature in 2009 revealed, involve members of the army, police forces and local politicians. To counter this at national level, Luiz Inácio Lula da Silva ('Lula'), President from 2003–2011, introduced programs, which endure beyond his tenure, including Fome Zero (Zero Hunger) and Programa Bolsa Família (PBF) (family support program) as well as the national biometric Identity Card. These are all biopolitical measures, aimed at creating a material and psychic sense of inclusion and are extraordinarily popular.

For those who are the generally willing subjects of such measures, particularly the most marginal, those living in informal settlements known as the favelas or morros of Rio there has also been long-term policy oscillation between deliberate state ignorance and problematization. However, with their continued growth, and immediate proximity to some of the wealthiest communities in the city, ignoring them has become impossible. Instead, the question is the nature of the problem they present: whether one of crime and security or one of social policy.

The former problematization sees the morros as a source of threat: the morros are controlled by armed drug gangs, and no-go areas for non-moradores (residents). And many morros are indeed controlled by gangs associated with the cocaine trade, affiliated either to the Commando Vermelho (Red Command, CV) or it offshoots, particularly the Amigos de Amigos (Friend of Friends, AdA). The answer in this case has been necropolitical: either obliteration, expulsion and/or rebuilding along official lines, or police invasion and retreat. The latter perspective in contrast works in a more conventional biopolitical manner by incorporating morros into wider society through the provision of infrastructure, education and social services.

The biopolitical problematization predominated in two particular periods: firstly, under Governor Lionel Brizola, who concentrated on the building of schools in morros, as well as the upgrading of some services, and more recently under Mayor César Maia, whose

Favela-Bairro (Favela to Neighbourhood) program was endorsed by major international institutions and donors from the UNDP to the World Bank. However, more recently the pendulum has swung back towards the necropolitical problematization, however instead of a purely hardline approach, there is a two-handed strategy. Both the City and State are united in an approach to the morros characterized by Mayor Eduardo Paes as 'choque de ordem' ('shock of order'). Favela-Bairro has been abandoned in favour of a new policy of 'pacification', in which special Military Police Unidades de Policía Pacificadores (Police Pacification Units, UPPs) occupy selected morros, driving out gangs and instilling state control. After the restoration of state authority, social programs are introduced, very much in the way that 'hearts and minds' programs are introduced in the aftermath of military invasions. When I visited in 2009, the program had just started with three communities, Santa Marta, Cidade de Deus (the 'City of God' of movie fame) and Jardím Batan. By December 2012, it covered 28 morros (see Table 1), including the biggest and most complex of all, Complexo do Alemão and Rocinha, and involved around 8,000 officers. According to the state security administration, by 2014 there would be more than 40 UPPs involving 12, 500 police officers.

It is clear from talking to those involved in the development of the UPP strategy that the long-term ambitions remain broadly biopolitical and inclusive. According to a former PM Colonel turned state security administrator, this is all about demographic transformation – it is a question of population management, "favelas in Rio de Janeiro these days have 150,000–200,000 residents, they're cities and they need support from visible policing integrated within the community". The result of this biopolitical understanding is that the forms of control that are envisaged as necessary cannot be the traditional military solution but need a combined approach that borrows from previous such experiences around the world from Colombia to Haiti, from Israel to New York City.

An important source of security knowledge is the travelling circus of the global city network: sports mega-events. Rio de Janeiro is hosting the 2016 Olympic Games and part of the 2014 FIFA World Cup. Sports mega-events are a significant element in economic competition between global cities and security is now part of the management of reputational risk that enables global cities to attract further investment. The securing of the games entailed the increasing insecurity of the marginalized residents of morros that are either perceived as significant threats to order – in that they harbour active and armed gangs – or who are occupying space required for event venues or are simply alongside major routes into the city where they might detract from visitors'

Year	No. of New UPPs	Running Total of UPPs	Communities Occupied by UPPs (Date Occupied)
2008	1	1	Santa Marta (19/12)
2009	4	5	Cidade de Deus (16/2), Jardim Batan (18/2), Babilônia e Chapéu Mangueira (10/6), Pavão-Pavãozinho e Cantagalo (23/12)
2010	8	13	Ladeira dos Tabajaras/Cabritos (14/1), Providência (26/4), Borel (7/6), Formiga (1/7), Andaraí (28/7), Salgueiro (17/9),Turano (30/10), Macacos (30/11)
2011	5	18	São João, Quieto e Matriz (31/1), Coroa, Fallet e Fogueteiro (25/2), Escondidinho e Prazeres (25/2), Complexo de São Carlos (17/5), Mangueira/Tuiuti (3/11)
2012	10	28	Vidigal (18/01), Fazendinha (18/04), Nova Brasília (18/04), Adeus/Baiana (11/05), Alemão (30/05), Chatuba (27/06), Fé/Sereno (27/06), Parque Proletário (28/08), Vila Cruzeiro (28/08), Rocinha (20/09)

Table 1 UPPs in Rio de Janeiro, December 2012

Source: Secretaria de Estado de Segurança do Rio de Janeiro (SESEG), 2012.

aesthetic experience and generate psychic insecurity (Gaffney 2010).

Finally, tying many other sources together is globalizing technocratic surveillance. By 2012, Rio had not one but two integrated control rooms under construction or in operation, and a vastly expanding network of cameras. The first is ostensibly not a police control room, but exactly what the GM officer had hoped for: the hub of a so-called 'Smart City' initiative, the Centro de Operações Rio da Prefeitura (Rio Prefectural Operations Centre) largely funded by IBM, to harness 'big data' to solve multiple problems of urban government. This Mayoral initiative links weather forecasting, traffic control and emergency services in a single large control room, with real time monitoring and response. Its focus is on anticipating, pre-empting or responding to emergencies, including in particular landslides caused by flooding, and traffic accidents. This is important for the security of residents of the morros, who tend to live on land that is very steep and/or vulnerable to flooding. This risk is often used as a basis for evictions and slum clearings instead of a source for humanitarian action.

FUTURE DIRECTIONS

The integration of Rio de Janeiro into global circuits of urban security knowledge looks set to accelerate and intensify. The Smart City control room is being joined by a new police control centre bringing together all the forces under the Security Secretary of the State of Rio's mandate. This promises much but has been continuously behind schedule. Part of this project involved acting as a command and control centre for the FIFA World Cup in 2014. Although sports mega-events have often been seen as a driver of security innovation in cities, it seems that in Rio, the Cup and the Olympic Games are additional drivers to an already expanding neoliberal security agenda.

Rio is not simply a passive receiver of sources of information circulating in global urban knowledge networks. Just as Favela-Bairro circulated as a social exemplar, so Rio's contemporary biopolitics is a discursive resource for other global cities. This is no accident: the City has relentlessly promoted the initiative through social media and slick online video

Plate 24 Luxury housing, Barra da Tijuca, Rio de Janeiro

Source: Roger Keil

marketing which describes the centre of operations as the "heart and soul of Rio", and Eduardo Paes himself has been presenting himself at 'progressive' public, like the TED talks, as a forward-thinking techno-savvy Mayor who can manage his city in real time from anywhere in the globe (Paes 2012). At the same time, other Latin American cities, in particular Buenos Aires in Argentina, have been eyeing the UPP model. And more generally, Brazil's example is also being eagerly promoted by the American state as the democratically acceptable face of urban securitization.

But Rio's security agenda is not going unchallenged, from three main sources. Firstly, the occupations of the morros have not ended gang violence. Secondly, pushing out drug gangs from the morros of Rio has resulted in a large upsurge of crime in places to which they have been displaced, in particular the neighbouring city, Niterói. And finally, there is the emergence of an anti-occupation and anti-surveillance politics in some UPP-occupied communities but has allies among middle-class human rights advocates and left-wing politicians. Whether successful or not, the Rio initiatives will continue to circulate within global urban networks of security knowledge. In an era of neoliberalizing government, failure merely generates a perceived need for further biopolitical innovation.

REFERENCES FROM THE READING

Abu-Lughod, J. L. (1991) *Before European Hegemony: The World System AD 1250–1350*, Oxford: Oxford University Press.

Caldeira, T. (2001) *City of Walls: Crime, Segregation, and Citizenship in São Paulo*, Berkeley, CA: University of California Press.

Coaffee, J., Murakami-Wood, D., and Rogers, P. (2008) *The Everyday Resilience of the City: How Cities Respond to Terrorism and Disaster*, Basingstoke: Palgrave Macmillan.

Ericson, R. V. (2007) *Crime in an Insecure World*, Cambridge: Polity Press.

Foucault, M. (1977) *Discipline and Punish: The Birth of the Prison*, London: Penguin.

Foucault, M. (2007) *Security, Territory, Population: Lectures at the College de France 1977–1978*, New York: Picador.

Gaffney, C. (2010) Mega-events and socio-spatial dynamics in Rio de Janeiro, 1919–2016, *Journal of Latin American Geography*, 9, 1: 7–29.

Garland, D. (2001) *The Culture of Control: Crime and Social Order in Contemporary Society*, Chicago, IL: University of Chicago Press.

Graham, S. and S. Marvin (2001) *Splintering Urbanism*, London: Routledge.

Holston, J. (2008) *Insurgent Citizenship: Disjunctions of Democracy and Modernity in Brazil*, Princeton, NJ: Princeton University Press.

Koonings, K. and Kruijt, D. (eds.) (2007) *Fractured Cities: Social Exclusion, Urban Violence and Contested Spaces in Latin America*, London: Zed Books.

Larner, W. and Laurie, N. (2010) Travelling technocrats, embodied knowledges: Globalising privatisation in telecoms and water, *Geoforum*, 41, 2: 218–26.

Mbembé, J-A. (2003) Necropolitics, trans. L. Meintjes, *Public Culture*, 15, 1: 11–40.

Murakami-Wood, D. (2015) *The Watched World: Globalization and Surveillance*, Lanham, MD: Rowman & Littlefield.

Paes, E. (2012) The four commandments of cities, ted. com, www.ted.com/talks/eduardo_paes_the_4_ commandments_of_cities.html.

Standing, G. (2011) *The Precariat: The New Dangerous Class*, New York: Bloomsbury.

Wacquant, L. (2008) *Urban Outcasts: A Comparative Sociology of Advanced Marginality*, Cambridge, MA: Polity.

THREE

28
"The virtual palimpsest of the global city network"

from Michele Acuto and Wendy Steele (eds.),
*Global City Challenges: Debating a Concept,
Improving the Practice* (2013)

Mark Graham

EDITORS' INTRODUCTION

Mark Graham is Associate Professor and Senior Research Fellow at the Oxford Internet Institute and an Associate in the University of Oxford's School of Geography and the Environment. His research is on information and communications technologies (ICT) and development, Internet, and information geographies. He has published widely in leading journals on many aspects of digital geographies. In this chapter, Graham investigates the global city network with the metaphor of a virtual palimpsest, by which he refers to cities being "incessantly made and remade, and layered with historical, contemporary, tangible, intangible, visible, invisible, material, and virtual elements." Graham points to the increasing importance of digital information as a source of additional layers in the palimpsest of globalizing cities. He identifies four main aspects of relational properties of globalizing cities in the digital sphere—virtual absences, virtual presences, ghettoization, and the ephemerality of local information, all of which have the potential to destabilize the existing global urban hierarchy.

FOR THE CARTOGRAPHIC ATTRIBUTES OF THE INVISIBLE[1]

Cities are comprised of bricks and mortar, concrete and glass, roads, rails, pipes and cables, people, plants and animals. The layers of cities also include the many histories, memories, legends, and stories that people ascribe to place (Crang 1996; Graham 2010). Yet cities have been going through two important transitions that have brought into being new dimensions that profoundly matter for the ways that we interact with our urban environments. Cities are no longer just confined to their material presences: they have become both digital and digitised. This chapter focuses on the many, often invisible and ephemeral, digital layers of cities.

The virtual elements of cities are immensely significant. Cities ooze data; they are structured by code and software; they cast innumerable digital shadows.

PALIMPSESTS

The notion of a palimpsest was first employed to refer to medieval writing blocks that could be reused while still retaining traces of earlier inscriptions. Because those earlier inscriptions could never fully be erased, every writing block was a composite containing traces of all inscriptions (Crang 1998). The notion of a palimpsest thus becomes useful for describing the urban environment. Cities are incessantly made and

Figure 1 The material/digital palimpsests of Manhattan
Source: Mark Graham

remade, and layered with historical, contemporary, tangible, intangible, visible, invisible, material, and virtual elements.

Digital urban layers can take myriad forms. The most visible of which are probably the digital maps that many people use to navigate through cities. Google, Yahoo!, Apple, OpenStreetMap, Baidu, and many other companies and organisations all host publicly accessible platforms that reflect cities. However, these services also become the platform for an almost unimaginable amount of additional content that both reflects the materiality of cities and augments it with additional content. This additional content is comprised of photographs, blogs, tweets, social media checkins, webcams, videos, and Wikipedia articles. These layers of digital representations are then further reproduced and repurposed in the ways that they annotate the urban environment (e.g. see Figure 1 for an example of the layering of digital content onto the materiality of the city).

The virtual content layered in and over cities is also not just of-the-moment, but also of-the-past: allowing for an 'unavoidable continuity' of the city's geography (Beauregard and Haila 2000). Some of these layers are relics of earlier events, places, and experiences (for instance old user-reviews that hover over particular places). Other layers are much more current (for instance the incessant posts in micro-blogging services that react to live events).

This peculiarity of the city as a material/virtual palimpsest means that the city itself can be experienced in entirely new ways. Not only do we no longer experience and interact with the city solely from our grounded offline positionalities, but we also experience many traces and data shadows of the city from earlier points in time. In other words, the enrolling of innumerable digital layers into the palimpsests of our

cities has fundamentally changed what the urban environment is and what it means to many of its inhabitants. Importantly, this leads us to wonder whether the material/virtual palimpsests of cities are able to encourage more generalised and less geographically unique digital layers, or if the selective reflections and representations of the city reinforced fragmented and splintered urbanisms. Furthermore, does this mean that the cities with the most visible digital presences are better able to project themselves and reinforce their global city statuses?

AUGMENTED REALITIES

We therefore need to ask how to conceptualise the ways in which the hybrid material/virtual layers of cities are experienced. It is here that the idea of 'augmented reality' is particularly useful. The term has been broadly applied as a way of describing experiences of place that are supplemented with digital information. The trend toward digital augmentation can be traced via three successive moments in the development of Internet practices and technologies: the growth of authorship, the emergence of a geospatial web (or Geoweb), and the rise of the mobile web. The rapidly increasing number of content contributors coupled with the ability to easily annotate places in the Geoweb has resulted in dense layers of virtual information augmenting some parts of the world. Furthermore, as ever more people use connected mobile devices (i.e. smartphones and tablets), layers of information over cities are no longer just accessible non-proximately. In other words, we are able to bring in all of this information into our portable, mobile devices. It can all augment the ways in which we bring place into being whilst we are in place, enacting place. We quite literally have access to all of the information augmenting our cities in the palms of our hands.

We know that representations of places are more than just lines and shapes on maps. Augmented information shapes what is known and what can be known about cities. This discoverable knowledge, in turn, influences the myriad ways in which urban spaces are (and can be) experienced. We should therefore be increasingly concerned with the ways in which augmented inclusions and exclusions, visibilities and invisibilities, will shape digital representations. As we experience not just the city, but the augmented city, digital geographic representations don't just influence how we think about places, but also, in a very real sense, influence how we move through, interact with, and enact place.

CODE/SPACE

Digital representations are not the only way that immaterial information mediates our urban experiences. Equally important is the role of software or code (i.e. formalised rules for information into other outputs and representations). Dodge and Kitchin (2004, 2005) distinguish between 'code/space', 'coded space', and 'background coded space'. 'Code/spaces are parts of our cities in which code or software in essential to the production or enaction of that space. A failure of code in such spaces inevitably results in a serious disruption of those spaces (ticket machines at a train station would be an example of this). Coded spaces also rely on code to function, but a failure of code would not render the space unusable (a busy traffic intersection would be an example). Background coded spaces are dormant until accessed (for instance with a mobile phone) and transformed into code/spaces or boded spaces.

Many code/spaces and coded spaces don't just alter the ways that cities are enacted and brought into being for their inhabitants, they also emit data shadows and information trails. These emitted shadows of 'big data' can take a variety of forms, but most are manifestations or byproducts of human/machine interactions in code/spaces and coded spaces. We now see hundreds of millions of connected people, billions of sensors, and trillions of communications, information transfers, and transactions producing unfathomably large data shadows (Anderson and Rainie 2012). Examples include geolocated mobile phone data showing spatial patterns of conversations, real-time electricity usage, and records of all underground journeys taken with smartcards (such as London's Oyster, Hong Kong's Octopus, and New York's MetroCard).

When using data about the city, policy makers and researchers no longer just take snapshots in time or samples from particular people and places. In other words, policy processes that shape global cities are likely to be ever more data-driven. Although many local governments and private sector entities do not fully know how best to utilise such a panoply of data, we are seeing an increasing number of examples of ways that urban data shadows can be fed back into algorithms and urban layers, and services; and ultimately have significant effects on how people move through the city.

DIGIPLACE

Because urban data shadows are now so unfathomably large, complex algorithms are needed to sort, rank, and order the virtual layers of cities. It is these algorithms that form an important part of our final concept that is useful for elucidating links between the virtual and the material: 'digiplace.' Digiplace is a heuristic for the subjective mixing of code, data, and material places (Zook and Graham 2007). It "encompasses the situatedness of individuals balanced between the visible and the invisible, the fixed and the fluid, the space of places and the space of flows, and the blurring of the lines between material place and digital representations of place" (Zook and Graham 2007: 1327). In other words, digiplace becomes useful as a way of bringing together the material and digital layers of the city, understandings of how the virtual can augment the material through technological mediations, and analyses of the power that code has to shape the ways that place is brought into being.

Importantly, digiplace is also a way of imagining the highly subjective and individualised experiences of the melding of cities and information. Cities (or places) have never been statically and objectively knowable. The London that I know is very different from the London that anyone else knows. This is not because of any particular insights into the workings of the city that I personally possess, but rather is because we know, experience, and enact fundamentally different cities. But, because of the complex and targeted algorithms used to mediate our interactions with and use of urban augmentations, digiplaces can potentially become further reified, segmented, and individualised. There is nothing inherently problematic about the fact that fundamentally different cities can be presented to us, but when combined with other issues of power, voice, and information inequalities, the opaque and selective digital mediations of our cities can start to become cause for concern. It is to these worries that the rest of this chapter now turns.

VIRTUAL CHALLENGES: BIG DATA AND THE EPHEMERAL, AUGMENTED CITY

This section highlights four main concerns related to the virtual geographies of global cities. First, the issue of information absences is addressed. Second is the issue of informational presences. Third and fourth are the interlinked issues of ghettoization and the ephemerality of local information. The targeted nature of urban information, code, and data shadows means that we all experience fundamentally different cities: a concern that has far-reaching implications for how we should think about cities.

Virtual absences

Mapping content indexed in Google Maps is a useful way to explore these layers of information about material places (because it is the world's most used website, and aggregates a huge variety of data sources). In Google, (a service with a stated aim to "organise the world's information and make it universally accessible and useful") we see that some cities are covered with highly dense virtual layers of content and information; many others have only very sparse virtual presences. The Tokyo metropolitan region (with a population of about thirty-five million people), for instance, is layered with more content that the entire continent of Africa (with a population of over one billion people). Examinations of other platforms for digital layers and representations of cities (e.g. Flickr, Wikipedia, etc.) display similar patterns of absences and presences. In other words, most data currently do seem to indicate that the density of information is highest in the usual suspects in the Global Cities literature while other cities are simply left off the map.

Not only are representations of urban environments characterised by highly uneven geographies, but the production of those layers is equally concentrated. Despite the fact that there are now over two billion Internet users, a very small minority tends to produce the bulk of digital information. Nowhere is this more apparent than in Wikipedia (a free encyclopedia with a mission to "contain the sum of all human knowledge"). Wikipedia in theory allows anyone with an Internet connection to make changes and contribute content. In practice, however, we see stark geographic and gendered inequalities.

Some of these informational absences undoubtedly come from disconnectivity and poor communications capabilities. Other absences also arise from informational fragmentation (e.g. in China and a handful of other countries, viable competing services exist to challenge global standards [Sina Weibo instead of Twitter, Baidu Baike instead of Wikipedia, etc.]) and censorship at either micro- (e.g. workplaces) or macro- (i.e. countries) levels.

But why should we care if London is covered by a denser cloud of information than Lagos? Why does it matter if some places are digitally mirrored by content that omits the voices of women, minorities, the oppressed, the invisible? Why do these augmentations really matter? The answer is that geographic representations don't just influence how we think about places, they also, in a very real sense, influence how we move through, interact with, and enact place. In other words, there are real, material consequences for cities and places absent from digital representations.

Virtual presences

The lack of information about, or contributions from, many cities in the world is only one way that certain people and places can find themselves at relative disadvantages. The massive data shadows (i.e. geocoded information or digital representations of place) over many parts of the world can cause a few important problems of their own.

As virtual layers of cities become ever more influential, there will be increasing needs and pressures to study, map, and understand them; and because the these virtual layers almost entirely exist in digital form, the data lend themselves more easily to certain types of analysis. In particular, this means that we are increasingly in an age of 'big data' in which samples, inferences, speculations, and hypotheses are no longer necessary. The data shadows that we emit about so many social and economic activities can be harvested across entire countries and populations in a relatively easy manner.

However, as we increasingly come to rely on 'big data' shadows, there is a lot that can be lost. Having more data about any particular process or practice does not necessarily equate to having more insights. Large-scale analyses of large-scale datasets necessarily cause us to ask only certain types of questions and derive only certain sorts of answers. Geodemographics have, for instance, been used by financial institutions from the urban to the national-scale for loan-scoring and credit-worthiness checks (e.g. Goss 1995). In an era of 'big data,' a lack of codified data can therefore mean lack of access to necessary financial resources. In other words, financial risk (amongst many other things) is intimately tied to codified and standardised information. Places with thin data shadows can therefore suffer from exclusion and barriers simply because of an absence of data rather than actual higher risk. The increasing reliance on 'big data' shadows can also start to self-reinforce the importance of those very shadows, with the viability of less automated methods of scoring and analysis (e.g. meetings, conversations) ever more undermined.

An especially relevant concern in many of the world's most wired cities is the issue of privacy and a decreasing ability for networked urban citizens to expect anonymity. It has long been possible for state security agencies to piece together detailed pictures about all but the most reclusive of citizens by triangulating sources of information like mobile phone calls, credit card purchases, and travel card journeys. However, private firms are now beginning to develop an equal amount of sophistication in the methods that they use to make inferences about urban inhabitants.

Ghettoization

Cities have never been stable, homogenous experiences. My experience of Manchester is unique, as is the urban experience of the two other million people who live there: we all inhabit the same geographic area, but ultimately encounter and enact very different cities. It might be assumed that the virtual dimension to cities allows people to move away from the grounded and material nature of urban splintering (e.g. Graham and Marvin 2001) to have more homogenous urban experiences due to the fact that we all access the same virtual representations. However, in many ways, accessing augmented digital information allows us to reify and reinforce the existing ways that we enact and understand cities. The rest of the section discusses two important ways in which digital ghettoization can happen.

Language is an obvious but crucially important way that different annotations and augmentations of the same place can be constructed. Representations and digital elements of cities can sometimes exist in hundreds of languages, all of which highlight, describe, critique, and order place in unique ways. A simple example from Bangkok can illustrate these very different linguistic shadows of place. In the maps below, Google Maps was queried for content about Bangkok containing the words "temple" or "วัด" (temple in Thai). The size of the circles in each map indicates the amount of content that Google indexes about that particular part of the city.

We know that that mapping references to "temple" in English unsurprisingly highlights high-profile tourist destinations. In contrast, Thai content highlights entirely different parts of the city. We see a focus on the Temple of Dawn and the many other temples on the eastern bank of the river. There are also clusters of content around the Golden Mount, Chinatown, and the temples in Sathon (e.g. Wat Yan Nawa). Again, these are locations in the city that are more likely to be known and frequented by Thais than foreigners. This case of temples indexed by Google in Bangkok doesn't tell us much about temples or Bangkok; but it does illustrate how fundamentally different virtual layers of the same city exist for English and Thai speakers.

In early 2012, Microsoft made headlines with a patent that proposed to offer tailored and selective representations of cities available to users of one of its apps. The app, which later was dubbed the 'avoid ghetto' app, was tailored to allow pedestrians to avoid certain 'undesirable' parts of cities (Keyes 2012). The proposal attracted much criticism, but illustrates just one of the ways that as personal data shadows become more encompassing and code becomes more sophisticated, the virtual dimensions of cities can become increasingly tailored and personalised.

Does it matter that different representations of a city are being produced and reproduced for different people? Urban inhabitants are accessing not just filtered bubbles of information, but also fundamentally different cities. Balkanised bubbles of augmented urban information have the potential to reinforce real, material, balkanized spaces. As a result, we need to ensure that urban research maintains a focus on mapping and theorising how the fluid and sometimes fleeting representations that exist on the internet might reinforce balkanized, hermetically sealed imagined spaces.

Ephemeral cities

Not only do the virtual layers of cities help us to enact fundamentally different places, but they also define the ephemerality of places. Doreen Massey (2005) argued that places are necessarily ephemeral temporary constellations. In other words, cities never were and never will be fixed and static entities. However, because of the increasingly hybrid material/digital nature of urban environments we now see a much more rapid reconfiguring of the raw materials of cities than ever before.

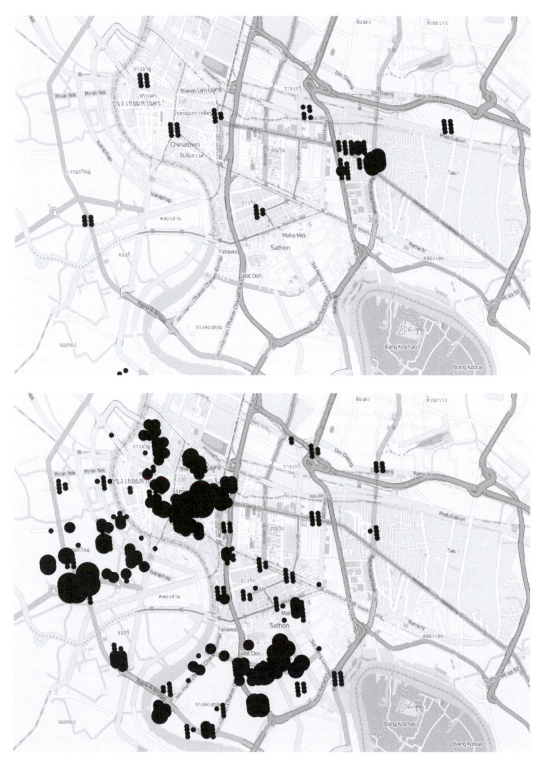

Figure 2 References to "Temple" in English (left) and Thai (right) in Bangkok

Source: Mark Graham

Almost every platform and repository of digital geospatial information is characterised by some degree of informational flux, but it is in information search that we can observe the unfixity of information at its most powerful. We see the increasing relevance of local search through the battles that are played out in order to be listed on the first page of results for search-terms, the multi-billion dollar industries of search advertising and search engine optimisation, the fact that the utility of large directories of information has long since faded, and the fact that many urban inhabitants now access and find local services through searches on mobile phones and local mapping services. Search engines and practices of local digital search thus play a central role in the ways in which much of the world accesses, enacts, and brings into being specific types of information. This local information, as mediated through search engines (a simple example of this would be a local search for a bicycle mechanic or public swimming pool) has become unfixed and destabilised.

First, every search relies on a vast 'ecosystem' of networked information that is both created and ordered by a crowd of contributors. The incessant online activity of millions of people means that not only is content itself constantly changing, but so are the preferences that feed into rankings that separate the visible from the invisible. A related implication is that large global cities covered by 'thick' layers of content will likely also be the places that are also characterised by the most significant amount of informational ephemerality.

Second, the sorting algorithms that mediate the enormous amount of crowd-sourced content have become spatially, socially, and temporally unfixed. Information accessed through search engines is thus decentred and destabilised. In other words, answers that search engines provide to our questions vary substantially based on our social, spatial, and temporal positionalities. Furthermore, each time that we are presented with (and influenced by) temporally, socially, and geographically targeted content, created and filtered by a crowd of millions, we reinforce the power of that content.

The ephemerality of urban information ultimately means that it becomes increasingly difficult for different people to discover authorial intent and to share bundles of space-time trajectories. Customisation, targeting, rapidly switching networks, and the temporal unfixity of information all increase the difficulty of talking about what is 'at' any particular place. Yet, it remains important to do just that, and understand how particular places are represented and brought into being.

NOTE

1. This phrase was coined by William Gibson in his 2008 novel, *Spook Country*.

REFERENCES FROM THE READING

Anderson, J. and Rainie, L. (2012) *The Future of Big Data*, Pew, http://pewinternet.org/Reports/2012/Future-of-Big-Data/Overview.aspx.

Beauregard, R. A. and Haila, A. (2000) The unavoidable continuity of the city, in Marcuse, P. and van Kempen, R. (eds) *Globalizing Cities*, London: Basil Blackwell: 22–37.

Crang, M. (1996) Envisioning urban histories: Bristol as palimpsest, postcards, and snapshots, *Environment & Planning A*, 28, 3: 429–52.

Crang, M. (1998) *Cultural Geography*, London: Routledge.

Crang, M. and Graham, S. (2007) Sentient cities: Ambient intelligence and the politics of urban space, *Information, Communication & Society*, 10, 6: 789–817.

Dodge, M. and Kitchin, M. (2004) Flying through code/space: The real virtuality of air travel, *Environment and Planning A*, 36, 2: 195–211.

Dodge, M. and Kitchin, M. (2005) Code and the transduction of space, *Annals of the Association of American Geographers*, 95, 1: 162–80.

Goss, J. (1995) We know who you are and we know where you live: The Instrumental rationality of geodemographic systems, *Economic Geography*, 71: 171–98.

Graham, M. (2010) Neogeography and the palimpsests of place, *Tijdschrift voor Economische en Sociale Geografie*, 101, 4: 422–36.

Graham, S. and Marvin, S. (2001) *Splintering Urbanism*, London: Routledge.

Keyes, A. (2012) This app was made for walking – but is it racist? *NPR*, www.npr.org/2012/01/25/145337346/this-app-was-made-for-walking-but-is-it-racist.

Massey, D. (2005) *For Space*, Thousand Oaks, CA: Sage.

Zook, M. and Graham, M. (2007). The creative reconstruction of the Internet: Google and the privatization of cyberspace and digiplace, *Geoforum*, 38: 1322–43.

29

"Relationality/territoriality: toward conceptualization of cities in the world"

from *Geoforum* (2010)

Eugene McCann and Kevin Ward

EDITORS' INTRODUCTION

Eugene McCann is Professor of Geography at Simon Fraser University, and Kevin Ward is Professor of Human Geography and Director of the Manchester Urban Institute at the University of Manchester. Both McCann and Ward have done comparative fieldwork in a large variety of locations to examine the global urban network of policy transfers. Based on the authors' individual and joint fieldwork, this chapter examines the relational and territorial properties of global urban networks. Focusing specifically on the increased worldwide mobility of urban policies, they point to the fact that in a globalizing world "policies and policy-making are intensely and fundamentally local, grounded, and territorial." Employing the mobilities approach (McCann and Ward 2011), this chapter highlights both fixities and mobilities in the global urban network through which policies are borrowed, adapted, and transformed.

INTRODUCTION

The policy world seems to be one in constant motion. In a figurative sense, policy-makers seem to be under increasing pressure to 'get a move on' – to keep up with the latest trends and 'hot' ideas that sweep into their offices, to convert those ideas into locally-appropriate 'solutions,' and 'roll them out', thus making the most of them before the next trend emerges. As waves of innovation arrive more frequently, a concordant 'churning' has been identified in urban policy, with new ideas and initiatives replacing old with increased regularity. Contemporary policy-making, at all scales, therefore involves the constant 'scanning' of the policy landscape, via professional publications and reports, the media, websites, blogs, professional contacts, and word of mouth for ready-made, off-the-shelf policies and best practices that can be quickly applied locally.

It is in this context of 'fast policy transfer' (Peck and Theodore 2001: 429) that figurative motion in the policy world becomes literal motion. Policy actors (a broadly defined category including politicians, policy professionals, practitioners, activists, and consultants) act as 'transfer agents' (Stone 2004), shuttling policies and knowledge about policies around the world through attendance at conferences, fact-finding trips, consultancy work, etc. These travels involve the transfer of policies from place to place, which, in some cases, seem to diffuse with lightning speed, e.g., welfare policies (Peck and Theodore 2001) and creative city policies (Florida 2002; Peck 2005). These travels and transfers involve local and national policy-makers in networks that extend globally, bringing certain cities into conversation with each other, while pushing others further apart. They create mental maps of 'best cities' for policy that inform future strategies – Austin for quality of life and creativity, Barcelona and Manchester for urban planning and regeneration, Curitiba for environmental planning, Portland for growth management, Porto Alegre for participatory budgeting and direct democracy. Thus, in a policy sense as in other ways, cities are constituted through their relations with other places and scales (Massey 2011).

Yet, while motion and relationality define contemporary policy-making, this is only half the picture. Policies and policy-making are also intensely and fundamentally local, grounded, and territorial. Even a cursory familiarity with the examples above confirms this point, since our ability to refer to complex policies through the use of a shorthand of city names indicates how tied they are to specific places. There is a 'Barcelona model' of urban regeneration, for example, which is contingent on the historical-geographical circumstances of that city and its relationship with other regional and national forms of decision-making. While other cities might be encouraged to learn or adopt that model, it is generally understood that, in doing so, adjustments will need to be made in order for it to work elsewhere. Furthermore, policy is fundamentally territorial in that it is tied up with a whole set of locally dependent interests, with those involved in growth coalitions being the most obvious. Therefore, while there are substantial literatures in urban studies that emphasize cities' relationality and fluidity and while there are other equally important literatures that emphasize their territoriality, we argue that urban policy-making must be understood as both relational and territorial; as both in motion and simultaneously fixed, or embedded in place. The contradictory nature of policy should not, however, be seen as detrimental to its operation. Rather, the tension between policy as relational and dynamic, on the one hand, and fixed and territorial, on the other, is a productive one. It is a necessary tension that produces policy and places.

Our purpose in this paper is to explore the implications of this tension for our understanding of urban policy and to use the study of the 'local globalness' of urban policy to inform the study of urban-global relations more generally. We discuss how

contemporary scholarship across the social sciences is exhibiting a remarkable convergence around questions of inter-scalar relations and around a conviction that specific cases of regulation, design, or policy-making, for example, must be understood in terms of processes stretching over wider geographical fields and in terms of imperatives that may not be immediately evident at the scale of, or on the face of the cases themselves. We argue that this is an important moment in which to consider global-urban relations since ongoing discussions about the relationships between cities and global processes (Robinson 2006; Taylor 2004) and about networked, relational, and territorial conceptualizations of social space (Allen and Cochrane 2007) indicate that cities are important nodes in a 'globalizing' world. Yet, scholars still do not understand, in a deep and detailed way, how those involved in urban politics and policy-making act beyond their own cities in order to practice or perform urban globalness and to articulate their cities in the world (but see the essays in McCann and Ward 2011). So, while we will outline a convergence of thought around the need for empirical detail on global political-economic relations, we will also suggest that the literature needs more empirical accounts of the struggles, practices and representations that underpin urban-global relations and that territorialize global flows. In doing so, we pay close attention to: (1) how urban policies are set 'in motion' globally and how global circuits of policy knowledge and the transfer of policy models influence the governance of specific cities; (2) how the 'making up' of policy is a fundamentally territorialized and political process, contingent on specific historical-geographical circumstances.

CONCEPTUALIZING GLOBAL-URBAN CONNECTIONS: RELATIONALITIES, TERRITORIALITIES, POLICIES

The relational and territorial geographies of urban policies

A great deal of critical geographical scholarship on cities examines the connections between urbanization and capitalism, the changing territorial forms of the state, and the production of new institutional arrangements for urban and regional governance, focusing on economic development and the 'new urban politics'. Yet, more analysis is needed on how – through what practices, where, when, and by whom – urban policies are produced in global relational context, are transferred and reproduced from place to place, and are negotiated politically in various locations. That said, a number of influential, although varied, and not always entirely compatible, theorizations have sought to understand the tensions and power relations central to these global-urban connections. Harvey's (1982) conceptualization of the dialectic of fixity and mobility in capitalism and the implications of investment and disinvestment for urban built environments is one of these. Massey's (see Ch. 57) notion of a global sense of place, in which specific places are understood to be open to and defined by situated combinations of flows of people, communications, responsibilities, etc. that extend far beyond specific locales, is another. The literature on spatial scale, much of which focuses on conceptualizations of territorialization and deterritorialization (Brenner 2004), and the world/global cities literature, with its focus on certain cities as powerful nodes in the networked geographies of finance capital, are two other established bodies of work. The burgeoning 'mobilities' approach, which seeks to conceptualize the social content of movements of people and objects from place to place at various scales and the immobilities and 'moorings' that underpin and challenge these dynamics, constitutes another worthwhile approach (Cresswell 2001; Hannam et al. 2006; Sheller and Urry 2006).

Each of these literatures seeks to conceptualize how cities are produced in relation to processes operating across wider geographical fields, while recognizing that urban localities simultaneously provide necessary basing-points for those wider processes. Each suggests that there can be no separation between place-based and global-relational conceptualizations of contemporary political economies. As Hannam et al. (2006: 5) put it: "[m]obilities cannot be described without attention to the necessary spatial, infrastructural and institutional moorings that configure and enable mobilities." Rather, Brenner (2004) suggests that territory must be seen as relationally produced rather than bounded and static. He argues that "the image of political-economic space as a complex, tangled mosaic of superimposed and interpenetrating nodes, levels, scales, and morphologies has become more [analytically] appropriate than the traditional Cartesian model of homogeneous, self-enclosed and contiguous blocks of territory" (Ibid, 66). The tensions and crises involved in this multi-scale urban

experience are objects of policy-making and politics. Harvey's (1989: 143) account of urban politics is particularly clear on this issue: While it is important to understand cities as always in a process of becoming, social relations, state policy, and politics shape and are shaped by urban regions, or territories, which exist "in the midst of a maelstrom of forces that tend to undermine and disrupt" their coherence.

Allen and Cochrane's (2007: 1171) discussion of (urban) regions resonates strongly with this viewpoint. They do not reject the importance of territory, only its traditional bounded connotation: "[T]here is little to be gained by talking about regional [and by inference, urban] governance as a territorial arrangement when a number of the political elements assembled are not particularly regional in any traditional sense, even if they draw on what might called the 'spatial grammar' of regionalism". They continue:

> Many are 'parts' of elsewhere, representatives of political authority, expertise, skills, and interests drawn together to move forward varied agendas and programmes. The sense in which these are [urban] 'regional' assemblages, rather than geographically tiered hierarchies of decision-making, lies with the tangle of interactions and capabilities within which power is negotiated and played out.

The urban region is, thus a social and political product that cannot be understood without reference to its relations with various other scales. Yet, to study how this social production gets done involves the study of a whole series of very specific and situated interactions, practices, performances, and negotiations.

Conceptualizing urban policy-making and politics through the productive tension between relationality and territoriality entails both the study of how urban actors manage and struggle over the 'local' impacts of 'global' flows and also the analysis of how they engage in global circuits of policy knowledge that are produced in and through a "relational geography focused on networks and flows" (Olds 2001: 6). These transfer agents seek, through this engagement, to take policy models from their own cities and promote them as 'best practices' elsewhere, or to tap into a global field of expertise to identify and 'download' models of good policy. This process of territorializing and deterritorializing policy knowledge is highly political in which "[zones] of connectivity, centrality, and empowerment in some cases, and of disconnection, social exclusion and inaudibility" (Sheller and Urry 2006: 210) are brought into being as struggles ensue over how policies get discursively framed as successes, while the insertion of new 'best practices' from elsewhere into specific cities can empower some interests at the expense of others, putting alternative visions of the future outside the bounds of policy discussion. The construction of 'models' of redevelopment and their circulation and re-embedding in cities around the world can have profoundly disempowering consequences. On the other hand, this process of policy transfer can also spur contest within cities where activists question the 'pre-approved' credentials of newly imported policy models or where activists are motivated to 'scan' globally for alternative policies as part of what Purcell (2008: 153) calls "fast resistance transfer."

From policy transfer to mobile policies

How might we think specifically about the movement of policies from a relational/territorial perspective? We might consider the already existing political science literature on policy transfer which studies how policies are learned from one context and moved to another with the hope of similar results. In one sense, this is a literature that is all about global relations and territories. While internally differentiated and heterogeneous, the literature shares some common features. It focuses on modeling how transfer works, creating typologies of transfer agents (Stone 2004), and identifying conditions under which transfer leads to successful or unsuccessful policy outcomes in the new location (Dolowitz and Marsh 2000). So, it is not without its insights.

Yet, while this literature is certainly about global relations and territories, it has exhibited less attention to the full range of social territoriality. It is limited in its definition of the agents involved in transfer, focusing largely on national and international elites largely working in formal institutions. It focuses solely on national territories – transfer among nations or among localities with single nations – without considering the possibility, or actuality, of transfer among cities that transcend national boundaries. Furthermore, it tends not to consider transfer as a socio-spatial process in which policies are changed as they travel (Peck and Theodore 2001).

These limits to the 'traditional' policy transfer literature (for a full critique, see McCann 2010) offer a

series of opportunities for further theorization from perspectives that understand, often in different ways, transfer as a global-relational, social, and spatial process which interconnects and constitutes cities. For Wacquant (1999: 321), the aim should be "to constitute, link by link, the long chain of institutions, agents and discursive supports" that constitute the current historical period while Peck (2003: 229) calls for more analyses of the circulation of policies in relation to "the transnational and translocal constitution of institutional relations, governmental hierarchies and policy networks."

Larner (2003: 510) also advocates a move in the same intellectual direction, towards a "more careful tracing of the intellectual, policy, and practitioner networks that underpin the global expansion of neoliberal ideas, and their subsequent manifestation in government policies and programmes." Explicitly interested in understanding both how and why governing practices and expertise are moved from one place to another, she advocates the 'detailed tracings' of social practices, relations, and embeddings. For example, her study of the global call center and banking industries and the place of New Zealand in the globalization of these economic activities shows the value of the detailed rendering of what might be seen as the banal or mundane practices of various actors who, individually and collectively, play an important role in constituting globalization.

Much of the mobilities work attempts to understand the details of a particular form of mobility, or a specific infrastructure that facilitates or channels mobilities, in reference to wider processes and contexts:

[It] problematizes both 'sedentarist' approaches in the social sciences that treat place, stability, and dwelling as a natural steady-state, and 'deterritorialized' approaches that posit a new 'grand narrative' of mobility, fluidity or liquidity as a pervasive condition of postmodernity or globalization ... It is a part of a broader theoretical project aimed at going beyond the imagery of 'terrains' as spatially fixed geographical containers for social processes, and calling into question scalar logics such as local/global as descriptors of regional extent ...

(Hannam et. al. 2006: 5).

For us, the language of the mobilities approach is a useful frame for our discussion of mobile policies because it emphasizes the social and the scalar, the fixed and mobile character of policies. We utilize 'mobilities' in the sense that people, frequently working in institutions, mobilize objects and ideas to serve particular interests and with particular material consequences.

We can, then, see convergences among scholars about the need to be alive to both the why and the how of policy transfer. This demands that we pay attention to how – through 'ordinary' and 'extra-ordinary' activities – policies are made mobile (and immobile), why this occurs, and the relationship between these mobilities and the socio-spatial (re)structuring of cities. The question remains how might we best frame these sorts of empirical discussions? Should we understand contemporary policy-making as primarily about territory, as primarily about relationality, or in terms of a both/and logic which recognizes that contemporary "global restructuring has entailed neither the absolute territorialization of societies, economies, or cultures onto a global scale, nor their complete deterritorialization into a supraterritorial, distanceless, placeless, or borderless space of flows" (Brenner 2004: 64)? We take the latter position.

CONCLUSION

The burgeoning literature on policy mobilities reveals the range of transfer agents involved in this process. Some with little reach, overseeing their introduction in a specific city. Others with a far longer reach, able to influence policy reform at a distance. In policy mobilization a process of translation is performed, both by those coming in from outside and by actors resident in each of the contexts. In these moments – whether they are literally 'performed' at conferences or workshops, or occur through circulated written publications – supply and demand come so close as to be almost indistinguishable.

While these empirical details are important, we want to conclude this chapter by noting a number of broader conceptual issues. First, we argue for a conceptualization of the making of urban policy through both its territorial and relational geographies and for an appreciation of how cities are assembled by the situated practices and imaginations of actors who are continually attracting, managing, promoting, and resisting global flows of policies and programs. Following Olds (2001: 8), we advocate "a relational geography that recognizes the contingent, historically

specific, uneven, and dispersed nature of material and non-material flows". Second, we critique the existing literature on policy transfer, while also acknowledging its contributions. In sketching out a new way of thinking about the mobility of policies and programs, our thinking draws on a number of different literatures in which there appears to have been convergence around documenting in detail the means through which policies are made mobile. Our way forward is to argue for a framework that includes a broad understanding of transfer agents, takes seriously the inter-urban policy transfer that links cities across national boundaries, and understands transfer as a socio-spatial process in which policies are subject to change as they are moved. Third, we offer some thoughts on what the current concern for thinking relationally might mean for doing of research. It means paying attention to the various spaces that are brought into being during the journey of a policy or program: a mixture of following the policy together with sensitivity to the particular territorial contexts at every step in the process of movement. Fourth, the approach we advocate has, at its core, sensitivity to both structure and agency. There is a set of macro supply and demand contexts in which some 'idea brokers' (Smith 1991), are structurally advantaged. Some, more than others, are likely to have their ideas and policies made mobile. And, of course, there is an interaction of a range of differently scaled forces in and through which these agents mobilize, broker, translate and introduce ideas in such a way as to make the territorial embedding of globally-circulating policies and programs not just possible but probable.

REFERENCES FROM THE READING

Allen, J. and Cochrane, A. (2007) Beyond the territorial fix: Regional assemblages, politics and power, *Regional Studies*, 41: 1161–75.

Brenner, N. (2004) *New State Spaces: Urban Governance and the Rescaling of Statehood*, Oxford and New York: Oxford University Press.

Cresswell, T. (ed.) (2001) Mobilities, *New Formations*: 43.

Dolowitz, D. and Marsh, D. (2000) Learning from abroad: The role of policy transfer in contemporary policy making, *Governance*, 13: 5–24.

Florida, R. (2002) *The Rise of the Creative Class*, New York: Basic Books.

Hannam, K., Sheller, M. and Urry, J. (2006) Mobilities, immobilities and moorings, *Mobilities*, 1: 1–22.

Harvey, D. (1982) *The Limits to Capital*, Chicago: University of Chicago Press.

Harvey, D. (1989) *The Urban Experience*, Baltimore, MD: Johns Hopkins University Press.

Larner, W. (2003) Guest editorial: Neoliberalism? *Environment and Planning D: Society and Space*, 21: 508–12.

Massey, D. (2011) A counterhegemonic relationality of place, in McCann, E. and Ward, K. (eds.) *Mobile Urbanism: Cities and Policy Making in the Global Age*, Minneapolis, MN: University of Minnesota Press: 1–14.

McCann, E. J. (2010) Urban policy mobilities and global relational geographies: Toward a research agenda, *Annals of the Association of American Geographers*, 101, 1: 107–30.

McCann, E. J. and Ward, K. (eds.) (2011) *Mobile Urbanism: Cities and Policy Making in the Global Age*, Minneapolis, MN: University of Minnesota Press.

Olds, K. (2001) *Globalization and Urban Change: Capital, Culture, and Pacific Rim Mega-Projects*, Oxford: Oxford University Press.

Peck, J. (2003) Geography and public policy: Mapping the penal state, *Progress in Human Geography*, 27: 222–32.

Peck, J. (2005) Struggling with the creative class, *International Journal of Urban and Regional Research*, 24: 740–70.

Peck, J. and Theodore, N. (2001) Exporting workfare/importing welfare-to-work: Exploring the politics of Third Way policy transfer, *Political Geography*, 20: 427–60.

Purcell, M. (2008) *Recapturing Democracy*, New York: Routledge.

Robinson, J. (2006) *Ordinary Cities: Between Modernity and Development*, London: Routledge.

Sheller, M. and Urry, J. (2006) The new mobilities paradigm, *Environment and Planning D: Society and Space*, 38: 207–26.

Smith, J. (1991) *The Idea Brokers: The Rise of Think Tanks and the Rise of the Policy Elite*, New York: Free Press.

Stone, D. (2004) Transfer agents and global networks in the 'transnationalisation' of policy, *Journal of European Public Policy*, 11: 545–66.

Taylor, P. (2004) *World City Network: A Global Urban Analysis*, London: Routledge.

Wacquant, L. (1999) How penal common sense comes to Europeans: Notes on the transatlantic diffusion of the neoliberal doxa, *European Societies*, 1: 319–52.

PART FOUR

Regulations

Plate 25 Hong Kong, Asia's world city

Source: Roger Keil

INTRODUCTION TO PART FOUR

In their original formulation, Friedmann and Wolff recognized the highly political nature of global city formation. In their contribution to this volume (see Ch. 3), they describe the inherent contradictions of the world city as a source of conflict and struggle. However, neither these authors nor most other early contributors to world cities research engaged systematically with the politics of the global city. Although Friedmann and Wolff insisted on the strategic importance of sociopolitical struggles within global city-regions, the question of how political institutions structure, and are in turn structured by, processes of global city formation, was left largely open in the global cities scholarship of the 1990s. We take up this concern in two separate sections. In Part 4, we approach the topic from the intersection of the state, policy and political economic structures. In Part 5, we highlight the many ways in which urban polices and projects have come under scrutiny and have been contested through social mobilizations from below.

In one particularly vivid passage of their classic text, Friedmann and Wolff (Ch. 3) evoke the image of the "citadel" and the "ghetto" to describe emergent patterns of sociospatial inequality in the global city. The metaphors powerfully illustrate the internal cleavages in social fabrics of the global city, which are expressed spatially in juxtaposition between the gleaming office towers of the new downtowns and the impoverished residential quarters of the increasingly internationalized urban proletariat. In this view, the ghetto's inhabitants are isolated like a "virus" by the corporate and political elites (Friedmann and Wolff 1982: 325); this marginalized urban space is then subdivided into ethnic and racial enclaves, which seem entirely separated from both each other and from the "citadel." Concomitantly, the political sphere of the world city appears to be differentiated clearly into two different entities—the first concentrates political and economic control capacities; and the second is inhabited by the politics of social reproduction and everyday survival. Meanwhile, national and local state institutions are seen as being unable to manage the proliferating regulatory problems associated with global city formation. Traditional welfarist policies are increasingly abandoned in favor of diverse boosterist strategies intended to attract and maintain investment by transnational corporations. The political conflicts that subsequently ensue within global cities are characterized by a local/global dialectic that pits globally mobile corporations ("economic space") against diverse, territorially circumscribed interests associated with "life space." This intense struggle between life space and economic space assumes a concrete form in proliferating battles over livelihood, diversity, and land use planning throughout the global city region (Newman and Thornley 2005).

Friedmann and Wolff's initial formulation was provocative and challenging to prevalent understandings of urban politics, but it ultimately proved unsatisfactory to later generations of researchers interested in urban politics in global city-regions. For, as Friedmann himself pointed out in subsequent publications, the emergence of a new global-local relationship could not be grasped adequately through metaphors of confrontation between two seemingly static poles (life space/economic space; ghetto/citadel); it needed to be examined, instead, as a result of material power relations (Beauregard 1995; Swyngedouw 1997). Moreover, not only did locally embedded communities take up the struggle against global capita, they also transformed themselves, both politically and institutionally, in and through such struggles. In this sense, the citadel/ghetto metaphor could not adequately illuminate the dynamics of sociopolitical contestation and institutional transformation in global city-regions; it also left open the question of which political structures and agents actually produce, define, and continually reorganize the global city.

Much research has been conducted on the political aspects of global city formation (see, for instance, Acuto 2012; Ancien 2011; Davis 1990; Hitz et al. 1994; Keil 1998). Some have studied the consolidation of local growth machines that are concerned to position the city strategically within supranational circuits of capital, generally through the consolidation of global command and control functions. Others have examined the interplay between the consolidation of global cities and various ongoing transformations of state power at national, regional, and local scales (Brenner Ch. 32; Taylor 1995; Sassen 1998). Additionally, scholars have examined the challenges of urban governance in cities that are attempting, simultaneously, to articulate themselves to the world economy and maintain internal sociospatial cohesion. Some have examined the supra-local interrelationships in the global urban network. Michele Acuto, in particular, has researched the geopolitical dimension of the global city and the shift from an international to a *global* politics of the global city (Acuto 2013a, 2013b). Others have begun to explore the role of major city political networks such as C40 Cities Climate Leadership Group and the role of mayors (Barber 2013; Betsill and Bulkeley 2004).

The resurgence of urban citizenship (more on this in Part 5) has been intertwined with diverse transformations of state space under contemporary conditions, which are increasingly relativizing the entrenched role of nationalized forms of economic regulation and political regulation (Brenner et al. 2003). In his brief prologue to this section, Warren Magnusson addresses such issues by arguing that inherited notions of nation-state sovereignty and nationalized models of state/society relations are no longer adequate. Magnusson perceptively argues that "the concept of the global city invites us to abandon a number of old distinctions: between the local ("the city") and the global; between the economic, the social, the cultural and the political; and between the static ("structures," "systems," "space") and the dynamic ("movements," "time")" (2000: 295). Magnusson emphatically calls for a "politicizing" of the global city which he considers as both "the venue and the product of our own struggles to become what we would like to be, and in the end there is no alternative but to take responsibility for what we have created" (2000: 304). In later work, Magnusson (2011) has provided more precision with regards to this conundrum by claiming that global politics now in general needs to "see like a city." Work in political theory has taken up this claim and studied the particular politics of the "transition narratives"–from localized, state-centered urbanization to emergent forms of global urbanization (Tedesco 2012; 2015).

Most world city researchers have conceived urban *regions* as the relevant scale on which the city's global functions are realized. Airports, residences, jobs, and educational facilities are not confined to the centralized, downtown financial districts of these cities, but can be found in strategic locations throughout the metropolitan region (Sassen 2000). This means that the politics of the global city are, in practice, actually metropolitan or regional politics. These dynamics of regionalization have also underpinned recent struggles over metropolitan institutional reform, regional economic governance, cross-border cooperation, and interurban networking in many western European, North American, and East Asian global city-regions (Brenner 2004; Friedmann 1997; Sassen 2001; Scott 2001 and Ch. 31). In addition, over the past decade, new work has examined two additional developments in globalizing urban regions. First, an emerging literature has begun to look specifically at megaregions as a possible new form of globalization which has deep implications for governance (Harrison and Hoyler 2015). Second, researchers have begun to investigate the particular patterns of suburban governance as a global phenomenon (Hamel and Keil 2015). Arguably, the two processes are related and their intertwined dynamics have begun to be studied (Harris and Keil 2016; Keil Ch. 63).

Often, transformations of global city regions are shaped by strategic state actions. Hill and Kim's study of Tokyo and Seoul (Ch. 33) suggests that urban development in East Asia represents a key spatial outcome of developmentalist state policies oriented towards both national *and* local economic growth. However, local social forces may also attempt, with varying degrees of success, to harness state institutions in order to block particular pathways of global city formation, protecting certain local interests while nonetheless facilitating (a modified form of) globalized urban development. One strand of this literature has focused on the rise of civil society. In his contribution to this section, Mike Douglass examines the changing state and civil society relations in major cities throughout East Asia (Ch. 34). In globalizing cities throughout the world, the role of civil society organizations has been greatly enhanced, as national and local states retreat from service delivery and corporate downsizing undermines job security. Under these conditions, civil society organizations have become important economic and social stabilizers of neoliberal urban governance. State institutions thus become a key political terrain in which the form, pace, and geography of globalized urbanization are fought out by diverse local and supralocal social forces.

In Chapter 35, Neil Smith explores how institutions of states have aligned with those of the market in driving gentrification. While critiquing the heavy focus of the global city theory on financialization, he examines specifically the changes in the 1980s and 1990s, in globalizing urban environments where strategic state action and market activity colluded to gentrify many central areas.

Two further contributions to this section explore the changes in the political geography of globalizing African cities. While Fourchard (Ch. 36) explains urban structures and processes in Africa's cities in relation to state building, Parnell and Pieterse (Ch. 37) examine the more recent experiences with local state formation in South Africa, based on rights-based politics in post-apartheid Cape Town. Both chapters challenge common perceptions of the geographies of theory that underlie conventional research on globalizing cities in Africa or any other region of what is commonly called the global South. Fourchard insists on seeing Africa as impacted by multiple global processes (rather than just one master-process). Parnell and Pieterse challenge the common assumptions about viewing an anti-neoliberal agenda as purely economic and adversarial, and they propose instead an added focus on rights which they understand need to be claimed in a multi-scale state formation.

In an increasingly connected world, urbanization also runs up against certain ecological limits (Keil 1995, 1998; Derudder 2003; Terlouw 2003). This has led to debates on sustainability and environmental justice in global cities and on the "urban political ecology" of global city formation (Gandy 2002; Desfor and Keil 2004). Researchers in this tradition study how globalized urbanization transforms human-nature interactions, leading to a variety of deeply rooted governance problems—pertaining, for instance, to the provision of public goods (water, electricity) and the conditions of everyday social reproduction (housing, pollution, public health, and transportation) in cities around the world. Timothy Luke (Ch. 38) explores some of the ways in which global cities are affected by, and are in turn influencing, emergent patterns of ecological interdependence and sustainability.

Taken together, the selections included in this section reflect some of the diverse ways in which urban researchers have attempted to decipher the many dimensions of political life and institutional restructuring in globalizing cities. The selections below are intended at once to provide readers with an accessible survey of some of the main lines of research and to demarcate a starting point for further inquiry into the problem of "governing complexity" (Keil 2003) within globalizing city-regions.

REFERENCES AND SUGGESTIONS FOR FURTHER READING

Acuto, M. (2012) Ain't about politics? The wicked power-geometry of Sydney's greening governance, *International Journal of Urban and Regional Research*, 36, 2: 381–99.

Acuto, M. (2013a) *Global Cities, Governance and Diplomacy: The Urban Link*, London: Routledge.

Acuto, M. (2013b) The geopolitical dimension, in M. Acuto and W. Steele (eds.) *Global City Challenges: Debating a Concept, Improving the Practice*, Houndmills, Basingstoke: Palgrave Macmillan: 170–87.

Ancien, D. (2011) Global city theory and the new urban politics twenty years on: The case for a geohistorical materialist approach to the (new) urban politics of global cities, *Urban Studies*, 48, 12: 2473–93.

Barber, B. (2013) *If Mayors Ruled the World*, New Haven: Yale University Press.

Beauregard, R. (1995) Theorizing the global-local connection, in P. Knox and P. J. Taylor (eds.) *World Cities in a World-System*, New York: Cambridge University Press: 232–48.

Betsill, M. and Bulkeley, H. (2004) Transnational networks and global environmental governance: The cities for climate protection program, *International Studies Quarterly*, 48: 471–93.

Bodnàr, J. (2001) *Fin de Millénnaire Budapest*, Minneapolis: University of Minnesota Press.

Brenner, N. (2004) *New State Spaces: Urban Governance and the Rescaling of Statehood*, Oxford: Oxford University Press.

Brenner, N., Jessop, B., Jones, M., and McLeod, G. (eds.) (2003) *State/Space: A Reader*, Boston: Blackwell.

Davis, M. (1990) *City of Quartz*, London: Verso.

Derudder, B. (2003) Beyond the state: Mapping the semi-periphery through urban networks, *Capitalism, Nature, Socialism*, 14, 4: 91–120.

Desfor, G. and Keil, R. (2004) *Nature and the City*, Tucson, AZ: University of Arizona Press.

Friedmann, J. (1997) World city futures: The role of urban and regional policies in the Asia-Pacific region, Occasional Paper no.56, Hong Kong Institute of Asia-Pacific Studies, The Chinese University of Hong Kong, New Territories, Hong Kong.

Friedmann, J. and Wolff, G. (1982) World city formation: An agenda for research and action, *International Journal of Urban and Regional Research*, 6, 3: 309–43.

Gandy, M. (2002) *Concrete and Clay*, Boston, MA: MIT Press.

Haila, A. (2016) *Urban Land Rent: Singapore as a Property State*, Oxford: Wiley Blackwell.

Hamel, P. and Keil, R. (eds.) (2015) *Suburban Governance: A Global View*, Toronto: University of Toronto Press.

Harris, R. and Keil, R. (2016) Globalizing cities and suburbs, in A. Bain and L. Peake (eds.) *Urbanization in a Global Context: A Canadian Perspective*, Oxford: Oxford University Press: 52–69.

Harrison, J. and Hoyler, M. (eds.) (2015) *Megaregions: Globalization's New Urban Form?* Cheltenham: Edward Elgar.

Hitz, H., Schmid, C., and Wolff, R. (1994) Headquarter economy and city-belt: Urbanization in Zurich, *Environment and Planning D: Society and Space*, 12, 2: 167–85.

Keil, R. (1995) The environmental problematic in world cities, in P. Knox and P. Taylor (eds.) *World Cities in a World System*, Cambridge: Cambridge University Press: 280–97.

Keil, R. (1998) *Los Angeles: Globalization, Urbanization and Social Struggles*, Chichester: Wiley.

Keil, R. (2003) Globalization makes states, in Brenner, N., Jessop, B., Jones, M., and McLeod, G. (eds.) (2003) *State/Space: A Reader*, Boston, MA: Blackwell: 278–95.

King, A. D. (2004) *Spaces of Global Cultures*, London and New York: Routledge.

Knox, P. L. and Taylor, P. J. (eds.) (1995) *World Cities in a World-System*, Cambridge and New York: Cambridge University Press.

Lefebvre, H. (1968) *Le Droit á la Ville*, Paris: Anthropos.

Magnusson, W. (2000) Politicizing the global city, in Isin, E. F. (ed.) *Democracy, Citizenship and the Global City*, London and New York: Routledge: 289–306.

Magnusson, W. (2011) *Politics of Urbanism: Seeing like a City*, London: Routledge.

Newman, P. and Thornley, A. (2005) *Planning World Cities*, Houndmills Basingstoke: Palgrave Macmillan.

Olds, K. (2001) *Globalization and Urban Change*, New York and Oxford: Oxford University Press.

Ronneberger, K. (2002) Contours and convolutions of everydayness: On the reception of Henri Lefebvre in the Federal Republic of Germany, *Capitalism, Nature, Socialism*, 13, 2: 42–57.

Sassen, S. (1998) *Globalization and its Discontents*, New York: The New Press.

Sassen, S. (2000) *Cities in a World Economy* (2nd edition), Thousand Oaks, CA: Pine Forge Press.

Sassen, S. (2001) Global cities and global city-regions: A comparison, in A. J. Scott (ed.) *Global City-Regions*, New York: Oxford University Press: 78–95.

Scott, A.J. (ed.) (2001) *Global City-Regions*, New York: Oxford University Press.

Swyngedouw, E. (1997) Neither global nor local: Glocalization and the politics of scale, in K. Cox (ed.) *Spaces of Globalization*, New York: Guilford: 137–66.

Taylor, P. J. (1995) World cities and territorial states: The rise and fall of their mutuality, in P. Knox and P. J. Taylor (eds.) *World Cities in a World-System*, New York: Cambridge University Press: 48–62.

Tedesco, D. (2012) The urbanization of politics: Relational ontologies or aporetic practices? *Alternatives: Global, Local, Political*, 37, 4: 331–47.

Tedesco, D. (2015) Begin again, return again: The transition narratives and political continuities of global urban politics, *International Political Sociology*, 9, 1: 106–9.

Terlouw, K. (2003) Semi-peripheral developments: From world systems to regions, *Capitalism, Nature, Socialism*, 14, 4: 71–90.

F
O
U
R

30 Prologue
"The global city as world order"

from *The Search for Political Space* (1996)

Warren Magnusson

One of the distinctive features of the city as a mode of order and domination is that it is not governed from a single centre. The principle of sovereignty does not work effectively within the civic domain. Of course, sovereigns often try to exercise control *over* cities, but they can rarely if ever work *through* cities. A city is in large degree a self-organizing system produced by a variety of cultural, social, and economic enterprises. It is where people come to do things outside the domain of sovereignty, in relative freedom from the dictates of church and state. The medieval proverb that "city air makes free" refers to more than the liberation of serfs, who could gain the status of free persons after a year and a day. It also alludes to the possibility for new enterprises that escape the dead hand of established authority. Such enterprises are not, in principle, contained within the territory of a particular city: they reach out to the surrounding countryside and to other cities in the world beyond. Urbanity in its fullest sense implies de-territorialized relations between people in different parts of the world. In this sense, the city is not fixed to a particular place the way a village is. Urbanity implies a kind of nomadism: a presence within a space of flows that connect and reconnect different places in the world. Obviously, the market structures many of these flows, and the logic of the market tends to determine which cities will expand and which will contract. However, the social and cultural flows that occur within urban space are not reducible to the logic of the market. They have more complex origins and many autonomous effects.

Present-day municipalities are lineal descendants of the early medieval corporations that were designed to contain and control urban development. Whether constituted by Royal Charter or formed from below by civic insurgents, the municipal corporations were intended to fix economic activity in particular places and to give urban life the form that people believed would be most rational. However, the municipal corporations were generally unable to contain what they were supposed to manage. This was partly because they lacked sovereignty, but more fundamentally because the activities that typified a city burst the bounds of any particular place. London could not be kept in its square mile, Paris could not be confined to the Ile de la Cité, and even Manhattan was not big enough for New York. This physical overflow was a sign of economic, social, and cultural spillage of a much more profound character. By the seventeenth century the whole world was in London's domain, and, although London ultimately became the leading world-city, it was by no means the only city that could boast of a global or near-global reach. Moreover, the leading cities were all linked to one another in patterns of dizzying complexity. None was "independent" of the others, nor sought to be. Although there were governments of a sort in particular cities – sometimes even national governments with ostensibly sovereign authority – these governments exercised only a shadow control over their urban domains. Urbanism itself – the system of cities – was under no one's direct control, followed no one's orders in its development, and could not be managed from any single center. On the contrary, as Braudel and others have reminded us, sovereigns of one sort or another have always depended heavily on the urban system that they pretend to govern. A productive and dynamic urbanism will produce the surpluses necessary to pay soldiers and make arms. It will also generate the ideas and the functionaries necessary for effective government or imperial expansion. Sovereignties will be sustained or overwhelmed by urban dynamism.

Arguably, the true heirs of the medieval cities are not the municipalities but the multinational corporations of the contemporary world. The corporate form as exemplified by IBM and ICI is a late mutation of the municipal corporation of medieval times, and is not connected (except nominally) to a particular place. It is slimmed down for the pursuit of profit, and projected into a truly global space. As organizations for economic enterprise, municipalities are largely anachronistic, because they are tied to particular territories and burdened with tasks of government. The free corporations of the present day occupy deterritorialized spaces that cannot be mapped onto the world like countries or provinces. Increasingly, they function within a cyberspace that is characterized by instantaneous electronic communication. This cyberspace is not stable, and in fact the major actors within it are constantly changing the systems of communication and control for their own purposes. Thus, the space within which these organizations act is largely the product of their own activities. To keep up with the most innovative actors is enormously difficult, and the evolution of cyberspace is not governed from any single center. In this respect, the emergent cyberspace is typical of the spaces created by urban activity.

There is, of course, more to urban space than cyberspace. Urbanism is characterized by the continual production and reproduction of spaces of habitation, work, recreation, cultural expression, and so on. There is a dynamism to these processes that again defies any static representation. A building is simply a momentary expression of people's ideas about the way their activities need to be spatialized. Fixed as it is, a building is reformed in use until it becomes almost unrecognizable to its original founders. The physical form of the city as a whole is even more plastic. Once we take into account the city's relationship to the countryside and to other cities, it becomes apparent that urban space is a dynamic presence in the world as a whole. People are within urban space in their airplanes going from airport to airport, and in their cars speeding along the motorways. The airport cafe in Honolulu, which serves passengers on their way from San Francisco to Sydney, is a part of the space of all three cities – and many others. Similarly, the restaurants in the Black Forest, the beaches of Mauritius, and the mountains of British Columbia are

extensions of the recreational space of urbanites in many parts of the world. Thanks to the means of transportation and communication that have been developed over the last thirty to thirty-five years, prosperous urbanites can inhabit any and all parts of the world in the routine course of their lives. It is a sign of parochialism – or poverty – if one fails to inhabit the whole world.

We might think of the wider habitat as a global "hyperspace," within which the "cyberspace" of the computer nets is a particular but important domain. Airports, motorways, offices, hotels, and boutiques are other elements in the most privileged domain of this hyperspace. Access to that domain is carefully controlled. In fact, cities are marked by exceedingly complicated strategies of territorialization. Among the obvious signs of this are the urban fortresses of the sort that were first established in the late medieval Italian cities and later replicated in many forms. These are not the fortresses of kings and bishops, but of burghers who seek security within the turbulent, expansive and absorptive space of the city. When people venture out from these fortresses, they carry with them their personal security systems. Much public and private enterprise is directed toward securing privileged people's routes and places of work, recreation, shopping, and cultural expression.

Whole cities (for example, Paris) and whole regions of the tropics are being re-formed to make them comfortable spaces for the public life of the prosperous. Beside, beneath, and often co-present with these spaces are other, more constricted spaces that provide for the less prosperous. Homeless people live in every crack and cranny, having been swept from the places of privilege with ruthless efficiency. The zones of exclusion and inclusion are subtly layered and dynamically articulated, so that they register in the consciousness only in the enactment (and often not even then). Nevertheless, these half-understood urban zones are generally of much greater significance in people's lives than the boundaries between states.

Although the world functions as a single city, it is not a "global village," as Marshall McLuhan once suggested. It lacks the fixity, community, and intimacy that is part of the image of the village. In fact, the global city is inherently complex, dynamic, and socially differentiated. It is at once expansive and inclusive, and it is more obviously marked by separations and exclusions than by the intimate relations of communal solidarity. The world as a city has many ethnic enclaves, rich neighborhoods and poor, a multinucleated central business district, suburban office centers, shopping precincts, and recreational complexes, overcrowded systems of public transportation, and vast slums on its fringes that flow over and through the better-ordered and more prosperous districts. The government of this whole is ineffective at best and nonexistent at worst – just as it is within particular metropolitan areas. And, just as in particular cities, there is intense competition among the different jurisdictions, intermittent co-operation among them, and a vague sense that the whole could be much better ordered. If it is held together as a whole, it is mainly by economic transactions, facilitated by a common physical infrastructure. There is some sense that there is a common environment to be maintained, and that violence has to be curbed; and there is something of a shared culture that is dependent on common media of communication. However, there is no effective sovereignty, and it is by no means clear that people would want such a mode of political organization if they could have it.

To assert their sovereignty over municipal authorities, states have had to strip those authorities of most of the powers they would need for effective governance. A self-governing city would have the power to regulate its own economy and determine its own foreign trade policy. It would have to control the flow of arms in and out of its domain, and break the power of the armed gangs that defy its authority. Even to deal with questions of public health, it would have to project its authority far beyond its immediate boundaries. In a sense, it would have to follow its particular connections throughout the world. Thus, a municipal government that was determined to protect the interests of its own citizens would need a kind of world reach and freedom of action that is clearly inconsistent with state sovereignty. On the other hand, without such powers, municipal governments are condemned to a sort of observer status within the cities they are supposed to govern. Ironically, this is the status to which most if not all national governments are presently being reduced. With respect to economic organization, public health, social services, and even "security," sovereignty seems increasingly like a "show" that offers a comforting illusion of national control over national destinies.

31
"Globalization and the rise of city-regions"

from *European Planning Studies* (2001)

Allen J. Scott

EDITORS' INTRODUCTION

Allen J. Scott is Distinguished Professor Emeritus of Public Policy and Geography at UCLA. Scott is one of the most prominent figures in the "LA School" of urban studies and has written a number of seminal books on industrialization, cultural industries, and regional development (Scott 1988; 1993; 2005; 2008). His work has explored the interplay between technology, inter-firm organization, labor markets, and agglomeration economies. More recently, Scott has examined the shift toward a cognitive and cultural economy and the consequences on world cities in his book, *A World in Emergence: Cities and Regions in the 21st Century* (Edward Elgar, 2012). Building upon his earlier studies of urban agglomeration, in this chapter Scott joins other global cities researchers in arguing for the enhanced significance of geographical proximity, and therefore of large-scale urban regions, under conditions of intensifying economic globalization. On this basis, Scott explores some of the new political spaces that have emerged in conjunction with the consolidation of globalizing city-regions. According to Scott, national state sovereignty has been partially undermined due to the accelerated "debordering" of national space-economies; meanwhile, new forms of political organization and regulatory experimentation have emerged on the subnational scale of urban regions. This chapter deciphers some of the emergent contours of this "new regionalism," at once with reference to local economic development initiatives, new forms of inter-firm coordination, land-use planning strategies, new forms of labor market regulation, and, more generally, intensifying struggles over the meaning of urban citizenship. Scott suggests that new forms of collective order and institutional organization have emerged in city-regions throughout the world economy, and that the latter serve increasingly important regulatory functions within the global political system. Due to the ongoing clash of neoliberal and social democratic models of capitalism, Scott argues, the institutional shape and political form of regional governance remain intensely contested. While some scholars have questioned certain aspects of Scott's analysis—for instance, his apparent contention that national state power is declining—his analysis represents a provocative, lucid, and empirically rich account of how global city-regions have become sites for the construction of new regulatory strategies and new models of political life.

INTRODUCTION

Contrary to many recent predictions (e.g. O'Brien 1992), geography is not about to disappear. Even in a globalizing world, geography does not become less important; it becomes more important because globalization enhances the possibilities of heightened geographic differentiation and locational specialization.

Indeed, as globalization proceeds, an extended archipelago or mosaic of large city-regions is evidently coming into being, and these peculiar agglomerations now increasingly function as the spatial foundations of the new world system that has been taking shape since the end of the 1970s (Scott 1998). The internal and external relations of these city-regions and their complex growth dynamics present a number of extraordinarily perplexing challenges to researchers and policy-makers alike as we enter the twenty-first century.

There is an extensive literature on "world cities" and "global cities" that focuses above all on a concept of the cosmopolitan metropolis as a command post for the operations of multinational corporations, as a center of advanced services and information-processing activities, and as a deeply segmented social space marked by extremes of poverty and wealth. I seek to extend this concept so as to incorporate the notion of the wider region as an emerging political-economic unit with increasing autonomy of action on the national and world stages. I refer to this type of region by the term "global city-region."

Global city-regions constitute dense polarized masses of capital, labor, and social life that are bound up in intricate ways in intensifying and far-flung extranational relationships. As such, they represent an outgrowth of large metropolitan areas – or contiguous sets of metropolitan areas – together with surrounding hinterlands of variable extent which may themselves be sites of scattered urban settlements. In parallel with these developments, embryonic consolidation of global city-regions into definite political entities is also occurring in many cases, as contiguous local government areas (counties, metropolitan areas, municipalities, etc.) club together to form spatial coalitions in search of effective bases from which to deal with both the threats and the opportunities of globalization. So far from being dissolved away as geographic entities by processes of globalization, city-regions are by and large actually thriving at the present time, and they are becoming increasingly central to the conduct and coordination of modern life.

GLOBALIZATION AND THE NEW REGIONALISM

In the immediate post-World War II decades almost all of the major capitalist countries were marked by strong central governments and relatively tightly bordered national economies. These countries constituted a political bloc within the framework of a Pax Americana, itself overlain by a rudimentary network of international arrangements (the Bretton Woods monetary system, the World Bank, the IMF, GATT, and so on) through which they sought to regulate their relatively limited – but rapidly expanding – economic interrelations. Over much of the post-war period, the most prosperous of these countries could be said to constitute a core zone of the world economy, surrounded in its turn by a peripheral zone of Third World nations, with a complex set of interdependences running between the two.

Today, after much economic restructuring and technological change, significant transformations of this older order of things have occurred across the world, bringing in their train the outlines of a new social grammar of space, or a new world system. One of the outstanding features of this emerging condition is the apparent though still quite inchoate formation of a multilevel hierarchy of economic and political relationships ranging from the global to the local. Four main aspects of this state of affairs call for immediate attention:

Huge and ever-increasing amounts of economic activity now occur in the form of long-distance, cross-border relationships. Such activity is, in important ways, what I mean by globalization as such, even though it remains far indeed from any ultimate point of fulfilment. Further, as globalization in this sense moves forward, it creates numerous conflicts and predicaments that in turn activate a variety of political responses and institution-building efforts. Practical expressions of such efforts include a complete reorganization of international financial arrangements as compared with the post-war Bretton Woods system, together with the restructuring and reinforcement of international forums of collective decision-making and action such as the G7/G8 group, the OECD, the World Bank, the IMF, and a newly streamlined GATT, now known as the World Trade Organization. While these political responses to the pressures of globalization remain limited in scope and severely lacking in real authority, they are liable to expansion and consolidation as world capitalism continues its predictable expansion.

In part as a corollary of these same pressures, there has been a proliferation over the last few decades of multination blocs such the EU, NAFTA, MERCOSUR,

ASEAN, APEC, CARICOM, and many others. These blocs, too, can be seen as institutional efforts to capture the benefits and control the negative externalities created by the steady spilling over of national capitalisms beyond their traditional political boundaries. They remain in various stages of formation at the present time, with the EU being obviously in the vanguard.

Sovereign States and national economies remain prominent, indeed dominant, elements of the contemporary global landscape, though they are clearly undergoing many sea-changes. On the one hand, individual States no longer enjoy quite the same degree of sovereign political autonomy that they once possessed, and under conditions of intensifying globalization they find themselves less and less able or willing to safeguard all the regional and sectional interests within their jurisdictions. On the other hand, national economies have been subject to massive debordering over the last few decades so that it is increasingly difficult, if not impossible, to say precisely where, say, the American economy ends and the German or Japanese economies begin. As a result, some of the regulatory functions that were formerly carried out under the aegis of the central State have been drifting to higher levels of spatial resolution; at the same time, other functions have been drifting downward.

Accordingly, and most importantly for present purposes, there has of late been a resurgence of region-based forms of economic and political organization, with the most overt expression of this tendency being manifest in the formation of large global city-regions. These city-regions form a global mosaic that now seems to be overriding in important ways the spatial structure of core-periphery relationships that has hitherto characterized much of the macro-geography of capitalist development.

Point (4) calls for some amplification. The propensity of many types of economic activity – manufacturing and service sectors alike – to gather together in dense regional clusters or agglomerations appears to have been intensifying in recent decades. This renewed quest for collective propinquity on the part of all manner of economic agents can in part be interpreted as a strategic response to heightened (global) economic competition in the context of a turn to post-Fordism in modern capitalism. Propinquity is especially important in this context because it is a source of enhanced competitive advantage for many

types of firms (Scott 1988; Storper 1997), and, as a corollary, large regional production complexes are coming increasingly to function as territorial platforms for contesting global markets. At the same time, the diminishing capacity of central governments to deal with all the nuanced policy needs of each of the individual regions contained within their borders means that many regions are now faced with the choice of either passive subjection to external cross-border pressures, or active institution-building, policy-making, and outreach in an effort to turn globalization as far as possible to their advantage. Regions that take the latter course are by the same token faced with many new and unfamiliar tasks of political coordination and representation. Special urgency attaches to these tasks not only because of their economic import, but also because large city-regions function more and more as poles of attraction for low-wage migrants from all over the world, so that their populations are almost everywhere heavily interspersed with polyglot and often disinherited social groups. As a consequence of this, many city-regions today are being confronted with pressing issues related to political participation and the reconstruction of local political identity and citizenship.

THE POLITICAL ORDER OF GLOBAL CITY-REGIONS

The world system is thus currently in a state of rapid economic flux, leading in turn to many significant adjustments in patterns of political geography. On the one side, the profound changes that have been occurring on the economic front are giving rise to diverse responses and experiments in regulatory coordination at different geographic levels from the global to the local. On the other side, the new regulatory institutions that are now beginning to assume clearer outline on the world map simultaneously reinforce the channeling of economic development into spatial structures that run parallel to the quadripartite political hierarchy described earlier. While the political shifts going on at each level in this hierarchy pose many perplexing problems, the level that is represented by the new global mosaic of city-regions is perhaps one of the least well understood. Moreover, precisely because the individual regional units at this level constitute the basic motors of a rapidly globalizing production system, much is at stake as they steadily sharpen their political identities and institutional foundations.

We may well ask, at the outset, how these regions are to be defined (in political-geographic terms) as territorial units with greater or lesser powers of coordinated action. In many instances, of course, the boundaries of given city-regions will tend to coincide with some pre-existing metropolitan area. But how will these boundaries be drawn when several different metropolitan areas lie in juxtaposition to one another, as, for example, in the case of the north-east seaboard of the US? And how far out into its hinterland will the political mandate of any city-region extend? The final geographic shape of any given global city-region must remain largely indeterminate in a priori terms. Even so, we can already perhaps see some of the outlines of things to come in the new regional government systems that have been put into place in a number of different European countries over the last couple of decades (Keating 2001), and in the maneuvering, some of which may bear fruit, some of which will certainly lead nowhere, that is currently gathering steam around prospective municipal alliances such as San Diego-Tijuana, Cascadia, the Trans-Manche region, Padania, Copenhagen-Malmo, Singapore-Johore-Batam, or Hong Kong-Shenzen. Note that a number of these alliances involve trans-border arrangements.

To an important extent, much of the political change going on in the world's large city-regions today represents a search for structures of governance capable of securing and enhancing their competitive advantages in a rapidly globalizing economic order. Agglomerated production systems are the arenas of both actual and potential region-wide synergies, but these synergies will always exist in some sub-optimal configuration so long as individual decision-making and action alone prevail in the economic sphere. These synergies have enormous relevance to the destinies of all the firms and workers in the immediate locality, and by the same token, they assume dramatic importance in a world where the continued spatial extension of markets brings each city-region into a position of vastly expanded economic opportunities, but also of greatly heightened economic threats from outside. The economies of large city-regions are thus intrinsically overlain by a field of collective order defined by these synergies, and this constitutes a crucial domain of social management. No matter what specific institutional form such management may assume (e.g. agencies of local government, private-public partnerships, civil associations, and so on), it derives its force and legitimacy from the positive role that coordinating agencies can play in regional economic systems by promoting and shaping critical increasing returns effects that would otherwise fail to materialize or that would be susceptible to severe misallocation. The possible shape and character of agencies such as these can be suggested by reference to strategies such as the fostering of agglomeration-specific technological research activities, the provision of high-risk capital to small start-up firms, the protection of certain kinds of infant industry, investments in upgrading workers' competencies, the cultivation of collaborative inter-firm relations, the promotion of distant markets for local products, and so on (Scott 1998). There is also, of course, a continued urgent need for more traditional types of urban planning to ensure that the negative effects of periodic land use and transportation breakdowns do not cut too deeply into local economic performance and social life.

The prospect of a mosaic of global city-regions, each of them characterized by an activist collectivity resolutely seeking to reinforce local competitive advantages, however, raises a further series of questions and problems. Rising levels of concerted regional activism can be expected to lead to specific kinds of destabilization and politicization of inter-regional relations, both within and across national boundaries. Consider, for example, the formation of regional alliances (such as the Four Motors for Europe Programme, or the recent (failed) linking of the London and Frankfurt stock exchanges) giving rise to complaints about unfair competition on the part of those excluded. Another example can be found in the currently prevalent attempts by the representatives of some regions to lure selected assets of other regions into their own geographic orbit, often at heavy social cost. Another can be deciphered in the development races that occur from time to time when different regions push to secure a decisive lead as the dominant center of some budding industry. Still another is evident in the expanding opportunities for multi-national corporations to play one region off against another in competitive bidding wars for new direct investments. In view of the likelihood that stresses and strains of these types will be magnified as the new regionalism takes deeper hold, a need for action at the national, plurinational, and even eventually the global levels of political coordination is foreseeable in order to establish a framework of ground rules for the conduct of interregional relations (including aid to failing regions) and to provide appropriate forums for interregional problem-solving. The European Committee of the Regions, established under the terms of the Maastricht Treaty, may conceivably represent an early, even if still quite fragile, expression in the transnational sphere of this dawning imperative.

As these trends and tendencies come more resolutely to the surface, a further question arises as to what macro-political or ideological formations will be liable to assert a role in defining the calibrating frameworks for the institution-building and policy-making projects that can now be ever more strongly envisioned at various spatial levels. Giddens (1998) has forcefully argued that two main contending sets of political principles appear now to be moving toward a war of position with one another in relation to recent events on the world stage, certainly in the more economically advanced parts of the globe. One of these is a currently dominant neo-liberal view – a view that prescribes minimum government interference in and maximum market organization of economic activity (and that is sometimes but erroneously taken to be a virtually inescapable counterpart of glo-balization). In light of the above remarks regarding the urge to collective action in global capitalism and its various appendages, neo-liberalism strikes me as offering a seriously deficient political vision. The other is a renascent social democracy or social market approach. On the economic front, social democracy is prepared to acknowledge and to work with the efficiency-seeking properties of markets where these are consistent with standards of social fairness and long-term economic wellbeing, but to intervene selectively where they are not. As such, social democratic politics would seem to be well armed to face up to the tasks of building the social infrastructures and enabling conditions (at every geographic level) that are each day becoming more critical to high levels of economic performance as the new world system comes increasingly into focus. At the city-region level, in particular, these tasks can be centrally identified with the compelling social need to promote those local levels of efficiency, productivity, and competitiveness that markets alone can never fully secure.

There is a further forceful argument in favour of a social democratic approach to the governance of global city-regions, one that is associated with, yet that

also goes well beyond, the need for remedial collective action in local economic affairs. Quite simply put, issues of representativeness and distributional impact are always in play in any political community, whether or not social management of the local economy is in some sense under way. In brief, the question of local democratic practice and how to establish effective forums of popular participation is inescapably joined to the more technocratic issues raised by the challenges of economic governance in global city-regions. This question takes on special urgency in view of the role of large global city-regions as magnets for low-wage migrants – many of them undocumented – from all over the world, so that often enough significant segments of their populations are made up of socially marginalized and politically dispossessed individuals. At the same time, and above and beyond any considerations of equity and social justice, enlargement of the sphere of democratic practice is an important practical means of registering and dealing with many of the social tensions that are especially prone to occur in dense social communities; and this remark in turn is based on the observation that the mobilization of voice in such communities is typically an important first step in the constructive treatment of their internal dysfunctionalities. Large city-regions, with their rising levels of social distress as a result of globalization, are confronted with a series of particularly urgent political challenges in this regard, not only because their internal conviviality is in jeopardy, but also because any failure to act is likely, too, to undermine the effectiveness of more purely economic strategies.

From all of this, it follows that some reconsideration of the everyday notion of citizenship is itself long overdue. An alternative definition of citizenship, one that is more fully in harmony with the unfolding new world system, would presumably assign basic political entitlements and obligations to individuals not so much as an absolute birthright, but as some function of their changing involvement and practical allegiances in given geographic contexts. In fact, traditional conceptions of the citizen and citizenship are vigorously and increasingly in question at every geographic level of the world system – for we are all rapidly coming to be, at one and the same time, participants in local, national, plurinational, and global communities – but nowhere as immediately or urgently as in the large global city-regions of the new world system (Holston 2001). Even though only a few tentative and pioneering instances of pertinent reforms in such regions are as yet in evidence (as in certain countries of the EU), more forceful experiments in local political enfranchisement will no doubt come to be initiated in the near future as city-regions start to deal seriously with the new economic and political realities that they face. In a world where mobility is continually increasing, it may not be entirely beyond the bounds of the conceivable that individuals will one day freely acquire title of citizenship in large city-regions many times over in conjunction with their movements from place to place throughout their lifetimes.

CODA

Globalization has potentially both a dark, regressive side and a more hopeful, progressive side. If the analysis presented here turns out to be in principle broadly correct, then those views that have been expressed of late in some quarters to the effect that any deepening trend to globalization must constitute a retrograde step for the masses of humanity can be taken as a salutory warning about a possible future world, but by no means as a representation of all possible future worlds. Insistent globalization under the aegis of a triumphant neo-liberalism would no doubt constitute something close to a worst-case scenario, leading to greatly increased social inequalities and tensions within city-regions and exacerbating the discrepancies in growth rates and developmental potentials between them. Alternative and realistic possibilities can be plausibly advanced, however, and I have tried to sketch out some of these in the preceding pages. Globalization, indeed, is the potential bearer of many significant social benefits. At this stage in history, its future course is still quite open-ended, and it will certainly be subject with the passage of time to many different kinds of political contestation, some of which will mold it in decisive ways. In particular, and as I have tried to indicate, globalization raises important new questions about economic governance or regulation at all spatial levels, and some form of social market politics seems to offer a viable, fair, and persuasive way of facing up to these questions.

Finally, while I have said little or nothing in this account about large cities in the less-developed countries of the world, it seems to me on the basis of both current trends and theoretical speculation – and with due acknowledgement of the enormous difficulties posed by the vicious circles in which they are often caught – that at least some of them might well be able to capitalize on the processes of urbanization and economic growth described above. These processes suggest that selected urbanized areas in a number of less-developed countries are likely eventually to accede as dynamic nodes to the expanding mosaic of global city-regions, just as places like Seoul, Taipei, Hong Kong, Singapore, Mexico City, São Paulo, and others have done, and are doing, before them.

REFERENCES FROM THE READING

Giddens, A. (1998) *The Third Way*, Cambridge: Polity Press.

Holston, J. (2001) Urban citizenship and globalization, in A. J. Scott (ed.) *Global City-Regions*, Oxford: Oxford University Press: 325–48.

Keating, M. (2001) Governing cities and regions: Territorial reconstruction in a global age, in A. J. Scott (ed.) *Global City-Regions*, Oxford: Oxford University Press: 371–90.

O'Brien, R. (1992) *Global Financial Integration: The End of Geography*, London: Pinter.

Scott, A. J. (1988) *New Industrial Spaces*, London: Pion.

Scott, A. J. (1993) *Technopolis*, Berkeley: University of California Press.

Scott, A. J. (1998) *Regions and the World Economy*, Oxford: Oxford University Press.

Scott, A. J. (ed.) (2001) *Global City-Regions*, Oxford: Oxford University Press.

Scott, A. J. (2005) *Social Economy of the Metropolis: Cognitive-Cultural Capitalism and the Global Resurgence of Cities*, Oxford: Oxford University Press.

Scott, A. J. (2008) *On Hollywood: The Place, The Industry*, Princeton: Princeton University Press.

Scott, A. J. (2012) *A World in Emergence: Cities and Regions in the 21st Century*, Cheltenham: Edward Elgar.

Storper, M. (1997) *The Regional World*, New York: Guilford Press.

32

"Global cities, 'glocal' states: global city formation and state territorial restructuring in contemporary Europe"

from *Review of International Political Economy* (1998)

Neil Brenner

EDITORS' INTRODUCTION

Neil Brenner is Professor of Urban Theory in the Graduate School of Design at Harvard University. This essay was written in the mid-1990s while he was a graduate student in the Department of Geography at UCLA and it was subsequently published in an interdisciplinary journal, the *Review of International Political Economy*. Brenner's intellectual starting point is the observation that most global city theorists have postulated, either implicitly or explicitly, a declining role for national states in the governance of economic life. This assumption of "state decline" has in turn led scholars either to focus entirely on *local* scales of regulation within global city economies and/or to bracket the ways in which (reconstituted) national state institutions have, in many cases, actively facilitated the process of global city formation. Against these intellectual tendencies, Brenner argues that state institutions at national, regional, and local scales have been instrumental in promoting the development of globally interlinked cities, both in the Western European context (his central empirical focus) and beyond. This claim, which has been developed by several other urbanists since the mid-1990s (see, for instance, Ch. 33 by Hill and Kim), leads Brenner to explore, more specifically, the ways in which state institutions themselves have been reorganized since the mid-1970s, in significant measure through their role in promoting urban restructuring. Brenner focuses upon the process of state rescaling in which established hierarchies of intergovernmental relations and political-economic regulation are being recalibrated so as to enable new forms of urban governance. While a significant part of Brenner's original article explored diverse empirical cases of global city formation and state rescaling in western Europe, this essay is devoted primarily to an elaboration of these arguments on a more conceptual level. In the context of this Reader, one of the contributions of Brenner's work is to explore the interplay between global city formation and various ongoing transformations of statehood—including patterns of state spatial organization and changing modes of state intervention.

World city theory has been deployed extensively in studies of the role of major cities such as New York, London and Tokyo as global financial centers and as headquarters locations for transnational corporations (TNCs). While the theory's usefulness in such research has been convincingly demonstrated, I believe that the central agenda of world city theory is best conceived more broadly, as an attempt to analyze the changing geographies of global capitalism in the late 20th century. From this point of view, the project of world cities research is not merely to classify cities within world-scale central place hierarchies, but, as Friedmann (1986: 69) has proposed, to analyze the "spatial organization of the new international division of labor." The key feature of this newly emergent configuration of world capitalism is that cities – or, more precisely, large-scale urbanized regions – rather than national territorial economies are its most fundamental geographical units. These urban regions are said to be arranged hierarchically on a global scale according to their specific modes of integration into the world economy.

But how is this emergent global urban hierarchy articulated with the geographies of national state territories? Insofar as world city theory is directly concerned with the "contradictory relations between production in an era of global management and the political determination of territorial interests" (Friedmann 1986: 69), an analysis of changing relations between world cities and national state spaces is arguably one of its most central theoretical and empirical tasks. Yet, in practice, the methodological challenge of analyzing the changing linkages between different spatial scales has not been systematically confronted by world cities researchers.

Much of world cities research has been composed of studies that focus primarily upon a single scale, generally either the urban or the global. Whereas research on the socioeconomic geography of world cities has focused predominantly on the urban scale, studies of changing urban hierarchies have focused largely on the global scale. To the extent that the national state has been thematized, it has usually been understood with reference to its local/municipal institutions or as a relatively static background structure. Indeed, like many other prevalent approaches to the study of globalization, the bulk of world cities research during the 1980s and early 1990s was premised upon the underlying assumption that intensified globalization entails an erosion of national state territoriality (see, for instance, Friedmann and Wolff 1982). This conception of globalization as a process of state decline has led world cities researchers to focus on the global scale, the urban scale and their changing interconnections while systematically neglecting the role of nationally configured political-economic dynamics. The privileging of the global/local dualism among world cities researchers has also been grounded upon what might be termed a zero-sum conception of spatial scales in which the global, the national and the urban scales are viewed as being mutually exclusive – what one gains, the other loses – rather than as being intrinsically related, co-evolving layers of territorial organization.

I argue here, by contrast, that national states are being rescaled and redefined in conjunction with processes of global city formation rather than being eroded. The resultant, rescaled configurations of national state space have come to figure centrally in mediating the processes of geoeconomic integration and urban-regional restructuring. While it is evident that the current round of geoeconomic restructuring has undermined Fordist-Keynesian forms of state regulation and economic governance, the narrative of state decline conflates the ongoing reconfiguration of the national state space with a withering away of national state power as such. Current transformations of state power may indeed herald the partial erosion of central state regulatory control over global flows of capital, commodities and labor-power, and they have also clearly entailed new fiscal and legitimation problems for national governments. Despite this, however, national states arguably remain central institutional matrices of political power and crucial geographical infrastructures for capital accumulation (Jessop 1994; Panitch 1994). More generally, as I argue in this chapter, national states have figured crucially in catalyzing and mediating the process of global city formation.

RESCALING URBAN SYSTEMS, RESCALED STATE SPACES

Cities are at once basing points for capital accumulation (nodes in global flows) and organizational-administrative levels of national states (coordinates of state territorial power). First, as nodes in global flows, cities operate as loci of industrial production, as centers of command and control over inter-urban, interstate and global circuits of capital and as sites of exchange within local, regional, national and global markets. Second, as coordinates of state territorial power, cities are regulatory-institutional levels within each national state's intergovernmental hierarchy. The term coordinate is intended to connote the embeddedness of cities within the national state's organizational matrix. These coordinates may be interlinked through various means, from legal-constitutional regulations, financial interdependencies, administrative divisions of labor and hierarchies of command to informal regulatory arrangements.

During the Fordist-Keynesian period (circa 1950 to 1970), these two dimensions of urbanization were spatially coextensive within the boundaries of the national territorial state. As nodes of accumulation, cities were enframed within the same territorial grids that underpinned the national economy. The cities of the older industrialized world served as the engines of Fordist mass production and as the urban infrastructure of a global economic system compartmentalized into nationalized territorial matrices. It was widely assumed that the industrialization of urban cores would generate a propulsive dynamic of growth that would in turn lead to the industrialization of the state's internal peripheries, and thereby counteract the problem of uneven geographical development. Likewise, as coordinates of state territorial power, Fordist-Keynesian regional and local regulatory institutions functioned primarily as transmission belts of central state socioeconomic policies (Mayer 1994). Their goal was above all to promote growth and to redistribute its effects on a national scale. To this end, redistributive regional policies were widely introduced to promote industrialization within each state's internal peripheries (Albrechts and Swyngedouw 1989).

Since the 1970s, however, these nationalized geographies of urbanization and state regulation have been profoundly reconfigured. The crisis of global Fordism was expressed in a specifically geographical form, above all through the contradiction between the national scale of state regulation and the globalizing thrust of postwar capital accumulation (Peck and Tickell 1994). Consequently, since the global economic crises of the early 1970s, the scales on which the Fordist-Keynesian political-economic order was organized – national regulation of the wage relation; international regulation of currency and trade – have been significantly reconfigured. While the deregulation of financial markets and the global credit system since

the collapse of the Bretton Woods system in 1973 has undermined the viability of nationally organized demand management policies, the increasing globalization of financial flows has diminished the ability of national states to insulate themselves from the world economy as quasi-autarchic national economic spaces (Agnew and Corbridge 1995). The intensification of global interspatial competition among cities and regions has also compromised traditional national industrial policies and led regional and local states to assume increasingly direct roles in promoting capital accumulation on subnational scales.

The central geographical consequence of these intertwined shifts has been to destabilize the most elemental building block of the postwar geoeconomic and geopolitical order – the autocentric national economy. Despite this, however, cities and national states have continued to operate as fundamental forms of territorialization for capital, even though this role is no longer tied primarily to the nationally configured patterns of urbanization and to nationally centralized strategies of economic governance. Since the crisis of North Atlantic Fordism in the early 1970s, new subnational and supranational patterns of urbanization and state regulation have been consolidated throughout the older industrialized world. Our task in the present context is to examine more closely the geographical-institutional interface between the rescaling of urbanization and the remaking of state spatiality.

First, as world cities researchers have indicated at length, the contemporary rescaling of urbanization must be viewed as a multidimensional reorganization of entrenched national urban systems in close conjunction with the consolidation of new world-scale urban hierarchies. To illustrate this ongoing rescaling of the urbanization process, Figure 1 depicts the ways in which the European urban hierarchy has been reconfigured since the crisis of the Fordist-Keynesian regime during the early 1970s.

This schematic representation of the contemporary European city-system (derived from Krätke 1995: 140–1) focuses upon the first dimension of urbanization, the role of cities as nodes of capital accumulation. Krätke's model describes contemporary transformations of the European urban hierarchy with reference to two criteria—the industrial structure of the city's productive base (Fordist vs. post-Fordist) and the spatial scale of its command and control functions (global, European, national, regional, non-existent). The arrows in the figure indicate various

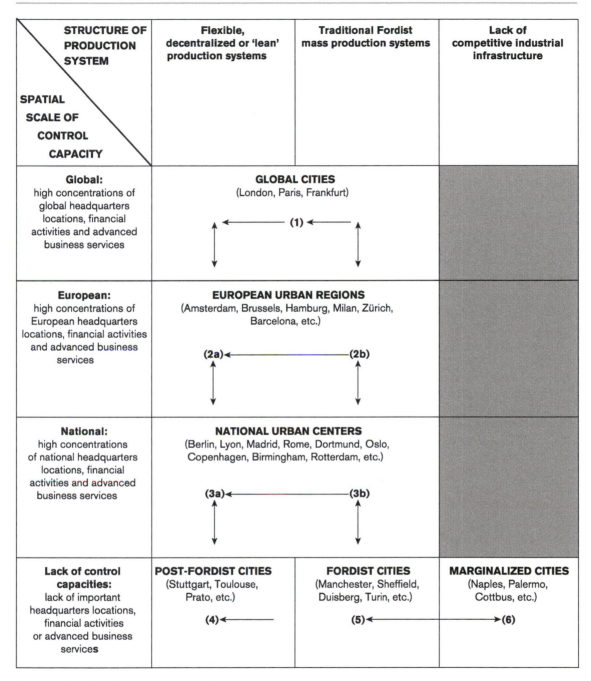

STRUCTURE OF PRODUCTION SYSTEM / SPATIAL SCALE OF CONTROL CAPACITY	Flexible, decentralized or 'lean' production systems	Traditional Fordist mass production systems	Lack of competitive industrial infrastructure
Global: high concentrations of global headquarters locations, financial activities and advanced business services	**GLOBAL CITIES** (London, Paris, Frankfurt) ↔ **(1)** ←		
European: high concentrations of European headquarters locations, financial activities and advanced business services	**EUROPEAN URBAN REGIONS** (Amsterdam, Brussels, Hamburg, Milan, Zürich, Barcelona, etc.) **(2a)** ← → **(2b)**		
National: high concentrations of national headquarters locations, financial activities and advanced business services	**NATIONAL URBAN CENTERS** (Berlin, Lyon, Madrid, Rome, Dortmund, Oslo, Copenhagen, Birmingham, Rotterdam, etc.) **(3a)** ← → **(3b)**		
Lack of control capacities: lack of important headquarters locations, financial activities or advanced business services	**POST-FORDIST CITIES** (Stuttgart, Toulouse, Prato, etc.) **(4)** ←	**FORDIST CITIES** (Manchester, Sheffield, Duisberg, Turin, etc.) **(5)** ←	**MARGINALIZED CITIES** (Naples, Palermo, Cottbus, etc.) → **(6)**

Figure 1 The changing European urban hierarchy

Source: Neil Brenner

possible changes in position among cities within the European urban hierarchy; and various cities have been listed to exemplify each of these levels. As the figure indicates, global city formation has entailed the emergence of a new global urban hierarchy, defined through the increasing scale of urban command and control functions, of inter-urban exchange relations and of inter-urban competition. As nodes of accumulation, therefore, cities are no longer enclosed within relatively autocentric national economies, but have been embedded more directly within transnational urban hierarchies and inter-urban networks. Although the cities currently positioned at the apex of the global, European, North American and East Asian urban hierarchies present the most dramatic evidence of this transformation, their newly acquired positions within the global urban system are indicative of a more general rescaling of urbanization processes across the world economy.

Most crucially here, the current wave of global spatial restructuring has also had important implications for the role of cities as coordinates of state territorial power. Despite its neglect of nationally scaled processes, the methodology of world cities research provides a useful starting point for investigating recent rescalings of state spatiality. Much like the place-based territorial infrastructures of global cities, I would argue, post-Keynesian state institutions can be viewed as crucial forms of reterritorialization for capital in the current period. Whereas the centralized, bureaucratized states of the Fordist-Keynesian era converged around the national scale as their predominant organizational-regulatory locus, the national states of the post-Keynesian, neoliberal era have been restructured substantially to provide capital with many of its most essential territorial preconditions and collective goods at other spatial scales, including both the supranational and the subnational.

Soja's (1992) concept of the exopolis provides a strikingly appropriate image for describing the transformed spatial form of these rescaled state apparatuses. Like the exopolis, the spatial expression of post-Fordist forms of capitalist industrialization in which inherited urban spaces have been turned "inside-out" and "outside-in" (Soja 1992), the geographies of post-Keynesian state institutions are polymorphic, multitiered and decentered; and they are likewise being simultaneously being turned inside-out and outside-in – inside-out insofar as they attempt to promote the global structural competitiveness of their major cities and regions; and outside-in insofar as supranational agencies and international agreements have come to play more direct roles in structuring their "internal" political spaces. This rescaling of state space is rearticulating inherited political geographies in ways that are deprivileging nationally organized regulatory arrangements while ceding new roles both to supranational and to subnational institutional forms. Thus understood, state institutions retain a critical role as forms of territorialization for capital,

but this role is no longer premised upon an isomorphic territorial correspondence between state institutions, urban systems and circuits of capital accumulation centered around national state boundaries.

This rescaling of state power has not only reshuffled entrenched political geographies and administrative hierarchies, but has also been associated with a profound transformation of the relationship between states, capital and territory. Territorial organization has long operated as a force of production under capitalism through its natural goods, its supplies of fixed capital and labor-power, its technological-institutional infrastructures and other place-specific externalities and collective goods. The state has arguably played a crucial role in the production, regulation and reproduction of these socio-territorial and productive ensembles throughout the long-run history of capitalism. During the Fordist-Keynesian period, most older industrial states deployed indirect forms of regulatory intervention oriented towards the reproduction of labor-power, industrial relocation and the promotion of collective consumption. Although the collapse of the Fordist-Keynesian regulatory regime has undermined the monolithic unity of national states as territorially self-enclosed containers of socioeconomic activities, this development has also arguably intensified the state's role in the territorialization of capital (Brenner 2004).

The rescaled state institutions have come to play essential roles in the production, coordination and maintenance of the customized, place-specific configurations of socioeconomic organization upon which global competitive advantages today increasingly depend, both in global city-regions and in other major capitalist cities as well. For, in contrast to the various incentive-based and indirect policies of the Fordist-Keynesian era, contemporary post-Keynesian modes of state intervention have entailed a more direct, unmediated involvement of state institutions in the territorialization of capital. Faced with the apparently increased mobility of capital, commodities and labor-power across national borders, post-Keynesian state apparatuses are orienting themselves above all towards the provision of immobile factors of production – that is, towards those externalities associated with capital's moment of territorialized fixity within major cities and city-regions. From public-private partnerships, labor retraining programs, science parks, conference centers, waterfront redevelopment schemes, technology transfer projects, information-sharing networks, venture capital programs and market research projects to large-scale investments in technopoles, innovation programs, enterprise zones and free trade areas, an immense range of state-organized economic development policies are being mobilized in order to enhance the territorially specific productive capacities of strategically delineated economic spaces. The overarching goal of these state strategies is to secure new locational advantages in international economic competition through the construction of territorially rooted immobile assets. In the current period, many if not all of the socially produced features of territorial competitiveness – such as human capital resources, cost efficiency, product quality, turnover time, flexibility and innovative capacities – have become central concerns of state institutions, at various spatial scales, in their governance of economic development at a range of spatial scales. And, even when such territorial assets are not directly produced by the state, a rapidly growing number of state agencies have become directly or indirectly engaged in financing, monitoring, coordinating and maintaining them.

More generally, by adopting new strategies of economic promotion and place-marketing, national state institutions have also come to play central roles in marketing their own territories (or strategic sites therein) as locational products on the world market. Under these conditions, the role of state institutions in economic governance is no longer merely to reproduce localized production complexes, but continually to restore, enhance, intensify and restructure their capacities as territorially specific productive forces. These developments lead Swyngedouw (1992: 431) to conclude that "the role of the state is actually becoming more, rather than less, important in developing the productive powers of territory and in producing new spatial configurations." The goal of creating place-specific or geographically immobilized competitive advantages may be pursued through both deregulatory and reregulatory political strategies. The balance between the latter is frequently a matter of intense sociopolitical conflict.

It is in this context, I would argue, that the enhanced role of subnational institutional forms in contemporary processes of socioeconomic governance is to be understood. It is above all through their role in securing, promoting, maintaining and advertising any number of place-specific conditions for capital investment that local and regional states, in particular, are gaining structural significance within each national

state's internal administrative hierarchy (Mayer 1994). Indeed, the process of state rescaling can be viewed in significant measure as a concerted political strategy through which political-economic elites at various levels of state power are attempting to propel major cities and regions upwards within the urban hierarchy depicted in Figure 1. Throughout Europe, local, regional and national governments are mobilizing diverse state strategies intended at once to revalorize decaying industrial sites, to promote industrial growth in globally competitive sectors and to acquire command and control functions in the world economy by providing various territorial preconditions for transnational capital, including transportation and communications links, office space, labor-power and other place-specific externalities (Hall and Hubbard 1996).

Figure 2 summarizes the ways in which the relations between urbanization patterns and forms of state territorial organization have been reconfigured since the Fordist-Keynesian period, highlighting at once the globalization of the world economy, the rescaling of state space, and the ramifications of these shifts for both dimensions of urbanization. As nodes of accumulation, global cities are embedded within flows of capital that no longer overlap coextensively with national economic space. As coordinates of state territorial power, global cities are strategic targets for rescaled state strategies oriented towards the continual enhancement of territorially specific competitive advantages and productive forces. In this sense, global cities are simultaneously spaces of global accumulation and coordinates of rescaled state spaces. In this sense, the governance of contemporary urbanization patterns entails not only the construction of "new industrial spaces" (Scott 1988) for post-Fordist forms of industrialization but, just as crucially, the consolidation of what might be termed new state spaces to enhance each state's capacity to mobilize the productive force of urban and regional spaces and to regulate the sociopolitical contradictions induced by such projects (Brenner 2004).

In the late 20th century, therefore, the state's own spatial and scalar configuration has become an important locational weapon in the interspatial competition between cities, regions and national states in the world economy. Under these conditions, a new "politics of scale" (Smith 1992) has emerged in which the scalar geographies of state power have become a direct object of sociopolitical contestation. If, as Friedmann and Wolff (1982: 312) have proposed, "world cities lie

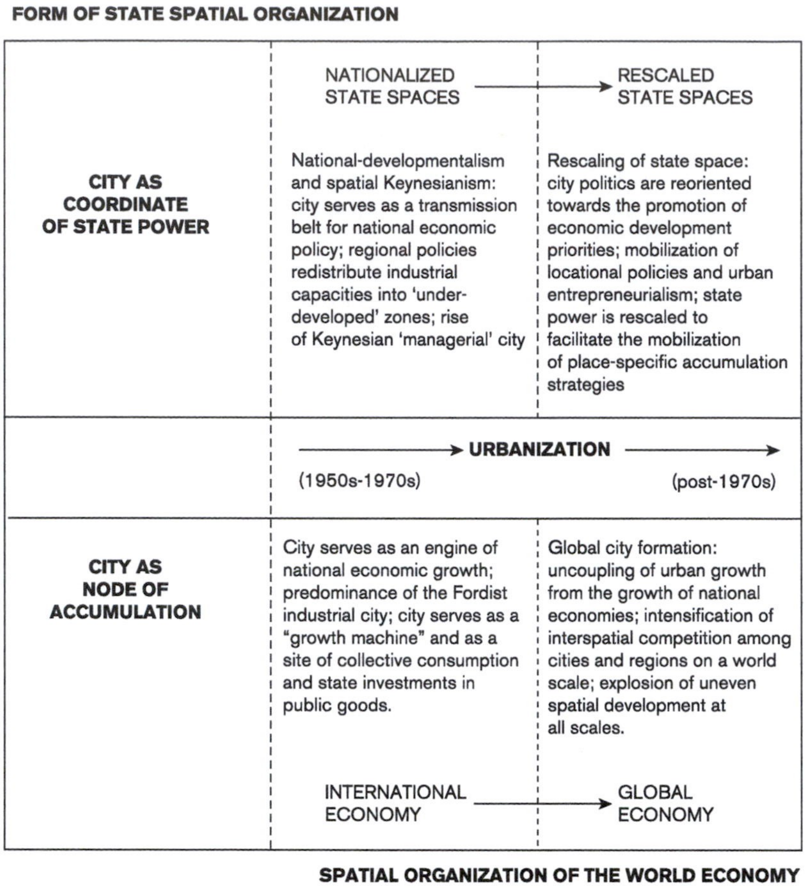

FORM OF STATE SPATIAL ORGANIZATION

	NATIONALIZED STATE SPACES ⟶	RESCALED STATE SPACES
CITY AS COORDINATE OF STATE POWER	National-developmentalism and spatial Keynesianism: city serves as a transmission belt for national economic policy; regional policies redistribute industrial capacities into 'under-developed' zones; rise of Keynesian 'managerial' city	Rescaling of state space: city politics are reoriented towards the promotion of economic development priorities; mobilization of locational policies and urban entrepreneurialism; state power is rescaled to facilitate the mobilization of place-specific accumulation strategies
	⟶ **URBANIZATION** ⟶	
	(1950s–1970s)	(post-1970s)
CITY AS NODE OF ACCUMULATION	City serves as an engine of national economic growth; predominance of the Fordist industrial city; city serves as a "growth machine" and as a site of collective consumption and state investments in public goods.	Global city formation: uncoupling of urban growth from the growth of national economies; intensification of interspatial competition among cities and regions on a world scale; explosion of uneven spatial development at all scales.
	INTERNATIONAL ECONOMY ⟶	GLOBAL ECONOMY

SPATIAL ORGANIZATION OF THE WORLD ECONOMY

Figure 2 Urbanization, state forms and the world economy, 1950–2000

Source: Neil Brenner

at the junction between the global economy and the territorial nation-state," then it seems appropriate to view the political-regulatory institutions of world city-regions as geographical arenas in which this new politics of scale are fought out with particular intensity.

CONCLUSION: THE URBAN QUESTION AS A SCALE QUESTION

Amidst the confusing and contradictory geographies of contemporary globalization, world cities represent a particularly complex "superimposition and inter-penetration" (Lefebvre 1991: 88) of social, political and economic spaces. Because urban regions occupy the contradictory interface between the world economy and the territorial state, they are embedded within a

multiplicity of political-economic processes organized upon a range of superimposed geographical scales. The resultant politics of scale within the political insti-tutions of major urban regions can be construed as a sequence of groping, trial-and-error strategies to manage these intensely conflictual forces through the continual construction, deconstruction and recon-struction of relatively stabilized configurations of ter-ritorial organization. The rescaling of urbanization leads to a concomitant rescaling of the state through which, simultaneously, urban and regional spaces are mobilized as productive forces and social relations are circumscribed within new political boundaries and scalar hierarchies. These rescaled configurations of state power in turn transform the everyday social con-ditions under which the urbanization process unfolds. Whether these disjointed strategies of reterritorializa-tion within European cities might eventually establish new spatial and scalar fixes for sustained capitalist growth on any geographical scale is a matter that can only be resolved through the politics of scale itself, through ongoing struggles for hegemonic control over the form, trajectory and territorial organization of the urbanization process.

REFERENCES FROM THE READING

Agnew, J. and Corbridge, S. (1995) *Mastering Space*, New York: Routledge.

Albrechts, L. and Swyngedouw, E. (1989) The challenges for regional policy under a flexible regime of accumulation, in L. Albrechts et al. (eds.) *Regional Policy at the Crossroads: European Perspectives*, London: Jessica Kingsley: 67–89.

Brenner, N. (2004) *New State Spaces: Urban Governance and the Rescaling of Statehood*, New York: Oxford University Press.

Friedmann, J. (1986) The world city hypothesis, *Development and Change*, 17: 69–83.

Friedmann, J. and Wolff, G. (1982) World city formtion: An agenda for research and action, *International Journal of Urban and Regional Research*, 6: 309–44.

Hall, T. and Hubbard, P. (1996) The entrepreneurial city, *Progress in Human Geography*, 20, 2: 153–74.

Jessop, B. (1994) Post-Fordism and the state, in A. Amin (ed.) *Post-Fordism: A Reader*, Cambridge, MA: Blackwell: 251–79.

Krätke, S. (1995) *Stadt, Raum, Ökonomie*, Basel: Birkhäuser.

Lefebvre, H. (1991) The production of space, trans. D. Nicholson-Smith, Oxford, UK and Cambridge, MA: Blackwell.

Mayer, M. (1994) Post-Fordist city politics, in A. Amin (ed.) *Post-Fordism: A Reader*, Cambridge, MA: Blackwell: 316–37.

Panitch, L. (1994) Globalization and the state, in R. Miliband and L. Panitch (eds.) *Socialist Register*, London: Merlin Press: 60–93.

Peck, J. and Tickell, A. (1994) Searching for a new institutional fix, in A. Amin (ed.) *Post-Fordism: A Reader*, Cambridge, MA: Blackwell: 280–315.

Scott, A. J. (1988) *New Industrial Spaces*, London: Pion.

Smith, N. (1992) Geography, difference and the politics of scale, in J. Doherty, E. Graham and M. Malek (eds.) *Postmodernism and the Social Sciences*, New York: St. Martin's Press: 57–79.

Soja, E. (1992) Inside exopolis, in M. Sorkin (ed.) *Variations on a Theme Park*, New York: The Noonday Press: 94–122.

Swyngedouw, E. (1992) Territorial organization and the space/technology nexus, *Transactions of the Institute of British Geographers*, 17: 417–33.

33

"Global cities and developmental states: Tokyo and Seoul"

from *Urban Studies* (2000)

Richard Child Hill and June Woo Kim

EDITORS' INTRODUCTION

In this contribution, Richard Child Hill and June Woo Kim (National University of Singapore) elaborate a comparative analysis of economic restructuring and urban development in Tokyo and Seoul. Whereas Hill's work on Detroit and Houston (see Ch. 13) applied certain key arguments of global city theory to the comparative investigation of those two cities, this analysis of Tokyo and Seoul adopts a more critical perspective on this approach. In a section of their article that could not be reproduced here, Hill and Kim develop a detailed critique of major global city theorists, such as Saskia Sassen and John Friedmann, for a purported tendency to overgeneralize the effects of globalization upon urban structures (Hill and Kim 2001). Against this alleged assumption of "convergence," Hill and Kim insist upon the highly variegated national and local pathways of urban restructuring that are crystallizing under contemporary capitalism. In this context, Hill and Kim also insist that the developmental trajectories of East Asian cities are profoundly shaped by national state institutions, which continue to channel significant resources into urban growth (see also Ch. 44 by Ren and Ch. 16 by Wu). For Hill and Kim, the existence of these activist, developmental states in East Asian cities differentiates them qualitatively from the supposedly paradigmatic cases of New York and London, and thus undermines the applicability of global city theory beyond the "market centered" urban systems of the Anglo-American world.

Tokyo and Seoul, according to Hill and Kim, diverge from the standard, New York- and London-centric model of global city formation due to the lack of extensive urban sociospatial polarization; the embeddedness of local financial institutions within the national industrial fabric; the persistence of manufacturing industries in the city; the continued role of activist national governmental policies and national political elites in guiding urban development; and the continued contribution of metropolitan economic dynamics to the national economy as a whole. Hill and Kim conclude by arguing that contemporary capitalism is composed of multiple, competing national and regional institutional systems, and not an overarching global regime.

The "world city paradigm" is the most important contribution by urbanists to the contemporary globalization literature. Oddly, however, the world city hypothesis has not generated the vigorous debate among globalists, statists and those attempting to bridge the two camps that so enlivens most current work on globalisation. John Friedmann and Saskia Sassen, the best-known architects of world city theory, take a globalist view—that is, they believe that a single global system is becoming superimposed on nation-states which are losing importance as a result. Globalisation produces a world city system that transcends national institutions, politics and culture, they argue. Such a view assumes a convergence in "economic base, spatial organization and social structure" among the world's major cities, especially New York, London and Tokyo (Sassen 1991: 4).

In fact, however, fundamental differences in "economic base, spatial organization and social structure" persist between major cities in the North Atlantic and East Asian regions. Most telling for the paradigm, Tokyo, center of the world's second-largest national economy and the world's largest urban agglomeration, departs from the world city paradigm on most salient dimensions. Seoul, center of East Asia's second OECD member and the region's second-largest metropolis, exhibits the same anomaly. An awareness of these differences has been creeping into the world city literature. John Friedmann (1995) has acknowledged that Tokyo does not fit the world city paradigm in some respects, but he does not address the implications for the world city hypothesis. Saskia Sassen (1991: 86), on the other hand, explains away Tokyo's differences as a temporary function of "Japan's uniqueness" and continues to assume that convergence among the world's major financial centers is the overall trend.

We disagree. Understanding Tokyo and Seoul necessitates a different conception of the world system from that underlying the globalist version of the world city argument. World cities differ from one another in many salient respects because they are lodged within a non-hegemonic and interdependent world political economy divided among differently organized national systems and regional alliances (Stallings and Streeck 1996).

GLOBAL CITIES AND DEVELOPMENTAL STATES

Western neo-classical economists enquire mostly into markets, and occasionally into organizational hierarchies. They recognize the state as a third means for economic governance but confine it to defining property rights, enforcing contracts, overseeing the general rules of competition and (sometimes) providing collective goods; that is, to setting the minimal conditions without which markets and hierarchies could not function. More state involvement in the economy than this would interfere with the market mechanism, and by deviating from efficiency and productivity, would ultimately give way to competitive pressures. In this view, capitalist societies are bound to converge upon a market-driven corporate model.

In contrast to their neo-classical brethren, Western development economists have long been interested in comparing national paths to development. In their view, latecomers to the industrialization process must forge their own development institutions and ideologies because they invariably face a different set of problems and possibilities from those of their technically more advanced predecessors (Gershenkron 1962). A poor country's fledgling firms, for example, confront formidable competition from transnational corporations (TNCs) possessing far greater economies of scale, advanced technologies and global networks. But less developed countries also have hidden reserves of labor, savings and entrepreneurship. Nations wishing to overcome the penalties and realize the possibilities of late development require a strong state. The real issue, development economists often conclude, is not whether the state should or should not intervene, but rather "the art of getting something done with intervention" (Amsden 1989: 140).

Japanese thinkers developed their own art of late development between the Great Depression and the end of World War II. In contrast to liberal capitalism, Japanese developmentalism addressed industrialization at the level of the nation-state. Strengthening national production was the top priority of industrial policy. The economy was viewed strategically with the aim of building an industrial structure that would maximize Japan's gains from international trade. State regulations and non-market governance mechanisms were designed to restrain competition in order to concentrate resources in strategic industries and maintain orderly economic growth. And the quest for short term profits was rejected to secure workers' co-operation in promoting productivity. The Japanese first institutionalized these principles between 1931 and 1945, but a number of studies (Gao 1997) suggest that these tenets continued to underlie Japan's postwar industrial policy, despite changes in Japan's political institutions and national purpose.

The transformation of Japanese developmentalism from militarism to trade was largely accomplished by the end of the 1960s. The state emerged with a strong capacity to sustain economic growth in contrast to the more free-wheeling role played by the market in Anglo-American capitalist economies. Japanese managers emphasized cooperative industrial relations in contrast

RICHARD CHILD HILL & JUNE WOO KIM

to their conflict-prone Western counterparts. Family-based *zaibatsu* business groups were reorganized into management-controlled *keiretsu* networks as a powerful weapon in market competition.

Given the differences between Anglo-American liberalism and East Asian developmentalism, it seems reasonable to expect related contrasts in the role each sphere's major cities play in the world economy. Table 1 offers such a contrast by hypothesizing two world city types: a market-centered, bourgeois type, modeled on New York City, and a state-centered, political-bureaucratic type, modeled on Tokyo. We provide a brief sketch of Tokyo to establish the contrasting type, and then focus our empirical investigation on another state-centered world city, Seoul, Korea.

The global economy is spatially imbedded in Tokyo, to be sure, but Tokyo is not primarily a global basing-point for the operations of stateless TNCs. Rather, Tokyo is mainly a national basing-point for the global operations of Japanese TNCs. Tokyo's relationship to the world economy is not driven in the first instance by market efficiency, but by a strategic concern to preserve national autonomy through global economic power (Johnson 1995). In Japan,

	Market-centred bourgeois	State-centred political-bureaucratic
Prototypical city	New York	Tokyo
Regional base	West Atlantic	East Pacific
Leading actors	TN capitalist class	State bureaucratic elite
Group Organisation	Finance TNCs	State Ministries tied to business networks via main banks
Economic ideology	Vertically integrated firms	
	Liberalism	Developmentalism
	Self-regulating market	Strategic national interest
Trade, investment and production	Market-rational	Plan-rational
Relation to world economy	Private wealth	National power
Prime objectives	Profit-maximising	Market-share, employment-maximising
Global control capability via	Private-producer service complexes	Government ministries
		Public corporations
		Policy networks
Industrial structure	Manufacturing HQs and production dispersed	Manufacturing HQs and high-tech production concentrated
	Services emphasised	Services de-emphasised
Occupational structure	Polarised	Compressed
(social and spatial)	Missing middle	Missing extremes
	High inequality	Low inequality
	High segregation	Low segregation
Foreign immigration	Weak controls	Strong controls
	High	Low
Culture	Consumerist	Productionist
	Yuppie, ethnic	Salaryman, officelady
City-central state relationship	Separation	Integration
Source of urban contradictions	Short term profit	State capital controls
	Market volatility	Overregulation
	Polarisation	Centralisation
Competitive advantages	Fluidity	Stability
	Mobility	Planning

Table 1 Two world city types

Source: Richard Child Hill and June Woo Kim

economic power is indexed by the world market shares held by the nation's industries, not by quarterly dividends and privately accumulated wealth.

Tokyo offers corporations global control capability, but the primary vehicle is not private financial and producer services clustered into complexes by market forces. Tokyo's global control apparatus resides in financial and industrial policy networks among public policy companies, banks and industrial enterprises, under the guidance of government ministries like the Ministry of Finance (MOF) and the Ministry of International Trade and Industry (MITI). Indeed, by emphasizing reinvestment and employment rather than high profits and individual consumption, Japanese policies have actively discouraged growth in the kinds of services distinguishing New York City.

The practice of global control in Tokyo has not resulted in a social regime characterized by massive loss in manufacturing jobs, high levels of foreign immigration, extreme wealth concentration and social and spatial polarization. One-quarter of Tokyo's labor force continues to work in manufacturing (as against less than 10 per cent in New York City), primarily in high-tech, research-intensive pilot plants, and the headquarters of Japan's major manufacturing companies continue to concentrate in Tokyo to be near government ministries. The state tightly controls foreign immigration with an eye to available employment, and the foreign-born represent a minuscule 1.8 per cent of Tokyo's population, compared to 28 per cent in New York City. In contrast to New York City's dualism, Tokyo's occupational structure is compressed around the median, the middle strata encompass most city residents, and the extremes in wealth concentration and impoverishment found in New York are missing.

Tokyo's commanding place in the world urban hierarchy is not determined by the city's ability to attract global investments, but by the ability of Tokyo companies to generate earnings from abroad. In 1990, for example, Japanese TNCs controlled 12 per cent of world FDI, while foreign investment in Japan represented only 1 per cent of the total world stock. The comparable figures in 1980 were 4 per cent and 1 per cent (Ostry 1996: 334). While Japan's *keiretsu* networks and the main bank system enabled the MOF to influence big-firm strategies via the supply and cost of capital to network banks, the system was also explicitly designed to protect Japanese companies

against foreign penetration and short-term profit pressures.

Control over investment equals power in Tokyo as in New York City, but Tokyo is under the sway of a political-bureaucratic elite, not a transnational capitalist class. Control in New York City is in the hands of a private investor class. The stock market is the barometer of New York's economy. Control in Tokyo is exercised through management-run corporate networks centering upon main banks which in turn are guided by government ministries. Employment is the barometer of Tokyo's economy. Tokyo's elite possesses a "productionist" not a consumerist ideology. The clash between classes in the state-centered world city is not between transnational and local capitalist classes but between bourgeois and political-bureaucratic elites.

Finally, Tokyo is not parting company with the Japanese nation and central state. Japan is a unitary state, and the relationship between the city of Tokyo and the central government is bureaucratically integrated in a myriad of ways, and especially through the Ministry of Home Affairs. Tokyo is, in effect, a national champion (Hill and Fujita 1995).

SEOUL

Seoul is Korea's command post for government planning and business management. While 24 per cent of the nation's population resides in Seoul city, virtually all of Korea's central government agencies (96 per cent) and top corporate headquarters (48 out of the top 50) are located there. Sixty-one per cent of Korea's business managers and 64 per cent of the nation's research scientists work in the city. The city of Seoul combines with surrounding satellites in Kyonggi-do Province to form the Seoul Metropolitan Region (SMR). Koreans often refer to the SMR as the "Seoul Republic" because it is so dominant over other regions of the country. With 17 million residents, the capital region contains 39 per cent of South Korea's population (Kim and Choe 1997: 2, 43).

Seoul is also Korea's window on the world. All but one of the nation's foreign embassies, and 15 out of Korea's 22 foreign consulates, are in Seoul. All of the nation's stock brokerages (76), foreign bank offices (66), offices of foreign media (25) and broadcasting networks (8) are in the city. Seoul hosts 71 per cent of Korea's overseas-based service industries, half the nation's international hotels and trading companies, and nearly all of its communication services (Hong 1996). Seoul now ranks 7th among the world's cities in the number of industrial Fortune 500 transnational corporations (TNCs) headquartered there, 13th in the number of TNC bank headquarters, 17th in number of international organizations and 23rd in frequency of international conferences (Jo 1992). The outward indicators certainly point to Seoul's world city status, but how well in fact does the Seoul Republic fit the world city paradigm?

Seoul's global base

With scarce natural resources and a small domestic market, the Korean state subsidized export-oriented industries and South Korea industrialized by exporting to overseas markets. Seoul, as home to the central government, was the place to be for all who desired contact with government ministries and exposure to international markets. Little wonder then that Korea's major firms and business associations were head-quartered in Seoul. Following the *chaebols,* related industries and supporting services also clustered in Seoul in pursuit of close contacts with the major business groups, central government agencies, and trade and industrial associations.

Seoul is certainly a basing-point for TNCs, as emphasized by world city theory. Seoul hosted four companies on the list of the world's 100 largest in 1997 – only six cities in the world had more. Ten of the global Fortune 500 companies are located in Seoul. But, contrary to world city theory, Seoul's TNCs are industrial not finance or producer service companies, and the contrast with New York City in Table 2 is revealing. Ranked by sales, 16 out of the top 20 New York City-based firms are in the finance and producer services sector, but only three out of the top 20 Seoul corporations are in that category; the rest are manufacturing and construction firms. Lest one think this difference is explained by New York City's more advanced economy, Table 2 also reveals that Tokyo, the capital of global capital according to the Global Fortune 500 list, resembles Seoul not New York City – just four out of Tokyo's top 20 firms are in the finance and producer service sector.

Contrary to the world city model, Seoul is primarily a *national* basing-point for the global operations of *Korean* TNCs. The Korean state has controlled the inward and outward flow of foreign investment until recently; the largest foreign holdings in Seoul as of 1990 were minority equity shares in major Korean corporations, mostly held by Japanese TNCs, rather than subsidiaries that were wholly or majority owned by companies headquartered abroad. Seoul hosts many fewer branches of foreign headquartered companies (161) than comparable world cities in the Americas, like Mexico City (266) or Sao Paulo (380) (Hoopes 1994).

Seoul's global control capability

Seoul certainly hosts an infrastructure for global control. But, contrary to world city theory, Seoul's global control apparatus is anchored in central government ministries and the state's continuous channels of communication with business leaders and organizations that monitor industrial performance, not in private finance and producer service firms. Because export marketing requires substantial fixed costs and externalities in the initial stages of any industry, the expanded credit made available by the state crucially enabled Korean exporters to fill foreign orders and explore foreign markets. Without government intervention in the allocation of credit, it is unlikely that Korea's rapid transformations in industrial composition and level of industrial development would have been possible. Firms which finance their investments primarily through bank credit and foreign loans, instead of through stock issues, accumulate heavy debt. Through its control over finance, the Korean government became a risk partner for industrialists, enabling new export ventures and entrepreneurship. And by controlling the banks, the government created incentives for firms to maximize their assets and growth, rather than to strive for immediate profitability. As long as they satisfied the government by expanding exports and constructing new plants, firms ensured their access to credit.

Economic organization

According to the world city paradigm, the decentralization of manufacturing and the associated shift to a service-based economy occasion a massive loss in a world city's manufacturing employment, the exodus of manufacturing headquarters and a downgrading in

New York		Seoul		Tokyo	
1.	Philip Morris, Inc.	1.	Samsung Corporation	1.	Mitsui & Co. Ltd
2.	**AT&T Corporation**	2.	Daewoo Corporation	2.	Mitsubishi Corporation
3.	**Citicorp**	3.	Hyundai Corporation	3.	**Nippon Tel. & Tel. Corporation**
4.	**Chase Manhattan Corporation**	4.	LG International Corporation	4.	Hitachi, Ltd
5.	**American International Inc.**	5.	Hyundai Motor Co. Ltd	5.	Sony Corporation
6.	**Merrill Lynch & Co., Inc.**	6.	Korea Electric Power	6.	**Dai-Ichi Mutual Life Insurance**
7.	**ITT Corporation**	7.	Yukong Ltd	7.	Toshiba Corporation
8.	**Travelers Group, Inc.**	8.	LG Electronics Co. Ltd	8.	Honda Motor Corporation
9.	**Loews Corporation**	9.	Hyundai Engineering & Construction	9.	**Bank of Tokyo-Mitsubishi**
10.	**American Express Co.**	10.	Kia Motors Corporation	10.	Tokyo Electric Power Co.
11.	RJR Nabisco Holdings Corporation	11.	Ssangyong Corporation	11.	NEC Corporation
12.	**Morgan J.P. & Co.**	12.	Hyundai Motor Service	12.	Fujitsu Ltd
13.	Bristol-Myers Squibb Co.	13.	Ssangyong Oil Refining	13.	Japan Tobacoo, Inc.
14.	**Lehman Brothers Holdings**	14.	**Korea Exchange Bank**	14.	Mitsubishi Motors Corporation
15.	**Nynex Corporation**	15.	Daewoo Electronics Co.	15.	**Meiji Mutual Life Insurance**
16.	**Morgan Stanley Group Inc.**	16.	**Hanil Bank Ltd**	16.	Mitsubishi Electric Corporation
17.	**Viacom, Inc.**	17.	**Cho-Hung Hank, Ltd**	17.	Kanematsu Corporation
18.	Pfizer Incorporated	18.	Korean Air Lines Co. Ltd	18.	Mitsubishi Heavy Industries
19.	**Chase Manhattan**	19.	Ssangyong Cement	19.	Nippon Steel Corporation
20.	**Time Warner, Inc.**	20.	LG Chemical Co. Ltd	20.	Ito-Yokado Co.

Table 2 The distribution and ranking (by net sales) of producer services in New York, Seoul and Tokyo

Source: Richard Child Hill and June Woo Kim
Note: Producer services firms are shown in bold
Source: Disclosure (1998)

the manufacturing jobs that remain in the city (Sassen 1991). But Seoul does not fit this profile, either. As we have seen (Table 2), the headquarters of Korea's major manufacturing companies continue to concentrate in Seoul, one-quarter of Seoul's labor force continues to work in manufacturing and the city's manufacturing base is the most advanced in the nation.

Social and spatial polarization

You can certainly find class-segregated residential areas in Seoul. Squatters, most of whom work in construction, as street vendors, housemaids and taxi drivers, or in small factories, have settled on the hillsides surrounding the city. But, contrary to world city theory, most are migrants from Korea's countryside not from abroad. Seoul's foreign residents numbered 39,246 in 1994, only 0.4 per cent of the city's population, much lower even than Tokyo's 1.8 per cent, let alone New York City's 28 per cent! (Crahan and Vour-volais-Bush 1997). Several wards south of the Han River, on the other hand, are well planned, mixed residential and business districts for the middle classes.

Still, Seoul has few of the plywood shanty towns visible in many Asian cities. And, contrary to world city theory, income disparities among Seoul's wards are hard to discern. In 1993, the average monthly household income of the poorest of Seoul's 26 wards was 97 per cent that of the wealthiest. By contrast, per capita income in the Bronx, New York City's poorest borough, was just 38 per cent that of Manhattan, the wealthiest. Indeed, even with the extreme concentration of capital in the hands of a few business groups, the distribution of income among Korean families is among the most egalitarian in the world.

Seoul's bureaucratic elite

Seoul is not under the sway of a transnational class, as posited by world city theory. Rather, the large political power of Korea's small bureaucracy continues to the present day. As in Japan, Korea's state officialdom, not the bourgeoisie, led the industrialization effort. And political bureaucrats controlled the state. Most parliamentary statutes originated with the bureaucracy, not with legislators, and administrative policies were also orchestrated within the bureaucracy.

Seoul's integration with the nation-state

Korea, like Japan, is a unitary state. Until recently, the city of Seoul was simply an appendage of the central government. There is no material basis for arguing that Seoul is severing economic ties with the rest of Korea, as world city theory would predict. Just the opposite is true. Seoul's gross product per capita was almost twice that of the nation in 1960, but the gap has steadily diminished, and by 1991 the city was about on a par with the nation as a whole.

CONCLUSION

Like Tokyo, Seoul does not conform to the world city model. Seoul, like Tokyo, is a national basing-point for the global operations of Korean transnational corporations, not a global basing-point for the global operations of borderless firms. Like Tokyo, Seoul's industrial policy and social structure are geared less to attracting investments from abroad than to facilitating the foreign trade and investments of Korean corporations. Command and control functions are concentrated in Seoul, but so too is industrial production, particularly knowledge-intensive manufacturing, as in Tokyo. Seoul, like Tokyo, has not experienced severe manufacturing decline, rapid expansion in producer service employment, extensive foreign immigration or much social and spatial polarization. Like Tokyo, Seoul is under the sway of a political bureaucratic elite, not a transnational capitalist class. And, as with Tokyo, it would be senseless to claim that Seoul is severing ties of mutual interest with the nation-state; if anything, the capital city is becoming even more integrated with the rest of South Korea.

Tokyo and Seoul challenge world city theory's assumptions about the nature of globalization and the role of world cities in the globalization process. How damaging are the East Asian anomalies to the validity of the world city paradigm? It depends upon how one interprets the discrepancies.

One could argue, for example, that Tokyo and Seoul do not fit the world city definition and therefore their anomalous characteristics have no bearing on the model. This resolution hardly seems satisfactory, however, for it would drastically reduce the geographical scope and empirical testability of the theory. In any case, Tokyo and Seoul do fit the world city definition since both cities provide an infrastructure that enables TNCs to control their global operations. However, both cities emphasize the global operations of indigenous not foreign companies, and their international infrastructure is primarily rooted in state ministries and bureaus, not in private finance and producer service firms.

One could also argue that while Tokyo and Seoul may not have conformed to the world city model in the past, they are being forced by global pressures to move in that direction today and will continue to do so in the future. However, it is premature to equate the very real crisis in the East Asian developmental state with the end of East Asian developmentalism let alone with the transition to liberal market capitalism. There is considerable popular support in Japan and Korea for more state decentralization, deregulation and policy transparency, but there is no similar groundswell support for market-driven capitalism. Neither Japan nor Korea has the historical, ideological or political underpinnings for Western neo-liberalism. Indeed, there is entrenched opposition to market liberalism in the state bureaucracies, business groups and trade unions. The current restructuring is more likely to result in a new phase of

East Asian developmentalism than anything approximating Anglo-American liberalism.

A third way to resolve the East Asian anomaly is to conclude that Tokyo and Seoul are a different type of world city from that conceptualized by Friedmann and Sassen. The world city paradigm makes sense in market-centered New York and London, but not in state-centered Tokyo and Seoul. But this resolution would put the state square in the center of world city analysis, and that clashes with two of the paradigm's central assumptions: that globalization diminishes the power and integrity of the nation-state, and that cities are replacing states as central nodes in the world economy. This approach is compatible, however, with comparative findings that national institutions, politics and culture mediate the impact of global processes to produce diverse urban outcomes.

We believe Tokyo and Seoul's divergence from the world city model reflects more than national variation within a common global context, however. Japan and Korea have developed a different kind of political economy from Western market capitalism, one nurtured, ironically, under the US geo-political umbrella during the Cold War. IMF pressure for financial reforms indicates "system friction" between Anglo-American and East Asian political economies, a kind of economic Cold War. Understanding Tokyo and Seoul necessitates a different conception of the world system from that underlying the globalist world city argument. Tokyo and Seoul differ from New York in so many salient respects because these cities are lodged within a non-hegemonic and interdependent world political economy divided among differently organized national systems and regional alliances.

Countries are attempting to open their markets to foreign competition *and* to pursue national and regional industrial policies *simultaneously*. Concepts like non-hegemonic interdependence (Stallings and Streeck 1996) better capture the trajectory of cities in today's world political economy than claims about cities "abandoning national ties" in order to embrace supranational alliances and "denationalized expertise." In short, the economic base, spatial organization and social structure of the world's major cities are strongly influenced by the national development model and regional context in which each city is embedded.

REFERENCES FROM THE READING

Amsden, A. (1989) *Asia's Next Giant*, New York: Oxford University Press.

Crahan, M. and Vourvoulais-Bush, A. (eds.) (1997) *The City and the World: New York's Global Future*, New York: Council on Foreign Relations.

Disclosure (1998) *World Scope CD-ROM*. Available from www.disclosure.com.

Friedmann, J. (1995) Where we stand: A decade of world city research, in P. Knox and P. Taylor (eds.) *World Cities in a World-System*, New York: Cambridge University Press: 21–47.

Gao, B. (1997) *Economic Ideology and Japanese Industrial Policy: Developmentalism from 1931 to 1965*, New York: Cambridge University Press.

Gershenkron, A. (1962) *Economic Backwardness in Historical Perspective*, Cambridge, MA: Harvard University Press.

Hill, R. C. and Fujita, K. (1995) Osaka's Tokyo problem, *International Journal of Urban and Regional Research*, 19: 181–93.

Hill, R.C. and Kim, J.W. (2001) Response to Friedmann and Sassen, *Urban Studies*, 38, 13: 2541–2.

Hong, S. W. (1996) Seoul: a global city in a nation of rapid growth, in F. Lo and Y. Yeung (eds.) *Emerging World Cities in Pacific Asia*, New York: United Nations University Press: 144–78.

Hoopes, D. (1994) *Worldwide Branch Locations of Multinational Companies*, Detroit: Gale Research, Inc.

Jo, S-J. (1992) *The World City Hierarchy and the City of Seoul*, Unpublished PhD Thesis, University of Delaware.

Johnson, C. (1995) *Japan: Who Governs?* New York: W.W. Norton.

Kim, J. and Choe, S. (1997) *Seoul: The Making of a Metropolis*, New York: John Wiley & Sons.

Ostry, S. (1996) Policy approaches to system friction: Convergence plus, in S. Berger and R. Dore (eds.) *National Diversity and Global Capitalism*, Ithaca, NY: Cornell University Press: 333–52.

Sassen, S. (1991) *The Global City*, Princeton, NJ: Princeton University Press.

Stallings, B. and Streeck, W. (1996) Capitalisms in conflict? The United States and Japan in the post-cold war world, in B. Stallings (ed.) *Global Change, Regional Response*, Cambridge: Cambridge University Press: 67–99.

34

"World city formation on the Asia Pacific Rim: poverty, 'everyday' forms of civil society and environmental management"

from M. Douglass and J. Friedmann (eds.), *Cities for Citizens: Planning and the Rise of Civil Society in a Global Age* (1998)

Mike Douglass

EDITORS' INTRODUCTION

Mike Douglass is Professor at the Lee Kuan Yew School of Public Policy at the National University of Singapore. Douglass is one of the pioneers of world city research and has been one of the most consistent observers of urbanization processes on the Pacific Rim and in Pacific Island nations. His work on urban development and planning in cities of the South is widely recognized as a leading contribution to the field. In this riveting text, which is an abbreviated version of a chapter for an influential book he co-edited with John Friedmann, Douglass examines the interrelationships of world city formation and social and environmental struggles in East Asian cities. Isolating three themes—globalization, localization, and community empowerment—Douglass examines the new forms of community politics that are emerging in the rapidly changing metropolises of East Asia. Focusing on diverse state-civil society relations in various East Asian developmental states, Douglass concludes that the project of empowering the urban poor remains extraordinarily difficult in the context of economic globalization and entrenched political authoritarianism. Nevertheless, Douglass places high hopes in the progressive potential of democratic, civil society-based activism and makes a strong argument for including the politics of the poor within the narrative of global city formation.

INTRODUCTION

Three major themes run through the discussion of the processes of global-local interaction along the Asian arc of the Pacific Rim, with particular emphasis on the ways in which world city formation is part of a process of emergent social struggles around the built and natural environment. The first is that the forces impinging on urban restructuring in Pacific Asia generate and reveal heightening tension between, on one hand, the (re)positioning of cities in a global system of cities and, on the other, citizen mobilization and demands for substantial improvements in daily life space. While linking up to and gaining higher positions in a system of world cities calls for increasing investments in mega-projects and the forced mobility of large numbers of people, the "discovery of civil society" associated with the rise of the urban middle class, organized labor, voluntary organizations and heightened political action from all quarters of society, including the poor, is making the implementation of such projects problematic as control over space becomes increasingly contested.

A second theme is that globalization is also a process of localization. Contrary to much literature on the tendency for capitalist penetration and Western cultural imperialism to lead towards the "rendering of the world as a single place" (King 1989: 5), variations in sociocultural, political and economic institutions continue to emerge to profoundly affect the ways in which global impulses are amalgamated into real historical settings. The array of possibilities for social mobilization and the enlargement of democratic spaces to create alternative development paths is greater than much of received world systems theories or mainstream economic theory allow. No matter how successful or unsuccessful, the future of Pacific Asia societies and their cities will be determined as much by localized socio-cultural and political processes as by global imperatives.

The third theme is that efforts by the poor to take command over community space are integral to a more embracing process of self-empowerment (Friedmann 1992). This includes not only securing land for housing, but also investing in efforts to gain access to and manage environmental infrastructure and resources. Again, contrary to conventional wisdom that the poor are "too busy being poor" to care about their habitat or the environment, research in low-income communities in Asia consistently shows that substantial amounts of household allocations of time, labor and resources are devoted to environmental considerations even among the very poor.

WORLD CITY FORMATION ON THE PACIFIC RIM

The appearance of "world city" as the new shibboleth of global achievement has not been missed by governments in Pacific Asia. The increasing strength of labor in these economies has shifted their comparative advantage away from labor-intensive manufacturing and towards higher-order production

and service industries, including global information and control functions. The intensive economic restructuring necessitates a parallel urban restructuring, with major cities in competition with each other to capture key global functions. Those gaining top positions would be on the cutting edge of high technologically driven production and producer services and would enjoy a position of power unprecedented in their history. There is little doubt that a major contributor to both the accelerated growth and the rising tensions over the built environment in Pacific Asian metropolitan regions is what Friedmann has summarized as world city formation.

STRUGGLES OVER THE BUILT AND NATURAL ENVIRONMENT

The enthusiasm by Asian governments for achieving world city status stands in contrast to the imagery presented by Western writers on the subject, most of which argue that the processes attending world city formation result in profound ethnic, racial and other social divides. Although a powerful vortex of global accumulation, the world city is seen as an arena of deepening social and political crisis. Direct confrontation with the requirements of world city formation occurs daily in the metropolitan centers of Pacific Asian countries in the form of struggles over slum demolitions, evictions of the poor, destruction of older petty capitalist business neighborhoods, conversion of rich agricultural lands to urban zones, loss of open spaces and longer commuting distances for the hapless wage worker. In the case of Bangkok, for example, by the mid-1980s there were more than 1,000 slum areas with a total population ranging up to 1.5 million (Kaothien and Rachatatanun 1991). Internationalization and the building boom during the same period saw high-rise commercial development displace more than 11,000 low-income housing units – more than 100,000 people – within a 10-mile radius from the center from 1984 to 1988 (Padco-LIF 1990). Many of these households were compelled to move to the metropolitan fringe. Most, however, chose to relocate in other slum areas in the inner city where job prospects are not only better but the petty economy of slums and neighborhood self-help relations are well established.

Low-income households that are able to maintain a hold on land and housing in the metropolis face another threat: severe urban environmental deterioration that has its greatest impacts in and around slum and squatter settlements. Not infrequently, the poor are able to stay in the city because certain quarters are so environmentally unsafe that private housing or commercial developers are not interested in them. These locations include sites adjacent to polluting industries, such as the remnants of the famous Diamond Hill slum in Hong Kong located below and downstream from textile dying factories (Chan et al.1994), along heavily polluted urban canals and waterways, on steep slopes that easily collapse during the rainy season, in low-lying areas subject to heavy flooding, along railroad tracks and, on a smaller scale, underneath bridges. These patterns support the thesis that the increasing vulnerability of Third World cities to natural environmental hazards is part of a circular process of residential occupation of marginal urban environments by households that, under land markets that increasingly favor the wealthy, are unable to secure land at less vulnerable sites (Main and Williams 1994).

But the poor are not the principal source of urban environmental deterioration in Asia. The crowding of low-income households into environmentally poor areas is taking place in great urban regions that are experiencing widespread environmental deterioration. Untreated sewage and industrial effluents have left major waterways – including the magnificent Chao Phrya river running through Bangkok – unable to support life; breathing the air has become hazardous to health; and heavy metals and other pollutants are destroying coastal ecologies and fishing industries. In Jakarta, urban expansion into upland areas is resulting in a loss of ground cover that intensifies both flooding and drought in the city and its hinterland. In this city, which has one of the world's highest levels of suspended particulate matter in the air, recent World Bank estimates are that infrastructure needed to begin to meet environmental standards would currently require annual investments of US $1 billion (World Bank 1993). In Hong Kong per capita municipal wastes continue to rise and water quality in Victoria Harbor, already very bad, continues to deteriorate steadily despite government White Papers and campaigns to clean up the environment.

The open competition by governments and corporate actors for world city status discloses the nature of choices which privilege economic growth over providing livable habitats for all citizens at the cost of environmental sustainability. Given the high visibility of this conflict in the day-to-day urban experience, the relative absence of explicit treatment of the built and natural environment in world city literature has meant that critical linkages between other features of Friedmann's world city hypothesis and with his subsequent writings on poverty and alternative development are difficult to make. Specifically, if the relationships between poverty and world city formation are to be explored fully in terms of relations of power, the question of decision making and control over the formation of the urban habitat needs to be included in the analysis.

POVERTY AS (DIS)EMPOWERMENT: LESSONS ON ENVIRONMENTAL MANAGEMENT FROM ASIAN SLUMS

Although variations are significant, if poverty is seen a condition of low levels of social, economic and political power, the globally attached processes of urban spatial restructuring and environmental deterioration have been a major contributor to it. Gaining access to land, housing, basic infrastructure and environmental resources occupies a substantial portion of the time and energies of low-income urban households throughout Asia. As earlier observed by Castells (1983: 312), the urbanization process in a capitalist world entails a commodification of the city itself that disrupts communities and cultures, leads to unbalanced patterns of growth and creates chronic crises in housing, services and collective consumption. While the more affluent classes respond with ever higher levels of consumption, often in fortified neighborhoods and air-conditioned cars that insulate them from the deteriorating social and physical scenes around them, the poor are crowded into environmentally degraded areas that are systematically denied basic infrastructure and services. Although some forms of environmental deterioration, such as air pollution, cannot totally be avoided even by the rich, they nonetheless affect the lives of the poor more severely.

CIVIL SOCIETY, THE STATE AND EMPOWERMENT

A great deal of energy is spent by the poor just on maintaining a place in the city and managing daily routines. Household divisions of labor, reciprocity among neighbors, community leadership, NGOs and government assistance can all help in reducing the time burden on poor households in carrying out these life-supporting routines. But even where those burdens seem overwhelming, events can bring to communities an extraordinary sense of urgency that is shared by enough households to lead to organized resistance and, further, mobilization for change. Galvanizing resistance is itself a complex phenomenon that, while emerging from civil society, invariably confronts the state either as a potential ally against landlords, for example, or as the targeted cause of the problem. As it broadens its inclusion, mobilization of citizens also entails raising the "moral high ground" to subsume a wide range of interests under a banner of righteousness (Apter and Sawa 1984). This capacity, too, is highly contingent on shared cultural and moral values that can be drawn on to overcome class and other social cleavages.

In Pacific Asia the struggles engaging social energies arise in a milieu of accelerated urbanization and globally integrated local economies that have been managed by highly interventionist states. Social mobilization as a form of empowerment has not been achieved by unending and entrenched anti-statist resistance, but rather by longer-term achievement of democratic reforms allowing for political association in civil society to be included in a territorially defined political community. Participation in political, social and economic affairs is the goal that translates social power into political power through, ultimately, collaboration with the state. Achieving this goal is, however, seen as a process of struggle that principally swells up from the grassroots with the support of mediating organizations also emerging from civil society. A central task of the (democratic) state is to sustain the territorial basis for civil society's claim for inclusion in the formation of political community.

While much has been written about regulation of the (international) economy by the "developmental state" in Asia, its more recent role has been to institute processes of political empowerment through the establishment of an inclusive democracy. Evidence from Asia, while mixed, has shown remarkable advances in this direction over the past decade, giving cautious support to the thesis that social mobilization can lead to fundamental political reform. Yet it can also lead, as in the case of Burma, to brutal suppression or, as in the case of Singapore, may not appear in any substantial way at all. Moreover, without progress in the other dimensions of empowerment, constitutional provisions for elected governments may fail to provide for the type of level playing field of inclusiveness. In the Philippines, for example, the strong patron-client relations rooted in rural landlordism and concentration of land ownership in the hands of powerful families reaches into the state and deeply erodes the potential of elected governments to carry out popular economic and social reforms. This "Latin America in Asia" syndrome has left this country, which after the Second World War was thought to be the brightest star in Pacific Asia, far behind other market-oriented countries of the region.

This brings to the fore the contrasting differences in state-civil society-economy relations in Asia. Table 1 presents a summary glimpse of some of the key contrasts among the Asian NIEs, a group of societies that are all too often lumped together as a single "four tigers" or "neo-Confucianist" development model, but are in fact markedly different in culture as well as in state-society relations (Berger and Hsiao 1990; Douglass 1994).

The table suggests that the struggle for inclusiveness through activating civil society involves many different strategies and tactics, depending on particular constellations of relations in a given setting.

A key distinguishing feature among the countries of the region is the degree to which the state allows collective association to emerge from civil society, which has a decisive effect on community-level strategies across the gamut of concerns. Where, for example, community power structures are replaced by the state or otherwise inhibited from emerging from within communities, community mobilization often moves away from self-provisioning efforts to forms of routinizing requests for state assistance, as in Hong Kong, or to street protests and political agitation, as in Korea.

In some instances, such as in Indonesia, the state permeates urban communities through networks of elected and appointed officials reaching all the way down to the neighborhood level in municipalities. Here the lines between community leadership and government officialdom are blurred. This has a tendency to foster a mini-bureaucratic process of implementing government programs in a "soft," top-down fashion. Neither spontaneous organization outside of official lines nor the establishment of non-government organizations is encouraged or facilitated

Dimension	Hong Kong	Korea	Singapore	Taiwan
Domestic capital	"Accommodationist," no explicit industrial strategy, indirect subsidies to firms via housing and infrastructure. Small-scale firms dominate	"Corporatist state" allied with *chaebol* – large-scale oligopolies created/regulated via nationalization of banks	"Extroverted corporatist," extreme reliance on TNCs and biased against domestic firms	"Entrepreneurial state:" state-owned enterprises used to promote major sectors. Domestic industry dominated by small-scale firms
Transnational capital (TNCs)	Same as above	Least reliance on direct foreign investment (FDI); use of military alliance with US to leverage investors	Virtually exclusive reliance on TNCs	High reliance on TNC FDI, but selectively used to avoid direct competition with domestic firms
Labor	Indirect control via housing, price controls and management of international labor migration. Small-scale firms dampen worker organization	Direct support of proletarianization and direct suppression of labor via use of police power. *Chaebol* and "Fordist" factory systems foster militant labor	Largely indirect via housing, international migration control, wage controls. Unions largely eliminated in 1960s	Domestic controls via prohibition of strikes, limited unionization under KMT Party. Small-scale firms inhibit worker organization
Civil society	Colonial, managed by public welfare and other agencies	Authoritarian use of police power to suppress social movements	Overt use of state power to regulate behavior and muzzle press in the name of "communitarianism"	Authoritarian use of police power to suppress anti-KMT movements by indigenous peoples

Table 1 Variations in State-Civil Society-Corporate Economy Relations in East Asia (Early 1990s)

Source: Mike Douglass

by this system. In Bangkok, relations between slum communities and government are perhaps among the least defined and, as a consequence, are much less routine or predictable. While community leadership is recognized when it appears, it is not a part of an official apparatus of the Indonesian type. The result is that government attention to the more than 1,000 slum areas of Bangkok is sporadic and partial, often on an issue-by-issue rather than a programmatic basis. At the same time, the absence of a strong state presence in the community may allow for more authentic grassroots organizations and leadership.

These differing experiences show that the currently fashionable policy of forging "partnerships" between state and community not only has many possible configurations, but is also fraught with lack of conceptual clarity about such key concepts as participation, citizen rights and the political versus technical role of non-governmental organizations. If a common pattern can be identified, it would be that extensive state bureaucratic involvement leads to increased reliance on the state to provide community infrastructure and services. Contrary to the commonly presented view that the poor are somehow naturally inclined to depend on the state for investments for collective consumption, case studies show that such dependence, where it does occur, is a product of a longer history of state-community and, on a larger scale, state-civil society relations that could have produced much more active community involvement.

The renewed interest in decentralization, democracy and citizen participation prevalent throughout Asia represents a potentially major shift in state-community relationships in cities in almost all settings. This has already been partially reflected in more open attitudes by governments toward communities, non-governmental organizations and political association outside of the state. Whatever concrete manifestations these changes may have will, of course, continue to differ among the various national and local contexts in Asia, but most governments have already come a long way from the anti-slum policies and eradication drives of the 1950s and 1960s. It is now more common for governments implicitly to recognize the existence of squatters and, to varying degrees, enter into dialogue about redressing the concerns of poorer communities and provide some forms of government assistance to at least a select number of locations. Whether they will now take the next step towards validating the legitimacy of poor people to reside in the city by recognizing their rights to have access to land and housing, to organize and select leaders and to become equal counterparts in political and planning affairs is one of the most important issues of the current decade.

As it reflects on the plight of poor people, the general evidence from this ongoing mobilization of households and communities from within civil society is that accountability of government, even through relatively weak and indirect forms of democracy, brings positive (if limited) attention to slum and squatter settlements. Where government officials must stand for election in poor sections of the city, palpable evidence of representing the voice of the people must be given. In Bangkok, for example, local elections led to promises of piped water and electricity to a squatter settlement that were later realized. This is not to say that democracy, particularly in its limited and often token and co-optive forms, is a panacea for eliminating poverty or improving the environment. Regimes in power still tend to be located within a range of authoritarian, paternalistic and non-democratic modes of governance. The ideal of authentic state-community partnerships that is currently being put forth in development plans throughout Asia is, in most instances, still in search of a real-world application. Even within the realm of democratic action, much also depends on class relations and how the emerging urban middle class in Asia will either support or move against slum and squatter settlements.

Much may also depend on whether the poor can tag their political agendas on to what is predominantly a middle-class movement for political liberalization. This is one of the most troubling features of democratization drives in situations of widening social inequalities that, in part, follow from the economic growth processes of Asia's miracle economies. In key aspects of the construction of (new) urban space, such as in housing, the implicit agenda of the middle class to live in posh neighborhoods near commercial skylines moves against the quest for empowerment of the poor.

To many the bridge across classes is manifested in the flowering of what are summarized as non-governmental organizations (NGOs), which are generally staffed by members of the middle class and disaffected elites and take on social justice and, currently, sustainable development concerns. Recognizing that even in the best circumstances, poor households and their communities have clear limitations on what they can accomplish on their own, the general proposition placing hope on NGOs is that, without some form of outside non-governmental support, sustaining community self-empowerment efforts will encounter severe, often insurmountable difficulties. As one dimension of collective association from within civil society, the NGO movement has generated a vast literature on their expected and real roles, typologies and orientations. There are NGOs that are genuinely involved in selfless endeavor for the poor, but there are also ones that are thin disguises for charlatans to exploit the poor. Some are completely autonomous from the state, while others are de facto arms of the state.

Generalizations about NGO-community relations must therefore be treated with care. Although the ideal type of NGO might be the "empowering" type described by Lee (1992) and identified by Friedmann, it can also be allowed that circumstances will dictate which approaches are viable in a giving setting. In several countries, for example, the emergence of NGOs is still severely limited by governments that remain entrenched in top-down, non-participatory approaches to development planning. The role of NGOs is also necessarily limited to small-scale, charitable activities rather than open advocacy of, for example, squatters' rights in these situations. In other settings, such as in the Philippines and Thailand, NGOs have appeared in almost astounding numbers and play a much wider variety of roles. As yet there are, however, few NGOs in Pacific Asia that express interest in urban environmental problems in low-income communities. What can be called the environmental NGOs still tend to focus on global and rural environmental issues, and when they do focus on cities, they do so at the urban rather than community level. Community-based NGOs, on the other hand, tend to focus either on immediate crises such as housing rights and eviction threats, or on health and education programs, and rarely focus on either income-generating or environmental management questions. Thus even in cities where NGOs are highly visible, the appearance of mediating organizations concerned with livelihood-environment questions is still rare.

CONCLUSIONS: THE GLOBAL-LOCAL NEXUS OF CIVIL SOCIETY AND EMPOWERMENT

The outstanding question raised by linking world city formation with an alternative development is whether the urban poor can gain power in the global-local nexus of economy and polity. While disempowering relations are the subject of their resistance, on a day-to-day basis the resistance is often muted, taking the form of non-compliance, avoidance and subtle disregard for rules established from above. It also takes the form of self-exploitation, of expending household energies to sustain the basic conditions for the reproduction of their own labor beyond levels made possible by selling labor power in the market alone.

Occasionally, resistance is galvanized into moments of collective action and, more rarely, urban social movements that span ethnic and class divides. As witnessed by the shifting patterns of direct foreign investments in response to political upheavals and political reforms, such moments reach outwards to the global scale as well. There is, however, no overriding imperative that global-local interaction will either deliver the poor from poverty or condemn them to conditions of immiseration. While the resolution of this question is a dynamic one that involves layers of interaction moving from the individual and household to the global scale, it fundamentally rests on the mobilization of the poor themselves. In the longer term, for the rise of civil society in East and Southeast Asia to bring an end to poverty, it must also transform adversarial relations between the state and the poor into collaborative ones. A change from confrontation to accommodation may seem improbable in the light of current political realities; yet events such as democratization taking root in the authoritarian states of Asia, which seemed equally unimaginable only a decade ago, suggest more possibilities for such accommodation than ever before.

REFERENCES FROM THE READING

Apter, D. and Sawa, N. (1984) *Against the State: Politics and Social Protest in Japan*, Cambridge: Harvard University Press.

Berger, P. and Hsiao, H. M. (eds.) (1990) *In Search of an East Asian Development Model*, London: Transactions.

Castells, M. (1983) *The City and the Grass Roots*, Berkeley: University of California Press.

Chan, C., Chang, F., and Cheung, R. (1994) Dynamics of community participation in environmental management in low-income communities in Hong Kong, *Asian Journal of Environmental Management*, 2, 1: 11–16.

Douglass, M. (1994) The 'developmental state' and the Asian newly industrialized economies, *Environment and Planning*, 26: 543–66.

Douglass, M. and Friedmann, J. (eds.) (1998) *Cities for Citizens: Planning and the Rise of a Civil Society in a Global Age*, New York: Wiley.

Friedmann, J. (1992) *Empowerment: The Politics of Alternative Development*, Oxford: Blackwell.

Kaothien, U. and Rachatatanun, W. (1991) *Urban Poverty in Thailand*, Bangkok: Government of Thailand, National Economic and Social Development Board.

King, A. D. (1989) Colonialism, urbanism and the capitalist world economy, *International Journal of Urban and Regional Research*, 13, 1: 1–18.

Lee, Y. F. (1992) Urban community-based environmental management, *Supercities International Conference on the Environment*, San Francisco State University, 26–30.

Main, H. and Williams, S. W. (eds.) (1994) *Environment and Housing in Third World Cities*, New York: John Wiley.

Padco-LIF (1990) *Bangkok Land Market Assessment*, Bangkok: NESDB/TDRI.

World Bank (1993) *Indonesia: Urban Public Infrastructure Services*, Washington, DC: World Bank.

35

"New globalism, new urbanism: gentrification as global urban strategy"

from *Antipode* (2002)

Neil Smith

EDITORS' INTRODUCTION

Neil Smith (1954–2012), who last held the position of Distinguished Professor of Anthropology and Geography at the Graduate Center of the City University of New York, was one of the most influential geographers and critical urban intellectuals of his generation. He wrote on a broad range of topics, including political ecology, labor markets, imperialism, and perhaps most importantly in relation to globalizing cities, gentrification. The excerpt below is arguably Smith's most important intervention into the global cities debate. Here he makes two salient arguments. First, global cities should be defined not only according to their command and control functions, but also by their participation in the global production of surplus value. Second, gentrification has become a competitive urban strategy in today's global economy. Revisiting the concept of the "revanchist city" (Smith 1996), he argues that the liberal urban policy of the post-1960s period, characterized by redistribution and anti-poverty legislation, has been replaced by revanchism, marked by policies against the poor and minorities. In the transition, the neoliberal state now acts as both a regulator and an agent of the market. This shift is particularly visible in the latest wave of gentrification underway since the 1990s. Unlike earlier waves of gentrification led by the middle class between the 1950s and the 1980s, the current process is spurred by a partnership of local governments and corporate capital and typically aims to promote urban regeneration. Geographically, the post-1990s gentrification has spread beyond the largest cities in the West to rust-belt cities such as Cleveland and Detroit; to small market towns such as Lancaster, Pennsylvania; and also to globalizing cities in the South, from Shanghai to Rio de Janeiro and Istanbul. The processes of global gentrification entail divergent pathways, as they are conditioned by local institutional and economic contexts (see Lees, Shin, and Lopez-Morales 2015, 2016), but as a global urban strategy, gentrification invariably represents the triumph of socioeconomic interest of the elite class.

NEW URBANISM

Saskia Sassen (Ch. 4 and Ch. 40) offers a benchmark argument about the importance of local place in the new globalism. Place, she insists, is central to the circulation of people and capital that constitute globalization, and a focus on urban places in a globalizing world brings with it a recognition of the rapidly declining significance of the national economy, while also insisting that globalization takes place through specific social and economic complexes rooted in specific places. In the end, Sassen's argument is a little vague about how places are, in fact, constructed. It does not go far enough. It is as if the global social economy comprises a plethora of containers—nation-states—within which float a number of smaller containers, the cities. Globalization brings about a dramatic change in the kinds of social and economic relations and activities carried on in these containers, a re-sorting of activities between different containers, and an increased porosity of the national containers, such that turbulence in the wider global sea increasingly buffets cities directly. I want to argue here that in the context of a new globalism, we are experiencing the emergence of a new urbanism such that the containers themselves are being fundamentally recast. "The urban" is being redefined just as dramatically as the global. The new concatenation of urban functions and activities vis-à-vis the national and the global changes not only the make-up of the city but the very definition of what constitutes—literally—the urban scale.

Let us now step back and look at the question of "globalization," because if we are talking about global cities presumably their definition is implicated in the processes thereof. What exactly is globalizing at the beginning of the twenty-first century? What is new about the present? A good case can be made that to the extent that globalization heralds anything new, the new globalism can be traced back to the increasingly global—or at least international—scale of economic production. As late as the 1970s, most consumer commodities were produced in one national economy either for consumption there or for export to a different national market. By the 1990s, that model was obsolete, definitive sites of production for specific commodities became increasingly difficult to identify, and the old language of economic geography no longer made sense. In autos, electronics, garments, computers, biomedical, and many other industrial sectors ranging from high tech to low, production is now organized across national boundaries to such a degree that questions of national "import" and "export" are supplanted by questions of global trade internal to the production process. The idea of "national capital" makes little sense today, because most global trade across national boundaries is now intrafirm: it takes place within the production networks of single corporations.

Two mutually reinforcing shifts have consequently restructured the functions and active roles of cities. In the first place, systems of production previously territorialized at the (subnational) regional scale were increasingly cut loose from their definitive national context, resulting not just in the waves of deindustrialization in the 1970s and 1980s but in wholesale regional restructuring and destructuring as part of a reworking of established scale hierarchies. As a result, production systems have been downscaled. The territorialization of production increasingly centers on extended metropolitan centers, rather than on larger regions.

The corollary is also taking place, as national states have increasingly moved away from the liberal urban policies that dominated the central decades of the twentieth century in the advanced capitalist economies. The point here is not that the national state is necessarily weakened or that the territoriality of political and economic power is somehow less potent. National states are reframing themselves as purer, territorially rooted economic actors in and of the market, rather than external compliments to it. Social and economic restructuring is simultaneously the restructuring of spatial scale, insofar as the fixation of scales crystallizes the contours of social power—who is empowered and who is contained, who wins and who loses—into remade physical landscapes.

Neoliberal urbanism is an integral part of this wider rescaling of functions, activities, and relations. It comes with a considerable emphasis on the nexus of production and finance capital at the expense of questions of social reproduction. It is not that the organization of social reproduction no longer modulates the definition of the urban scale but rather that its power in doing so is significantly depleted. Public debates over suburban sprawl in Europe and especially the US, intense campaigns in Europe promoting urban "regeneration," and the emerging environmental justice movements all suggest not only that the crisis of social reproduction is thoroughly territorialized but, conversely, that the production of urban space has also come to embody that crisis. A connection exists between the production of the urban scale and the efficient expansion of value, and a "mis-scaled" urbanism can seriously interfere with the accumulation of capital.

The leading edge in the combined restructuring of urban scale and function lies in the large and rapidly expanding metropolises of Asia, Latin America, and parts of Africa, where the Keynesian welfare state was never significantly installed, the definitive link between the city and social reproduction was never paramount, and the fetter of old forms, structures, and landscapes is much less strong. These metropolitan economies are becoming the production hearths of a new globalism. No one seriously argues that the twenty-first century will see a return to a world of city-states—but it will see a recapture of urban political prerogative vis-à-vis regions and nation-states.

Finally, the redefinition of the scale of the urban in terms of social production rather than reproduction in no way diminishes the importance of social reproduction in the pursuit of urban life. Quite the opposite: struggles over social reproduction take on a heightened significance precisely because of the dismantling of state responsibilities. However, state abstention in this area is matched by heightened state activism in terms of social control. The transformation of New York into a "revanchist city" (Smith 1996) is not an isolated event, and the emergence of more authoritarian state forms and practices is not difficult to comprehend in the context of the rescaling of global and local geographies. According to Swyngedouw (1997: 138), the substitution of market discipline for that of a hollowed-out welfare state deliberately excludes significant parts of the population, and the fear of social resistance provokes heightened state authoritarianism. At the same time, the new urban work force increasingly comprises marginal and part-time workers who are not entirely integrated into shrinking systems of state economic discipline, as well as immigrants whose cultural and political networks—part of the means of social reproduction—also provide alternative norms of social practice, alternative possibilities of resistance.

In summary, my point here is not to argue that cities like New York, London, and Tokyo lack power in the global hierarchy of urban places and high finance. The concentration of financial and other command functions in these centers is undeniable. Rather, I am trying to put that power in context and, by questioning the common assumption that the power of financial capital is necessarily paramount, to question the criteria according to which cities come to be dubbed "global." If there is any truth to the argument that so-called globalization results in the first place from the globalization of production, then our assessment of what constitutes a global city should presumably reflect that claim.

URBAN REGENERATION: GENTRIFICATION AS GLOBAL URBAN STRATEGY

Let me now shift scales and focus toward the process of gentrification. If one dimension of neoliberal urbanism in the twenty-first century is an uneven inclusion of Asian and Latin American urban experiences, especially at the forefront of a new urbanism, a second dimension concerns what might be called the generalization of gentrification as a global urban strategy. At first glance these surely seem like two quite different arguments, the one about luxury housing in the centers of global power, the other about new models of urbanism from the integrating peripheries. They certainly express contrasting experiences of a new urbanism, but that is precisely the point. Neoliberal urbanism encompasses a wide range of social, economic, and geographical shifts, and the point of these contrasting arguments is to push the issue of how varied the experience of neoliberal urbanism is and how these contrasting worlds fit together.

Most scholars' vision of gentrification remains closely tied to the process as it was defined in the 1960s by sociologist Ruth Glass. Here is her founding 1964 statement (Glass 1964: xviii), which revealed gentrification as a discrete process:

> One by one, many of the working-class quarters of London have been invaded by the middle classes—upper and lower. Shabby, modest mews and cottages—two rooms up and two down—have been taken over, when their leases have expired, and have become elegant, expensive residences. Larger Victorian houses, downgraded in an earlier or recent period—which were used as lodging houses or were otherwise in multiple occupation—have been upgraded once again . . . Once this process of "gentrification" starts in a district it goes on rapidly until all or most of the original working-class occupiers are displaced and the whole social character of the district is changed.

Almost poetically, Glass captured the novelty of this new process whereby a new urban "gentry" transformed working-class quarters. Today such developments are couched in the language of urban renaissance and the scale of ambitions for urban rebuilding has expanded dramatically. The current language of urban regeneration, particularly in Europe, is not one-dimensional, but it bespeaks, among other things, a generalization of gentrification in the urban landscape. Whereas, for Glass, 1960s gentrification was a marginal oddity in the Islington housing market—a quaint urban sport of the hipper professional classes unafraid to rub shoulders with the unwashed masses—by the end of the twentieth century it had become a central goal of British urban policy. Whereas the key actors in Glass's story were assumed to be middle- and upper-middle-class immigrants to a neighborhood, the agents of urban regeneration thirty-five years later are governmental, corporate, or corporate-governmental partnerships. A seemingly serendipitous, unplanned process that popped up in the postwar housing market is now, at one extreme, ambitiously and scrupulously planned. That which was utterly haphazard is increasingly systematized.

Most importantly, perhaps, a highly local reality, first identified in a few major advanced capitalist cities such as London, New York, Paris, and Sydney, is now virtually global. Its evolution has been both vertical and lateral. On the one hand, gentrification as a process has rapidly descended the urban hierarchy; it is evident not only in the largest cities but in more unlikely centers such as the previously industrial cities of Cleveland or Glasgow, smaller cities like Malmö or Grenada, and even small market towns such as Lancaster, Pennsylvania or Cˇeské Krumlov in the Czech Republic. At the same time, the process has diffused geographically as well, with reports of gentrification from Tokyo to Tenerife, São Paulo to Puebla, Mexico, Cape Town to the Caribbean, Shanghai to Seoul.

Of course, these experiences of gentrification are highly varied and unevenly distributed, much more diverse than were early European or North American instances of gentrification. They spring from quite assorted local economies and cultural ensembles and connect in many complicated ways to wider national and global political economies. The important point here is the rapidity of the evolution of an initially marginal urban process first identified in the 1960s and its ongoing transformation into a significant dimension of contemporary urban-ism. Whether in its quaint form, represented by Glass's mews, or in its socially organized form in the twenty-first century, gentrification portends a displacement of working-class residents from urban centers.

In the context of North America and Europe, it is possible to identify three waves of gentrification (Hackworth 2000). The first wave, beginning in the 1950s, can be thought of as sporadic gentrification. A second wave followed in the 1970s and 1980s as gentrification became increasingly entwined with wider processes of urban and economic restructuring. Hackworth (2000) labels this the "anchoring phase" of gentrification. A third wave emerges in the 1990s; we might think of this as gentrification generalized. Of course, this evolution of gentrification has occurred in markedly different ways in different cities and neighborhoods and according to different temporal rhythms. And yet, to differing degrees, gentrification had evolved by the 1990s into a crucial urban strategy for city governments in consort with private capital in cities around the world. By the end of the twentieth century, gentrification fueled by a concerted and systematic partnership of public planning with public and private capital had moved into the vacuum left by the end of liberal urban policy. Elsewhere, where cities were not governed by liberal urban policy during much of the twentieth century, the trajectory of change has been different, yet the embrace of a broadly conceived gentrification of old centers as a competitive urban strategy in the global market leads in a similar direction. In this respect, at least, turn-of-the-century neoliberalism hints at a thread of convergence between urban experiences in the larger cities of what used to be called the First and Third Worlds.

The generalization of gentrification has various dimensions. These can be understood in terms of five interrelated characteristics: the transformed role of the state, penetration by global finance, changing levels of political opposition, geographical dispersal, and the sectoral generalization of gentrification. Let us examine each of these in turn.

First, between the second and third waves of gentrification, the role of the state has changed dramatically. In the 1990s, the relative withdrawal of the national state from subsidies to gentrification that had occurred in the 1980s was reversed with the intensification of partnerships between private capital and the local state, resulting in larger, more expensive, and more symbolic developments, from Barcelona's waterfront to Berlin's Potsdamer Platz. Urban policy no longer aspires to guide or regulate the direction of economic growth so much as to fit itself to the grooves already established by the market in search of the highest returns, either directly or in terms of tax receipts.

The new role played by global capital is also definitive of the generalization of gentrification. From London's Canary Wharf to Battery Park City—developed by the same Canadian-based firm—it is easy to point to the new influx of global capital into large mega-developments in urban centers. Just as remarkable, however, is the extent to which global capital has percolated into much more modest, neighborhood developments. The reach of global capital down to the local neighborhood scale is equally a hallmark of the latest phase of gentrification.

Third, there is the question of opposition to gentrification. From Amsterdam to Sydney, Berlin to Vancouver, San Francisco to Paris, gentrification's second wave was matched by the rise of myriad homeless, squatting, housing, and other antigentrification movements and organizations that were often loosely linked around overlapping issues. These rarely came together as citywide movements, but they did challenge gentrification sufficiently that, in each case, they were targeted by city politicians and police forces. Apart from anything else, the heightened levels of repression aimed at antigentrification movements in the 1980s and 1990s testified to the increasing centrality of real-estate development in the new urban economy. Cities' political regimens were changing in unison with their economic profile, and the dismantling of liberal urban policy provided as much a political opportunity as an economic one for new regimes of urban power. The emergence of the new revanchism was explicitly justified in terms of making the city safe for gentrification. The new authoritarianism both quashes opposition and makes the streets safe for gentrification.

The fourth characteristic of this latest phase is the outward diffusion of gentrification from the urban center. This is far from a smooth or regular process, but as gentrification near the center results in higher land and housing prices, even for old, untransformed properties, districts further out become caught up in the momentum of gentrification. The pattern of diffusion is highly variable and is influenced by everything from architecture and parks to the presence of water. Above all, it is geared to the historical patterns of capital investment and disinvestment in the landscape.

Finally, the sectoral generalization that typifies this most recent phase goes to the heart of what distinguishes the new gentrification. Whereas urban renewal in the 1950s, 1960s, and 1970s sought a full-scale remaking of the centers of many cities and galvanized many sectors of the urban economy in the process, it was highly regulated and economically and geographically limited by the fact that it was wholly dependent on public financing and therefore had to address issues of broad social necessity, such as social housing. In contrast, the earliest wave of gentrification that followed urban renewal proceeded with considerable independence from the public sector. Despite considerable public subsidy, the full weight of private-market finance was not applied until the third wave.

What marks the latest phase of gentrification in many cities, therefore, is that a new amalgam of corporate and state powers and practices has been forged in a much more ambitious effort to gentrify the city than earlier ones.

Retaking the city for the middle classes involves a lot more than simply providing gentrified housing. Third-wave gentrification has evolved into a vehicle for transforming whole areas into new landscape complexes that pioneer a comprehensive class-inflected urban remake. These new landscape complexes now integrate housing with shopping, restaurants, cultural facilities, open space, employment opportunities—whole new complexes of recreation, consumption, production, and pleasure, as well as residence. Just as important, gentrification as urban strategy weaves global financial markets together with large- and medium-sized real-estate developers, local merchants, and property agents with brand-name retailers, all lubricated by city and local governments for whom beneficent social outcomes are now assumed to derive from the market rather than from its regulation. Most crucially, real-estate development becomes a centerpiece of the city's productive economy, an end in itself, justified by appeals to jobs, taxes, and tourism. In ways that could hardly have been envisaged in the 1960s, the construction of new gentrification complexes in central cities across the world has become an increasingly unassailable capital accumulation strategy for competing urban economies. Herein lies a central connection to the larger outline of a new urbanism.

The strategic appropriation and generalization of gentrification as a means of global interurban competition finds its most developed expression in the language of "urban regeneration." The language of regeneration sugarcoats gentrification. Not only does "urban regeneration" represent the next wave of gentrification, planned and financed on an unprecedented scale, but the victory of this language in anesthetizing our critical understanding of gentrification represents a considerable ideological victory for neoliberal visions of the city.

The point here is not to force a one-to-one mapping between regeneration and gentrification strategies, or to condemn all regeneration strategies as Trojan horses for gentrification. Rather, I want to insist that gentrification is a powerful, if often camouflaged, intent within urban regeneration strategies and to mount a critical challenge to the ideological anodyne that sweeps the question of gentrification from sight even as the scale of the process becomes more threatening and the absorption of gentrification into a wider neoliberal urbanism becomes more palpable. Gentrification as global urban strategy is a consummate expression of neoliberal urbanism. It mobilizes individual property claims via a market lubricated by state donations.

CONCLUSION

In this paper, I present two rather different arguments. On the one hand, I challenge the Eurocentric assumption that global cities should be defined according to command functions rather than by their participation in the global production of surplus value. On the other hand, I want to highlight the ways in which gentrification has evolved as a competitive urban strategy within the same global economy. The post-1990s generalization of gentrification as a global urban strategy plays a pivotal role in neoliberal urbanism in two ways. First, it fills the vacuum left by the abandonment of twentieth-century liberal urban policy. Second, it serves up the central- and inner-city real-estate markets as burgeoning sectors of productive capital investment: the globalization of productive capital embraces gentrification. The emerging globalization of gentrification, like that of cities themselves, represents the victory of certain

economic and social interests over others, a reassertion of (neoliberal) economic assumptions over the trajectory of gentrification. Even where gentrification per se remains limited, the mobilization of urban real-estate markets as vehicles of capital accumulation is ubiquitous. Urban real-estate development—gentrification writ large—has now become a central motive force of urban economic expansion, a pivotal sector in the new urban economies.

REFERENCES FROM THE READING

Glass, R. (1964) *London: Aspects of Change*, London: Centre for Urban Studies and MacGibbon and Kee.

Hackworth, J. (2000) The Third Wave, PhD dissertation, Department of Geography, Rutgers University.

Lees, L., Shin, H. B., and Lopez-Morales, E. (eds.) (2015) *Global Gentrification: Uneven Development and Displacement*, Bristol: Policy Press.

Lees, L., Shin, H. B., and Lopez-Morales, E. (2016) *Planetary Gentrification*, Cambridge: Polity Press.

Smith, N. (1996) *New Urban Frontier: Gentrification and the Revanchist City*, London: Routledge.

Swyngedouw, E. (1997) Neither global nor local: "Glocalization" and the politics of scale, in K. Cox (ed.) *Spaces of Globalization: Reasserting the Power of the Local*, New York: Guilford: 137–66.

36

"Between world history and state formation: new perspectives on Africa's cities"

from *Journal of African History* (2011)

Laurent Fourchard

EDITORS' INTRODUCTION

Laurent Fourchard is a senior researcher with the French Foundation of Political Science at the research institute 'Les Afriques dans le Monde' at Sciences Po Bordeaux, France. He is one of the leading international researchers on urban Africa. Fourchard's work is both theoretically influential in current debates and empirically thorough, ranging across the continental diversities of urban Africa. In the following chapter, Fourchard intervenes in theoretical debates about the reorientation of global urban studies from a western centric view towards a more pluralist and decentralized conceptual framework. First, Fourchard points to the invisibility of cities in Africa in the global cities debate and their perennial treatment as outliers of development. Then he critiques the scholarship on urban Africa that oscillates between localism (i.e., perceiving urban Africa as essentially different from other urban regions) and the global city paradigm (i.e., focusing on the most globalized cities in Africa and ignoring the rest). To move beyond these localizing and homogenizing tendencies, Fourchard proposes to study Africa's cities from the perspective of social history and in relation to state-building and world history. He makes a convincing argument that the notion of "African cities," as often used in the academic literature, is inappropriate, as it assumes a common history of the whole continent. He suggests using "Africa's cities" instead, and urges us to pay attention to how cities in Africa have been differently shaped by heterogeneous influences of global economy and geopolitics from a historical perspective.

INTRODUCTION

Global city research theorists contend that there is no global city in Africa, with Johannesburg arguably the only exception. However, a less normative vision has recently suggested that cities should not be seen only in terms of what they lack but rather on the basis of what they are and how they arrived at their contemporary configurations. Analyzing Africa's cities within the larger framework of the global South serves the dual purpose of avoiding their construction as exceptional and reinserting them in broader academic debates that are not confined to the continent's history. According to this argument, it is imperative to include Africa's cities in a larger context in order to look at them comparatively and move beyond a perception of urban Africa as being essentially different from the rest of the world (see Robinson Ch. 9 and Simone Ch. 45).

Plate 26 Johannesburg, a world-class African city

Source: Roger Keil

AFRICA'S CITIES IN WORLD HISTORY

Various labels have marked the last thirty years of urban studies and the literature on urban history in Africa. From the criticized Islamic and colonial city paradigms to more recent global city theory, Africa is alternatively perceived at the core or at the periphery of a process of academic categorization that concerns most of the cities of the south. In a way, these approaches have been developed to either include the continent's cities in, or exclude them from, a world history perspective, though neither has been entirely able to escape an oversimplifying vision of Africa's urban change. Concentration on large-scale, anonymous structures and processes has neglected the life experiences of ordinary residents that are at the core of social history.

FROM ISLAMIC CITY PARADIGM TO THE GLOBAL CITY THEORY

From the 1930s to the 1990s, the Islamic city paradigm enjoyed considerable popularity especially in North Africa. The Islamic city (that is, a set of political, economic, social, and cultural characteristics supposedly

shared by towns and cities of the successive Arab and Ottoman Empires between the seventh and nineteenth centuries), was mainly defined negatively on the basis of a number of elements it lacked according to Max Weber's prototypical city: the regularity and institutions of the Classical city; the political autonomy of the medieval town; urban planning; links with the countryside. In the last thirty years, several researchers have denounced the absurdity of using Islam as a conceptual framework to account for the urban phenomena of countries with such varied historical traditions (Abu-Lughod 1987). In sub-Saharan Africa, indigenous forms of urbanism predated Islamic expansion and had no links with Arab trade networks (Aksum, Zimbabwe, Benin City, Old Oyo, Ile Ife, Jenne-Jeno). On the East African coast, Swahili towns were polyglot, multiethnic frontiers and composed of various population groups (Arabic, Indian, African) while (Sahelian) cities ruled by Islamic elites (Timbuktu, Djenne´, Gao, Katsina, Kano) were equally places of mutual influence between Northern and Western African societies.

The 'colonial city' is another widespread label used both within and beyond the African continent to refer to the city under colonial – and by extension – apartheid rule. It may refer to a particular moment during which colonialism was portrayed as a power demarcating, racialising, and ordering urban space. The overemphasis on colonial control, segregation schemes, and the 'sanitation syndrome' has indisputably played a role in the development of the colonial city paradigm. Yet, to view colonial cities as 'dual cities' is misleading since colonial societies cannot be understood merely in terms of a 'European versus Indigenous' dichotomy. It omits the agency of African societies, their capacity to overcome such divisions, to ignore them or even to imagine them differently.

A similar ambiguity revolves around the postcolonial city label, commonly used to qualify cities which have been developing since the end of the colonial period. This approach implies the comprehension of an enduring common colonial legacy that unites Africa's cities and other cities of the South in a postcolonial framework (see Robinson Ch. 9). The colonial situation, instead of being analysed as a total social project, is limited to urban planning, technologies of control, and the civilizing mission. Sources of conflicts are said to emanate mainly from the spatial separation between colonizers and the colonized, while the contemporary city is perceived as a fluid and more variegated space shaped by conflicts of a more complicated nature. This latter approach does not always avoid the trap of introducing forms of binary discourse that surprisingly tend to classify colonial and postcolonial cities into ontologically distinct categories. Seeing Africa's cities only in terms of their colonial and postcolonial relationships may preclude a fuller understanding of the multifaceted ways in which they have engaged with the larger world.

While the various attempts to integrate Africa in large historical frames have sometimes led to an essentialist vision of its cities, global city research has left the African continent aside. The globalizing city notion might be more useful as it construes globalization as a process, not a state that reifies and classifies cities of the world (Grant 2009). However, many scholars agree that Africa constitutes a site of extremely uneven globalizing processes. Such unevenness stems from the movement of capital across national borders, in the process linking particular and dispersed sites of global relevance thereby leaving huge regions simply by-passed (Ferguson 1999). The notion of a globalizing city might be useful to define post-1980s Accra, but the self-prophetic dimension of this construct is of little help in understanding the social and economic heterogeneity of the continent.

In all likelihood, all these labels will be unable to fully reintegrate Africa's cities into world trends without oversimplifying and homogenizing their multiple histories. One single notion does not appear to be sufficient for grasping the uneven integration of Africa's cities into world history.

HISTORICIZING WORLD IMAGINARIES

To move beyond the excessively narrow economic approach of globalization, more innovative research has looked at the everyday social practices of inhabitants shaped by 'globalised imaginaries' to highlight the worldliness of contemporary African life forms. As mentioned by AbdouMaliq Simone (2001: 28), African urban residents have developed a 'worlding' from below as they are able to operate at larger scales in a broader world through the mobilization of religious practices, modes of dress, food, and musical taste. These perspectives share a vision of Africa's cities in which world cultural repertoires are both locally produced and imagined, and propose a comprehensive reading of the city which avoids both localism and the global city paradigm.

Historicizing various competing world imaginaries among city dwellers elucidates the messy process by which people get connected to worldwide influences and helps to identify groups (who might be defined along social, gendered, or generational lines) who claim to share a form of cosmopolitan life. It also sheds light on the very processes by which becoming cosmopolitan was equivalent with becoming urban: both marked a distinction from the countryside. Heterogeneity and historical contingency shaped how each group or individual, in each town or city, came to imagine themselves as belonging to several worlds at the same time.

Becoming cosmopolitan and becoming urban were contested historical processes everywhere, though especially in regions with longer established urban traditions (for example, the Swahili coast, hinterland West Africa, North Africa). In these places, Islam played a central role in the process of dissociating oneself from a rural, 'backward' environment and being associated with real or imaginary prestigious external links. There is variety in what it has meant to be urban and to be cosmopolitan in Africa at different moments in history. It was important how these strategies, when pursued by elites, could be intimately tied to processes of state building.

RECASTING CITY/STATE RELATIONSHIPS

John Peel (1980: 273) recommended that 'a satisfactory African urban history must be about politics . . . which directs attention to the larger unit of the state in which West African urban analysis must be set'. In some of the literature, 'African states are often identified as failed not by what they are, but by what they are not, namely, successful in comparison to Western states' (Péclard and Hagmann 2010: 541). This analytical framework is thus at the core of the 'failed state' perspective frequently criticized by anthropologists and political scientists but which is nevertheless influential in the ways cities in Africa are often understood: A long historical urban past in Africa might be seen as either marginal to state-building processes or as the manifestation of the inherent weaknesses of the African state in general.

Some authors have suggested less essentialist ways of exploring state-building in Africa. The historicity of the state, the embeddedness of its bureaucratic organizations in society, the interactions between state officials and non state actors, the material dimension of statehood, and the importance of accumulating basic legitimacies are all key dimensions that apply to African states as well as to many states in the world. Scaling down the focus of analysis from the level of the territorial state to local arenas or to socio-economic areas (such as mining or border areas) has been suggested as another possible way of evaluating the manner in which the state is built from the margins or from local situations (e.g. Lund 2006). Two other analytical possibilities are advanced here. First, understandings of the 'state' and the 'town' have, for too long, engaged with these as separate political entities even though several historical studies have recently emphasized the necessity of analyzing them within a common analytical framework. Second, cities have been privileged places in which multiple authorities compete over state functions and, as such, should contribute more to our understanding of the history of the state in Africa than is the case at present.

MAKING THE TOWN, MAKING THE STATE

There is a clear correlation between state formation and processes of urbanization in various regions of the world and Africa is no exception. This is in spite of the examples of stateless though urbanized societies, for instance Jenne-Jeno, and states without towns but with mobile court capitals, such as Burundi, Rwanda, and Buganda until the nineteenth century. Most African states needed the revenues generated by trade activities concentrated in towns and ports in order to strengthen their power over a territory and to finance war.

Warfare was an integral part of state formation and was a crucial stimulus not just for social and political change but also town growth. In many parts of the continent, warfare led to the multiplication of city walls which, in turn, led to a greater concentration of people seeking protection. This emergent settlement pattern allied to the practice of warfare also facilitated taxation and the control of trade while at the same time imparting prestige to rulers. Significantly, in the nineteenth century, the two most urbanized regions in Africa south of the Sahara, Hausaland and Yorubaland, were also the densest groupings of African walled cities.

As state formation and town growth are aspects of the same historical process, it is near impossible to disentangle the terms for 'state' and 'town' in several African languages. The foundational concept of Yoruba political sociology is the term ilu, commonly translated as 'town' or community. An ilu is both a town and a polity and Yoruba people do not make a conceptual distinction between the two. In Dahomey, urbanity was not only defined by size and the concentration of people, but by political autonomy and the role of the town as an administrative seat: *to* referred to a town or settlement of any size while *togan* (the chief of the two) was a provincial governor within the kingdom. In the Ga language (southern Ghana), the term *man* most commonly denotes 'town' and has wider social and political connotations such as people, nation, or state. There are also a large number of names that distinguish townspeople enjoying full civic rights (mambii) from people of the bush. The very things that made Accra a state – organized political, military, legal and religious institutions –are also the things that made Accra distinctively "urban". These examples suggest that networks of towns directly shaped precolonial states and militate against the idea that the latter developed independent of the former.

In several parts of Africa, a wide range of negotiations, compromises, and conflicts came to define complex relationships between European officials and local elites. Precolonial urban-based institutions were often important in making the early colonial state as many colonial territories came into existence through them. The old and new networks of towns became the architecture of emerging bureaucratic colonial states (with expanded though limited administration and public services). It is also possible – though this remains to be confirmed – that embryonic town services (municipal police, town planning, hygiene and welfare services) shaped the emerging central services of the colonies. This might represent a similar experience to that of Europe, where the emergence of the welfare state was often extrapolated from municipal experiments and the personnel involved in their operation. While ad hoc town initiatives might have shaped the central bureaucratic state, it is also clear that colonial officials, like African political leaders, dramatically transformed urban networks. In the long term, public investments gave rise to different patterns of networks. 'Primacy' as a colonial heritage of centralization still dominates Africa more than any other continent but has begun to decline over the last three decades.

CONTESTING THE STATE AND COMPETING AUTHORITIES OVER STATE FUNCTIONS

Post-1970s structural adjustment policies (SAPs) together with the reduction of state services and decentralization of government administration have facilitated the return of local power centres, increased the number of non-state actors involved in delivering services, and multiplied the patron-client ties and personal networks upon which people must rely to survive. In itself, this process does not necessarily imply the weakening or privatization of the state. Scaling down the focus of analysis to local arenas has recently helped to renew our understanding of the state in Africa. The multiplication of parallel structures and alternative sites of authority (for instance, chiefs, political factions, hometown associations, neighbourhood groups, and vigilante organizations) has managed to 'bring the state back' into local arenas (Lund 2006: 688). Looking at 'cities at war' in post-conflict situations, or towns beyond the control of government bureaucracies, helps elucidate how the state can be built from the margins through unofficial agents imposing taxes on populations, building

embryonic administrations, capturing the resources of humanitarian aid (for example, Goma, Lumumbashi, Juba), or developing financial centres for money laundering, receiving migrant remittances, and recycling the goods of the regional economy (for example, Nador [Morocco], Benguerdane [Tunisia], Touba/Mbacke´ [Senegal]).

An historical approach that examines the contingent relationships constituted through conflicts and negotiations between state officials and multiple competing authorities in local and urban areas reveals processes of state construction and deconstruction in Africa. To oppose the rural and the urban here is probably of little help. Juxtaposing urban Africa as a privileged site of emergent civil society and of challenges to colonial and authoritarian regimes with tribal and despotic power in rural areas (as suggested by Mamdani 1996: 289–93) overstates the role of urban political forms and undervalues the strategic place of rural constituencies and the importance of urban-rural links in the making of African politics. Nevertheless, urban areas should remain a significant focus for exploring day-to-day interactions between state and non-state actors.

Of radical importance are the effects of urban unrest and large-scale mobilization on the shape of the state. Heightened political tensions in cities have led to public actions and reinforced the role of bureaucracy as a means of reasserting the authority of the state over turbulent populations. Large scale mobilizations are probably not the dominant form of political and social confrontation in most of Africa's cities. Less visible, though no less important, are a set of fragmented confrontations and negotiations between the urban poor and the state. Poor people fostered what Asef Bayat (2000) calls quiet encroachment, characterized by largely atomized and prolonged mobilization with episodic actions by individuals and families to acquire the basic necessities of their lives, and to get access to collective resources (land, shelter, piped water, electricity) and to public space (street pavements, intersections, street parking places). The contemporary urban landscape of the poor in several large metropolises seems dominated, on the one hand, by the apparent lack of public policies and, on the other, by the everyday bargaining of citizens with state officials and local political leaders.

In Lagos, Nairobi, Casablanca and Cairo, the poorest neighbourhoods are, however, not necessarily abandoned by the central or the local state, as the 'failed state' literature contends, but instead remain strongly tied to it through networks of intermediaries, caïds, and political or union leaders, relationships that result in multiple forms of political belongings. Ultimately, the government's inability to ensure security for its citizens is usually considered the most important indicator of contemporary state failure in Africa. Historically, however, coercive practices including collecting taxes, recruiting labour, implementing sanitary regulations, and policing towns and cities have often been delegated to 'Native authorities', foreign companies and local communities. The colonial establishment of trained and professional police forces was slow and uneven, and never met the needs of the population. Local leaders and elders frequently demanded the provision of security by indigenous authorities, self-defence groups, and vigilante and community organizations in areas perceived as threatened by youth gangs. Such provisioning became a cheap way for the state to unload the expense of policing onto local communities. Nevertheless, the results of this

delegation of security functions were unpredictable. In the mid-1980s, no one would have guessed that street committees, neighbourhood watches or vigilante organizations would, twenty years later, become part of the South African police service through community police schemes and reserve police forces.

To cast the multiplication of actors performing or subverting state functions as the 'privatization of the state' or as 'the end of the postcolonial state' is to analyze such practices in terms of a one-way affair and ignore the capacity of local authorities to operate in the twilight between state and society, public and private. Looking at sites of authority that are linked to state officials and perform state functions might allow us to see new forms of authority being generated in apparently failing states. Achille Mbembe suggests that private actors who perform functions such as imposing taxes and create moral or political order through violence and other constraints should also be understood as doing so for their own personal material gain. Both arguments are probably valid, depending on circumstances.

Plate 27 Downtown Johannesburg

Source: Roger Keil

CONCLUSION

Considering Africa's cities as dysfunctional, chaotic, failed, informal, or not globalized works to retain the Western city as the paradigmatic model against which all others are to be assessed. The heterogeneity of urban situations and the extremely uneven influence of other continents in Africa ultimately suggest that it is more appropriate to speak about 'cities of Africa' rather than 'African cities'. Simultaneously, social history examinations have revealed how the specificities of localized situations have greatly shaped the everyday practices of ordinary residents. Given the sheer diversity of these situations, a comprehensive portrait of urban Africa thus remains a difficult challenge to achieve.

In this chapter, exploration of cities as places situated within country, state, and world networks has been suggested as a way to move beyond localism and unhelpful categorization, and to avoid fetishizing scales of analysis. Much recent historical and social scientific scholarship has insisted on the necessity of thinking about the city and the state, and the city and the world, simultaneously. Being urban, modern, civilized, or cosmopolitan appear, in a number of cases, to be overlapping processes. Making towns and making states were simultaneous historical processes in many African societies prior to colonial rule, while negotiations and contestations between a range of local and imperial actors shaped the early colonial state in the nineteenth century. Reasserting the importance of cities in history will likely help us to understand better processes of state formation in Africa today. Exploring, in more empirical detail, the making of the city and the state and their uneven connections to the wider world will make the banality of Africa's urban past part of the history of the urban world.

REFERENCES FROM THE READING

Abu-Lughod, J. (1987) The Islamic city: Historical myth, Islamic essence and contemporary relevance, *International Journal of Middle Eastern Studies*, 19: 155–76.

Bayat, A. (2000) From "dangerous classes" to "quiet rebels": Politics of the urban subaltern in the global South, *International Sociology*, 15, 3: 533–57.

Ferguson, J. (1999) *Expectations of Modernity: Myths and Meanings of Urban Life on the Zambian Copperbelt*, Berkeley, CA: Berkeley University Press.

Grant, R. (2009) *Globalizing City: The Urban and Economic Transformation of Accra, Ghana*, Syracuse: Syracuse University Press.

Lund, C. (2006) Twilight institutions: Public authority and local politics in Africa, *Development and Change*, 37, 4: 673–84.

Mamdani, M. (1996) *Citizen and Subject: Contemporary Africa and the Legacy of Late Colonialism*, Princeton, NJ: Princeton University Press.

Péclard, D. and Hagmann, T. (2010) Negotiating statehood: Dynamics of power and domination in Africa, *Development and Change*, 41: 4.

Peel, J. D. Y. (1980) Urbanization and urban history in West Africa, *Journal of African History*, 21, 2: 269–77.

Simone, A. (2001) On the worlding of African cities, *African Studies Review*, 44, 2: 15–41.

FOUR

37

"The 'right to the city': institutional imperatives of a developmental state"

from *International Journal of Urban and Regional Research* (2010)

Susan Parnell and Edgar Pieterse

EDITORS' INTRODUCTION

In this chapter, Susan Parnell and Edgar Pieterse, both based at the University of Cape Town, argue that under conditions of globalization large cities present unique challenges and opportunities for poverty reduction and the realization of rights. The "rights to the city" refer to specifically four generations of rights: from basic individual rights such as voting and health (1st generation), household services such as housing (2nd generation), to neighborhood- and city-level amenities (3rd generation), and finally, security from warfare and environmental risks (4th generation). Using Cape Town, South Africa as an example, they argue that it is imperative to expand residents' socioeconomic rights in addition to their political rights, in order to lift the marginalized from poverty. Such a universal rights agenda can only be fulfilled with participation of all levels of the state, and the authors pay particular attention to local (i.e., municipal) governments. Historically, the weak local state capacity in South Africa has been the primary factor leading to the compromised socio-economic rights, such as poor delivery of urban services to households and neighborhoods. Parnell and Pieterse argue that regulatory reforms at the local level are needed in order to clear the institutional barriers that define the current, unequal terms of citizen rights regimes. Instead of focusing the discussion on the negative effects of neoliberal governance, Parnell and Pieterse urge us to debate on how to achieve effective regulatory reforms so that the state can play a stronger role in redistribution and development.

INTRODUCTION

Under conditions of globalisation large cities present unique challenges for poverty reduction and the realisation of rights. The inexorable urbanisation of society places the city at the core of the developmental agenda of the twenty-first century. The urbanisation of poverty also underscores the imperative of downscaling the emerging debate about the developmental state to the city scale. The arguments we make in this chapter start from the proposition that a universal rights agenda can and should be fulfilled as much at the city-region scale as it is at a national scale, or for migrants who move between places. Thus, a commitment to the roll out of universal rights implies not only that all people should be afforded minimum rights, but that these rights should be protected by governments regardless of the scale or type of settlement that people occupy. Indeed, it is from recognition of the locationally specific impediments to the realisation of rights and the multi-scalar nature of the state's actions that are necessary for the full realisation of human rights, that the identification of an agenda for sustainable urban poverty reduction action emerges. Only more recently have enduring debates on urban poverty, including issues of land use management, housing, work, network infrastructures, environmental protection and services been cast in an overtly rights based frame. We probe the issue of the universal right to the city as the moral platform from which the developmental role of the state should be defined, and from which alternatives to neoliberal urban managerial positions should be articulated. Our focus is on large fairly well resourced places that nevertheless have very large concentrations of chronically poor people who are institutionally excluded from the government support structures that are necessary for their well being.

Using the realisation of rights as the litmus of urban poverty reduction changes the understanding of the nature and scale of government's interventions that are required to achieve poverty reduction targets. The Millennium Development Goals (MDGs), with their diversified emphasis on varied aspects of poverty, including living in slums and without services, are a clumsy and very crude recognition of the link between poverty reduction and the different settlement-based expressions of the denial of human rights. If, however, the rights implied by the MDGs (especially goal 7 target 11 of slum eradication) are to be achieved in cities across the global South, it will only be possible if full consideration is given to the three overlapping concerns of this paper: the role of the sub-national state in urban poverty reduction; the imperative of government targeting the household and neighbourhood in addition to the individual in its roll out of urban services; and the imperative of understanding the role of location and scale in the roll out of settlement-based rights.

PUTTING HUMAN RIGHTS INTO AN URBAN PERSPECTIVE

This chapter takes issue with two conventional reference points in the literature on urban poverty and calls for a wider but more nuanced understanding of what a rights-based approach to development might entail in urban settlement policy and practice. Our first objection is to the tendency of the urban poverty reduction literature to focus exclusively and simplistically on the realisation of democratic not socio-economic rights. We argue that democratic deepening must be linked to rights-based advocacy to achieve better socio-economic outcomes in our cities.

To some extent the exclusion of socio-economic rights is the result of historically weak sub-national state capacity to engage the complexity of sustainably addressing urban poverty issues, a situation compounded by neo liberalism, but it is also a product of the individualised understanding of how human rights are realised in an urban context. Clearly the demand for political representation at the local or municipal level and the affirmation of the right to food, water and shelter are crucial for individual and household advancement in the city. But the pre-occupation with these basic or 1st generation human rights drives international support for transparency in local government elections and for basic infrastructure provision, at the expense of defining a more nuanced and demanding agenda of urban transformation in which more complex rights can be addressed for increasingly large numbers of people who live in the cities of the South.

Few in apartheid South Africa would disagree that the primacy afforded to establishing democracy and a universal right to vote, including at the local level, was well placed. Similarly the post-democracy popular demand for basic services was (and is) unambiguous, not least because of the incomplete delivery by government on its promises of providing affordable services for all (McDonald 2007). But the ongoing focus on electoral and participatory democracy as well as on protecting other individual rights (freedom from discrimination, freedom of expression, etc.) may marginalise new efforts to advance 2nd-generation socio-economic rights. These 2nd-generation rights are achieved through the sustained delivery of affordable urban services to households and neighbourhoods (not individuals) and through viable service administration and finances not just through infrastructure investment. How this ongoing service delivery is achieved will vary greatly between urban and rural contexts. We identify a further gap in political commitment and action at the urban scale to provide '3rd generation' rights—defined as including the right to the city or a safe environment, to mobility or to public spaces. While the right to freedom of movement, safety, environmental protection and economic opportunity are recognised in both the South African Constitution and the International Declaration on Human Rights, the urban planning and enforcement mechanisms that protect or enable these rights are poorly understood. This is largely because these rights are exercised or denied collectively not individually,

1st generation rights	Focus on the individual (e.g. the vote, health, education)
2nd generation rights	Focus on household services like housing, water, energy and waste
3rd generation rights	Focus on neighborhood or city scale entitlements such as safety, social amenities, public transport, etc.
4th generation rights	Focus on freedom from externally induced anthropogenic risk, such as war, economic volatility or climate change

Table 1 The right to the city

Source: Susan Parnell and Edgar Pieterse

and at various geographical scales across the city region (individual, house, property, neighbourhood, municipality and city region). What we term 3rd generation rights form part of the public good and are more easily claimed in places that are free of environmental risk and economic and social exclusion. Implementation of these 2nd and 3rd generation rights (one might add 4th generation rights to climate secure cities) rests on robust and capable sub-national structures – contrary to neoliberal imperatives for lean and fragmented institutional state arrangements. Embryonic postcolonial local state structures, unfunded decentralisation and privatisation all militate against strong urban government in the global South. Consequently, despite obvious wealth being concentrated in large urban areas, the poor are trapped in second class strata of the city that might one day provide for universal 1st generation rights, but will never facilitate full urban citizenship.

We suggest that the concept of the rights based city offers innovative ways of advancing debate about the developmental state and places a more empowering agenda on the table in contrast to the neoliberal governmentality agenda. The intangible area of 'rights-supporting place-making' holds the key to meaningful urban poverty reduction, especially in middle income contexts. We argue that, especially for the chronically poor of the city, putting the emphasis on 3rd generation rights, that are generally realised through stronger state capacity to provide inclusive development planning and to enforce land use management, plus the economic and environmental regulations that advance the interests of the poor is essential to a sustainable model of urban poverty reduction based on job creation and economic growth. A developmental state or effective pro-poor planning at the city scale provides a real alternative to local area or interest based livelihood projects, the narrow extension of emergency support or even of grant based social

safety nets. In the pursuit of such an agenda we suggest that a narrow focus on good governance or municipal planning is inadequate to upholding 2nd, 3rd and 4th generation urban rights (Table 1). The right to the city is framed by a strong ethical base and (interlocking) actions to reduce inequality across the local, city, city-region, national and international scale. This includes the notion that complementary and strategically articulated poverty reduction actions can and should take place at each scale to ensure the realisation of the right to the city.

RIGHTS AND THE CITY

The achievement of a rights based city will not happen with political will alone. Paradigmatic shifts and an institutional revolution in city management are required if an enabling environment for implementing the multi generational rights of the urban poor is to be realised. Such shifts are unlikely to emerge in the absence of sustained political pressure and contestation from progressive interest groups. However, grassroots pressure that remains stuck in an oppositional mode without explicit propositional demands tied to concrete institutional reforms is unlikely to effectively displace neoliberalism in practice at the local scale. Our argument is based on the case of Cape Town where there is on the ground action to advance the agenda of developmental local government and the realisation of human rights, although with limited direct engagement from progressive civil society groups.

Cape Town's is a flawed beauty, marked by extreme, enduring and highly concentrated poverty. For hundreds of households government has yet to deliver on its Constitutional promise of universal rights for all (McDonald 2007). What makes the Cape Town example interesting is that there is explicit and high level commitment from all spheres of government (national, provincial and local) to poverty reduction and the realisation of rights. It is also a place that enjoys relative affluence and thus there is the means to do something about poverty in the city. Within the South African context the Western Cape Province is the second wealthiest and within the province the greater Cape Town city region is the dominant node of economic power. Cape Town has seen examples of developmental transformation from across the intergovernmental spectrum, which highlights the fact that it is not only local government that is responsible for urban poverty reduction measures or the realisation of rights in the city. What we seek to highlight here is how we might think or imagine the institutions that are necessary for establishing a rights-based city in the South. Our broader contention is that this important line of elaboration is left unattended by an exclusive focus on the negative effects of neoliberal governmentality.

CONTESTING NEOLIBERALISM THROUGH A RIGHTS AGENDA

We seek to emphasize the potential of consolidating more empowering forms of local governance by paying closer attention to the underlying institutional and regulatory urban systems that secure those rights which underpin public interest goods as opposed to individual allocations. By placing certain expressions of neoliberal governmentality at the centre of analysis strategic opportunities are being missed to ensure through effective regulation that the state plays a stronger redistributive and developmental role. The challenges associated with effective land-use management, and avoiding the structural economic exclusion that stems from driving development in poor areas, can only be recognised if the state is not treated as an uncontested and monolithic force intent on guaranteeing neoliberal outcomes, but as replete with contestation and contradiction which presents a number of opportunities for advancing more radical policy and political projects if there is sufficient and strategic articulation with savvy actors within civil society who are alert to such opportunities. In the case of Brazil participatory budgeting systems allow such opportunities, often mediated by the 'PT', which allows social movement activists and government officials to interact and engage outside of the formal participatory

forums as well as inside those arenas of stylised contestation. In South Africa there are ample opportunities within the emergent and highly uneven local democratic systems to do the same but this territory remains under explored in large measure, we argue, because the local state is too often written off as simply an extension of a national abdication to neoliberal policy imperatives.

At this point it is important to clarify our confidence in the role that regulatory reform can play as an element in a larger radical political project. We are assuming that a primary commitment to citizen empowerment through effective democratic processes and institutions is in place. Citizen empowerment is most likely to flourish if there is a simultaneous engagement with the state on the basis of the democratic autonomy of such organisations. Strategic and tactical engagement of civil society with the state can then identify when and how the interests of the poor can be embedded in the routine functioning of the state, especially in as far as the state delimits resource allocation, sequencing of public investment priorities and articulating a series of actions to achieve higher order developmental outcomes such as growth, economic inclusion and urban spatial restructuring. By contrast, citizen action that relies exclusively on an oppositional logic or a political stance of perpetual resistance is unlikely to achieve reforms in the mundane functioning of the state, which we know from the Cape Town experience are a precondition for cumulative changes that can transform the political economy of opportunity and provide the institutional access to resources that enable the realisation of anything other than the most basic 1st generation rights in cities.

It is precisely because the progressive agenda for citizenship empowerment lends itself to claims that need a city-wide perspective that it can benefit from a rights-based discourse of urban management. For example, arguments can be marshalled to demonstrate how, because access to more strategic urban opportunities are good for overall long-term economic performance, a more radical and inclusive approach to urban land-use and control is imperative. Alternatively, organisations representing the urban poor could explore the importance of universal minimum infrastructure standards at the household, neighbourhood and city regional scale. Because these investments in the poor require redistribution a city-wide approach is imperative. These examples of an alternative politics and claim-making do not come into view when the left is preoccupied with the manifestation of neoliberal managerialism as it expresses itself in very low levels of subsidy for basic services for the urban poor. It is not that the issues of service cost and access are not important, but we would argue that it is easier to drive home an urban rights based agenda for more public investment and bigger subsidies in a context where state institutions work and have universal application for all residents of the city.

Achieving rights-based city management necessitates radical local governance transformation that cannot only be achieved by protests of the urban poor. Specifically, negotiating land use reform requires a fine-grained engagement with the systems and procedures of the local state and state agencies that are rooted at regional and national scales. This implies building interest-based coalitions across diverse institutional sites, which include the state, the professions, non-governmental actors, and even selective business interests. This is not the same as the partnership discourses characteristic of many neoliberal inspired urban management agendas. Rather, it invokes an appreciation of the necessity of strategic articulation to advance a transformative agenda in an urban context sutured by capitalist modernity. What this suggests to us is the need for a propositional politics as opposed to a defensive one against the vicissitudes of neoliberalism.

In practical terms a propositional politics to advance a comprehensive rights-based agenda means building up the planning and institutional systems to ensure the effective provision of various public goods and services such as infrastructure (ideally through environmentally sustainable technologies), stable and sufficient income through wage employment (or at least some form of welfare grant), health, including environmental health, education, housing and land, public space for assembly, culture and sport (Turok 2006). Most of these public goods require an urban development framework that is both future-oriented (sensitive to different life-cycles) and city-regional in scope. Regionalism matters for achieving the right to the city because area-based and sectoral interventions can only be planned effectively in temporal (return-on-investment cycles) and spatial terms (e.g. environmental catchment territories for water-borne infrastructures) at the larger scale to arrive at meaningful calculations about affordability and long-term operational maintenance.

A rights based city with a focus on poverty reduction and economic inclusion requires a fundamental conceptual and administrative transformation of the urban. Failure to recast the developmental focus of urban planning means it is virtually impossible to marshal a comprehensive rights based-policy agenda, rooted in a coalition of progressive forces, to address the underlying drivers of poverty and economic exclusion. Suffice to note that while the basic needs of the poor may be provided, inequality will remain unchallenged and the poor will be institutionally, socially, economically and environmentally excluded from full urban citizenship.

CONCLUSION

This chapter has shown that the issue and implications of a full right-based agenda for the city is poorly understood, largely because of the dominance of support for or opposition to a particular brand of neoliberalism. The issues of voting and basic service provision deflect from substantive issues of urban citizenship, especially the kind of rights, such as effective and equitable town planning, that can only be conferred by legitimate and resourced subnational government. Even in a city like Cape Town, where there is nominal commitment to the agenda of increased inclusion the poor are marginalised institutionally and not everyone enjoys the same land use rights. Implementing a rights based agenda at the subnational scale thus necessitates a radical critique of the instruments as well as values of the local state and will require a massive process of state rebuilding and institutional reform, without which everyday practices of urban management remain unchallenged and exclusionary.

South Africa has an especially strong tradition of state led development, including the notion of municipal engagement in social and economic advancement through 'developmental local government'. This focus on city government is part of a wider resurgence of interest in what the sub-national state can do through better city scale planning for the poor in the achievement of a rights based urban agenda.

Large cities with significant concentrations of poverty are a feature of middle income countries. Mega cities or large city regions are the key to emerging economies global positioning. Johannesburg, São Paulo or Shanghai are typical examples (Segbers

2007). These cities, while poor by the standards of London or Frankfurt have some disposable income and face the imperatives of responding to poor communities. Making cities of the South work better purely in terms of becoming economic nodes in the global systems of trade, production and consumption is not going to help the poor in those city regions. But failing to make these emerging global nodes work for all their residents may hinder their global progress. Either way, a radical programme of urban citizenship is required. The fundamental assumptions of how government operates institutionally (across the devolution of powers and functions, including legal, regulatory and enforcement functions and between departments and sectors) has to be challenged so that rich and poor are equally visible to the state and so

that the state is empowered to achieve a transformatory rights based agenda.

REFERENCES FROM THE READING

McDonald, D. (2007) *World City Syndrome: Neoliberalism and Inequality in Cape Town*, London: Routledge.

Segbers, K. (ed.) (2007) *The Making of Global City Regions*, Baltimore, MD: Johns Hopkins Press.

Turok, B. (2006) The connections between social cohesion and city competitiveness, in *OECD: Competitive Cities in the Global Economy*, Paris: Organisation for Economic Co-operation and Development.

38

"Global Cities vs. 'global cities': rethinking contemporary urbanism as public ecology"

from *Studies in Political Economy* (2003)

Timothy W. Luke

EDITORS' INTRODUCTION

Timothy Luke is University Distinguished Professor in Political Science at Virginia Polytechnic Institute and State University in Blacksburg, Virginia. He is one of the leading radical political theorists in North America. The interface of Luke's work with the topic of this section is politics, particularly environmental politics or urban political ecology. In this innovative chapter, which is an excerpt from a collection of essays on political ecology published by the Canadian journal *Studies in Political Economy*, Luke discusses the importance of ecological processes in the dynamics of global urbanization. Counterposing the limited number of "Global Cities" to the overall process of urban restructuring within "global cities" around the world, Luke directs our attention to the process of globalized urbanization. Reminiscent of Lefebvre's notion of "urban society" (see Ch. 68) and Friedmann's (2002) concept of "complete urbanization," Luke's insistence on the pervasiveness of the urbanization process worldwide is a reminder that globalization does not just occur in a select number of distinct places. Pointing towards the intricate links between socio-technological as well as bio-metabolic processes in the current phase of urbanization, Luke demands that we explore the political ecology of these relationships. Criticizing the tendency for urban ecological issues to be channeled into "subpolitical" processes of "private ecologies," Luke argues for a democratically open urban political process that is capable of dealing with the global and local inequalities of the current phase of globalized urbanization. Luke's essay is a powerful statement for a political ecology of the urban instead of the city (see Angelo and Wachsmuth 2015), by emphasizing on the transnational political constitution of the environmental *problematique* in an era of globalizing cities (Keil 2011). It also presents a utopian vision of a future world of cities linked together in sustainable and creative ways of production and consumption (Taylor and Derudder 2016).

A POINT OF DEPARTURE

Human beings always have proven destructive to their natural environments. Until the twentieth century, however, this damage was either limited and local or it was more broadly widespread in only a handful of large conurbations centered in the imperial economies of the planet's northern hemisphere. Today, however, the inhabitants of hundreds of large cities all over the world are relentlessly reshaping the traditional and modern economies of every continent as they exert global and local demands in "glocalized" spaces for energy, foodstuff, information, labor, and material through world markets. Hence, this analysis responds to the call for paying more attention "to urban ecologies and the policies developed around them as part of the formation of world cities" (Keil 1995: 293) by exploring the environmental impact of generalized urbanism or "global cities."

Because of this unchecked proliferation of such citified spaces in the twentieth century, it is no longer as clear that "Nature" is what surrounds humans in cities. Instead, one must ask if the world-wide webs of energy, information, material, and population exchange flowing between cities around the world now are infiltrating Nature so completely that this new artificial ecology will undercut entirely the survival of human and nonhuman beings?

Plainly, there are a handful of major metropoles, like London, Frankfurt, Hong Kong, Tokyo, or New York, where the command, control, communication, and intelligence functions of transnational commerce are highly concentrated. Many researchers have investigated the peculiar qualities of urban life in these Global Cities, and they are, in many ways, the limit cases of global urbanism. In many other ways, however, focusing upon such extraordinary Global Cities misses another qualitatively different transformation unfolding behind the quantitative proliferation of urbanized living in all "global cities." While the work of Global Cities leads into the spread of "global cities," it is at the latter sites, rather than the former, where the rising level of a globalized urbanization is overwhelming the Earth's natural ecology to the point of threatening the sustainability of the entire planet's human and nonhuman life.

Today's "global cities," then, are entirely new built environments tied to several complex layers of technological systems whose logistical grids are knit into other networks for the production, consumption, circulation, and accumulation of commodities. Along with sewer, water, and street systems, cities are embedded in electricity, coal, natural gas, petroleum, and metals markets in addition to timber, livestock, fish, crop, and land markets. All of these links are needed simply to supply food, water, energy, products, and services to their residents. Thus, "global cities" leave very destructive environmental footprints as their inhabitants reach out into markets around the world for material inputs to survive, but the transactions of this new political ecology also are the root causes of global ecological decline.

In 1900, only 10 per cent of the world's 1.6 billion people lived in cities. During 2000, just over 50 per cent of the world's 6 billion people lived in cities. And, by 2050, 67 per cent of a projected population of 10 billion people supposedly will live in cities. Today's premier Global Cities, plainly, are intriguing, but the more ominous numbers posed by all "global cities" taken together are far more threatening. Urbanism on this scale is creating a set of contested regions where command and insubordination, control and resistance, communication and confusion, and intelligence and incomprehension must all be rejiggered daily as transnational commerce dumps an ever-accelerating turnover of goods and services into the global economy. With over 50 per cent of humanity now residing in urban areas, the quantitatively growing logistical pull of all global cities together constitutes a new political ecology whose demands have acquired such operational mass that they are qualitatively more distinct and interesting.

Cities do have ecologies, and the ecological impact of all global cities as a system of biopolitics is building up into this wholly new built and unbuilt environment. The collective of global cities begins to constitute a "world of near complete internationalized urbanization" (Keil 1996: 42). Consequently, the larger ecology implied by the aggregate collectives of global cities is "an array of urban ecologies: "environments" in the plural" that must address "global populations, globalized everyday practices, and internationally diversified gender relations, as well as images and uses of nature" (Ibid).

Many discussions of Global Cities, then, typically approach them in "metageographical" terms, trying to gauge how much they fully subsist in a space of flows or still persist in the space of territories. A cartography of states driven by Westphalian logic of territorialized sovereignty still occludes the charts and maps that tracking flows would require. Many traditional efforts at Global Cities analysis, as the Loughborough University Globalization and World Cities (GaWC) group illustrates, maps mostly these intercity linkages to see how Global Cities provide the high-level command, control, and communication services needed by global economy through inter-city linkages among themselves. This work on Global Cities and their interconnected networks is interesting, but these static views of the links, magnitudes of interaction, and primacies in various industries among Global Cities is not the whole story. All of these links taken together now also constitute a vast megalogistical collective whose aggregate ecological effects are redirecting the world's built and unbuilt environments. The "global cities" approach must ask how this new "organizational architecture for cross-border flows" affects all local and global ecologies, fixed territorial sites, and streams of commodity circulation (Sassen 2002).

The ecologies between, beneath, or behind the Global Cities' organizational architectures are increasingly public, but subpolitical; largely artificial environments, but rooted in many layered unbuilt ecologies; globally flow-based, but locally frozen in particular territorialized material sites and spaces. Global Cities do make possible a new metageography, but this spatial frame has its own metaecological realities that become manifest in the logistics of all the many global cities being infiltrated and influenced by the high-end command, control, and communication dictates of a few Global Cities.

Rather than focusing upon that handful of Global Cities which serve as the core nodes in networks for global capitalism, we need to ask instead about the collective impact of all "global cities." As a planetary system of material production and consumption, these built environments constitute much of the world-wide webs of logistical flows which swamp over the conventional boundaries between the human and the natural with a new biopolitics of urbanism. Here critical environmentalism must fuse the concerns of public health with the goals of sustainable ecology in public ecology. Decisions that are made, and patterns that become fixed, in a subpolitical fashion must be identified, addressed, and corrected in a more political register. Public ecology is a strategy for opening these discussions and effecting these changes.

Global Cities are the usual suspects in the line-up of world cities. They are, typically, presented in a fairly conventional manner. They are seen as limited in number, tightly interconnected in function, located at the center or semiperiphery of the global economy, and formed by abstract forces. A global cities perspective tries instead to address how ecological sustainability, municipal politics, and global citification interconnect in "local social struggles that try to keep damaging consequences of globalization by bay" (Keil 1996: 38).

In this respect, Keil is quite right, "the world city is a place where the global ecological crisis manifests itself concretely" (Keil 1995: 282). Looking at "global cities" takes this insight, and then combines it with Friedmann's and Wolff's recognition that world cities are the control centers of the global economy by looking at world cities as control centers as well as controlled centered sites. Moving from a perspective that all counts and measures all Global Cities, like the work done by the GaWC project, to one that gauges the overall impact of the general citified formations marked by "global cities" is much more important. Here ecology does not stand outside of, and apart from, urbanism, which permits a critical analysis that emphasizes "the social nature of nature and the natural basis of society," and leads to a point where "finding a strategy to solve ecological problems leads potentially to a democratization of society, economy, and the state" (Keil 1995: 285).

Focusing upon the "metageography" of Global Cities, and studying these urban forms as core nodes in commercial networks, is highly useful for understanding how they shape and steer the world's megalogistical systems as a "private ecology." Yet, the work of these five, ten, or twenty Global Cities now has led to over a half of the world's human population living in urban settlements. Global Cities are small in number, but "global cities" are many in number. The impact of the carrying charges against the Earth's ecological carrying capacity for these hundreds of settlements should force everyone to see that privatized ecology of Global Cities as, in fact, a highly public ecology, which must be repoliticized, resocialized, and relocalized by environmental activists in many everyday struggles. The costs of allowing large corporate formations to privatize urban ecologies are unacceptable, and the metaeconomic questions about how humans should live with all other nonhuman beings and things must be addressed by public ecology (Harvey 2000). Public ecology can, of course, be discussed apart from "global cities," but the environmental challenges posed by sustaining the logistical grids that Global Cities have propounded for "global cities," as well as global towns and countrysides, provide a critical point of departure for this discussion.

URBANISM AS LOGISTICS

As the art of moving war material and quartering troops, logistics is about organizing and sustaining supply chains, but it also suggests the bigger issues of lodging, accommodation, and shelter for people as well as moving whatever materials are needed to sustain those activities. In many ways, cities essentially are concretions of logistics past, articulations for logistics present, and speculations about logistics future. And understanding these enduring elements in the creation of urban civilization is captured best by reconsidering the ecological links they create between, with, and after human beings and nonhuman things. From small changes in the daily traffic of materiale in human collectives, major ecological outcomes occur later, in succession to, and after such modifications in the movement of things and people.

Immense logistical spaces, then, are always carved out beyond, beneath, or behind the flows of urban existence. They help produce the permanent quarters of urban space, which fixes the conditions for quartering of city residents. These spaces also materially concretize all the arts and sciences of the broader civilization underpinning global cities. Here one sees the complex codes, collectives, and commodities of global commerce creating products and by-products out of global society's logistical exchanges, which are all dramatically recontouring the world's economies and environments around a highly privatized transnational, but still mostly transurban, trade. Cities remain pivotal sites at which the everyday exchanges between built and unbuilt environments occur, but they also are where much of what is regarded as international relations between different spatially divided economies, governments, and societies transpire.

Clearly, any single Global City is a particular site where larger forces burrow into a given place in specific, but also varied, ways, localizing the global in some determinate fashion (Keil 1998). Yet, all of these determinate formations in the aggregate now also add up in the collective megalogistics of global cities which a handful of Global Cities has made much more possible.

The grids of global cities simultaneously are works in the present for what is hoped to be greater future logistical efficiency as well as past products of what was once believed to be efficient logistical greatness. Citification has led to rich civilizations, but those cultural advancements typically were highly localized, rarely permanent, and still subject to decay and collapse. Only with the advent of global capitalism and industrial production over the past five hundred years have cities become much more than huge agricultural villages. Medieval London held fewer than 60,000 people, and its core area was only about

700 acres. By 1800, it had nearly 1 million people, and a large network of roads, horse-drawn public transport, and latter railways were needed to move people and things within an urbanized region composed of many scores of square miles. Until 1800, the cities were by and large not unlike they were in 800 A.D. or 80 A.D. Only about 2.5 per cent of the world's population lived in cities in 1800, but this quadrupled to 10 per cent in 1900, and then quintupled again to 50 per cent by 2000. In 1800 only two cities in the world – London and Edo – held a million people, in 1900 ten did, but in 2000 almost 300 did. The world's urban population, in turn, grew from around 225 million in 1900 to right at 3 billion in 2000. While large cities covered about 0.1 per cent of the world's land in 1900, this figure grew ten-fold by 2000 to about 1.0 per cent.

To comprehend fully the destructive demands of today's transnational urbanism, one must accept how globalization is operating now in 2005. This acceptance is important if one hopes to understand how fully the reticulations of power and knowledge work in most locales through what Baudrillard has identified as "the system of objects" in culture, urbanism, and globalization on a local, national or global level. All of these terms, however, are quite mutable in their meanings, and they constantly are evolving every day in new objectifications of the systems at play in objects – capitalism, nationalism, technology, urbanization – within globalization (Baudrillard 1981; 1996).

FINDING PUBLIC ECOLOGY IN POLITICAL ECOLOGY

Articulating the many ambiguous interconnections between urbanization and the environment requires one first to come to terms with privatized ecologies. Only then can globalization and the nature of modern urbanism, the boundaries of the political and the subpolitical, the nature of personal and public health, and the social formations that link health and the environment be connected as a public ecology. The recurring motif that emerges from this reconsideration is inequality, so to cope with globalism's ambiguities this discussion speaks in favor of funding a new normative discourse, or a public ecology, to guide critical thinking and political activity to improve life in both the unbuilt and built environment for human beings as well as nonhuman life. While recognizing that the concept of "public" brings a great deal of baggage with it, this notion still provides a useful point of departure for rethinking our collective ecological future in terms of sustainability, equality, and justice.

While there perhaps is no single hegemonic power today, the dilemmas posed by preserving the health of the world environment suggest there is this hegemonic form of globality at work. Moreover, many globalists are willing to push certain conditions of consumption to advance globalization into where it does not yet exist, even though there are many resistances. Most political rulers as well as most corporate managers are all working quite openly to perfect this new private ecology. It is being built to bring many more global goods and services to consumers as a part of, first, their on-going programs to advance globalization, second, as an implicit sign of their globality, and, third, as a marker, complicitly, of their shared submission if only for now, to globalism.

The privatization of collective ecological goods, as it is celebrated in the quest for greater performativity, is not advancing everyone's welfare. On the contrary, these practices are leading down paths that often are producing greater and greater malfare. Not only is the earth's "natural ecology" being degraded, but so too is the "social ecology" being neglected. Corporate visions of private ecology often discount healthy built environments, health care systems, and health-centered lifestyles to advance global growth; hence, these utilities are not being maintained or not being developed at all. Likewise, a provision of potable water, edible food, safe housing, efficient sewerage, reliable hospitals, and effective medical care as mandatory features of many built environments has never been done.

There are severe inequalities at work today in global affairs. Some are very old, and well known. Some are quite old, and only now being recognized. Some are new, and just now being felt. Most of them however, can be tied back to unequal levels of access, power, status, and wealth, which are becoming so quantitatively unbalanced on a global scale that they are turning into something qualitatively different. The analytical tools in both global studies and environmental studies are perhaps not adequate to the tasks of interpreting what is now unfolding. Instead, too many of these existing tools occlude what needs to be analyzed, who needs to be criticized, and what must be done to overcome these trends toward powerlessness and inequality.

All too often, global studies is relegated to the realm of "Society" and its analysis is assigned to only the cultural and social sciences, while environmental studies are shuttled off to the domain of "Nature" and its consideration is given over exclusively to the biological and physical sciences. To really get at what is happening today, however, we need to focus on hybridities of Nature/Society at sites which intermix the natural and the social, like the "built environment," "natural history," or "political ecology" in privatized ecology. These amalgams of Nature/Society are what sustain and/or degrade overall levels of health and environmental quality for both human and nonhumans, and they materially manifest themselves in patterns of urban settlement, industrial ecology, and natural economy.

A public ecology must fuse the administrative concerns of civic public health with the activist engagements of a critical political ecology (See the Public Ecology Project at: http://www.cnr.vt.edu/publiceecology). Public ecology should mix the insights of life science, physical science, social science, applied humanities, and public policy into a cohesive conceptual whole. Public ecology should work at the local and global level to develop "pre-pollutant" or "noncontaminant" approaches to environmental problems by using political pressures to work back up the commodity chain to lessen ecological damage by mobilizing solutions drawn from collaborative management, green engineering, industrial ecology, or vernacular design. Public ecology must show how private ecology has turned the built and unbuilt environments into a formation that is one and of a piece, not two and wholly separable.

CONCLUSION: RESISTING INEQUALITY

What surrounds one in Dallas is not what surrounds one in Delhi, but those different surroundings have high economic, political, and social costs inside and outside of both environments. A persistent feature of all global societies today is toxic wastes. Like weather, water, and wildlife, such waste is to be found everywhere in the planetary environment, making this by-product a new fixed characteristic of the Earth's ecology as it is being transformed by modern agricultural, industrial, and technological development. Nonetheless, many mechanisms in the world's political economy permit Dallas more than Delhi to dump more toxic wastes outside specific locales, boost their concentrations beyond permissible thresholds, raise exposures so intensively as to threaten health, and disperse effects indiscriminately across space and time. These irrationalities in the private ecology of global cities come from a subpolitical realm, but they now are negatively affecting every political system on a global scale as transnational environmental problems. All of this, in turn, exposes the key meta-economic issues raised by the metageography of Global Cities.

In the realm of the subpolitical, ordinary processes of democratic legitimation fail, because modern industrial revolutions with all of their profitable products and toxic by-products are highly technified economic actions. Each always "remains shielded from the demands of democratic legitimation by its own character" inasmuch as "it is neither politics nor non-politics, but a third entity: economically guided action in pursuit of interests." Because of property rights and expert prerogatives, most occupants of this planetary subpolis have yet to realize fully how "the structuring of the future takes place indirectly and unrecognizably in research laboratories and executive suites, not in parliament or in political parties. Everyone else – even the most responsible and best informed people in politics and science – more or less lives off the crumbs of information that fall from the tables of technological sub-politics." This elaborate subpolis evolves in the reified dictates of industrial ecologies, whose machinic metabolism, in turn, entails the planned and unintended destruction of nonhuman and human lives in many different environments (Beck 1992: 222–3).

Beck worries about how to face this modernity as he recognizes how fully "the possibilities for social change from the collaboration of research, technology, and science accumulate" in new loci of social order and disorder when real power and knowledge "migrates from the domain of politics to that of subpolitics" (Ibid). In the subpolis, activities that often may begin at an individual level as a rational plan combine at a collective level into the irrational, unintended, and unanticipated. It is difficult to resist these outcomes inasmuch as the workings of modern technics and markets are "institutionalized as "progress," but remain subject to the dictates of business, science, and technology, for whom democratic procedures are invalid" (Ibid).

The subpolis shapes, and then is itself shaped, in the global market's imbrication of the polis for humans and the subpolis of things. Modernity becomes an inegalitarian mechanism whereby the few who know-how and own-how maintain domination over the many who do not know-how or own-how. The illusion of progress through greater education and broader opportunity, in fact, always belies grittier realities of exploitative avarice fostered by growing disinformation and greater dispossession. Consequently, the subpolitically structured inequalities in global cities need to be more closely policed in public policy and political practice to correct the inequalities of overall health and environmental quality behind today's economic crises and political contradictions. The notion of a public ecology offers a set of values and practices to push such decisions out of the subpolitical domain.

Environmentalism, urban studies, and public health are among some of the last remaining discourses available to provide some ethical consideration or political reflection about the effects of inequality and technique and property. Private ecology degrades the overall civic life of society as the privileged millions still benefit from the international misery of billions. We cannot continue on this track if the Earth's ecologies are ever to be mended. Here one must not only talk about the iterated ecological links of the top ten Global Cities. Instead, we must consider the imposition of productive material extraction systems and material consummative discharge grids by all the settlements occupied by over half the world's human population. Ecology here becomes inescapably public, and thereby it becomes political. A public ecology can begin to develop a more formalized discourse about the unecological conduct of and anti-environmental practice of living in global cities, which will help determine what must be done.

REFERENCES FROM THE READING

Angelo, H. and Wachsmuth, D. (2015) Urbanizing urban political ecology: A critique of methodological cityism, *International Journal of Urban and Regional Research*, 39, 1: 16–27.

Baudrillard, J. (1981) *For a Critique of the Political Economy of the Sign*, St. Louis: Telos Press.

Baudrillard, J. (1996) *The System of Objects*, London: Verso.

Beck, U. (1992) *The Risk Society: Towards a New Modernity*, London: Sage.

Friedmann, J. (2002) *The Prospect of Cities*, Minneapolis: University of Minnesota Press.

Harvey, D. (2000) *Spaces of Hope*, Berkeley: University of California Press.

Keil, R. (1995) The environmental problematics in world cities, in P. L. Knox and P. J. Taylor (eds.) *World Cities in a World-System*, Cambridge: Cambridge University Press: 280–97.

Keil, R. (1996) World city formation, local politics, and sustainability, in R. Keil, G. R. Wekerle and D. V. Bell (eds.) *Local Places in the Age of the Global City*, Montreal: Black Rose Books: 37–44.

Keil, R. (1998) *Los Angeles: Globalization, Urbanization, and Social Struggles*, Chichester and New York: J. Wiley.

Keil, R. (2011) Transnational urban political ecology: Health and infrastructure in the unbounded city, in G. Bridge and S. Watson (eds.) *The New Blackwell Companion to the City*, Oxford: Blackwell: 713–25.

Sassen, S. (2002) *Global Networks, Linked Cities*, London: Routledge.

Taylor, P. J and Derudder, B. (2016) *World City Network: A Global Urban Analysis* (2nd Edition), London and New York: Routledge.

PART FIVE

Contestations

Plate 28 Demolitions, Shanghai

Source: Xuefei Ren

INTRODUCTION TO PART FIVE

The sociospatial restructuring in global cities since the 1980s has contributed to the articulation of new political claims on citizen rights. Against this background, various urban social movements for economic, social, and environmental justice began to characterize themselves as active political participants in the transformation of globalizing cities. The local politics contesting capitalism thus appropriated the discourses of anti-globalization, as social movements groups struggle to promote grassroots empowerment and new forms of sociospatial justice. Under these conditions, diverse class factions and territorial communities of the global city clashed directly with globally oriented forces such as transnational corporations, real estate developers, and the boosterist local state apparatus. In these struggles, civil society-based agents such as trade unions, neighborhood organizations, and environmental groups have been actively shaping the politico-institutional dynamics within globalizing cities.

The chapters included in this section draw upon a diverse body of scholarship on urban conflicts and contestations in the post-2000 period. They provide both analytical frameworks and concrete examples for better understanding the contested urban politics across the global North and South. The first two chapters by Sassen (Ch. 40) and Harvey (Ch. 41) offer a conceptual lens to read urban politics in globalizing cities. Sassen challenges us to rethink what "local" and "global" entail when used to describe urban social movements, conflicts, and contestations in the context of economic restructuring. Facilitated by the Internet and social media, many of the "local" political actors, such as immigrant groups, environmental activists, and human rights groups can mobilize internationally, thus entering the stage of "global" politics. In other words, "global" politics can be invariably traced to "local" actors and organizations in globalizing cities. Harvey (Ch. 41) discusses "the right to the city," which is a popular discourse by now and has been appropriated by various political groups. Harvey argues that the right to the city is not limited to individual rights to access urban resources, but should also be understood as rights to change the city according to people's needs. Today, both of these understandings—local actors engaging in global politics, and the right to the city—have become key themes in studies on contentious urban politics.

We have also assembled in this section texts that examine urban contestations in different political regimes and socio-institutional settings. Some chapters focus on cities in the global South such as Shanghai, São Paulo, Johannesburg, and Dakar (Senegal), while others focus on cities in Europe and North America; some readings discuss urban conflicts in democracies, while others turn to contestations in authoritarian regimes such as Egypt and China; also, some chapters examine urban contestations in the context of weak local states such as in Africa, while others focus on urban struggles in the presence of strong local states, such as in China.

The qualitative differences in the wealth, regime types, and local state capacity have shaped both the substance of claims making and the strategies employed by local political groups. To elaborate, first, the wealth of cities and its distribution can strongly influence the dynamics of urban conflicts. In wealthier countries and cities, urban conflicts are often responses to sociospatial transformations brought about by the erosion of the welfare state and by the implementation of entrepreneurial policies (see Mayer Ch. 42 and Schipper et al. Ch. 47). By contrast, in poorer cities, urban conflicts are often about the expansion of socio-economic rights (Parnell and Pieterse Ch. 37), and residents employ a wide range of tactics to gain access to resources. By doing so they also remake their cities by altering the regulatory environment (Simone Ch. 45).

Second, democratic regimes allow more space for citizen participation, but even under non-democratic regimes, citizens are not invisible, voiceless victims of state power and they employ various strategies to assert their rights. Since the 1990s, some cities in Europe, North America, and Latin America have created particular participatory schemes to "invite" citizens to express their opinions on urban policymaking. Such schemes of participatory democracy are often state initiatives in response to demands from urban political groups. They are by no means perfect, but they do open up new possibilities for deepening democratic practices. Comparatively, in non-democratic contexts, both the scope and the institutional infrastructure for citizen participation in urban politics are limited and, as a result, urban conflicts often escalate to direct confrontations with the state, in the form of protests and direct agitations. However, one should not assume that citizens are powerless groups in authoritarian regimes, as the essays on Cairo (Prologue by Elshahed) and Shanghai (Ren Ch. 44) show, strong repression often leads to even stronger resistance. Some rapidly globalizing cities under authoritarian regimes have become fertile ground for large-scale urban social movements.

Third, local state capacity also shapes the form and content of urban political struggles.

A strong local state can define the terms and scope of citizen participation—either by widening it or narrowing it, while a weak local state leads to a power vacuum at the local level, which in turn, can give rise to political entrepreneurs. Here Chinese and African cities offer two contrasting examples. The strong local government in Shanghai took the lead in the city's urban renewal project from the late 1990s, and it decided the pace and magnitude of urban demolitions and repressed oppositions from residents (Ren Ch. 44). While in Dakar, Senegal, the local state has failed to provide basic welfare and services so that residents have to be entrepreneurial in developing social networks to access resources; in a paradoxical way, the weak local state has led to a flourishing of associational life in the city (Simone Ch. 45).

In addition to the variations, some common themes can also be identified in the scholarship on urban contestations, for instance, labor disputes, the struggle for citizenship rights, the dialectic between state and civil society, and the visibility of urban struggles in globalizing cities.

As predicted early on by Friedmann and Wolff (Ch. 3) and writers such as Robert Ross and Kent Trachte (1990), the arena of work and class struggles has been of great relevance in global city formation. Steven Tufts (Ch. 46) takes a closer look at the role of trade unions in the process.

In the field of global cities research, the question of citizenship has generated much theoretical debate and empirical research. This strand of research explores the question of whether new rights claims and demands for state action are being engendered in globalizing urban spaces. While most theories of urban citizenship are rooted in the experiences of the western world, there is also now a growing recognition that distinctive forms of urban citizenship have emerged in the global South (Holston and Appadurai 2003: 304). Caldeira (Ch. 43), for instance, examines social protests that broke out in São Paulo and other Brazilian cities in 2013. She notes that many of the participants are youths from the periphery of the city, which has been long neglected by the state as compared to the cosmopolitan urban center. With graffiti, music, and social media, the youths from the periphery were challenging the city's residential segregation and claiming their right to be included. Compared to Brazil, China offers an even more extreme case of unequal citizenship, which is institutionalized with the country's *hukou* system. This national household registration system divided the population into rural and urban, with each sector entitled to different economic and social rights. Migrant workers live and work in cities but have limited access to social welfare services, and thus, Chinese cities today have become contested terrains for reformulating citizenship rights (Bandurski 2016).

The dialectic relation between the state and civil society is another central theme in the scholarship on urban contestations. We have discussed how state capacity and political regimes can affect mobilization strategies and claims making. Here it is necessary to add that urban social movements and contestations can also force the state to acknowledge and expand citizen rights of the marginalized. In China, widespread demonstrations and homeowners' activism eventually forced the central government to revise its draconian regulations over demolitions. Different from the 1990s when demolitions could happen without residents' consent, today at least 80 percent of residents have to give their consents before any redevelopment project can take place (Ren 2011). Similarly, in India, housing rights activism has forced the state to revise its slum redevelopment policies, in order to entitle more residents for compensation in the case of evictions (Weinstein 2014).

Finally, studies on urban contestations have focused on the visual aspects of urban struggles taking place in global cites. One strand of the scholarship examines contestations over the built environment. Economic, cultural, and demographic globalization has dramatically reshaped downtowns and suburban peripheries from Shanghai to Singapore, from Mumbai to Cape Town (Bodnàr 2001; Olds 2001; King 2004; Haila 2016). Consequently, a distinct politics of the built environment has emerged in globalizing cities. Ren (Ch. 44) examines Shanghai's strategy of global city building via both urban renewal and historic preservation of colonial-era buildings. In Shanghai, the ambition to become Asia's global city is invoked visibly in the city's built environment, which in turn has been heavily contested by civil society groups. Another strand of scholarship focuses on the visibility of globalizing cities themselves and the implications on urban social movements. Many have noted that not all contestations taking place in global and globalizing cities are "urban" in their focus, but these cities are chosen by activists to stage their protests precisely because of the media exposure that global cities can bring. From Frankfurt to Hong Kong, activists blocked roads and tried to shut down the Central Business District, not because their claims target local economic policies, but because global cities offer high-visibility platforms for claims making. Taken together, the chapters in this section capture variations and similarities in the sociopolitical mobilization in a wide range of globalizing cities.

REFERENCES AND SUGGESTIONS FOR FURTHER READING

Appadurai, A. (2000) Spectral housing and urban cleansing: Notes on Millennial Mumbai, *Public Culture*, 12, 3: 627–51.

Bandurski, D. (2016) *Dragons in Diamond Village and Other Tales from the Back Alleys of Urbanizing China*, Melbourne: Penguin Books Australia.

Bodnàr, J. (2001) *Fin de Millénnaire Budapest*, Minneapolis: University of Minnesota Press.

Davis, M. (1987) Chinatown, part two: The 'internationalization' of downtown Los Angeles, *New Left Review*, 164: 65–86.

Gualini, E., Mourato, J. M., and Allegra, M. (eds.) (2015) *Conflict in the City: Contested Urban Spaces and Local Democracy*, Berlin: Jovis.

Haila, A. (2016) *Urban Land Rent: Singapore as a Property State*, Oxford: Wiley Blackwell.

Harvey, D. (2000) *Spaces of Hope*, Berkeley and Los Angeles: University of California Press.

Holston, J. and Appadurai, A. (2003) Cities and citizenship, in N. Brenner et al. (eds.) *State/Space: A Reader*, Boston, NY: Blackwell: 296–308.

Keil, R. (1998) *Los Angeles: Globalization, Urbanization and Social Struggles*, Chichester: Wiley.

Keil, R. (2003) Globalization makes states, in N. Brenner et al. (eds.) *State/Space: A Reader*, Boston, NY: Blackwell: 278–95.

King, A. D. (2004) *Spaces of Global Cultures*, London and New York: Routledge.

Olds, K. (2001) *Globalization and Urban Change*, New York and Oxford: Oxford University Press.

Ren, X. (2011) *Building Globalization: Transnational Architecture Production in Urban China*, Chicago: University of Chicago Press.

Ross, R. and Trachte K. (1990) *Global Capitalism: The New Leviathan*, Albany, NY: State University of New York Press.

Sandercock, L. (1998) *Towards Cosmopolis*, Chichester: Wiley.

Sandercock, L. (2003) *Cosmopolis II*, London and New York: Continuum.

Weinstein, L. (2014) *The Durable Slum: Dharavi and the Right to Stay Put in Globalizing Mumbai*, Minneapolis: University of Minnesota Press.

39 Prologue
"From Tahrir Square to Emaar Square: Cairo's private road to a private city"

from The Guardian (2014)

Mohamed Elshahed

In mid-February 2014, the developer behind the world's tallest building, the Burj Khalifa in Dubai, signed a deal with the Egyptian defense ministry. The agreement clears the way for the construction of Emaar Square, the centerpiece of a mixed-use development – exclusive residential units, a golf course, open-air shopping for international luxury brands – that is part of UAE-based Emaar's exclusive Uptown Cairo complex.

The defense ministry, which owns the massive tract of land where Uptown Cairo will be built, is Egypt's largest landowner and manager. In 1997, a presidential decree gave the military the right to manage all undeveloped non-agricultural land – 87% of the country, by one estimate. In the city of Cairo, this translates to massive, walled plots of land in lucrative locations, monitored from watchtowers. Signs forbid photography and identify them as military zones, but no military activity takes place here. They are vacant, awaiting their turn to be transformed into hotels, housing for military officers or upmarket malls.

Emaar Square is the latest and biggest of these military-secured developments. A private road will link it to Cairo's road network, likely requiring the "cleansing" of the poor homeowners in the Jabal al-Ahmar area. Egypt's military has a lengthy track record of forcibly evicting residents, sometimes using lethal force, in favor of private interests – but nearly always citing security as the reason. This is a private road to a private city – the chants for "bread, freedom, social justice" in Tahrir Square three years ago will not be heard in Emaar Square. The deal between the defense ministry and Emaar went unnoticed in the news; three years since the revolution, in a way it is business as usual. So how did we get from Tahrir Square to Emaar Square?

The Egyptian revolt is not usually discussed as an urban struggle, one that not only takes place in cities but also seeks to undo the very mechanisms that have produced Egypt's uneven urban environment. The events of Tahrir Square were a protest against the extreme unevenness of development in Egypt, in which the state neglects the urban majority while providing concessions to Gulf investors and local entities linked directly to the military state apparatus.

In the Mubarak years, the state and its business cronies began a development plan called Cairo 2050. This would have led to the mass eviction of thousands of families, in order to transform the city into pockets of high-end residential development, golf courses and shopping centers. Much of the investment was to come from Kuwait, Saudi Arabia and the UAE. Egypt's former regime was intent on the "Dubai-sation" of Cairo.

The revolution caused an unwanted turn: the Muslim Brotherhood was backed by Qatar rather than Saudi, Kuwait and UAE. But since the military re-took control in July last year, the battle over Egypt's economy has tilted in favor of the old regime and its supporters, and the old Cairo 2050 projects are back on track, including the Maspero Triangle and Uptown Cairo.

Those who celebrate Dubai's urban model and wish for its expansion across the region make the unethical choice of ignoring the fact that the Gulf cities emerge out of a very specific relation between political power and capital: namely, that they are usually one and the same. The expansion of this model into cities such as Cairo, where the military has unchallenged access both to politics and capital (land, resources, construction), would have a disastrous impact on the urban majority – who will be marginalized, moved out of the way when necessary and put to work under unacceptable conditions, with no power to mobilise and with little pay.

So why is this interesting? First, when the military, the biggest land owner and free from civilian oversight, makes a direct deal with a developer to build an exclusive and gated community in the heart of the capital, this is not a free market. It may be framed by the government as part of "building Egypt" and attracting investment, but it merely creates more opportunity for buyers to be locked into gated developments with no access to democratic municipal management: the residents of Emaar Square won't pressure the government for services, they will deal with the company instead. Such deals also give the growing "republic of retired generals" cushy jobs.

Second, this is not a democracy, and certainly not revolutionary. The Emaar Square deal included the ministries of housing, local development, investment and the Cairo authorities. This co-operation is what will allow Emaar to build its private road. And yet these state institutions are failing to solve Egypt's mounting urban problems, many of which are directly caused by the lack of accountability and the incompetence of these very institutions. Since the 1970s, the state has fallen short of providing services, of creating effective systems of urban management and of producing expansion plans that allow local private capital to grow while protecting the sanctity of the common – the public sphere, and its manifestation in shared public spaces. Instead, the institutions of the state came together to sign a deal for a private highway to a private city.

The protests that began in 2011 showed how cities are stages for a struggle not only to shape urban space but also to create new forms of democratic representation. They shed light on how everyday spaces can become sites of resistance, revolution and transformation. This struggle is directly linked to the way power and capital have produced socially and economically unjust urban experiences. In Egypt, the military bear weapons in civilian spaces and have direct access to capital and assets (such as land and building materials) that directly shape cities and their development. This means that if people stand in the way of military bulldozers coming to remove their "illegal" homes or huts, those residents risk being tried in military court for obstruction.

For the majority of Egyptians, cities have lost their vital role as places of economic possibility. Land is just one of many commodities monopolized by the military. There are also contracts worth billions of dollars of state money going directly to military contractors without a proper bidding process – from highway construction to slum clearance to redevelopment.

The recent $40 billion partnership with the UAE's Arabtec Construction to build one million low-income housing units on land controlled by the military is just the latest example of how a small cabal shapes the urban environment. Companies associated with former or current military generals are part of a network that produces the essential building materials (chiefly cement and brick) used for everything from luxury condos in gated communities to new residential buildings that expand unilaterally into agricultural land. The spatial confrontations, often violent, in Egyptian squares between protesters and conscripts are in many ways vivid illustrations of Egypt's struggle over its politics, economy and space – in other words, a struggle towards a more even urban development.

After the Egyptian government suddenly resigned last month, the former housing minister Ibrahim Mehlib was named premier; he appointed Mostafa Madbouly, a leading figure in the Cairo 2050 vision, to replace him. Will these experts in housing and urban development serve the interests of the majority of Egyptians living in cities? Or will they continue to serve the interests of a particular network of political and financial power? Emaar Square is all the indication you need.

40

"Local actors in global politics"

from *Current Sociology* (2004)

Saskia Sassen

EDITORS' INTRODUCTION

This text is part of Saskia Sassen's book *Territory, Authority, Rights* (2006), in which she discusses how the exclusive authority of the nation state over people and territory has been unbundled, and how this new space of global politics is now being filled by a variety of local actors. For Sassen, global cities are strategic sites not only for worldwide economic flows, but also for grassroots mobilization. Non-state actors–individuals, community organizations, and NGOs–here engage in various forms of activism in pursuit of priorities such as gender equality, better employment benefits, housing rights, and environmental justice. According to Sassen, new information technologies have actively facilitated the entry of local actors and organizations into global politics. Through examples from the public demonstrations of Turkish Kurds in major European cities to housing rights movements in Mumbai, Sassen argues that social media can make visible the poor and the less powerful, especially those located in global cities. In this way, Sassen suggests a rethinking of the traditional local/global dualism. For Sassen, the local is not simply a physical entity nested within the global scale, but is a multiscalar site of contestation.

Globalization and the new ICTs (information and communication technologies) have enabled a variety of local political actors to enter international arenas once exclusive to national states. Multiple types of claim-making and oppositional politics articulate these developments. Going global has been partly facilitated and conditioned by the infrastructure of the global economy, even as the latter is often the object of those oppositional politics. Further, the possibility of global imaginaries has enabled even those who are geographically immobile to become part of global politics. NGOs and indigenous peoples, immigrants, and refugees who become subjects of adjudication in human rights decisions, human rights and environmental activists, and many others are increasingly becoming actors in global politics.

That is to say, non-state actors can enter and gain visibility in international forum or global politics as individuals and as collectivities, emerging from the invisibility of aggregate membership in a nation-state exclusively represented by the sovereign. One way of interpreting this is in terms of an incipient unbundling of the exclusive authority over territory and people that we have long associated with the national state. The most strategic instantiation of this unbundling is probably the global city, which operates as a partly denationalized platform for global capital and, at the same time, is emerging as a key site for the coming together of the most astounding mix of people from all over the world. The growing intensity of transactions among major cities is creating a strategic cross-border geography that partly bypasses national states. The new network technologies further strengthen these transactions, whether they are electronic transfers of specialized services among firms or Internet-based communications among the members of globally dispersed diasporas and civil society organizations. These new technologies, especially the public access Internet, have actually strengthened this politics of places, and have expanded the geography for civil society actors beyond the strategic networks of global cities, to include peripheralized localities. This has enabled a politics of places on global networks.

A key question organizing this essay concerns the ways in which such localized actors and struggles can be constitutive of new types of global politics and subjectivities. The argument is that local, including geographically immobile and resource-poor, actors can contribute to the formation of global domains or virtual public spheres and thereby to a type of local political subjectivity that needs to be distinguished from what we would usually consider local. The new ICTs are important. But, as I will discuss, they are so under two conditions. One is the preexistence of social networks, and it is here that the cross-border geographies connecting places, especially global cities, provide a conducive environment. The other qualifier is that it took a lot of organizing and work to develop adequate technical infrastructure and software to make this happen. Civil society organizations and individuals have played crucial roles. The result has been that particular instantiations of the local can actually be constituted at multiple scales and thereby construct global formations that tend towards lateralized and horizontal networks rather than the vertical and hierarchical forms typical of major global actors, such as the IMF and WTO.

I examine these issues through a focus on various political practices and the technologies used. Of particular interest is the possibility that local, often resource-poor organizations and individuals can become part of global networks and struggles. Such a focus also takes the analysis beyond the new geographies of centrality constructed through the network of the forty plus global cities in the world today. It accommodates the possibility that even rather peripheralized locations can become part of global networks.

THE ASCENDANCE OF SUB- AND TRANSNATIONAL SPACES AND ACTORS

Cities and the new strategic geographies that connect them and bypass national states can be seen as constituting part of the infrastructure for global domains, including global imaginaries. They do so from the ground up, through multiple micro-sites and micro-transactions (Hamel et al. 2000). Among the actors in this political landscape are a variety of organizations focused on trans-boundary issues concerning immigration, asylum, international women's agendas, alter-globalization struggles, and many others. While these are not necessarily urban in their orientation or genesis, they tend to converge in cities. The new network technologies, especially the Internet, ironically have strengthened the urban map of these trans-boundary networks. It does not have to be that way, but at this time cities and the networks that bind them function as an anchor and an enabler of cross-border struggles.

Global cities are, then, thick enabling environments for these types of activities, even though the networks themselves are not urban per se. In this regard, these cities help people experience themselves as part of global non-state networks as they live their daily lives. They enact some version of the global in the micro-spaces of daily life rather than on some putative global stage.

A key nexus in this configuration is that the weakening of the exclusive formal authority of states over national territory facilitates the ascendance of sub- and transnational spaces and actors in politico-civic processes. Among these are spaces that tended to be confined to the national domain and can now become part of global networks, and they are spaces that have evolved as novel types in the context of globalization and the new ICTs. The loss of power at the national level produces the possibility of new forms of power and politics at the sub-national level and at the supranational level. The national as container of social process and power is cracked. This cracked casing opens up a geography of politics and civics that links subnational spaces. Cities are foremost in this new geography. The density of political and civic cultures in large cities localizes global civil society in peoples' lives. We can think of these as multiple localizations of civil society that are global in that they are part of global circuits and trans-boundary networks.

The organizational side of the global economy materializes in a worldwide grid of strategic places, uppermost among which are major international business and financial centers. We can think of this global grid as constituting a new economic geography of centrality, one that cuts across national boundaries and increasingly across the old North-South divide. It has emerged as a transnational space for the formation of new claims by global capital but also by other types of actors. The most powerful of these new geographies of centrality at the inter-urban level bind the major international financial and business centers: New York, London, Tokyo, Paris, Frankfurt, Zurich, Amsterdam, Los Angeles, Sydney, Hong Kong, among others. But this geography now also includes cities such as Sao Paulo, Shanghai, Bangkok, Taipei, and Mexico City. The intensity of transactions among these cities, particularly through the financial markets, transactions in services, and investment, has increased sharply, and so have the orders of magnitude involved.

The new urban spatiality thus produced is partial in a double sense: it accounts for only part of what happens in cities and what cities are about, and it inhabits only part of what we might think of as the space of the city, whether this be understood in terms as diverse as those of a city's administrative boundaries or in the sense of the public life of a city's people. But it is nonetheless one way in which cities can become part of the live infrastructure of global civil society.

The space constituted by the worldwide grid of global cities, a space with new economic and political potentialities, is perhaps the most strategic, though not the only space for the formation of transnational identities and communities. This is a space that is both place-centered in that it is embedded in particular and strategic cities, and trans-territorial because it connects sites that are not geographically proximate yet are intensely connected to each other. It is not only the transmigration of capital that takes place in this global grid but also that of people, both rich—i.e., the new transnational professional workforce—and poor—i.e., most migrant workers; and it is a space for the transmigration of cultural forms, for the re-territorialisation of 'local' subcultures. An important question is whether it is also a space for a new politics, one going beyond the politics of culture and identity while likely to remain at least partly embedded in it. One of the most radical forms assumed today by the linkage of people to territory is the loosening of identities from their traditional sources, such as the nation or the village. This unmooring in the process of identity formation engenders new notions of community of membership and of entitlement.

Global capital and immigrants are two major instances of transnationalized actors that have cross-border unifying properties internally and find themselves in conflict with each other inside global cities.

The leading sectors of corporate capital are now global in their organization and operations. And many of the disadvantaged workers in global cities are women, immigrants, people of color—men and women whose sense of membership is not necessarily adequately captured in terms of the national, and indeed often evince cross-border solidarities around issues of substance. Both types of actors find in the global city a strategic site for their economic and political operations. We see here an interesting corespondence between great concentrations of corporate power and large concentrations of 'others.'

Large cities in both the global South and the global North are the terrain where a multiplicity of globalization processes assume concrete, localized forms. A focus on cities allows us to capture, further, not only the upper but also the lower circuits of globalization. These localized forms are, in good part, what globalization is about. Further, the thickening transactions that bind cities across borders signal the possibility of a new politics of traditionally disadvantaged actors operating in this new transnational economic geography. This is a politics that arises out of actual participation by workers in the global economy, but under conditions of disadvantage and lack of recognition, whether as factory workers in export-processing zones or as cleaners on Wall Street.

PEOPLE'S NETWORKS: MICRO-POLITICS FOR GLOBAL CIVIL SOCIETY

The cross-border network of global cities is a space where we are seeing the formation of new types of 'global' politics of place that contest corporate globalization, environmental and human rights abuses, and so on. The demonstrations by the alter-globalization movement signal the potential for developing a politics centered on places understood as locations on global networks. This is a place-specific politics with global span. It is a type of political work deeply embedded in people's actions and activities but made possible partly by the existence of global digital linkages. These are mostly organizations operating through networks of cities and involving informal political actors—that is, actors who are not necessarily engaging in politics as citizens narrowly defined, where voting is the most formalized type of citizen politics. Among such informal political actors are women who engage in political struggles in their condition as mothers, anti-globalization activists who go to a foreign country as tourists but to do citizen politics, undocumented immigrants who join protests against police brutality.

These practices are constituting a specific type of global politics, one that runs through localities and is not predicated on the existence of global institutions. The engagement can be with global institutions, such as the IMF or WTO, or with local institutions, such as a particular government or local police force charged with human rights abuses. Theoretically these types of global politics illuminate the distinction between a global network and the actual transactions that constitute it: the global character of a network does not necessarily imply that its transactions are equally global, or that it all has to happen at the global level. It shows the local to be multiscalar computer-centered technologies have also here made all the difference; in this case the particular form of these technologies is mostly the public access Internet. The latter matters not only because of low-cost connectivity and the possibility of effective use (via e-mail) even with low bandwidth availability, but also and most importantly, because of some of its key features. Simultaneous decentralized access can help local actors have a sense of participation in struggles that are not necessarily global but are, rather, globally distributed in that they recur in locality after locality. In so doing these technologies can also help in the formation of cross-border public spheres for these types of actors, and can do so a) without the necessity of running through global institutions, and b) through forms of recognition that do not depend on much direct interaction and joint action on the ground. Among the implications of these options is the possibility of forming global networks that bypass central authority, and, further, especially significant for resource- poor organizations, that those who may never be able to travel can nonetheless be part of global struggles and global publics.

Such forms of recognition are not historically new. Yet there are two specific matters which signal the need for empirical and theoretical work on their ICT enabled form. One is that much of the conceptualization of the local in the social sciences has assumed physical/geographic proximity and thereby a sharply defined territorial boundedness, with the associated implication of closure. The other, partly a consequence of the first, is a strong tendency to conceive of the local as part of a hierarchy of nested scales, especially once there are national states. To a very large extent these conceptualizations hold for most of the instantiations of the local today, more specifically, for most of the

actual practices and formations likely to constitute the local in most of the world. But there are also conditions today that contribute to destabilize these practices and formations and hence invite a reconceptualization of the local that can accommodate a set of instances that diverge from dominant patterns. Key among these current conditions are globalization and/or globality as constitutive not only of cross-border institutional spaces but also of powerful imaginaries enabling aspirations to transboundary political practice even when the actors involved are basically localized.

The city is a far more concrete space for politics than the nation. It becomes a place where non-formal political actors can be part of the political scene in a way that is more difficult, though not impossible, at the national level. Nationally politics needs to run through existing formal systems, whether the electoral political system or the judiciary (taking state agencies to court). To do this you need to be a citizen. Non-formal political actors are thereby more easily rendered invisible in the space of national politics. The space of the city accommodates a broad range of political activities—squatting, demonstrations against police brutality, fighting for the rights of immigrants and the homeless—and issues, such as the politics of culture and identity, gay and lesbian and queer politics. Much of this becomes visible on the street. Much of urban politics is concrete, enacted by people rather than dependent on massive media technologies. Street-level politics make possible the formation of new types of political subjects that do not have to go through the formal political system.

It is in this sense that those who lack power and are "unauthorized" i.e., unauthorized immigrants, those who are disadvantaged, outsiders, discriminated minorities, can gain *presence* in global cities, vis-à-vis power and vis-à-vis each other. A good example of this is the Europe-wide demonstrations of largely "Turkish" Kurds in response to the arrest of Ocalan: suddenly they were on the map not only as an oppressed minority but also as a diaspora in their own right, distinct from the Turks. This signals, for me, the possibility of a new type of politics centered in new types of political actors. It is not simply a matter of having or not having power. These are new hybrid bases from which to act. There are a growing number of organizations that are largely focused on a variety of grievances of powerless groups and individuals. Some are global and others national. While powerless, these individuals and groups are acquiring presence in a broader politico-civic stage.

One of the characteristics of the types of organizations discussed here is that they engage in 'non-cosmopolitan' forms of global politics. Partly enabled by the Internet, activists can develop global networks for circulating not only information (about environmental, housing, political issues, etc.) but also can engage in actual political work and execute strategies. Yet they remain grounded in very specific issues and are often focused on their localities even as they operate as part of global networks. There are many examples of such a new type of cross-border political work. For instance, SPARC (the Society for Promotion of Area Resource Centers), started by and centered on women, began as an effort to organize slum dwellers in Bombay to get housing. Now it has a network of such groups throughout Asia and some cities in Latin America and Africa. The focus is local, and so are the participants and those whom they seek to reach, usually local governments. The various organizations making up the broader network do not necessarily gain power or material resources from this global networking, but they gain strength for themselves and vis a vis the agencies to which they make their demands.

This is one of the key forms of critical politics that the Internet can make possible: a politics of the local with a big difference in that these are localities connected with each other across a region, a country, or the world. Although the network is global this does not mean that it all has to happen at the global level.

THE FORGING OF NEW POLITICAL SUBJECTS

All of this facilitates a new type of cross-border politics, one centered in multiple localities yet intensely connected digitally. Adams (1996), among others, shows us how telecommunications create new linkages across space that underline the importance of networks of relations and partly bypass older hierarchies of scale. Activists can develop networks for circulating place-based information (about local environmental, housing, political conditions) that can become part of political work and strategies addressing a global condition—the environment, growing poverty and unemployment worldwide, lack of accountability among multinationals, etc. The issue here is not so much the possibility of such political practices; they have long existed even though with other mediums and with other velocities. The issue is

rather one of orders of magnitude, scope and simultaneity: the technologies, the institutions and the imaginaries that mark the current global digital context inscribe local political practice with new meanings and new potentialities.

The mix of focused activism and local/global networks represented by the organizations creates conditions for the emergence of at least partly transnational identities. The possibility of identifying with larger communities of practice or membership can bring about the partial unmooring of identities referred to in the first section. While this does not necessarily neutralize attachments to a country or national cause, it does shift this attachment to include trans-local communities of practice and/or membership. This is a crucial building block for a global politics of localized actors, that is to say, a politics that can incorporate the micro-practices and micro-objectives of people's daily lives as well as their political passions. The possibility of transnational identities emerging as a consequence of this thickness of micro-politics is important for strengthening global politics, even as the risk of nationalisms and fundamentalisms is, clearly, present in these dynamics as well.

The types of political practice discussed here are not the cosmopolitan route to the global. They are global through the knowing multiplication of local practices. These are types of sociability and struggle deeply embedded in people's actions and activities. They are also forms of institution-building work with global scope that can come from localities and networks of localities with limited resources and from informal social actors. They do not have to become cosmopolitan in this process, they may well remain domestic and particularistic in their orientation and remain engaged with their households and local community struggles, and yet they are participating in emergent global politics.

REFERENCES FROM THE READING

Adams, P. C. (1996) Protest and the scale politics of telecommunications, *Political Geography*, 15: 419–41.

Hamel, P., Lustiger-Thaler, H., and Mayer, M. (eds.) (2000) *Urban Movements in a Globalizing World*, London: Routledge.

Sassen, S. (2006) *Territory, Authority, Rights: From Medieval to Global Assemblages*, Princeton: Princeton University Press.

41
"The right to the city"
from *New Left Review* (2008)

David Harvey

EDITORS' INTRODUCTION

David Harvey, Distinguished Professor of Anthropology and Geography at the Graduate Center of the City University of New York, is one of the most influential Marxist urban geographers and the author of numerous books, including *Social Justice and the City* (1973), *The Condition of Postmodernity* (1989), *A Brief History of Neoliberalism* (2005), and more recently, *Rebel Cities* (2012) and *The Ways of the World* (2016). Many of his concepts have become foundational theoretical reference points for global urban studies–these include, among others, "urban entrepreneurialism," "accumulation by dispossession," and the theme of this essay–"the right to the city." Since the mid-2000s, urban scholars have explored the right to the city in a range of settings–housing, water, clean environment, and public space. For Harvey, the right to the city represents far more than a right of individual access to urban resources; rather, he insists, "it is a right to change ourselves by changing the city." The right to the city is thus a collective struggle to rework the urbanization process itself. In developing this argument, Harvey compares urban renewal in Haussmann's Paris in the 1860s with postwar American suburban sprawl, mass consumption, inter-state highway construction, and with more recent forms of urbanization in China, India, Korea, and in the Gulf States. He sees similar patterns of property speculation worldwide, in which surplus finance capital is directed towards property speculation, leading to the displacement of the poor. Looking ahead, Harvey argues that a global struggle against finance capital is needed to control and reorient the processes of urbanization. Like Sassen, Harvey is optimistic about the organizing capacity of local actors to engage in this global struggle. The theme of the right to the city is further examined in other essays in this section–for instance, the discussion of housing rights activism in Shanghai (Ren, Ch. 44), and urban protests in São Paulo (Caldeira, Ch. 43) and Frankfurt (Schipper et al., Ch. 47).

We live in an era when ideals of human rights have moved center stage both politically and ethically. A great deal of energy is expended in promoting their significance for the construction of a better world. But for the most part the concepts circulating do not fundamentally challenge hegemonic liberal and neoliberal market logics, or the dominant modes of legality and state action. We live, after all, in a world in which the rights of private property and the profit rate trump all other notions of rights. I here want to explore another type of human right, that of the right to the city.

The question of what kind of city we want cannot be divorced from that of what kind of social ties, relationship to nature, lifestyles, technologies and aesthetic values we desire. The right to the city is far more than the individual liberty to access urban resources: it is a right to change ourselves by changing the city. It is, moreover, a common rather than an individual right since this transformation inevitably depends upon the exercise of a collective power to reshape the processes of urbanization. The freedom to make and remake our cities and ourselves is, I want to argue, one of the most precious yet most neglected of our human rights.

From their inception, cities have arisen through geographical and social concentrations of a surplus product. Urbanization has always been, therefore, a class phenomenon, since surpluses are extracted from somewhere and from somebody, while the control over their disbursement typically lies in a few hands. This general situation persists under capitalism, of course; but since urbanization depends on the mobilization of a surplus product, an intimate connection emerges between the development of capitalism and urbanization. Capitalists have to produce a surplus product in order to produce surplus value; this in turn must be reinvested in order to generate more surplus value. The result of continued reinvestment is the expansion of surplus production at a compound rate—hence the logistic curves (money, output and population) attached to the history of capital accumulation, paralleled by the growth path of urbanization under capitalism.

URBAN REVOLUTIONS

Consider, first, the case of Second Empire Paris. The year 1848 brought one of the first clear, and European-wide, crises of both unemployed surplus capital and surplus labor. It struck Paris particularly hard and issued in an abortive revolution by unemployed workers and those bourgeois utopians who saw a social republic as the antidote to the greed and inequality that had characterized the July Monarchy. The republican bourgeoisie violently repressed the revolutionaries but failed to resolve the crisis. The result was the ascent to power of Louis-Napoleon Bonaparte, who engineered a coup in 1851 and proclaimed himself Emperor the following year. To survive politically, he resorted to widespread repression of alternative political movements. The economic situation he dealt with by means of a vast program of infrastructural investment both at home and abroad. In the latter case, this meant the construction of railroads throughout Europe and into the Orient, as well as support for grand works such as the Suez Canal. At home, it meant consolidating the railway network, building ports and harbors and draining marshes. Above all, it

entailed the reconfiguration of the urban infrastructure of Paris. Bonaparte brought in Georges-Eugène Haussmann to take charge of the city's public works in 1853.

Haussmann clearly understood that his mission was to help solve the surplus-capital and unemployment problem through urbanization. Rebuilding Paris absorbed huge quantities of labor and capital by the standards of the time and, coupled with suppressing the aspirations of the Parisian workforce, was a primary vehicle of social stabilization. The system worked very well for some fifteen years, and it involved not only a transformation of urban infrastructures but also the construction of a new way of life and urban persona. Paris became 'the city of light', the great center of consumption, tourism and pleasure; the cafés, department stores, fashion industry and grand expositions all changed urban living so that it could absorb vast surpluses through consumerism. But then the overextended and speculative financial system and credit structures crashed in 1868. Haussmann was dismissed; Napoleon III in desperation went to war against Bismarck's Germany and lost. In the ensuing vacuum arose the Paris Commune, one of the greatest revolutionary episodes in capitalist urban history, wrought in part out of a nostalgia for the world that Haussmann had destroyed and the desire to take back the city on the part of those dispossessed by his works.

Fast forward now to the United States. The suburbanization of the United States was not merely a matter of new infrastructures. As in Second Empire Paris, it entailed a radical transformation in lifestyles, bringing new products from housing to refrigerators and air conditioners, as well as two cars in the driveway and an enormous increase in the consumption of oil. It also altered the political landscape, as subsidized home-ownership for the middle classes changed the focus of community action towards the defense of property values and individualized identities, turning the suburban vote towards conservative republicanism. Debt-encumbered homeowners, it was argued, were less likely to go on strike. This project successfully absorbed the surplus and assured social stability, albeit at the cost of hollowing out the inner cities and generating urban unrest amongst those, chiefly African-Americans, who were denied access to the new prosperity.

By the end of the 1960s, a different kind of crisis began to unfold; Moses, like Haussmann, fell from grace, and his solutions came to be seen as inappropriate and unacceptable. Traditionalists rallied around Jane Jacobs and sought to counter the brutal modernism of Moses's projects with a localized neighborhood aesthetic. But the suburbs had been built, and the radical change in lifestyle that this betokened had many social consequences, leading feminists, for example, to proclaim the suburb as the locus of all their primary discontents. If Haussmannization had a part in the dynamics of the Paris Commune, the soulless qualities of suburban living also played a critical role in the dramatic events of 1968 in the U.S. Discontented white middle-class students went into a phase of revolt, sought alliances with marginalized groups claiming civil rights and rallied against American imperialism to create a movement to build another kind of world—including a different kind of urban experience.

Along with the '68 revolt came a financial crisis within the credit institutions that, through debt-financing, had powered the property boom in the preceding decades. The crisis gathered momentum at the end of the 1960s until the whole capitalist system crashed, starting with the bursting of the global property-market bubble in 1973, followed by the fiscal bankruptcy of New York City in 1975. As William Tabb argued, the response to the consequences of the latter effectively pioneered the construction of a neoliberal answer to the problems of perpetuating class power and of reviving the capacity to absorb the surpluses that capitalism must produce to survive.

GIRDING THE GLOBE

Fast forward once again to our current conjuncture. International capitalism has been on a roller-coaster of regional crises and crashes—East and Southeast Asia in 1997–98; Russia in 1998; Argentina in 2001—but had until recently avoided a global crash even in the face of a chronic inability to dispose of capital surplus. What was the role of urbanization in stabilizing this situation? In the United States, it is accepted wisdom that the housing sector was an important stabilizer of the economy, particularly after the high-tech crash of the late 1990s, although it was an active component of expansion in the earlier part of that decade. The property market directly absorbed a great deal of surplus capital through the construction of city-center and suburban homes and office spaces,

while the rapid inflation of housing asset prices—backed by a profligate wave of mortgage refinancing at historically low rates of interest—boosted the U.S. domestic market for consumer goods and services. American urban expansion partially steadied the global economy, as the U.S. ran huge trade deficits with the rest of the world, borrowing around $2 billion a day to fuel its insatiable consumerism and the wars in Afghanistan and Iraq.

But the urban process has undergone another transformation of scale. It has, in short, gone global. Property-market booms in Britain and Spain, as well as in many other countries, have helped power a capitalist dynamic in ways that broadly parallel what has happened in the United States. The urbanization of China over the last twenty years has been of a different character, with its heavy focus on infrastructural development, but it is even more important than that of the U.S. Its pace picked up enormously after a brief recession in 1997, to the extent that China has taken in nearly half the world's cement supplies since 2000. More than a hundred cities have passed the one-million population mark in this period, and previously small villages, such as Shenzhen, have become huge metropolises of 6 to 10 million people. Vast infrastructural projects, including dams and highways—again, all debt-financed—are transforming the landscape. The consequences for the global economy and the absorption of surplus capital have been significant: Chile booms thanks to the high price of copper, Australia thrives and even Brazil and Argentina have recovered in part because of the strength of Chinese demand for raw materials.

Is the urbanization of China, then, the primary stabilizer of global capitalism today? The answer has to be a qualified yes. For China is only the epicenter of an urbanization process that has now become genuinely global, partly through the astonishing integration of financial markets that have used their flexibility to debt-finance urban development around the world. The Chinese central bank, for example, has been active in the secondary mortgage market in the U.S. while Goldman Sachs was heavily involved in the surging property market in Mumbai, and Hong Kong capital has invested in Baltimore. In the midst of a flood of impoverished migrants, construction boomed in Johannesburg, Taipei, Moscow, as well as the cities in the core capitalist countries, such as London and Los Angeles. Astonishing if not criminally absurd mega-urbanization projects have emerged in the Middle East in places such as Dubai and Abu Dhabi, mopping up the surplus arising from oil wealth in the most conspicuous, socially unjust and environmentally wasteful ways possible.

PROPERTY AND PACIFICATION

As in all the preceding phases, this most recent radical expansion of the urban process has brought with it incredible transformations of lifestyle. Quality of urban life has become a commodity, as has the city itself, in a world where consumerism, tourism, cultural and knowledge-based industries have become major aspects of the urban political economy. The postmodernist penchant for encouraging the formation of market niches—in both consumer habits and cultural forms—surrounds the contemporary urban experience with an aura of freedom of choice, provided you have the money. Shopping malls, multiplexes and box stores proliferate, as do fast food and artisanal market places. We now have, as urban sociologist Sharon Zukin puts it, "pacification by cappuccino". Even the incoherent, bland and monotonous suburban tract development that continues to dominate in many areas now gets its antidote in a 'new urbanism' movement that touts the sale of community and boutique lifestyles to fulfill urban dreams. This is a world in which the neoliberal ethic of intense possessive individualism, and its cognate of political withdrawal from collective forms of action, becomes the template for human socialization. The defense of property values becomes of such paramount political interest that, as Mike Davis points out, the home-owner associations in the state of California become bastions of political reaction, if not of fragmented neighbourhood fascisms (Davis 1990).

DISPOSSESSIONS

Surplus absorption through urban transformation has an even darker aspect. It has entailed repeated bouts of urban restructuring through 'creative destruction', which nearly always has a class dimension since it is the poor, the underprivileged and those marginalized from political power that suffer first and foremost from this process. Violence is required to build the new urban world on the wreckage of the old. Haussmann tore through the old Parisian slums, using powers of expropriation in the name of civic improvement and renovation.

Consider the case of Seoul in the 1990s: construction companies and developers hired goon squads of sumo-wrestler types to invade neighborhoods on the city's hillsides. They sledgehammered down not only housing but also all the possessions of those who had built their own homes in the 1950s on what had become premium land. High-rise towers, which show no trace of the brutality that permitted their construction, now cover most of those hillsides. In Mumbai, meanwhile, 6 million people officially considered as slum dwellers are settled on land without legal title; all maps of the city leave these places blank. With the attempt to turn Mumbai into a global financial center to rival Shanghai, the property-development boom has gathered pace, and the land that squatters occupy appears increasingly valuable. Dharavi, one of the most prominent slums in Mumbai, is estimated to be worth $2 billion. The pressure to clear it—for environmental and social reasons that mask the land grab—is mounting daily. Financial powers backed by the state push for forcible slum clearance, in some cases violently taking possession of terrain occupied for a whole generation. Capital accumulation through real-estate activity booms, since the land is acquired at almost no cost.

Will the people who are displaced get compensation? The lucky ones get a bit. But while the Indian Constitution specifies that the state has an obligation to protect the lives and well-being of the whole population, irrespective of caste or class, and to guarantee rights to housing and shelter, the Supreme Court has issued judgments that rewrite this constitutional requirement. Since slum dwellers are illegal occupants and many cannot definitively prove their long-term residence, they have no right to compensation. To concede that right, says the Supreme Court, would be tantamount to rewarding pickpockets for their actions. So the squatters either resist and fight, or move with their few belongings to camp out on the sides of highways or wherever they can find a tiny space. Examples of dispossession can also be found in the U.S., though these tend to be less brutal and more legalistic: the government's right of eminent domain has been abused in order to displace established residents in reasonable housing in favor of higher-order land uses, such as condominiums and box stores. When this was challenged in the U.S. Supreme Court, the justices ruled that it was constitutional for local jurisdictions to behave in this way in order to increase their property-tax base.

In China millions are being dispossessed of the spaces they have long occupied—three million in Beijing alone. Since they lack private-property rights, the state can simply remove them by fiat, offering a minor cash payment to help them on their way before turning the land over to developers at a large profit. In some instances, people move willingly, but there are also reports of widespread resistance, the usual response to which is brutal repression by the Communist party. In the PRC it is often populations on the rural margins who are displaced, illustrating the significance of Lefebvre's argument, presciently laid out in the 1960s, that the clear distinction which once existed between the urban and the rural is gradually fading into a set of porous spaces of uneven geographical development, under the hegemonic command of capital and the state. This is also the case in India, where the central and state governments now favor the establishment of Special Economic Zones—ostensibly for industrial development, though most of the land is designated for urbanization. This policy has led to pitched battles against agricultural producers, the grossest of which was the massacre at Nandigram in West Bengal in March 2007, orchestrated by the state's Marxist government. Intent on opening up terrain for the Salim Group, an Indonesian conglomerate, the ruling CPI(m) sent armed police to disperse protesting villagers; at least 14 were shot dead and dozens wounded. Private property rights in this case provided no protection.

What of the seemingly progressive proposal to award private-property rights to squatter populations, providing them with assets that will permit them to leave poverty behind? Such a scheme is now being mooted for Rio's favelas, for example. The problem is that the poor, beset with income insecurity and frequent financial difficulties, can easily be persuaded to trade in that asset for a relatively low cash payment. The rich typically refuse to give up their valued assets at any price, which is why Moses could take a meat axe to the low-income Bronx but not to affluent Park Avenue. The lasting effect of Margaret Thatcher's privatization of social housing in Britain has been to create a rent and price structure throughout metropolitan London that precludes lower-income and even middle-class people from access to accommodation anywhere near the urban center. I wager that within fifteen years, if present trends continue, all those hillsides in Rio now occupied by favelas will be covered by high-rise condominiums with fabulous views over the idyllic bay, while the erstwhile favela dwellers will have been filtered off into some remote periphery.

FORMULATING DEMANDS

At this point in history, this has to be a global struggle, predominantly with finance capital, for that is the scale at which urbanization processes now work. To be sure, the political task of organizing such a confrontation is difficult if not daunting. However, the opportunities are multiple because, as this brief history shows, crises repeatedly erupt around urbanization both locally and globally, and because the metropolis is now the point of massive collision—dare we call it class struggle?—over the accumulation by dispossession visited upon the least well-off and the developmental drive that seeks to colonize space for the affluent.

One step towards unifying these struggles is to adopt the right to the city as both working slogan and political ideal, precisely because it focuses on the question of who commands the necessary connection between urbanization and surplus production and use. The democratization of that right, and the construction of a broad social movement to enforce its will, is imperative if the dispossessed are to take back the control which they have for so long been denied, and if they are to institute new modes of urbanization. Lefebvre was right to insist that the revolution has to be urban, in the broadest sense of that term, or nothing at all.

REFERENCES FROM THE READING

Davis, M. (1990) *City of Quartz: Excavating the Future in Los Angeles*, London and New York: Verso.

Harvey, D. (1973) *Social Justice and the City*, Athens, GA: University of Georgia Press.

Harvey, D. (1989) *The Condition of Postmodernity*, Malden, MA: Blackwell.

Harvey, D. (2005) *A Brief History of Neoliberalism*, Oxford and New York: Oxford University Press.

Harvey, D. (2012) *Rebel Cities: From the Right to the Cities to the Urban Revolution*, London and New York: Verso.

Harvey, D. (2016) *The Ways of the World*, London: Profile Books.

42
"Urban social movements in an era of globalization"

from *Urban Movements in a Globalizing World* (2000)

Margit Mayer

EDITORS' INTRODUCTION

Margit Mayer is Senior Fellow at the Center for Metropolitan Studies in Berlin and Professor Emerita of Political Science in the John F. Kennedy Institute of the Free University Berlin, Germany. She has also taught at many American and Canadian universities. Her work in the United States and Germany has provided her with a unique perspective from which to explore the globalization of urban social movements on both sides of the Atlantic. Mayer is a leading social movement researcher who works predominantly in the New Social Movements tradition, which both acknowledges the wider socioeconomic contexts of movements and focuses on the identity changes that occur in and through social movement struggles. Mayer's recent work has concentrated on the interactions of urban social movements with local states, particularly in the spheres of homelessness, nonprofit organizations, and employment policy. In the contribution below, she examines the intersection of urban social movement activity with processes of worldwide interurban competition. Mayer discusses the erosion of traditional welfare rights and the restructuring of urban governance under these circumstances. Mayer's analysis underscores the contradictions and conflicts associated with urban social movements, as well as the opportunities for social transformation they appear to open. In so doing, she indicates the key role of urban social movements in the process of global city formation.

This chapter discusses first the changes in local politics which recent urban research has identified as reactions to globalization practices, and then relates today's prevailing movements to these shifts in urban politics in order to, finally, discuss strategic implications and options for local actors. The focus is mostly on major cities, which, regardless of historical tradition, geographical location, or general level of economic development, are increasingly tied into global flows and networks in very similar ways.

NEW TRENDS IN LOCAL POLITICS AND OPPOSITIONAL MOVEMENTS

Reviewing the literature on recent developments in urban governance, there is consensus that three trends in particular are significant and novel on the level of local politics:

- the new competitive forms of urban development,
- the erosion of traditional welfare rights,

- and the expansion of the urban political system, also described as a shift from "government" to "governance."

Each of these trends has provoked or influenced new or existing urban movements, which I will present now in the context of their respective urban political setting.

The contemporary forms of urban growth and development consist primarily of the efforts of cities to upgrade their locality in the international competition for investors, advanced services, and mega-projects. Local political actors everywhere emphasize economic innovation, seek entrepreneurial culture, and implement labor market flexibility in order to counter the crisis of Fordism and to meet the intensified international competition. Other policy areas are increasingly subordinated to these economic priorities. The higher the position of a city within the global competitive structure of the new economy, the more important the role of advanced services in the central business district, and the more intense the restructuring of urban space. For global cities in particular, which compete as much for foreign investment and the economic megastructures of internationally oriented growth as they do for world class culture, everything from the production of the built environment to the priorities of the municipal budget has become subject to the function of the city as command center and its corresponding service industries. But other major cities have also seen the rebuilding and expansion of their downtowns into producer-oriented service centers. The intense tertiary development in the central business districts and the construction of huge new infrastructure projects have had undesired consequences for large parts of the resident population, because their effects have been gentrification and displacement, congestion and pollution, and often the loss of traditional amenities. While the city centers are being turned into luxury citadels, other neighborhoods are turned into preferential sites for unattractive functions, yet others are given up to abandonment. Further, the rebuilding of the central city has entailed an urban expansion into peripheral "green pastures."

Opposition movements have formed both in the cities and at their peripheries. They have either built on existent (latent) networks or organizations, or have sprung up anew, and they range from defensive and pragmatic efforts to save existing quality of life or privileges (which are sometimes progressive, environmentally conscious, and inclusive, but other times selfish, anti-immigrant, or racist) to highly politicized and militant struggles over whose city it is supposed to be (as in anti-gentrification struggles or movements against other growth policies).

Social movement research has produced most work on those mobilizations that are to protect the home environment – of too much traffic, too much development, or any other project which people don't like to have "in their own backyard." These often middle class-based quality-of-life movements frequently succeed in averting an unwanted facility (NIMBY), with the effect that then a poor or minority neighborhood is targeted.

Case studies show that such groups quickly become skilled at a variety of tactics and repertoires such as petition drives, political lobbying, street confrontations, and legal proceedings. Researchers tend to lament the fact that social justice orientations, which used to characterize the goals and practice of such citizens initiatives during the 1970s, have been replaced by particularist interests and/or a defense of privileged conditions. But there are also case studies of local movements composed of working class and middle class participants mobilizing against highway construction plans, traffic congestion, or housing shortages, and, particularly in U.S. minority/working class communities, against polluting industries and hazardous facilities, with which they are disproportionately burdened. The action repertoire of such groups goes well beyond that of the defensive NIMBY movements: beside direct action (demonstrations, blockades, corporate campaigns) to put public pressure on polluting firms, they also undertake independent analysis of urban problems, and they demand representation on relevant decision-making boards.

Frequently, movements against urban growth policies and gentrification are directly triggered by what have become increasingly used instruments of big city politics. These include large, spectacular urban development projects, such as London's Docklands or Berlin's Potsdamer Platz; festivals such as the Olympics, World Expo, international garden shows, or 1,000 year birthdays; or the attraction of mega-events, sports entertainment complexes, theme-enhanced urban entertainment centers – all of which depend on the packaging and sale of urban place images. The movements have attacked the detrimental side effects of and the lack of democratic participation inherent in these strategies of restructuring the city and of raising funds, and they criticize the spatial and temporal concentration of such development projects, as they prevent salutary effects for the city overall. The concentration on prestige projects tends to detract attention and finances from other urban problems and to restrict investments in other areas. Thus, protest campaigns against these forms and instruments of city marketing raise questions of democratic planning that urban elites concerned with intra-regional and international competitiveness like to downplay. Furthermore, they have the potential of bringing otherwise scattered local movement groups together in broad coalitions (as happened, for example, in the NOlympia Campaign in Berlin in 1991–3). A leading actor in such campaigns are often radical, so-called autonomous movements, who consciously seize on the importance image politics have gained in the global competition of cities, and seek to devise image damaging actions to make their city less attractive to big investors and speculators, to creatively prevent the takeover of the city by "global capital."

The trend of the eroding local welfare state has been another trigger of the structural change in the profile of urban social movements. It has two elements: First, there is the dualization of labor markets, the expansion of precarious and informal jobs, and the shift in social policies, which produced a new marginality, the most visible manifestation of which are the tens of thousands of homeless inhabiting major cities. Other, less visible forms of social exclusion and new poverty also concentrate in urban areas, even if their causes are increasingly identified in global processes. Secondly, since the image of cities is playing such an important role in attracting supra-local investment, stern anti-homeless and anti-squatter policies have been drafted, and regular raids are carried out at the showcase plazas of all major cities. This kind of regulation of public space has been observed since the early 1990s, even in cities with "progressive" governments, which have also adopted laws that prohibit people to sit or lie on sidewalks in business districts. In order to drive out beggars, homeless people, or "squeegee merchants" from the center of the cities (where they concentrate for a variety of reasons), these groups are being constructed as "dangerous classes" or "enemies of the state." Social policies have been abandoned in favor of punitive and repressive treatments.

In reaction to the combination of these trends, new poor people's movements have sprung up and actions by supporter groups and advocacy organizations, frequently also anti-racist initiatives. Research findings

on the forms of self-organization by the new poor are most scarce, which is – among other things – due to the fact that most authors assume this population to be not just poor and without resources but also disempowered and passive. In fact, the resources of these groups consist primarily of their bodies and time, so that their protest activities tend to be episodic and spontaneous, local in nature, and disruptive in strategy. At best, their disruptive tactics block normal city government operations and threaten the legitimacy of local policies of exclusion. This was the case, for example, when the homeless in Paris defended their right to the city in a campaign around the slogan "droit au logement" ("The Right to Housing") during the mid-1990s. In general, though, such new poor people's movements face an increasingly recalcitrant and punitive state and it is only under rare conditions that their struggle against efforts to drive them out of the downtowns, their setting up of encampments, holding public forums, and making demands on the city, allows them to develop solidarity, political consciousness, and organizational infrastructures – i.e. the elements of which social movement research assumes that they are preconditions for the emergence of mobilization. However, when resource-rich political advocacy groups dedicate themselves to the problems of the homeless or when professional activists make their resources available to such organizations, durable and effective mobilizations can be achieved.

Next to churches, political activist groups, and local coalitions, in Germany another relevant support network is provided by autonomous movements and anti-racist initiatives. These latter groups scandalize the production of new poverty and homelessness while also mobilizing against their own eviction from squatted buildings and "liberated areas" of the city centers. In Berlin, where the city government has marked 14 so-called "danger zones," from where individuals looking "suspect" may be deported, newly formed initiatives stage protest demonstrations, provide legal aid, and put public pressure on the local government. In June 1997, a "Downtown Action Week" took place in 19 German and Swiss cities to create public awareness and pressure about this widespread practice of driving out the new marginality from the core areas.

Another new movement form has emerged in the context of housing need and new poverty, though its members do not see themselves as a "poor people's movement." The majority of the so-called "Wagenburgen," i.e. groups of people squatting on vacant land, living in trailers, circus wagons, or other mobile structures, see their action as a form of resistance against the dominant pattern of political, social, and economic relations in German cities. There are about 70 to 80 such sites in Germany; of the 15 in Berlin most have been in the downtown area and thus threatened by eviction or already displaced to other locations. Their political orientations cover a wide spectrum: while some use the freedom this lifestyle allows them for political activism (such as sheltering refugees without legal status), others are content to explore alternative ways of living. But evictions and the threat of evictions have brought them together in campaigns to pressure city governments to tolerate the sites, delay construction, or provide other acceptable locations.

The new conditions on the labor market and the shift from social welfare to more punitive workfare policies have impacted on the urban movement scene in further ways. Not only have hundreds of new organizations sprung up, nonprofits run by and for the homeless, the unemployed, and the poor, but also the number and variety of institutions and projects "servicing" the marginalized has exploded; many of them function within municipal programs that harness the reform energy of community-based groups. Their labor seeks not just to "mend" the disintegration processes which traditional state activities cannot address; frequently they develop innovative strategies acknowledging the new divisions within the city.

Finally, the third novel trend in urban politics is that the local level of politics has gained renewed significance (and in the process has transformed itself), simply because the concrete supply-side conditions making for structural competitiveness can be provided neither by multinationals' strategies nor by uniform national policy (Mayer 1992). These conditions can only be identified and implemented at the local level of politics. Local politics, which seek to make local economies competitive in the world economy, are increasingly organized in partnership with an extended range of non-governmental stakeholders holding relevant resources of their own (such as private finances, local knowledge, community-based or locality-specific expertise). Besides investors and chambers of commerce, also education bodies, research centers, and local unions have become such partners, as well as voluntary sector groups and associations, including former social movement organizations (Mayer 1994). This trend, which political science has described as a

shift from government to governance arrangements, means that the state's involvement, besides that of a plurality of other actors, is becoming less hierarchical and more moderating than directing (Jessop 1995).

The opening up of the urban political system to non-governmental stakeholders and the strategy of many municipalities to employ former social movement organizations in the development and implementation of (alternative) social services, cultural projects, housing, and economic development has been a new and important force shaping the trajectory of urban movements since the 1980s. By including and funding third sector organizations the municipalities hope to achieve political vitalization as well as financial relief, though these goals frequently conflict with each other.

For the movements, this trend posits a particularly ambivalent opportunity structure, because it makes parts – but only parts – of the urban movement sector into "insiders." This development is rather more advanced in North America and Great Britain than on the European continent. Community-based and client-oriented groups now play a polyvalent function, in more and less developed forms, in and for cities all over North America and Western Europe. The establishment of alternative renewal agents and sweat-equity programs and the funding of self-help and social service groups was in most places a long and contested process, but since the late 1980s municipal social and employment programs everywhere have been making use of the skills, knowledge, and labor of such movement groups. Similarly, many cultural projects have become part of the "official" city, and youth and social centers play acknowledged roles in integrating "problem groups" and potential conflict. The participation of community and movement groups in different policy fields has, in other words, become routinized, especially in fields where both groups of the alternative sector and the political administration are keenly interested in solutions, such as urban renewal, drugs, immigrant integration, AIDS, or unemployment.

The bulk of the research focusing on these novel forms of institutionalization of social movements within the shifting relations of welfare systems and provision emphasizes the "contestatory character of their constituency" and the counterweight they pose to "conventional views of local economic planning" (Lustiger-Thaler and Maheu 1995: 162, 165). Whether in the economic development sector, the field of alternative services, or that of women's projects, the work of the groups is generally found to be an innovative and progressive challenge to public policy, as improving access to the local political system, and providing potentially more active citizenship.

Closer examination, particularly of the more recent developments, reveals, however, that these (former) movement organizations that have inserted themselves into the various municipal or foundation-sponsored funding programs play a rather complicated role within the urban movement scene. On the one hand, they enhance organization building and lend stability to the urban movement infrastructure and thus to the conditions for continuing mobilization. But on the other hand, the widening and the growing internal differentiation within the movement sector has led to new conflicts and antagonisms. The movement organizations now participating in the new governance arrangements are subject to the danger of institutional integration, "NGOization," and of pursuing "insider interests," and their own democratic substance is far from guaranteed. Especially since these organizations find themselves threatened by cuts and are faced with the reorientation of public sector programs toward labor market flexibilization, competition among them for funding has intensified, and the groups engage more in private lobbying strategies to secure jobs and finances instead of creating public pressure. Furthermore, some of the alternative renewal agents and community-based development organizations, who are busy developing low-income housing or training and employment opportunities for underprivileged groups, find themselves criticized and attacked by other movement actors who do not qualify for the waiting lists or who prefer squatting or other non-conventional forms of action.

While such attacks serve to illustrate new polarization tendencies and antagonisms within the movement sector, a series of indicators points toward an interpretation that the inclusion of movement groups in revitalization and other partnerships means, for many, that they become tied up with managing the housing and employment problems of groups whose exclusion by normal market mechanisms might otherwise begin to threaten the social cohesion of the city. This kind of instrumentalization of (former) movement groups thus harbors the real danger that their reform energy evaporates in the processing of urban disintegration tendencies or might even be used for the smooth implementation of state austerity policies.

But at the same time it is also the case that the increasing dependence of city governments on such (former) social movement organizations for processing the complex antagonisms within contemporary cities does also enhance the chances for tangible movement input. While this dependence is meanwhile institutionalized with the routinized cooperation between the local state and the former social movement organizations with regard to community economic development, client-based social services, and women's centers, these new partnership relations are also beginning to influence interaction between the local state and movements described under the first two categories, i.e. no-growth and anti-poverty movements. The eroding local competence and dwindling resources which many city governments are suffering from increase the pressure on the local political elites to negotiate and bargain with movement representatives within the channels and intermediary frameworks generated by the wave of routinization of alternative movement labor in the context of municipal (employment or revitalization) programs. Thus, today's movements making a stand on the use value of the city, such as environmental and poor people's movements, now may also expect to profit from the new culture and institutions of non-hierarchical bargaining systems, forums, and round tables. It is true though that these new structures of governance are open to the less progressive, xenophobic, and antisocial movements as well.

THE ROLES AND OPTIONS FOR URBAN MOVEMENTS IN GLOBALIZING CITIES

The specific socio-spatial context which cities provide for social movements, as well as its consequences for the dynamic and development potentials of movements, needs to be further differentiated for a contemporary assessment. Due to the position of cities within globally and regionally restructuring hierarchies a variety of city types, with different conflict patterns, is emerging, which means that the homogeneous pattern of conflicts and movements of the Fordist era is dissolving. Metropolitan regions at the top of the global hierarchy develop particularly pronounced conflict patterns along the internationalization of their working classes and neighborhoods, their precarious labor relations (made use of especially by migrants), and their eroding municipal powers. At the same time,

FIVE

large metropoles facilitate the emergence of a critical mass, which is precondition for the building up of movement milieus and the construction of collective projects and identities. This is where movements against central city development in favor of global headquarters as well as new poor people's movements proliferate and may expect – because of the presence of handed down movement cultures and institutions (such as community organizing in North America and leftist political organizations in Western Europe) – both support and instrumentalization.

Old, deindustrializing cities on the other hand feature struggles over plant closures and new employment possibilities, and, depending on the profile with which the city seeks to reposition itself in the new urban hierarchy, more or less intense cooperation between the municipality and community groups. Cities trying to make their fortune as "innovation centers" frequently provoke environmental protest and slow growth movements with this strategy. Cities transforming themselves into module production places are particularly dependent on cheap and flexible labor and thus provide a difficult terrain for movements struggling for social citizenship rights and sustainable development. Such different movement activities would need to be analyzed in order to systematically explain the heterogeneous picture of urban movements in the 1990s.

Even though such differences among urban movement milieus are far from adequately researched, the stocktaking of some of the changes movements have undergone as presented in the preceding section allows some preliminary statements about their current role and possibilities for action. The argument here is that just like the movements of the phases of the 1970s and 1980s have contributed to shaping and changing the forms of governance as well as the structure of the city, the movements active in and around the city today play a role, if a contradictory one, in contributing to and challenging the shape and regulation of the city. While their practice with innovative urban repair and their inclusion in municipal governance structures may well feed into the search for (locally adequate) post-Fordist solutions and arrangements (making the movements appear "functional" and co-optable), their challenge of undemocratic and unecological urban development schemes may yet contribute to a more participatory and more sustainable first world model of city. In order to realize this potential, however, the new problems confronting

urban movements of today have to be addressed head-on.

One of these new problems is the new antagonisms within the movement sector, which are also a product of the restructuring of the urban polity that has expanded and now includes some but not others in its governance arrangements. Besides this tendency to produce new forms of marginalization and new "losers," a second new problem demands attention, that is the evidence that the inclusion of movement groups in revitalization and other partnerships has meant, for many, that they become tied up with managing the housing, employment, or survival problems of groups whose exclusion by normal market mechanisms might otherwise begin to threaten the social cohesion of the city. And finally, a third problem arises with the pressures to entrepreneurialize the social and community work of these groups, as funding support for them is increasingly only available through workfare programs and microcredit arrangements. These structurally new constellations have to be acknowledged and their specific constraints – as well as the opportunities peculiar to them – have to be identified.

At the same time, the conquered positions and new institutional avenues described above offer opportunities that allow tackling the new problems. The growing role of local politics, even within global contexts, and the simultaneous inclusion of a variety of non-governmental stakeholders, including former movement organizations, into local politics have made new avenues available for those forces amongst the urban social movements that can seize them and that can tease out their ambivalence. But rather than doing so only for particular defensive spaces or individual threatened privileges, the challenge consists of making use of these avenues in the complex struggle for a democratic, sustainable, and social city.

Some urban theorists see this struggle as one between global elites and local communities, reduced to the simple antagonism between distant powerful forces (such as global capital) and local victims "retrenched in their spaces that they try to control as their last stand against the macro-forces that shape their lives out of their reach" (Castells 1994). Such an idealized view of local movements would already have been problematic for the 1960s and 1970s, when the majority of urban movements were still part of a larger social struggle against broadening forms of domination. Today's local movements certainly cannot in their entirety be listed on the positive side of the ledger, since they are such highly differentiated products of recent shifts in urban politics (their proliferation as well as their fragmentation can be shown to result from the three trends described above). They themselves are contradictory and complex agents in the shaping of globalizing cities and have to deal with the new fragmentation within the movement sector as well as with massive social disintegration processes increasingly characteristic of urban life. The institutionalized, professionalized, or entrepreneurial movements which now benefit from routinized cooperation with the local state frequently want nothing to do with younger groups of squatters and cultural activists. Because of their preoccupations due to the new funding structures, they are often at quite a distance from the growing marginalized and disadvantaged social groups. But since the latter's organizations and forms of resistance do not automatically lead to mobilization or widespread support, it becomes crucial that those parts of the movement sector that enjoy some stability, access, resources, and networks devote part of their struggle to creating a political and social climate where marginalized groups can become visible and express themselves.

Movement actors will thus need to acknowledge and make transparent their new dependencies (both on state and market) in order to identify the opportunities that exist under contemporary conditions. The new and difficult task consists in transforming the funds and the stability of the resource-rich movements into support for precarious movement groups.

Existing opportunities, whether workfare programs or poverty initiatives, need to be seized and used to attack and to restrict marginalization and discrimination at the root of the new form of poor people's movements. Urban movements need to politicize the new inequality and they can exploit the new access structures and the dependency of the new negotiation frameworks on local residents' input for this purpose.

REFERENCES FROM THE READING

Castells, M. (1994) European cities, the informational society, and the global economy, *New Left Review*, 204: 18–32.

Jessop, B. (1995) The regulation approach, governance and post-Fordism: Alternative perspectives on economic and political change? *Economy and Society*, 24, 3: 307–33.

Lustiger-Thaler, H. and Maheu, L. (1995) Social movements and the challenge of urban politics, in L. Maheu (ed.) *Social Movements and Social Classes*, London: Sage: 151–68.

Mayer, M. (1992) The shifting local political system in European cities, in M. Dunford and G. Kafkalas (eds.) *Cities and Regions in the New Europe*, London: Belhaven Press: 255–76.

Mayer, M. (1994) Post-Fordist city politics, in A. Amin (ed.) *Post-Fordism: A Reader*, Oxford: Basil Blackwell: 316–37.

F
I
V
E

43

"São Paulo: the city and its protest"

from *Open Democracy* (2013)

Teresa Caldeira

EDITORS' INTRODUCTION

Teresa Caldeira, Professor of City and Regional Planning at the University of California, Berkeley, focuses on spatial segregation and social discrimination in cities in the global South, especially Brazil. Her award-winning book *City of Walls: Crime, Segregation and Citizenship in São Paulo* (2000) offers a striking portrayal of a divided city—between the wealthy inner core flush with amenities, and the poor periphery that for generations has housed the poor. More than a decade later, this description still holds true, and sharp social-spatial segregation still very much characterizes São Paulo today. In 2013, major Brazilian cities erupted in protests, which Caldeira in this essay considers within the framework of "rights to the city." The initial trigger of the protests was an increase in bus fares—the majority of the working poor in São Paulo rely on public transportation. The protests reflected the citizens' unmet demands for their right to mobility in a city notorious for traffic congestion, which stems in part from the local government's emphasis on promoting private automobile ownership rather than investing in public transit. But, as Caldeira explains, the significance of the protests runs deeper. Most of the protesters are young people from the peripheries who now use social media, graffiti, and tagging to gain visibility. These recent protests represent, for Caldeira, new expressions of democratization, though she is quick to warn that the new democratic articulation must go beyond posters, hashtags, and graffiti on the walls.

In June 2013, a series of large demonstrations throughout Brazil shook its main cities and political landscape. They also perplexed politicians and analysts alike, many of whom found themselves without solid references to interpret the novelty and oscillated between silence and old discourses. It is always risky to interpret emerging processes. Minimally, we risk following secondary paths or, even worst, framing new events with the vocabulary made available by old interpretative models, exactly the ones that the new events are trying to displace. However, in order to reveal what is emerging it is necessary to risk, search for new hints, and follow signs already available. Several references that can guide us to interpret the June events have been around for quite a while; others are new, but we can trace their lineage and contextualize them.

GLOBAL FOOTPRINTS

The demonstrations in Brazil share some characteristics with others around the globe in recent times. Analyses of movements such as those of the Arabic Spring, Occupy, the *Indignados* of Spain, and, more recently, the protests in Istanbul, have already revealed features that point to the existence of a new type of public mobilization. They include: a symbiotic relationship with the internet and the social media; the diffused and spontaneous organization around networks; the capacity of attracting thousands of participants in a short period of time; the heterogeneity of the participants, who may or may not form coalitions; the hand-made quality of posters and banners; a high participation of young people. The demonstrations also disregard established political institutions such as political parties and unions and clearly indicate a shift in the way in which political languages are produced, circulate, and guide practices. They reveal that previous monopolies in the production of information and forms of organization have broken down.

The role of the internet is evident. It gives to all and everyone – and not only to those who control institutionalized means of expression and organization – the autonomy to formulate and distribute messages and interpretations, to select and distribute information, and to form networks. It also accustoms its users to political criticism and engagement, making the signing of petitions, for example, a frequent tool of protest. But the internet's power also crystallizes on the streets, where those who Brazilians call *internautas* (internauts) congregated carrying posters with sayings and ideas that have been in the making and circulating in social media and blogs for quite a while. Moreover, from the streets participants feed back the social media, describing, documenting, interpreting, and thus amplifying the events and re-articulating networks. This exchange between the internet and the streets was palpable in São Paulo and other Brazilian cities, where events such as police repression were posted and reproduced immediately in Twitter and Facebook, convincing many who had stayed at home to join the crowds.

THE CITY AND ITS SEGREGATION

Issues related to the city and its quality of life also framed the protests. There is a global dimension to this. Movements demanding rights to the city are nowadays articulated in the most diverse cities around the world and their experiences are exchanged frequently through networks of activists. It is also probably meaningful that São Paulo's demonstrations started just after those in Istanbul, which clearly articulated the issue of rights to the city. "Turkey is here" was a huge banner in the streets of São Paulo.

But it is obvious that there is also a local history to this. The demonstrations started in São Paulo triggered

by an increase in R$0.20 (US$ 0.10) in the fare of public transportation. It did not take long for protesters to insist that the demonstrations were not only about the 20 cents and for a long list of demands to appear on the streets. Moreover, the demonstrations persisted after the increase was cancelled. Nevertheless, it is undeniable that the city itself, its pattern of spatial segregation, and especially its everyday dynamics of stalled traffic and eternal difficulties of moving around are at the core of the protests.

São Paulo is a dispersed city in which long distances separate the center from the peripheries, workers from their jobs. As in many other metropolises in the global south, migrants who arrived by the hundreds of thousands a year to work on what was becoming an important industrial center could not find housing. They have thus relied on auto-construction. Starting in the 1940s, they bought cheap pieces of land in non-urbanized areas in distant peripheries and then started a long-term process of building and expanding their residences, transforming them year after year until it would become a nice house of their own. This process typically takes one to two decades to complete and simultaneously urbanizes the city.

As a consequence of this process, however, by the 1970s São Paulo was clearly segregated according to a center-periphery pattern. The better-off lived in well urbanized and equipped central areas of high population density, while the working poor inhabited the precarious, distant, and low density peripheries that they had built on their own. The provision of infrastructure under this pattern was difficult and very expensive. But by the 1980s, it started to arrive. Crucial for this to happen was the action of urban social movements, which spread throughout the peripheries claiming rights to the city they had built and in which they paid taxes. These movements helped to democratize Brazilian society and to significantly transform the quality of the urban space and of the public services in the peripheries.

In sum, São Paulo was urbanized by its citizens through a process that generated deep inequalities, long distances, high expectations of consumption, and a process of democratization that formed citizens and accustomed them to demand the improvement of their spaces and their participation in the polity. If in the past water, electricity, and sewage were the catalysts of social movements, now circulation becomes a core issue. Mobility through the immense metropolitan region has not improved, but rather worsened continuously over the last decades. The frustrations and indignities of the everyday life in a congested city in which people have to travel long distances found expression in the June demonstrations. However, they have been expressed in the internet, in new forms of political movements, and in the cultural production from the peripheries for quite a while.

THE RIGHT TO MOVE AROUND

The experience of moving around in traffic is painful for all, and the indignities of using absolutely packed public transportation – buses, vans, and subway – are a constant complaint of the millions who commute every day. The internet has been functioning for a long time as the space to express and spread the feelings of irritation. Anyone who follows Facebook and Twitter on a daily basis knows that people sitting in the immense traffic jams use their cell phones to post messages such as: "in the damn bus: stopped for 15 minutes!;" "Will be late for work AGAIN;" "Oh no! Now it stopped to get a handicapped person: will be

even more delayed." And so it goes, a breeding site for frustrations and a space for the expression of prejudices and intolerance, sometimes in cruel and vulgar terms.

Traffic is also associated with an intensified consumerism and with the fact that the government has chosen to emphasize it as a mode of promoting the social mobility of the poor. The individual automobile is central to this policy. Instead of developing public policies that could ameliorate the quality of life in cities by improving urban infrastructure and investing in public transportation, the Brazilian government, especially under the PT, has opted to subsidize the acquisition of individual automobiles. As a result, São Paulo, a city of around 11 million inhabitants, had in 2011 more than 7 million registered vehicles, 5 million of them automobiles. The inevitable outcome is a permanently congested city.

Traffic and mobility are at the core of the new type of social movement that made the call to the streets in June: the *Movimento Passe Livre* (MPL or Free Fare Movement). It describes itself as "an autonomous, a-political (a-party), horizontal, and independent movement, which fights for true public transportation,

Plate 29 São Paulo

Source: Xuefei Ren

free for all, and away from the private initiative." It is not a new movement, since it was founded in 2005, but it had not been taken seriously by the instituted political organizations. It is a movement that knew how to articulate the everyday frustrations of the city and the aspirations of its young people, and that dared to look for democratic and radical alternatives for the problem of urban mobility – such as the notion of zero fare, a proposal it has supported with studies of alternative modes of taxation able to finance a free public transportation of quality for all.

For years, the motto of the movement has been: *"a city only exists for those who can move around it."* This is a demand for a fundamental right to the city. In this regard, it is meaningful to notice that São Paulo's demonstrations have been mobile and occupied simultaneously several spaces of the city, instead of being identified with a fixed territory, such as Taksim Square or Tahrir Square. The demonstrations exercised the right to circulate throughout several parts of the city, including some where usually people don't walk, such as expressways. They have congregated both in the center and the peripheries and, symbolically, largely bypassed Praça da Sé, the central square that has been the iconic space of protests and political gatherings during the democratization process in the 1980s.

But to affirm that a city only exists for those who can move around it is more than a demand for a fundamental right to the city. This affirmation also expresses the practice and the values of an immense number of young people and of an intense cultural and artistic production that proliferates especially in the peripheries. In São Paulo, the majority of the participants in this production were born in the peripheries and are the children or grandchildren of the migrants who first moved to the middle of nowhere to build their houses. However, there is a striking generational difference in their ways to look at the city. When the young cultural producers were born, the social movements were already fading away and the spaces of their neighborhoods had improved considerably. They grew up in a world of democracy, NGOs, relative access to education, and increasing availability of technologies of communication. In spite of their poverty, the young artists from the peripheries are plugged into globalized circuits of youth culture, whose styles they reinterpret and adopt, and into an equally globalized and quite expanded consumption market.

By contrast, the generation that built the peripheries had one central project of consumption and social mobility: their houses and their appliances. For this

they made incredible sacrifices and accepted a significant immobility in the city. Today the new generation is either less interested in owning a house of their own or believes it to be untenable. Instead, they value the possession of a large range of items, from clothes to the cell phones, all electronic and communication-related items, and motorcycles and cars. They want to move around the whole city and to do so in a fashionable way. Their artistic production and their practices of circulation express clearly this desire, either through graffiti, pixação (São Paulo's style of tagging), rap, breakdancing, skating, parkour, or literary events.

But this production also expresses their indignation toward the difficulties they face to use the city and its resources. Some of their production aggressively tries to take over the city. *Pixação*, São Paulo's style of tagging, is the most evident and transgressive of them. For over a decade, *pixadores* have inscribed all types of spaces and articulated the desire for mobility and visibility of young men from the peripheries. A city in which pixação is omnipresent should have already understood both these desires, the aggressiveness of the interventions that attempt to materialize them, and the mounting indignation associated with the continuous difficulties to move around and become part of the whole city.

But this new cultural production expresses several other indignations related to everyday life in the city, most importantly the anger toward a police force which has never refrained from using violence in the peripheries. It is not surprising, thus, that social media exploded and the streets became full of people as soon as the demonstrations in mid-June started to be treated with an unusual level of police violence (tear gas and rubber bullets shot at the crowd, plus beating). It should also not be a surprise that posters incessantly reproduced on Facebook remarked: "The PM (Military Police) is doing on Paulista (a central avenue) what they have never stopped doing in the peripheries."

The class tension that marks the city and that is expressed by its spatial segregation became palpable in social media as the protest unfolded. An image that went viral juxtaposed two photographs: on one side, a middle-class young man held a poster with the words "The people woke up"; on the other side, a bus burning in some area of the periphery with the saying "I'll tell you a secret: the periphery has never slept." Posters indicating the peripheries' sharp awareness abounded. Thus, those who had been articulating new imaginaries and a deep indignation in alternative spaces for quite a while finally arrived to the streets and made sure to attribute any feelings of surprise to others. They knew. They have long understood and experienced injustice; it was the others who were finally waking up and experiencing surprise. These included the political parties that have not listened to them, the governments that have disrespected them continuously, and the middle classes that arrived late to the streets to join in the indignation.

Thus, in the same way as the spontaneity of the demonstrations and the new format of the MPL indicate a break from monopolies, authorities, and modes of political organization, the new cultural production of the peripheries and its circulation via internet or via the walls and streets of the city have been breaking monopolies in the production of representations and interpretations and displacing authorships and authorities. This disperse cultural production does not conceive itself in terms of political movements, but has certainly been producing new imaginaries that circulate in autonomous and non-regulated ways. Imaginaries that crystallized on the streets and that express some of the great inequalities and social tensions that constitute the metropolis.

BEYOND THE CITY: DEMOCRACY

But it is obvious that this was not the only thing that crystallized on the streets and that participants in the protests were not only young people from the peripheries. After the demonstrations exploded, all manner of simmering irritations and anger from across all social groups found their expression on the streets, most notably: the exasperation with politicians and their corruption; frustration with government at all levels and branches, whose buildings were attacked in several cities; the sense of absurdity of the expense of mega-events such as the World Cup contrasted with the disregard of basic social rights such as education and health (we want "FIFA standard" schools and clinics, said the posters); annoyance with political parties (the PT flags were burned on the streets); perplexity with the attempt of some in Congress to undermine LGBT rights; and the revolt at continuous police violence. This proliferation of protests on the streets, preceded by exchanges on the internet and years of cultural production and interventions into the space of the city, indicates the incapacity of established organizations and institutions to maintain a hegemony in the production of interpretations and practices. This can be very positive and liberating, opening new paths, breaking old monopolies, and revealing new articulations. But it also indicates risks and the need for a new democratic articulation that may go beyond posters, hash-tags, and inscriptions on the walls; an articulation able to contain authoritarian and violent impulses and create alternative political spaces without losing novelty.

44

"Global city building in China and its discontents"

from *Building Globalization: Transnational Architecture Production in Urban China* (2011)

Xuefei Ren

EDITORS' INTRODUCTION

Xuefei Ren, Associate Professor of Sociology and Global Urban Studies at Michigan State University, has written extensively on the political economy of urban transformation in contemporary China, including the politics of urban planning, historical preservation, housing rights activism, cultural industries, and environmental governance. In addition to her role as one of the editors of this volume, she is the author of *Building Globalization* (2011) and *Urban China* (2013), and has been working on a number of comparative projects, including a new book on urban governance and citizen rights in China and India; a set of articles comparing the politics of informal settlements in Mumbai, Guangzhou, and Rio de Janeiro; and a photo-documentary on Detroit and rust-belt China.

This chapter is excerpted from *Building Globalization* (2011), which analyzed urban social change in contemporary China in the context of architectural globalization. Departing from past studies that emphasized China's institutional reforms in land and housing sectors, Ren spotlights the Chinese state's role in harnessing transnational forms of architectural production to build global cities. The book demonstrates how the spatial articulation of built environments has become a major force of capital accumulation in China's globalizing cities, and it also illuminates the deepening inequality and contradictions that have subsequently emerged. The selection included here focuses on urban renewal in Shanghai in the mid-2000s, when tens of millions of residents were displaced from inner-city neighborhoods to make space for property speculation. Compared to a decade ago, large-scale demolitions have become less common in Chinese cities today, and the center of contention over rights to housing and land has shifted from property owners to migrants in urban villages (Al 2014) and landless farmers (Ren 2017). The protests against demolitions in Shanghai documented in this chapter marked the emergent phase of housing rights movements in China that will continue into the next decade.

In the early 1980s appeared the first research concerning the emergence of a new type of city—the global city. The original theoretical articulation by John Friedmann and Saskia Sassen (Ch. 3 and Ch. 4) attributes the formation of global cities to structural changes in the world economy. In Sassen's formulation, for example, one key dynamic is the decentralization of manufacturing jobs and service outlets, which has brought about a reconcentration and expansion of economic command-and-control functions. Global cities are thus strategic nodes in the world economy because of their concentration of specialized producer services firms. Urban researchers have conducted individual case studies, historical and comparative analyses, and relational network analyses to examine the changing power relationship in the global urban hierarchy and the different pathways of global city formation. It is widely agreed that global city formation is a long and historically accumulated process abundant with uncertainties and contingencies, and that economic polarization and spatial segregation are universally observed negative consequences in global cities.

Policy makers and academics in China got hold of the concept of global cities in the mid-1990s, when government officials started to speak of building China's "international metropolises." In the early discourse on global cities, government officials and academics argued that China needed to develop its own global cities in order to be "integrated into the world." In spite of the policy of "strictly controlling the growth of large cities, while developing medium and small-sized cities," proposed by the central government in 1990, the latter half of the 1990s saw a surge of global city discourse and policy initiatives. Local governments competed with one another and declared their intention to remake themselves into global cities. By the end of the 1990s, more than 43 Chinese cities had announced plans to become global cities; this group included not only booming commercial centers such as Beijing and Shanghai but also small trading posts such as Manzhouli and Heihe on the northern border with Russia. Government officials and academics argued that only by developing global cities could China effectively tap into international investment, attract multinational firms, and realize its economic takeoff. They also identified the deficiencies of Chinese commercial capitals as compared with other established global cities, such as low levels of productivity and economic output, underdeveloped tertiary industry, the high percentage of manufacturing in the urban economy, and, most important, the lack of infrastructure and architectural projects that could change the image of postsocialist Chinese cities.

The image being invoked in the global city discourse is of a city with skyscraper buildings, impeccably modern infrastructure, and a large number of specialized business services firms. Two features characterizing the specificities of Chinese global cities can be identified from this discourse. The first is the emphasis on infrastructure, signature buildings, and the built environment—namely, on the visual image of

a global city. In other words, for Beijing and Shanghai to become truly global cities, policy makers believe that these cities have to first adopt a "global city look" by constructing state-of-the-art infrastructure and flagship architectural projects. They argue that if the government took the lead in "building the stage" (a metaphor for developing infrastructure and remaking built environment), then "the performers" (referring to foreign investors and companies) would come (Gu and Sun 1999). This policy orientation has led to a rush of building expressways, bridges, airports, and skyscrapers in many large and small cities across the country.

The second signal feature of the global city discourse in China is its emphasis on the role of the state in directing the development of global cities. The project of global city building in China is essentially a state project. The state regulates the magnitude and pace of market-driven urban development by legal and administrative means. Meanwhile, as an economic actor, the state also actively participates in the race of urban development, through its agencies such as state-owned enterprises (SOEs) and quasi-SOEs. This political logic produces certain practices peculiar to global city building in China—practices that are centered on urban property development because all land belongs to the state and the state has more control in the property sector than in others. Thus, the two emphases in the global city discourse—the visual image of a city and the role of the state—reinforce one another, and together they explain why, in the Chinese case, construction of architectural megaprojects has become a core strategy in the state project of building global cities.

URBAN RENEWAL IN SHANGHAI

The ubiquitous Chinese character 拆, painted in white on buildings slated for demolition, has become a popular symbol synonymous with the machinery of urban renewal and displacement. Since 1990, when local governments started to renew neighborhoods with "old and dangerous houses," a large number of historic buildings have vanished under the wrecking ball. The scale of destruction and displacement has prompted widespread criticism, and in response, the government has passed preservation laws, increased conservation funds, and tightened control over developers to ensure fair compensation for the displaced. But in spite of these progressive measures, the pace of demolition and residential displacement has yet to slow down. On the contrary, in Beijing and Shanghai, the 2008 Olympics and the 2010 World Expo have further fueled the machinery of urban renewal.

Shanghai experienced massive demolition in the 1990s. In 1992, Deng Xiaoping's tour of southern China marked the real beginning of the pragmatic market reform. In the same year, at the Sixth Communist Party Congress of Shanghai, the Shanghai municipal government announced the famous 365 Plan, declaring that by the year 2000, the city would finish demolishing 365 hectares of "dangerous houses." The plan was based on a survey conducted in the previous year, which confirmed 15 million square meters of old-style housing in the city and identified 365 hectares as being in dangerous condition. The term "dangerous," however, was never clearly defined. Neighborhoods that promised good investment returns, such as those in central locations and with low residential density, were deemed to have "dangerous housing" and slated for demolition, even if the housing stock was in good condition. In comparison, many neighborhoods in desperate need of housing improvement, but with less promising investment prospects, were carefully bypassed by developers and local governments.

It is apparent that demolition of "dangerous houses" has become the most important instrument used by city governments to intervene in the land and housing markets. By demolishing old low-rise houses in the inner city, where land prices are the highest, the government can requisition land, transfer land-use rights to developers, and boost municipal revenues from land-leasing fees. Each wave of demolition creates tens of thousands of displaced residents, who then look for housing elsewhere in the city. Thus, massive demolition directly creates huge demand for more housing, especially in the suburbs, where property prices are lower, but also in the secondhand housing market. The city government has control over the pace of urban renewal and demolition, making it more aggressive and faster in some years than in others, to boost the real estate market during recessions and prevent the market from getting overheated in boom times. In 2007, when property prices peaked in Shanghai, the government slowed down the pace of demolition to control the demand for housing, while after 2008, in the midst of the global economic recession, the government sped up urban renewal

again and set the goal to "improve the living conditions of 200,000 households" and renew eight million square meters of old houses in five inner-city districts. The scale of demolition and relocation in Shanghai is unprecedented. From 1990, when the 365 Plan was announced, to 2008, the city demolished 70 million square meters of houses and relocated 1.2 million households from inner-city districts (Shanghai City Government 2009).

FROM DEMOLITION TO PRESERVATION

Since the 1990s, the Shanghai city government has shown steady movement toward historical preservation, by passing a series of preservation laws and landmarking a large number of historical buildings. The city has rapidly expanded the list of historical buildings designated for preservation from a mere 50 initially to more than 700 locations. The buildings listed for preservation are predominantly examples of Western-style architecture from the colonial period. Buildings from the following socialist era are mostly regarded as worthless for preservation and eligible for demolition as they age. Only recently, in response to criticism about the lack of preservation for architecture from the socialist era, were a few buildings built after 1949 added to the preservation list.

What marked the transition from demolition to preservation in Shanghai was the flagship redevelopment project at Xintiandi, where two blocks of *shikumen* houses, Shanghainese tenements built by Western landlords for Chinese tenants in the colonial period, were turned into a posh entertainment quarter by international developers and architects, with support from local governments. Although it is questionable what is actually preserved at Xintiandi, the commercial success has sent a clear message to mayors and developers across the country that history can sell, and very well. Since Xintiandi opened, Shanghai has begun to preserve much more architecture than it ever did before. As most financial institutions moved to modern office towers across the river in Pudong, the neoclassical buildings on the Bund were renovated as spaces for galleries, boutiques, and upscale restaurants and clubs. Various preservation and redevelopment plans followed for other buildings on the Bund. Old warehouses along Suzhou Creek and factory workshops in Yangpu district were turned into artists' studios, galleries, and offices for media firms.

Han Zheng, the mayor of Shanghai, delivered a new slogan in 2004: "Building new is development, preserving old is also development." The new slogan summarizes well the nature of historical preservation efforts in Shanghai—preservation is another instrument, a more sophisticated one than demolition, to promote urban growth.

DISPLACEMENT AND THE POLITICS OF COMPENSATION

The glamour of the space of conspicuous consumption at Xintiandi is multiplied when contrasted with the poor neighborhoods nearby. Across Taicang Road, the northern boundary of Xintiandi, was a cluster of old *shikumen* houses scheduled to be pulled down. In

the mid-2000s, about three hundred families refused to accept the compensation offered by the developer to resettle elsewhere. A year later, only 30 families were still remaining. The remaining families were those with fewer resources and special needs. To oust these residents, demolition companies tore down most of the empty buildings and left construction debris uncollected on the ground. On some occasions, water and electricity supplies were cut off. The families had to live amidst this environment of dirt, noise, and waste. And just across the street they could see the fancy shops and wealthy customers at Xintiandi.

The negotiation process for relocation compensation in China is an informal but coercive one. First of all, developers in general do not engage in any dialogue with the local community. For example, according to the agreement reached between the

Plate 30 A Starbucks at Xintiandi, Shanghai

Source: Xuefei Ren

district government and the developer, it is the district government's responsibility to relocate residents. The developer hired specialized demolition companies to clear the site. The demolition companies were paid a lump sum by the developer, which covered compensation for residents and demolition costs. Second, the negotiation between the district government and residents was an informal process, carried out on an ad hoc basis. Together with monetary compensation, the district government used a combination of other incentives to persuade residents to move out, such as giving job offers to certain family members, providing ready apartment units in exchange for the current ones, or offering retirement pensions and medical insurance for elderly family members. Specific terms were negotiated according to each family's needs, and resettlement plans varied case by case. Third, although the negotiation with residents was a flexible and informal process, it was also a coercive one. If a family had members working in the public sector—for example, in government branches, work units, or schools—the district government would put pressure on these public employees, asking them to persuade their families to move out. The unfair compensation is the single most important factor explaining why some families have refused to move. The remaining families at Xintiandi were mostly lower-income residents living in extremely crowded conditions. Since the calculation of compensation is based on living area (i.e., square meters) instead of number of residents, it was impossible for these families to buy an apartment in the city that was large enough to accommodate all family members.

HOUSING RIGHTS ACTIVISM

Overall, incidents of housing activism in Shanghai have increased over the course of urban renewal as residents have begun to combine various tactics to fight for their rights (Weinstein and Ren 2009). The combination of a non-electoral political system and a booming property market has led to new forms of housing activism in Shanghai. Instead of street actions and direct confrontations, residents tend to use other, more subtle tactics in their resistance. Some residents register extra family members and build extensions to gain better compensation; others mobilize the media to bring attention to their cause; still others engage in "rightful resistance" by framing their struggles in terms of legal rights (O'Brien and Li 2006).

A small group of housing activists has emerged in Shanghai, most of them lawyers advising residents who have lost their homes on how to negotiate with the government for fair compensation. However, as the state does not tolerate organized demonstrations for fear of social instability, many leading activists have been persecuted by the government. Zheng Enchong, a lawyer in Shanghai, had his law license revoked and was sentenced to a three-year jail term. He was charged with "revealing state secrets" after communicating with human rights organizations abroad about housing demolition in China. Zheng Enchong's sentence sent a reminder to other lawyers about the risks of representing demolition cases. Organized street protests are generally less successful, as organizers must first obtain authorization from the government to stage a legal protest. Many organizers are jailed when applying for authorization, and the police are quick to arrest and disperse protesters. Mass petitioning to the central government in Beijing is equally unsuccessful, as petitioners are often stopped by government security on their way to Beijing and then sent back. Increasingly, residents file lawsuits in local People's Courts, despite the fact that the courts are often not independent from the government.

Many residents try to delay construction efforts by standoff, simply staying in their apartments and refusing to leave even when demolition crews arrive, thus becoming so-called *dingzihu* (nail households). *Dingzihu* refuse to be evicted even after courts' final decisions. Some *dingzihu* also register extra family members and build extensions to their houses in order to negotiate for better compensation. The Internet has become a powerful medium for residents to communicate with one another, to seek legal advice, and to get their stories out. Compared with the tight censorship over conventional media, the government has less control over the content circulated on the Internet. Stories, photos, and even live videos of forced demolitions can be easily found online. Numerous websites offering legal services to evicted residents have been created in recent years. Even with the risks of defending demolition cases, an increasing number of lawyers are now more willing to take such cases, partly impelled by a sense of justice, but also partly driven by the lucrative market for demolition lawsuits.

CONCLUSION

China's urban transformation in the first three decades of reform has clearly exhibited a top-down pattern of decision-making, as the political, economic, and cultural elites formulate key spatial strategies for global city formation. This essay has examined both the efficacy and fragility of this top-down model of city making, a model that has produced vast discontents and inequalities. The strategy and practice of building global cities in China have been characterized by heavy investment in infrastructure, privatization, and deregulation endorsed by the state, and the symbolic articulation of urban space by the elite. The policy priority placed on urban infrastructure investment has helped channel surplus capital accumulated in the manufacturing and export sectors to the built environment. Real estate construction and property development have become a major industry driving national GDP growth. Meanwhile, the spectacular urban growth could not have been achieved without the introduction of a series of new legal and administrative instruments by the state, including the legalization of land transactions, protection of private property, and the deregulation of foreign producer services firms. These legal measures have opened up, if only partially, the domestic property market to global investors and firms. The rise of a service-based urban economy, the official promotion system emphasizing "physical achievements," i.e., urban construction and property development, and the dual land market have made conditions ripe for the urban elite to pursue symbolic spatial strategies in the competitive race of urbanization. This essay has examined how historical preservation is mobilized to promote Shanghai as China's global city, and how redevelopment and displacement have triggered a surge of housing rights activism. Shanghai's transformation into a global city demonstrates that urban space is not a container in which broader social processes take place, but a strategic terrain where symbolic spatial production leads to capital accumulation and where older power logics are rescaled to produce new dynamics.

REFERENCES FROM THE READING

Al, S. (ed.) (2014) *Villages in the City: A Guide to South China's Informal Settlements*, Hong Kong: Hong Kong University Press.

Gu, C. and Sun, Y. (1999) Jingji quanqiuhua yu zhongguo guojixing chengshi jianshe (Economic globalization and building China's international metropolises), *Urban Planning Forum*, 3: 1–6.

O'Brien, K. and Li, L. (2006) *Rightful Resistance in Rural China*, New York: Cambridge University Press.

Ren, X. (2011) *Building Globalization: Transnational Architecture Production in Urban China*, Chicago: University of Chicago Press.

Ren, X. (2013) *Urban China*, Cambridge: Polity Press.

Ren, X. (2017) Land acquisition, rural protests, and the local state in China and India, *Environment and Planning C*, 35, 1: 22–38.

Shanghai City Government (2009) *Suggestions on How to Advance Urban Renewal*, Shanghai: Shanghai City Government.

Weinstein, L. and Ren, X. (2009) The changing right to the city: Urban renewal and housing rights in globalizing Shanghai and Mumbai, *City and Community*, 8, 4: 407–32.

45

"Between ghetto and globe: remaking urban life in Africa"

from *Associational Life in African Cities* (2001)

AbdouMaliq Simone

EDITORS' INTRODUCTION

AbdouMaliq Simone is Research Professor at the Max Planck Institute for the Study of Religious and Ethnic Diversity, Visiting Professor of Sociology at Goldsmiths College, University of London, and Visiting Professor at the African Centre for Cities, University of Cape Town. For three decades, Simone has written extensively on urban informality and everyday life in African and Southeast Asian cities. In this essay, Simone draws upon his ethnographic work to examine the flexibility, informality, and resourcefulness of urban inhabitants in two African cities, Dakar and Johannesburg. He poses a central question: when the state is unable to deliver basic needs—jobs, infrastructure, housing, what do ordinary residents do to get by and get things done? In contrast to much of social movement scholarship, Simone considers the poetics of everyday negotiations to access multiple sources of income. With ethnographic sensitivity and precision, he portrays the resilience of urban African residents, and he argues that an understanding of the future of Africa's cities will first require some sense of their residents' needs and actions.

Within the global urban context, the move toward "normalization" of governance and economic management in most African cities opens up spaces where efforts to remake everyday life—households, schooling, religious devotion, morality, practices of accumulation, etc.—become increasingly significant. None of these "re-makings" have crystal-clear trajectories. At times, the efforts communities make to compensate for the absence of effective urban government converge, and at times they conflict, with new forms of social action made possible through broadened local democracies. The relationship of these "re-makings" with the new regulatory environments is uneasy, often volatile, resulting in a wide range of tensions.

Ways of organizing survival, economy and politics outside formal institutions are conventionally seen as either compensations for inefficiency and exclusion,

instruments for circumventing the "rules," or necessary but "hidden" domains so that formal institutions can "really" function (Sanyal 1996). Rarely are such ways of organizing viewed as historical "outgrowths" in their own right, which defy "formalization" either because it is perceived as disadvantageous or because existing formal institutions have largely served to repress them.

Decentralization provides greater political maneuverability for urban governments. But at the same time, these governments face shortages of funds, which render their maneuverability largely fictitious. The combination of these factors has produced greater diversity of organizational practices at the local level. The institutional boundaries between religion, politics, culture and social economy have become increasingly blurred, giving rise to a proliferation of ad hoc initiatives. Often, formally organized civil society organizations (CSOs) are either out of touch with such initiatives or attempt to control them. This does not mean that CSOs are not playing important roles, but their emphasis on targeting specific populations, identities, sectors or themes often means that they cannot extend themselves to work among other actors or relate to different processes (Paerregaard 1998). Given these dynamics and the proliferation of ways of organizing at local levels, urban development work must critically assess what various ways of organizing might be able to achieve and how different forms can be articulated into complementary instruments of action (Simon 1998).

TALES OF TWO CITIES: DAKAR AND JOHANNESBURG

The discussion in this essay focuses on two disparate forms of urban associational life in Dakar and Johannesburg. By focusing on diffuse and largely "invisible" ways in which participation and collaboration are mobilized, it is possible to discern tensions and contested development trajectories of the urban arena.

In Dakar and Johannesburg, as well as in most African cities, population shifts resulting from migration, demographic change, dispersion of economic resources and relocation of growth mean that localities encounter instability. This instability is exacerbated by the proliferation of survival strategies pursued by local residents forced into opportunistic and *ad hoc* behaviour. The adherence of African states to prevailing economic wisdom—the need to create an entrepreneurial class, small businesses, commercial agriculture, etc.—has provided opportunities for land appropriation, circumvention of rules regarding the hiring of labor, proliferation of political clientelism into the provision of social welfare and technical assistance, and hiding assets from the state (Bayart et al. 1999). For the most part, such informalization stems from opportunities for individuals to access multiple sources of income by securing niches in various sectors. These multiple positions are then used to cover up, divert, or circumvent scrutiny by regulatory agencies. In most of the developing world, no household is fixed in any single economic position, but is rather operating in-between several. This operation "in-between" changes the ways in which economic

positions are linked, constituted and remade as they evolve or dissipate over time.

These processes tend to promote a notion that all aspects of everyday life can be negotiated. Instead of adhering to an unequivocal set of rules and procedures, anything and everything is negotiable. Such negotiability introduces greater flexibility and potential for problem solving, but also raises the question as to when negotiations end and how many different actors can be negotiating partners. If everything can be negotiated, then predicting the outcomes of any given transaction is uncertain. It is thus difficult for individuals to assess what implications their actions will have for others or to ascertain what it is feasible for them to do. Uncertainty is also increased by a tendency either to limit the range of social exchanges and reduce the number of those with whom one negotiates, or to constantly change "negotiating partners."

Since some local groupings are often precluded from collaborating with each other in the formal public sphere, but cannot ignore each other, a series of informal processes comes into being to regulate exchanges. Given the diminution of authority in local public institutions—in part occasioned by an inability to comprehend the social changes under way, as well as their lack of capacity and resources— these informal processes increasingly take precedence. As localities are reasserted as a collection of parochial identities and lifestyles, often marked by extreme disparities in terms of access to opportunities and resources, there is also an intensifying sense of cosmopolitanism, especially in the larger African cities. In a globalized world, inter-penetration of all kinds across economic, cultural and political terrains throws into question the integrity of any identity. Any effort to achieve unequivocal clarity with regard to identities of race, religion, gender, nation or ethnicity finds itself "interrupted" and intersected with influences, "pollution" and exceptions beyond its control. In a world of incessant mobility, travel, communication and exposure, any culture or grouping has little choice but to take a vast range of "others" into consideration. By doing so, that culture or grouping implicitly becomes some of what those "others" are. Investments in association, then, must attempt to both bridge the need to maximize the opportunities involved in crossing borders of all kinds, and to solidify often highly localized arrangements of social support and solidarity.

SOIREE IN SOUMBEDIOUNE

Soumbedioune is a historic Lebu fishing village, incorporated as an extension of the Greater Medina in Dakar. In Soumbedioune, whom one goes to see to straighten out small disputes with neighbors or matters of cleanliness on the street, or organizing social events, varies, depending on whom amongst one's neighbors has acquired newly important connections, whether elections or religious events are upcoming, or on what kinds of deals have been made between particular actors and networks at any given time. In fact, there seems to be an incessant precipitation of problems, in excess of those usually associated with people living in dense quarters, simply to find out just who has what status and to ascertain what opportunities present themselves at any given time. Such processes give rise to a sense of fluidity about how local power is organized and deployed. Because this process can be tedious, many people end up doing what they want to do, while others end up seldom doing anything different at all.

One recent example of these dynamics concerns a situation where a project to manage the completion of a large housing estate was being put out to tender by one of Dakar's top companies. The company was obligated to invite tenders, even though it clearly preferred and had every intention of keeping the entire project "in-house". An architectural firm was to be hired, as well as a general contractor to secure materials and labor. In a cursory gesture, to abide by the rules, an announcement was issued to a nearly empty conference center on a Thursday evening to the effect that tenders would close at seven the next morning. The selection of this particular weekday was significant because Thursday evening is a time when many people participate in special worship in the mosques.

Nevertheless, and subsequent to this announcement, the night was busy with a proliferation of visits among architects, engineers, laborers, politicians, dealmakers, *marabouts*, soccer clubs, youth organizations, artisan guilds and scores of their relatives who all thought they could discover pathways of influence to the person who was to award the contract. Although the

Plate 31 Street scene, downtown Johannesburg

Source: Roger Keil

official who was to award the contract no longer resides in Soumbedioune, a sister occupies the family compound. While she received many visits, most of the activity pervaded the surrounding area, as neighbors, friends and purported social connections were sought out and discussions held. Hundreds of thousands of CFA, the local currency, changed hands, sexual scenarios were played out, as were prayers, family injunctions, pleas, reiterated obligations and "chance" encounters. A broad array of rhetorical strategies was involved: memories of past infractions and promises, personal knowledge, cultural etiquette, political procedure, technical know-how, sorcery—all circulating in various constellations marking gaps and hybrid conjunctures.

While professional firms unofficially joined the fray, there were also people looking for any opportunity to affiliate with the project. Even if they could not design, build, or supply relevant inputs, they knew others who could and thus hoped for possible in-kind "finder's fees" or opportunities to parlay information into other opportunities.

By morning, there were no official submissions of tenders but it had become clear to the company that there were hundreds of options to choose from, and theoretically, hundreds of different housing estates could have been built from this nocturnal deliberation. Dakar has roughly 100 architects to choose from and a lesser number of contracting firms capable of handling a project of this size. The political and religious loyalties of these players are commonly known. But with a flurry of activity crossing boundaries of all kinds and producing various constellations of alliances, workforce compositions, inputs and cash flows, it was not clear what the possible implications of any particular choice might have been. People known to be unwilling to work together expressed their willingness, best friends parted company for the moment to broaden their chances. There was a sudden opportunism and a proficiency for cutting across social divides, networks and alliances that could not be absorbed by any formal representation. The company asked for more time to weigh the decision that probably had already been made a long time ago.

ON THE OUTSIDE IN THE JOHANNESBURG INNER CITY

Johannesburg has long been Africa's most developed city. Its status was acquired through an oppressive political system, which rigidly regulated access to urban space and services and imposed two distinct development trajectories: a city with all the Western amenities for whites, and impoverished peri-urban labor reserves for blacks. The race-based zoning of urban residential communities kept blacks out of the inner city for several decades. Accelerating white movement to the suburban areas, coupled with economic recession, pushed up vacancy rates in the neighborhoods of Hillbrow, Bertrams, Joubert Park, Berea and Yeoville. Although officially illegal until 1991, blacks had begun moving to what was known as "grey areas" in the mid-1980s (Tomlinson et al. 1995).

The accelerated turnover of populations has itself provided a feasible cover, if not necessarily a major motivation, for the sizeable immigration of foreign Africans to Johannesburg. This migration, in turn, has substantially shaped the nature of inner-city life and commerce, further contributing to a process of internationalization. Because the inner city is one of the most circumscribed and densely populated urban spaces on the continent, with neighborhoods such as Hillbrow made up of row after row of high-rise apartment blocks, this socio-cultural reconfiguration has taken place with a large measure of invisibility.

In response to these changes, a large measure of xenophobia prevails, where foreign Africans are blamed for an overcrowded informal trading sector, the growth of the narcotics trade and general deterioration of the inner city. Many South African residents believe that it is because of such a foreign presence that government authorities and the private sector are unwilling to make investments in upgrading and service provision.

Although migrant labor from the sub-region had played a major historical role in the South African economy, the nature of migration radically changed once conduits of transport and a context suitable for individualized exchange of labor and pursuit of entrepreneurship were opened up. While the bulk of migration continues to originate in Southern African states, the inner city has become a staging area for individuals from Francophone countries, Ethiopia, Somalia, Nigeria, Ghana and Kenya.

Much of the initial African migration corresponded to shifting policies pursued by the South African state. Political favors offered by African states, as well as their co-operation in circumventing sanctions and creating trade opportunities and havens for capital flight, were often rewarded with relaxed entry and stay requirements. At different times, visas were waived for Congolese, Ivorians, Malians and Senegalese. Changing circumstances for potential migrants in other countries also contributed to a move southwards. For example, interviews with long-term Malian residents indicate that the bulk of the early Malian community was largely derived from diamond smuggling syndicates in Zambia, which faced crackdowns during the early days of the Chiluba regime. In another instance, long-term co-operation between South African multinationals and the Congolese ruling hierarchy produced an almost wholesale transfer of much of Zaire's ruling elite to South Africa, following the advent of a protracted political crisis.

Although the overall fluidity of the inner city provides an opportunity for putting in motion or improvising upon many of the standard practices of African entrepreneurship, conflicts between local South African small entrepreneurs and foreign Africans have bubbled to the surface. Because South Africans have not participated substantially in the circuits of movement and exchange dominating the rest of the continent, they know little about other African cultures and practices. Local South Africans frequently conclude that the success of foreign hawkers and well-heeled businessmen is at their expense. Foreign Africans, long accustomed to the relative domesticity of public space, have become easy targets for muggers and con men. Frequent assaults, coupled with cultural misunderstandings and disputes over space and access, have produced an increased defensiveness among the diffuse foreign African communities.

Local South Africans at times attribute an almost monolithic solidity to particular foreign national groupings, such as being "Nigerian" or "Senegalese"—characteristics in contradistinction to which they rally due to their own lack of history in and relative newness to the city. Just as Mozambicans, for example, may have their special hangouts, some local South Africans claim buildings, streets or certain bars as "their places," as a means of adding consistency to their social affiliations and movements. A frequent invocation by South Africans is that the French-speaking Africans never deal with their English-speaking counterparts, that Nigerians never deal with Cameroonians, etc., so why should they be compelled to deal with any of them.

At the same time, foreign Africans try to up-end these assumptions with moves to "Africanize" their associations in the widest sense, i.e. attempting to build their personal networks from many different national communities. While ethnic and national groupings are maintained, many traders, hawkers and businesspeople are beginning to elaborate collaborations that cut across these divides, from acquisition of buildings, coalescence of hawking and service groups, to cross-border trade. Once "business discussions" in the bars of the renowned immigrant hotels, e.g. the Mariston, Protea Gardens, the Mark and the Sands, were strictly between those of common national identity. But in recent years I have observed Ethiopians, Congolese, Mozambicans and Zambians, for example, talking about ways to consolidate their individual agendas and networks to provide greater scope and flexibility. Thus, local pressures, which act to "ghettoize" and constrict the operations of foreign migrants, are also prompting new forms of organization, which enable individuals of distinct nationalities to maximize their maneuverability and reach outside the confines of the Johannesburg inner city.

CONCLUSION: GOING FORTH INTO THE WORLD

This essay does not address viability or sustainability. It is not clear what is likely to happen in cities such as Dakar and Johannesburg, if adequate forms of governance fail to emerge. While such considerations are important in terms of what happens to the lives of individuals who are "condemned" to remain in cities unable to provide employment, legality, shelter, etc., the urbanites retain the possibility of circumventing such considerations.

Based on these stories from Dakar and Johannesburg, a critical future area of research would be to follow the trajectories and movements of informal and provisional associational dynamics as they intersect with the functioning and reach of specific institutions, politics and economic activities. If the city is characterized by increasing segregation and fragmentation, it may be important to chart the complicity, co-operation, boundary-crossing, interpenetration, affiliation and divergence which "come and go" across the city, its neighborhoods and its facets.

In a fundamental way the question of where African cities are going could be addressed by assessing—in a broader sense—where African urban residents are going, within their different time frames and in terms of their respective objectives. While urban poverty is indeed growing worse, as is a concomitant disarticulation among urban social spheres and a narrowing in the terms of belonging, urban residents are not standing still. There is something going on, efforts are being made to come up with new ways of earning a living, of helping others out and of trying to create interesting cities. An awesome sense of responsibility is being displayed, as well as an awesome sense of irresponsibility shown by the rest of the world to the future of African cities.

REFERENCES FROM THE READING

Bayart, J-F., Ellis, S., and Hibou, B. (1999) *The Criminalization of the State in Africa*, Bloomington: Indiana University Press.

Paerregaard, K. (1998) Alleviating poverty in Latin America: Can local organizations be of any help? in N. Webster (ed.) *In Search of Alternatives: Poverty, the Poor and Local Organizations*, Copenhagen: Centre for Development Research.

Sanyal, B. (1996) Intention and outcome: Formalization and its consequences, *Regional Development Dialogue*, 17: 161–78.

Simon, D. (1998) Rethinking (post)modernism, postcolonialism and posttraditionalism: South-north perspectives, *Environment and Planning D*, 16: 219–45.

Tomlinson, R., Hunter, R., Jonker, M., Rogerson, C., and Rogerson, J. (1995) *Johannesburg Inner-City Strategic Development Framework: Economic Analysis*, Greater Johannesburg Transitional Metropolitan Council.

46
"World cities and union renewal"

from *Geography Compass* (2007)

Steven Tufts

EDITORS' INTRODUCTION

The global city literature has always acknowledged the polarizing labor conditions in globalizing cities: at one pole, the high-earning professional class drawn to the strong presence of specialized business service industries, and at the other pole, short-term, low-wage workers who toil in low-end service industries. The growth of the two classes often feeds one another, as many of the lifestyle demands of the global professional class have to be met by the manual labor of low-wage workers, such as those working in restaurants, hotels, and the domestic sector. But we know little about how the polarizing labor conditions in global cities can change organizational structures and strategies of labor unions. Steven Tufts, Associate Professor of Geography at York University, fills this gap by examining the changing operations of labor unions in globalizing cities. He finds that the political economy of global city formation has reshaped unions' organizational structures and coalition building strategies, leading to new forms of labor internationalism and also to new modes of cooperation between labor and management. On this basis, Tufts emphasizes the strategic importance of global cities for the changing strategies of labor organizing, due to the critical mass and visibility of the working class. He calls, however, for researchers to expand their focus beyond the global city in order to consider other places and scales of labor organizing.

In most advanced capitalist economies, workers continue to struggle against the concerted attack on their working conditions and institutions. While union membership continues to increase in absolute terms in some countries, labor union density – defined as the percentage of workers belonging to labor unions – has declined or stagnated throughout most Organization for Economic Cooperation and Development countries. Indeed, it remains relatively high only in a few Northern and Western European countries, while in the USA, where the decline has been pronounced, the union membership rate has decreased from 20.1% in 1983 to 12.5% in 2005 (Bureau of Labor Statistics 2005). The overall decline is the result of a number of pressures, including global competition from low-wage regions, changing technology and labor processes, increasing employment in traditionally non-union sectors such as consumer services and neoliberal state policies limiting the ability of unions to organize workers in order to increase flexibility.

Yet, organized labor has responded to threats of irrelevance. Since the early 1990s, a number of researchers have noted labor's response and speak of the possibilities of a revitalized labor movement. A now immense literature continues to explore union renewal, manifest in new organizing tactics, campaign strategies and union structures necessary for organized labor's success in regulating labor markets. Geographers continue to contribute to the study of union renewal, although the emphasis has been largely on which scale to best mobilize labor against increasingly mobile capital.

In this article, I wish to add to this discussion through a brief investigation of how the 'urban' scale is reflected in union renewal discourse and practice. In order to focus the discussion, I attempt to identify how the 'world city' thesis is embedded in accounts of labor revitalization. I argue that the account of the world city hypothesis – initialized by Friedmann (1986) and developed by Sassen (1991) and other postmodern urban scholars (e.g. Davis 1998; Dear 2000; Soja 2000) – has a significant presence in the accounts and contemporary practices of union renewal. While this understanding of the changing role of cities in the global economy has directed the union renewal project, the extent to which it uncritically forms analysis and strategy can be questioned in the same way world city research has been criticized and refined for over two decades (see Knox and Taylor 1995; McCann 2002).

THE WORLD CITY'S ASSAULT ON LABOR

The world city hypothesis has influenced a number of commentators of urban change who have built models of how cities increasingly are shaped by global flows of capital and people. Indeed, the strength of such accounts is the way forces operating at multiple scales (such as neoliberalism, trade liberalization, global patterns of (dis)investment and changing technology) transform the urban landscape. In particular, researchers based in southern California (often termed the LA School) have developed taxonomies of the post-metropolis, an urban form largely driven by integration into global networks of capital (Dear 2000; Soja 2000). An uncritical examination of the way in which a hierarchy of world cities, key centers of accumulation and intense interaction for multinationals, reportedly command global production for transnational capitalist identifies several important implications for workers. Such widely recognized cities are where emergent working-classes (e.g. immigrants, young people, racialized minorities) are assaulted by the forces of post-industrialism and post-Fordist accumulation, victimized by intense inter-urban competition and increasingly disciplined by states restructuring people's access to and through urban spaces.

The world city thesis also emphasizes the intensification of a hierarchical network of cities with increasingly heterogeneous labor markets. Economic strength is gained by successfully tapping international flows of capital, information and the transnational elites that run through the networks of global metropolitan centers. The end result is cycles of *inter-city competition* among labor markets for investment. Regional competition can occur in several areas. In terms of competition for manufacturing jobs, local labor markets are disciplined by pressures for lower labor costs and/or highest productivity. In the race, to deregulate labor markets and establish a numerically and functionally flexible workforce organized labor is a primary target. Class interests are often subjugated to the needs of particular places to reproduce themselves and attract new investment.

Lastly, contemporary cities are increasingly *revanchist* as social regulation and political governance fortify capital against any instability and criminalize those on the economic margins (Davis 1998). Gated communities and urban (re)development projects sanitize, privatize and commercialize space in ways to minimize threats to production and consumption at the expense of publicly owned spaces and political dissent (Soja 2000). Since 11 September 2001, the nation state's efforts to secure borders and cities against 'terrorist threats' have had contradictory and complimentary effects on urban economic development. While trans-border flows of people, goods and services are disrupted, the state has taken the opportunity to discipline dissent further and restructure work. Undocumented workers are further exploited in environments with heavy-handed immigration policies and enforcement. The voices and organizing efforts of marginal workers are disciplined as a result.

LABOR UNION RENEWAL IN THE WORLD CITY

Labor union renewal is a broad concept involving many interrelated components that I have categorized as changes in organizational structure, leadership and democracy, intensive organizing, coalition building, new labor internationalism and labor-management cooperation. I highlight how some of organized labor's most important experiments with (re)organizing workers in the city are linked to the specific understanding of changing urban landscapes.

Changes in organizational structure

At the heart of recent experiments with union structures is the inability of most unions to adapt to the new geographies of small workplaces that are difficult to organize one employer at a time, especially in competitive low-wage sectors. A response has been experiments with geographical unionism, new structures that allow bargaining with many small employers in one area. The call for these new structures has been in response to the rise of a marginalized service class (e.g. office cleaners, taxi drivers) that co-exist alongside high-wage professional classes of world cities.

The archetypical example of such a structure is JfJ campaign in the USA as an example of how a union (the Service Employees International Union) used a range of tactics to organize thousands of cleaners scattered over multiple worksites (i.e. office buildings) employed by dozens of subcontractors. The campaign in Los Angeles – documented by Savage (1998) – identifies important changes that the SEIU had to endure, the most important of which was to view the 'workplace' as several unique geographical districts in the city that required 'local sensitivity' to different groups of workers and the specifics of the local industry.

Another form of geographical unionism is the Living Wage Movement. In the early 1990s, a union-community coalition in Baltimore campaigned for a local city ordinance that would force subcontractors working for the city to pay workers a minimum wage. Over the next decade, dozens of ordinances and well over a hundred campaigns have been established across the USA and abroad (Luce 2004). There is significant research indicating positive economic impacts derived from an increase in benefits for workers.

Implicit in all of these experiments and the accounts of their supporters is an appreciation for the rescaling of economic and political power to large centers identified by world city and other urban theorists. The campaigns target the growing numbers of post-industrial jobs with limited capital mobility, a source of power for less mobile labor. Walsh (2000) emphasizes, the metropolitan scale is a powerful 'spatial fix' for labor in its struggle to organize service sectors where capital mobility is qualitatively different from manufacturing. The campaigns are also based in large cities with a critical mass of workers,

Plate 32 Trade union headquarters, London

Source: Roger Keil

and community groups with which coalitions can be formed. More important, the changes in union structure allow unions to address local needs and regulate labor markets attacked by economic restructuring (e.g. subcontracting) in locally specific ways (e.g. with variations in minimum standards).

Intensive organizing

The focus on organizing has now moved beyond how to shift union resources to questions of how to *intensify* campaigns. If organizing drives are to be successful, a sense of ownership must be given to both present and future members. Committees of workers are established to make decisions about the campaign as opposed to 'top-down' coordination by paid organizers. Union members are engaged with non-union workers through intensive 'house visits' where workers can speak about union issues away from the workplace. The goal is to build organizational capacity before a workplace is unionized in order to build power for the negotiation of a first agreement.

An example of such a campaign is the *Hotel Worker's Rising* launched in 2006 by UNITE-HERE (see www.hotelworkersrising.org). The campaign is the culmination of a decade long struggle for common expiry dates of major collective agreements with hotels in over a dozen North American cities. For the first time, hotel workers are able to bargain collectively with multinational employers in large numbers. The goal of the current round of bargaining is to raise the standards of hotel workers through immediate improvements in wages and working conditions for union members and for *future* members by securing neutrality agreements with large employers to allow for easier union recognition and greater union density across the sector. While the campaign has an international component, a great deal of effort is being invested in building support for workers at the local level through community coalitions.

COALITION BUILDING

The city provides a base for coalitions given the number and diversity of groups that form the social fabric of the contemporary metropolis. 'Community Rising' has reached out to everyone from church groups to settlement agencies to progressive academics. On the production side, the city's local labor

council is supporting the workers with a concurrent campaign 'A Million Reasons to Support Hotel Workers'. The campaign to build local labor support was derived from the larger 'A Million Reasons' campaign aimed to raise the wages of one million Toronto workers beyond the low-income level (see www.laborcouncil.ca/amillionreasons). The theme is therefore about placing the hotel workers' struggle in the context of raising economic standards in marginalized communities throughout Toronto. By scaling the struggle beyond workplace issues, the campaign resonates with the broader community.

The above coalition building activity is part of a longer trajectory identified by scholars who witnessed dissent and resistance increasingly occurring beyond points of production in the post-war period. A key tenet of the world city thesis is that dominant classes are increasingly confronted by diverse groups from the global margins. Furthermore, while there are exceptions, the world city has been proven an effective springboard for local unions as they jump scales and build international alliances through global urban networks.

New labor internationalism

Recently, there is a move towards building transnational 'global union' campaigns that combine the strategic research resources of several unions to tackle multinational employers such as Wal-Mart. Similarly, a recent protest by the West London Citizens in calling for a Living Wage for Hilton hotel workers was directly linked to the Hotel Workers Rising campaign in North America as UNITE-HERE orchestrates international pressure on the multinational hotel chain. SEIU has worked with Australian unions to export the JfJ campaign (it is called Justice for Cleaners in Australia) and continues global initiatives as it expands the JfJ campaign to other countries. As Herod (1998) reminds, global networks may be built by labor as well as capital, and here we see evidence of transnational links of elite labor activists orchemobilizing across world cities.

Labor management cooperation

Among the most controversial renewal strategies adopted by unions is engagement in new labor-management partnerships as new tradeoffs are made between increasing productivity and maintaining job security. Through 'social partnerships,' it is argued that labor can strengthen their position by 'trapping' capital investment in local markets through 'high road' strategies that emphasize training and increased productivity in the workplace.

Labor continues to actively cooperate with capital in lobbying for state subsidies to support economic development initiatives (e.g. auto assembly plants, training) that have been devolved from nationwide to local strategies placing cities and regions in direct competition. In the sense that economic development is path dependent and some 'winners' may be able to maintain jobs and leverage employers after initial large (re)investments in specific human and capital assets, social partnerships may allow some form of union renewal. The success may be short-lived, however, and such partnerships may create as many 'losers' as they do 'winners'.

CONCLUSION

The above discussion is only a snapshot of union renewal and the ways the world city framework is implicated. There are numerous accounts of renewal projects that happen on a daily basis being established in rural areas (e.g. migrant farm workers) and in no way do I discount these interconnected struggles. Yet, even in these struggles, the world city is implicated as organizers wage consumerist campaigns in global cities as a means of effecting regulatory changes with the state granting more rights to migrant workers or leveraging pressure on employers at different points in the commodity chain.

The shadow the world city hypothesis cast on aspects of the union renewal literature and practices is evident, but will require further analysis and research. The importance of post-industrial/post-Fordist industries, the heterogeneity and marginalization of world city labor markets, the harsh discipline of neoliberal states and the growing global networks of capital and labor bases flowing through such cities are, in many ways, embedded in discussions of labor union renewal. At one level, the world city contributes significantly to our understanding of the changes facing workers in a global economy and account for the sectors that are at the center of the union renewal project (i.e. service sectors with less

capital mobility). New union structures are being developed that are more compatible with these fragmented sectors. New strategies ranging from intensive campaigning to coalition building are being used to reach workers marginalized by class, race, gender and citizenship. At the same time, the re-scaling of economic activity and governance to the urban level has created opportunity for union renewal in the city. As cities compete in a global economy, labor can exercise leverage by supporting (or undermining) efforts to increase tourism activity and restructure work processes in ways that attract targeted investment for specific sectors (e.g. tourism, manufacturing clusters). Labor unions are changing labor market regulation through coalitions that enforce local minimum wage requirements while scaling up organizational capacities with international networks connecting activist of in large cities. The JfJ, Living Wage and Hotel Worker's Rising campaigns are largely focused on metropolitan spaces but have made links between 'global' cities (as in the case of the West London Citizens' support for hotel workers in London).

By attaching union renewal efforts to world city frameworks, not only are different ways and scales of organizing restricted, but we reinforce the very framework that has dislocated and oppressed workers in and outside of world cities. If unions only see renewal in terms of their response to overwhelming global forces and the rescaling of capital, such forces are legitimated and the only way out will continue to be an international unionism organized at a comparable scale with capital. While global unions are undoubtedly part of union renewal, it may be time to re-imagine the project in terms that view globalization as a complex process that is continuously shaped by *labor* and capital acting at a number of interlocking scales. Imagining union renewal in a manner that does not immediately privilege the force of global capital and cities will allow the emerging multiscalar approach to flourish.

REFERENCES FROM THE READING

Bureau of Labor Statistics (2005) *Union membership in 2004*, Media Release, January 17, Washington, DC: US Department of Labor.

Davis, M. (1998) *Ecology of Fear: Los Angeles and the Imagination of Disaster*, New York: Vintage.

Dear, M. (2000). *The Postmodern Urban Condition*, Oxford: Blackwell.

Friedmann, J. (1986) The world city hypothesis, *Development & Change*, 17, 1 (January): 69–83.

Herod, A. (1998) Of blocs, flows and networks: The end of the Cold War, cyberspace, and the geo-economics of labor at the fin de millénaire, in A. Herod et al. (eds.) *An Unruly World? Globalization, Governance and Geography*, London: Routledge: 162–95.

Knox, P. and Taylor, P. (eds.) (1995) *World Cities in a World System*, Cambridge: Cambridge University Press.

Luce, S. (2004) *Fighting for a Living Wage*, Ithaca, NY: Cornell University Press.

McCann, E. J. (2002) The urban as an object of study in global cities literatures: Representational practices and conceptions of place and scale, in A. Herod and M. Wright (eds.) *Geographies of Power Placing Scale*, Oxford: Blackwell Publishers: 61–84.

Sassen, S. (1991) *The Global City: New York, London, Tokyo*, Princeton, NJ: Princeton University Press.

Savage, L. (1998) Geographies of organizing: Justice for Janitors, in A. Herod (ed.) *Organizing the Landscape: Geographical Perspectives on Labor Unionism*, Minneapolis, MN: University of Minnesota Press: 225–52.

Soja, E. (2000) *Postmetropolis: Critical Studies of Cities and Regions*, Oxford: Blackwell.

Walsh, J. (2000) Organizing the scale of labour regulation in the United States: Service-sector activism in the city, *Environment and Planning A*, 32: 1593–610.

47

"Blockupy fights back: global city formation in Frankfurt am Main after the financial crisis"

Sebastian Schipper, Lucas Pohl, Tino Petzold, Daniel Mullis, and Bernd Belina

EDITORS' INTRODUCTION

Sebastian Schipper is a Professor at the Institute of Geographical Sciences at Freie Universität Berlin. His research interests include neoliberalism, the political economy of housing, gentrification, and urban social movements. This essay is coauthored with his colleagues at the Goethe-University Frankfurt, including Lucas Pohl and Daniel Mullis (PhD candidates in Human Geography), Tino Petzold (researcher in Human Geography), and Bernd Belina (Professor of Human Geography). This chapter examines the role of Frankfurt as one of the chief financial centers in Europe and its implications for urban social movements. Frankfurt shifted from being West Germany's (later the united Germany's) main site for advanced producer services and one of Europe's major hubs for air travel and logistics, into a true financial center of continental proportions during the Euro era due to the presence of the European Central Bank (ECB) in the city's east end (see also Beaverstock et al., Ch. 22). The chapter details the emergence of a diverse anti-capitalist movement called "Blockupy" that has used Frankfurt strategically as a stage, target, and symbolic site for a series of protests and educational actions after the world financial crisis in 2008. As the seat of the ECB, Frankfurt is not just a place with common "urban" grievances, leading to the formation of "urban social movements," but a strategic node in financialized neoliberal capitalism and the European austerity regime. The chapter explores the specific conditions for the contestations of the Blockupy movement, with a nod to Frankfurt's history of social struggles since the 1970s.

In 2012 and 2013 Frankfurt am Main witnessed major social protests under the name of Blockupy against the European crisis management and its devastating impacts on the livelihoods of people all over Europe. The city was chosen as the site of protest by a national coalition of radical and as well as more modest civil society groups due to its economic, political and symbolic functions as Germany's most important global city, the capital of continental European financial industries and the location of the *European Central Bank* (ECB) which is one of the key European institutions executing neoliberal austerity politics. The slogan for the 2013 protests was chosen accordingly: "Resistance in the heart of the European crisis regime" (Blockupy 2013).

The Blockupy coalition's most visible activities so far were the 'days of action' in May 2012 and May 2013 that aimed at shutting down Frankfurt's central business district (CBD) via a social strike. The days of action included mass rallies in the CBD with about

30,000 demonstrators in the first year and about 20,000 in 2013 as well as smaller and decentralized actions directed at specific sites associated with neoliberal capitalism and the European austerity regime. In 2013, Blockupy also included a three-day protest camp with up to 1,000 national and international activists, who met for workshops and plenary sessions to discuss the situation in Europe.

Blockupy as a social movement is not easily classified. It is an urban social movement in the sense that its visible actions are concentrated in Frankfurt, using the symbols and the functions of this global city for mobilization and as the site of an urban social strike. Blockupy is, however, not primarily concerned with specifically urban issues and grievances—the defining aspect of urban social movements. While urban issues like gentrification and the housing question became part of Blockupy's agenda over time, the movement is centrally concerned with the more abstract, less immediately urban issues of financialized neoliberal capitalism and the European austerity regime. In this it resembles the summit-hopping alter-globalisation movement of the 1990s and 2000s that mobilized huge mass rallies in the cities and places where summits of transnational organisations were held. Blockupy, however, using existing local social infrastructures, did not follow the symbols and control functions of neoliberal capitalist globalisation, but attacked them where they are fixed in place: in Frankfurt's CBD.

WHY FRANKFURT? THE STRATEGIC RELEVANCE OF THE URBAN

Cities function as strategic sites for both politicization and policing (Uitermark 2004). With regard to politicization, Uitermark et al. (2012) argue, for instance, that urban centers are constitutive for counter-hegemonic social movements and are particularly robust spaces for driving mobilizations for at least three reasons. First, cities are privileged for breeding contention because contradictions within capitalist urbanization tend to produce a wide variety of grievances among its inhabitants. Cities are, second, due to their density, size and diversity, more likely to become laboratories where new ties between a broad and diverse range of activists and political groups are forged. As cities also concentrate (immobile symbols of) power relations, they represent, thirdly, "a privileged point of attack" (Ibid: 2550) as social movements can claim public

space exactly where key institutions of political and economic power are located. Especially in the case of global cities where a set of powerful institutions are spatially fixed and where symbols of power and prestige are concentrated permanently, movements have the opportunity to address global issues by confronting their concrete, local expressions.

Three aspects help understand the strategic relevance of Frankfurt for the Blockupy protests: First, the emergence of Frankfurt as Germany's leading international banking center and global city during the early 1980s resulted in a high concentration of powerful financial institutions including the headquarters of private banks like *Deutsche Bank* and *Commerzbank* as well as the *German Central Bank* and the ECB (Schipper 2014). Due to the global city formation, Frankfurt is, second, nowadays faced with a relatively high degree of socio-economic inequality (Bock and Belina 2012) accompanied by ongoing gentrification processes that have fuelled considerable grievances among the local population in recent years. Beyond that, Frankfurt has, third, a long-lasting tradition of urban based social struggles that could, at least temporarily, overcome the frequently witnessed 'othering' and criminalization of radical protests by successfully establishing alliances of a broad range of political actors across civil society (Mullis and Schipper 2013). This tradition enables local (Blockupy) activists today to build on already established ties and to enter broader coalitions with 'respectable' civil society actors. In the following, we briefly explain how these three aspects are historically interwoven.

The foundations for becoming Germany's leading banking center were laid after WWII when Frankfurt was selected as the location for the predecessor of the *German Central Bank*. The subsequent concentration of national financial institutions in Frankfurt and the resulting growth of the Central Business District (CBD) led to a radical transformation of the urban socio-spatial structure and planted the seeds for the following globalizing process that started in the early 1980s. However, the early state-led expansion of the CBD and the resulting partial demolition of the centrally located residential neighborhood *Westend* were met with fierce resistance. During the 1970s, lower and middle-class tenants, migrant workers, and radical students joined forces to protest against ruthless urban development projects in the interest of real estate speculators and the growing financial industries. With strong support among the local population, the militant movement organized rent strikes and mass rallies, and squatted, for the first time in German post-war history, a number of residential buildings that were supposed to be demolished in order to construct high-rise office towers. For several years, regular clashes with the police, mostly in the context of force-ful evictions of squatted buildings, grew into severe, violent and, up until then, unprecedented street riots (Stracke 1980).

During the 1980s a final shift from a national to an international financial center occurred, as Frankfurt's transformation into a global city was reflected in the further expansion of financial and producer service industries that materialized in new generations of high-rise buildings and an airport extension pushed through aggressively against local resistance. Especially resis-tance to a third runway called *Startbahn-West* was overwhelmingly supported by the local population including both respected citizens and left wing radicals. From 1980 to 1984, mass rallies with up to 120,000 people, an informal protest village in the forest close to the airport, site occupations barricaded highways and other tactics of sabotage and civil disobedience happened in the region. (Ronneberger and Keil 1995).

In addition to protests against the airport expansion, struggles against gentrification and the push of the office economy into residential neighborhoods con-tinued throughout the 1980s but became increasingly irrelevant politically in the 1990s (Keil and Ronneberger 2000). During the same time social protests 1) shifted from the center of the city to the (northern) periphery which has evolved into a prime location for the post-Fordist economy; and 2) became more populist, supported also by conservative farmers and middle-class suburbanites. These new types of turmoil against large scale urban infrastructure and mass housing projects represented "a clear departure from the notion that urban social movements in Frankfurt were hegemonized by the left wing or green milieu alone" (ibid.: 245). Since the 1990s, this populist kind of (partly NIMBY-) anti-growth protest has continued in different forms until today. One of the most remark-able mass protests erupted, for instance, over the opening of another runway called *Nordwestbahn* at the Frankfurt airport in 2011 since the aircraft noise of the new flight path now also affects upper and middle-class neighborhoods mainly in the south of Frankfurt. While the airport expansion had for years been proclaimed as a vital strategy for increasing the global competitiveness of the city by political elites

from conservative, green and social democratic par-ties, its realization still led to strong anti-growth unrest heavily supported by traditionally more conservative parts of the local population.

Frankfurt's selection as the location for the newly established ECB in 1998 reflected, first of all, the leading position of Frankfurt for European continental financial industries and, secondly, aggressive lobbying by the national and local governments. In 2014–15, the ECB relocated its headquarters from CBD into a newly built high-rise in the eastern part of the city called *Ostend*, putting that neighborhood under further heavy gentrification pressure. Sanctioned by the muni-cipality, large-scale luxurious urban redevelopment projects to gentrify this working-class neighborhood and the surrounding old industrial areas located along the axis of the Main River had been ongoing for years. Against this background, the political decision to permit the ECB construction of a high-rise building outside of the existing city-master-plan heightened the pressure on the housing market for working-class families. As the ECB employs more than 1,500 high-wage professionals often looking for upscale, centrally located housing, it is more than likely that the reloca-tion will, intended or not, lead to the further displace-ment of low-income households from one of the last remaining affordable and centrally located areas within the city.

As an effect of the global city formation based on financial industries and an international airport economy, the social polarization in Frankfurt increased significantly. A rapid deindustrialization process since the early 1990s, resulting from increasing land values, and a substantial loss of unionized blue-collar employ-ment opportunities went along with the growth of high paid jobs in the financial industries on the one side and low-wage jobs in the service sector on the other one (Schipper 2014:241). Therefore, the metro-politan region of Frankfurt is, compared to most other German urban centers, faced with a disproportionally high social inequality (Bock and Belina 2012). Further-more, this polarized income structure is, nowadays, confronted with a return of the housing question and a new wave of state-led and crisis-induced gentrifi-cation processes. Between 2010 and 2013, average housing prices increased by 35% (City of Frankfurt 2014: 22), while the closely linked rent level rose by 20%. As centrally located, former working-class neighborhoods face most severe rises of rents and housing prices, low-income and also middle-class households are displaced and pushed further to the periphery of the metropolitan region (Heeg and Holm 2012). As a result, a new wave of urban social protests, in many ways similar to the left wing housing struggles of the 1980s, has emerged, contesting escalating housing prices and the gentrification of centrally located neighborhoods. These protests are organized and carried out by grassroots networks of trade unionists, tenants' associations, local left wing politicians, neighborhood initiatives and more radical housing activists including a lively squatter movement that has occupied more than ten vacant buildings since 2011— all of them subsequently evicted by the police.

The social movements illustrated above are examples of politicization that emanated not least from the global city formation of Frankfurt. As Frankfurt is a place where the contradictions of the capitalist economy are concentrated, growth periods have, during the last fifty years, "always been linked to forms of local resistance" whereby "[u]rban social conflicts usually erupted where the new phase of expansion manifested itself most visibly" (Keil and Ronneberger 2000: 242). So far the struggles tackled processes primarily on the local (anti-gentrification, squatter movement, anti-growth protests) and regional (airport expansions: *Startbahn-West, Nordwestbahn*) scales. In contrast, Blockupy is the first attempt to not only articulate opposition to the processes evoked by global city formation but to use Frankfurt as a local anchor point to challenge globalized, finance-market driven capitalism and the European crisis regime.

THE BLOCKUPY PROTESTS IN FRANKFURT 2012 AND 2013

Blockupy is a broad German national political alliance with transnational network ties that was founded in reaction to the German and European crisis management in early 2012. The alliance consists of activists from left wing parties, trade unions, the Occupy movement, associations of the unemployed, the environmental and peace movements, anti-racist and anti-fascist initiatives and different other groups of the radical and more moderate Left. Inspired by the Arab spring, the US-Occupy-movement and the wave of mobilizations in Spain and Greece, the idea was born to organize five days of protest in the financial district of Frankfurt in May 2012 including demonstrations, workshops and lectures in public places all over the

city, and a number of creative, non-violent actions of civil disobedience. Part of the latter was the strategy to block and shut down the financial district and the ECB for several days by setting up protest camps in the city center and by peacefully blocking the main entrance routes with thousands of people. While Frankfurt was, originally, only one of several possible options, the city was selected due to its global city functions as well as its long-standing tradition of social struggles. For instance, the Blockupy Call for Action for 2013 states: "We carry our protest, our civil disobedience and resistance to the residence of the profiteers of the European crisis regime to Frankfurt am Main" (Blockupy 2013).

However, in contrast to the expectations of the organizers and in contrast to former experiences with similar forms of protest in Germany, the City of Frankfurt legally prohibited all demonstrations, camps, public lectures and workshops and banned any kind of protest for the week of the events. Arguing that the Blockupy protests could be used by militant radicals to riot in the city, the local government legally produced a space of exception where basic democratic rights were suspended (Petzold and Pichl 2013). Supported by a panic discourse that "our city" was "threatened", shops and banks in the city center were closed and boarded up, and subway stations and public institutions like the Goethe University were shut down. The banks turned off all ATMs in the city center and with them the access to cash money in Germany's financial capital and the region's most important commercial district. Further stirring the panic, the local police as well as several businesses advised employees to take a day off or to wear casual clothes instead of suits due to the alleged threat of being identified as bankers.

Although any protest had been legally banned, the aim of shutting down the financial district was surprisingly successful – yet not in the way the organizers had imagined. In order to prevent protesters from entering the city center, an overwhelming force of 5,000 police officers including hundreds of cruisers, mounted police, water cannons and even armored vehicles blocked the CBD of Frankfurt for several days (Plate 33).

However, the police also detained more than 1,400 people who tried to gather in small groups. In sum, this dialectic of control and contestation resulted in an urban space in which everyday life was thoroughly disturbed and through which the CBD was shut down

Plate 33 ECB headquarters

Source: Stefan Rudersdorf

in a much more efficient way than any blockades by protesters could have ever achieved. Setting a strong symbol against this authoritarian state of emergency, 30,000 people joined the demonstration on the final day of the Blockupy mobilisation—which was the only protest event that had been legalized by court order in a last-minute decision.

One year later, in May 2013, the Blockupy alliance mobilized again for days of action in Frankfurt. Like in 2012, it was agreed to perform mass actions of civil disobedience combined with a larger demonstration at the end to disturb the regular business flows within the global city. The non-violent but consistent resistance against the repression in 2012 had subverted the legitimacy and possibilities of state authorities to follow a similar strategy of criminalization. Thousands of police officers were concentrated in the CBD once more and acted on their order, this time, to be more reserved and let the protest take its course. On the morning of Friday, May 31, around 3,000 activists

gathered in front of the ECB in order to block its entrances. The blockades were held until noon when a 'second wave' of protest was launched. These more decentralized actions of civil disobedience took place at the Frankfurt International Airport to challenge its role of being Germany's main hub for the deportation of refugees; at the entry of the *Deutsche Bank* headquarters to protest against land grabbing and financial speculation on food prices; in front of real estate investors to claim the right to the city which has increasingly been denied to the urban poor due to escalating rents and housing prices; and on the main city center shopping street where several malls and multinational retailers were blocked to protest the working conditions in textile industries in the Global South. In sharp contrast to the situation in 2012, the days of action in 2013 could take place more or less without police interference. While in 2012 the dialectic of control and contention resulted in an authoritarian shut down of the city center, this

Plate 34 Police in front of the ECB

Source: Timmy Lichtbild

time around protesters were in the driver's seat. However, on the following day the final demonstration of around 20,000 people would not get far. After about 800 meters, police forces in full riot gear stopped the peaceful demonstration, cut off and encircled the first 1,000 participants for more than nine hours, and injured another several hundred with tear gas and batons (Plates 34 and 35). Subsequently, the media and the public roundly condemned the police strategy and the unjustified violence against a peaceful and legal demonstration. The public media outrage culminated in a tumultuous press conference of the Minister of the Interior of Hesse and the local police president, when mainstream journalists publicly denounced the police strategy as "a shame for Frankfurt." As a result, over 10,000 local citizens took to the streets a week later to demonstrate against police-violence and the second attempt to criminalize social protests against the European crisis management.

CONCLUSION

Since 2012, Blockupy has, even if not primarily focusing on urban issues and grievances, become part of Frankfurt's long-standing tradition of urban-based social movements by using the historical and ongoing global city formation to articulate its protest against

Plate 35 Police using tear gas

Source: Strassenstriche

the German and European austerity regime. In line with Uitermark's (2004: 710) argument that, "some points in space are obviously more vulnerable than others," the ECB within the global city of Frankfurt in particular became a symbol of neoliberal capitalism, austerity politics and European power relations. By using the ECB as a symbol, Blockupy opened up for the first time a path to connect transnational social struggles against German and European crisis management with local protests against the neoliberal urban restructuring within the global city of Frankfurt. This place-based and cross-scalar networking was made possible by and turned against the very networks that formed the global city Frankfurt in the first place. The global city, one might conclude, in peculiar ways produced the possibilities to attack the root causes of its devastating social, economic and political realities.

REFERENCES FROM THE READING

Blockupy (2013) Call for action: Blockupy Frankfurt! Resistance in the heart of the European crisis regime. https://blockupy.org/en/287/call-blockupy frankfurt accessed on October 10, 2014.

Bock, S. and Belina, B. (2012) Armut und Reichtum in der Rhein-Main-Region, Rosa-Luxemburg Stiftung. http:www.rosalux.de/fileadmin/rls_uploads/pdfs/sonst_publikationen/studie-armut_reichtum_rheinmain_2012-7.pdf accessed on October 10, 2014.

City of Frankfurt (2014) *Immobilienmarktbericht 2013 für den Bereich Frankfurt am Main des Gutachterausschuss für Immobilienwerte*, Frankfurt am Main.

Heeg, S. and Holm, A. (2012) Immobilienmärkte und soziale Polarisierung in der Metropolregion Frankfurt Rhein-Main, in J. Monstadt et al. (eds.) *Die diskutierte Region. Probleme und Planungsansätze der Metropolregion Rhein-Main*, Frankfurt am Main: Campus: 211–30.

Keil, R. and Ronneberger, K. (2000) The globalization of Frankfurt am Main: Core, periphery and social conflict, in P. Marcuse and R. van Kempen (eds.) *Globalizing Cities: A New Spatial Order?* Oxford: Blackwell: 228–48.

Mullis, D. and Schipper, S. (2013) Die postdemokratische Stadt zwischen Politisierung und Kontinuität. Oder ist die Stadt jemals demokratisch gewesen? *Sub/urban: zeitschrift für kritische stadtforschung*, 1, 2: 79–100.

Petzold, T. and Pichl, M. (2013) Räume des ausnahmerechts: Staatliche raumproduktion in der krise am beispiel der Blockupy-Aktionstage 2012, *Kriminologisches Journal*, 45, 3: 211–27.

Ronneberger, K. and Keil, R. (1995) Ausser Atem – Frankfurt nach der Postmoderne, in H. Hitz, U. Lehrer and R. Keil (eds.) *Capitales Fatales: Urbanisierung und Politik in den Finanzmetropolen Frankfurt und Zürich*, Zürich: Rotpunktverlag, 286–353.

Schipper, S. (2014) The financial crisis and the hegemony of urban neoliberalism: Lessons from Frankfurt am Main, *International Journal of Urban and Regional Research*, 38, 1: 236–55.

Stracke, E. (1980) Stadtzerstörung und Stadtteilkampf, Innerstädtische Umstrukturierungsprozesse, Wohnungsnot und soziale Bewegungen in Frankfurt am Main, Köln: Pahl-Rugenstein.

Uitermark, J. (2004) Looking forward by looking back: May Day protests in London and the strategic significance of the urban, *Antipode*, 36, 4: 706–27.

Uitermark, J., Nicholls, W., and Loopmans, M. (2012) Cities and social movements: Theorizing beyond the right to the city, *Environment & Planning A*, 44, 11: 2546–54.

PART SIX

Culture

Plate 36 Graffiti-covered streets, São Paulo

Source: Xuefei Ren

INTRODUCTION TO PART SIX

Culture as practice, talk about culture, cultural politics, cultural production, and products are all–and have always been–an integral part of the formation of world and global cities. British social theorist Raymond Williams (1976: 77) famously noted: "culture is one of the two or three most complicated words in the English language." Culture is a multifaceted concept that includes notions of civilization, product, anthropological practice, and so forth. Williams distinguishes culture as *material* production and culture as a *symbolic* system. In the most colloquial and widespread usage of the term, "culture is music, literature, painting and sculpture, theater and film" (Williams 1976: 80). Additionally, culture in global city scholarship can also refer to place-making strategies for urban revitalization, and to the symbolic representation of historical, political, and economic undercurrents in a social formation. The chapters included in this section discuss both material and symbolic elements of culture in globalizing cities by examining cultural production, cultural strategies, and cultural politics of representation.

The most direct and intuitive connection to culture made in the literature on global cities has been in work on multiculturalism. Often without making use of the term "culture," Friedmann and Wolff and other early analysts of global city formation noted that the arrival of migrants from the global periphery enhanced social polarization within major urban regions (Ch. 3). Yet, the reality of cultural diversity in world cities remained fairly underexplored in most accounts throughout the 1980s and 1990s. Many observers in the 1980s declared that "the Third World has come home"; many argued, further, that the new diasporic populations were not coherently integrated into social formations that were frequently hostile to immigration from non-European countries (Amin 2002). Subsequently, as world cities emerged as more clearly recognizable hubs of global migration during the 1980s, their multicultural character attracted greater academic attention (Davis 1987).

Several contributions to this section examine cultural production in globalizing cities, for example, in the form of music, architecture, nightlife, and also in the genre of media and creative industries. Watson et al. (Ch. 55) discuss the cultural geography of music production within and between global cities, and they underscore the importance of place in music creativity. As they point out, the multicultural, ethnically diverse neighborhoods in London are hotspots for music talent, and these places are woven into a global network of cultural production with travels of artists, record production, and music sharing on the web. Farrer and Field (Ch. 53) examine clubbing cultures in Shanghai's entertainment districts, and they identify the emergence of an ethnosexual field in which foreign visitors, expatriates, and locals negotiate their sexual subjectivity on dance floors.

Cultural dynamics also figure crucially in the production of built environments, urban form, and sociospatial divisions around the world, from the global cities of the West to the expanding metropolises of the global South. As King (Ch. 49) shows, a culture of distinction associated with gentrified built environments has now reached the rapidly expanding exurbs of Indian and Chinese world cities such as Bombay/Mumbai and Shanghai. As King (1996: 3) explains elsewhere, "the increasingly differentiated cultural discourse on the city is itself self-generating and part of the accumulation of cultural capital in the city. It is linked to the massive expansion of the cultural industries in (especially) the western metropolis–to publishing, education, the media, the symbolic and representational realms of advertising, and the 'aestheticisation' of social life." Similarly, Sklair (Ch. 52) views global cities as breeding grounds and markets for the production of architectural design. He identifies key components of the transnational capitalist class engaged in architectural production, and argues that the era of "starchitects" and megaprojects of the 2000s is inseparable from the rise of global cities and formation of global urban networks. The role of cultural strategies in shaping the built environment is further explored at length in Ute Lehrer's chapter (Ch. 51). Lehrer argues that the "willing" of the global city in post-unification Berlin rests on a distinctively culturalist discourse of "spectacularization." Lehrer emphasizes the discourse of urban design for the status and competitive strategies of global cities. While some cultural strategies are legitimate, others are ambiguous and they blur the line between legal and illegal. McAuliffe (Ch. 54) explores the moral geography of graffiti in Sydney and he presents a rather delicate picture of how place-boosters, policy makers, and artists negotiate the boundary of unauthorized graffiti/vandalism and legally sanctioned street art in their attempt to make Sydney into a creative global city.

In recent years, public and academic debates on urbanism have been strongly influenced by the notion of creativity, which is frequently invoked as the key to industrial renewal and urban revitalization. The idea of "creative cities" frequently presupposes that particular types of urban images must be mobilized in order to project the identity of particular places into the global space of flows (Florida 2002). Florida's thesis has been subject to intense criticism as the creative city formula for urban regeneration inevitably excludes communities at the margins of the creative economy. The consolidation of creative industries, such as media and cultural products across the network of global cities, has been explored systematically by Stefan Krätke (see Ch. 50). As Krätke explains, various clusters and cliques of media production centers, so called "global media fields," have emerged in close conjunction with the world city network (see also Scott 2000; 2005).

As we move from cultural production and strategies to the realm of representation, the influence of postcolonial theory in global city scholarship is more apparent. Postcolonial studies are focused on questions of knowledge and power, "issues of agency, representation and especially, the representation of culture(s) under asymmetrical political and social conditions" (King 2004: 48; see also King Ch. 49). For instance, Anthony King builds upon the notion of postcolonialism, a complex and contested term which he defines as a combination of material practices and discursive formations that emerge "after the colonial" (King 2004: 45). Beyond this relatively straightforward use of the term, the notion of postcolonialism also denotes the type of knowledge produced in postcolonial studies, a distinct theoretical approach that has emerged in close conjunction with the migration of diaspora academics to metropolitan universities. From the postcolonial perspective, Varma (Ch. 56) explores the cultural politics of representation of Bombay/Mumbai, and she finds the irony that the period from the 1990s when Mumbai became recognized as a global city was also the period of rising ethnic tensions and communal violence, which she calls "provincializing" tendencies. The troubling trends of ethnic intolerance are captured in film and literature, and these cultural representations both constitute and complicate the discursive construction of Mumbai as a global city.

In many ways, global cities research has itself been permeated by a form of cultural imperialism insofar as it has been grounded predominantly upon Western cultural forms and intellectual paradigms (see Robinson Ch. 9). Reflexively grappling with these issues, the contributions to this section point towards productive new ways of thinking about globalized urbanization that transcend conventional Eurocentric categories, assumptions, and narratives by viewing the global city as a site of multicultural production, intervention, and representation.

REFERENCES AND SUGGESTIONS FOR FURTHER READING

Amin, A. (2002) Ethnicity and the multicultural city: Living with diversity, *Environment & Planning A*, 34: 959–80.

Davis, M. (1987) Chinatown, part two: The 'internationalization' of downtown Los Angeles, *New Left Review*, 164: 65–86.

Florida, R. (2002) *The Rise of the Creative Class*, New York: Basic Books.

King, A. (ed.) (1996) *Re-Presenting the City*, London: Macmillan.

King, A. (2004) *Spaces of Global Cultures*, London and New York: Routledge.

Krätke, S. (2011) *The Creative Capital of Cities: Interactive Knowledge Creation and the Urbanization Economies of Innovation*, Oxford: Wiley-Blackwell.

Sassen, S. (1998) *Globalization and its Discontents*, New York: The New Press.

Scott, A. (2000) *The Cultural Economy of Cities*, London: Sage.

Scott, A. (2005) *On Hollywood*, Princeton, NJ: Princeton University Press.

Williams, R. (1976) *Keywords*, Glasgow: Fontana.

48 Prologue
"High culture and hard labor"

from *New York Times* (2014)

Andrew Ross

Across a narrow sea channel from Abu Dhabi's sleek towers, construction on Saadiyat Island is proceeding at a pace that's extreme even by the standards of this Persian Gulf boomtown.

Planned as the mother of all luxury property developments, Saadiyat's extraordinary offer to the buyers of its opulent villas is that they will be able to stroll to the Guggenheim Museum, the Louvre and a new national museum partnered with the British Museum. A clutch of lustrous architects—Frank Gehry, Jean Nouvel, Zaha Hadid, Rafael Viñoly and Norman Foster—have been lured with princely sums to design these buildings. New York University, where I am on the faculty, will join the museums when its satellite campus opens later this year. But there is a darker story behind the shiny facades of these temples to culture, arts and ideas.

On Saadiyat, and throughout the gleaming city-scapes of Abu Dhabi and Dubai, the construction workforce is almost entirely made up of Indian, Pakistani, Bangladeshi Sri Lankan and Nepalese migrant laborers. Bound to an employer by the kafala sponsorship system, they arrive heavily indebted from recruitment and transit fees, only to find that their gulf dream has been a mirage. Typically, in the United Arab Emirates, the sponsoring employer takes their passports, houses the workers in substandard labor camps, pays much less than they were promised and enforces a punishing regimen under the desert sun.

In its 2006 report "Building Towers, Cheating Workers," Human Rights Watch issued the first of its several critiques of the kafala system. The official response has been mixed: some reforms have been made to labor law, but representatives from Human Rights Watch have been barred from entry, and almost a thousand migrants have died in neighboring Qatar while building infrastructure for the 2022 World Cup.

Saadiyat is supposed to be a model exception. The government's Tourism Development and Investment Corporation has installed a well-equipped worker village (though it still has the feel of a detention camp), along with employment policies that look good on paper. But the policies are not adequately enforced. Employers are supposed to pay off their workers' recruitment fees, though very few do, and many contractors house their workers more cheaply in poor facilities elsewhere. Every independent investigator who has visited these off-island locations has turned up multiple violations of the employment codes.

Earlier this month, I interviewed workers employed on Saadiyat projects, accompanied by my colleagues from Gulf Labor, a coalition of artists and writers convened three years ago to persuade the Guggenheim and the Louvre to raise labor standards. Gulf Labor has led an international boycott of the museum's Abu Dhabi branch by more than 1,800 artists, writers, curators and gallery owners—many of them respected names whose work the Guggenheim would like to acquire for its Saadiyat collection.

On our trips through the archipelago of labor camps that encircles Abu Dhabi and Dubai, we stopped at a makeshift Punjabi restaurant in the industrial area known as Al Quoz. There, in the early morning hours, we spoke with a number of workers including one named Ganesh, who has worked on buildings for N.Y.U. and the Louvre. Slightly built with a dazzling smile, he switched between Hindi and English to explain his predicament. Owed a year's wages by the recruitment company that brought him from Nepal, he is unable to leave the U.A.E., more than 10 years later, because his sponsor has his passport (and his back pay). His labor visa has expired, and he is surviving on canteen credit and illegal work stints.

Paying off recruitment debts consumed his first two years of hard labor in the U.A.E. During that time, his family's subsistence farm in the Himalaya foothills had been at his creditor's disposal. "Three or four out of 10 lose their land," he said, "when they can't repay on time." His next decade in the U.A.E. was spent scratching out thin remittances to send to his wife and children. On some work projects, he was housed three hours from the construction site. To put in a mandatory 12-hour shift, "I had to wake up at 4 a.m.," he said, "and then had to cook my dinner after I returned at 10 p.m."

Last month, a Gulf Labor offshoot (the Global Ultra Luxury Faction) occupied the Guggenheim Museum in New York, protesting labor conditions in Abu Dhabi. In response, the museum's director, Richard Armstrong, claimed that the Guggenheim's Abu Dhabi expansion, designed by Mr. Gehry, is not yet under construction. Yet the extensive foundation pilings and much of the surrounding infrastructure have already been laid.

It's not too late for the museum to break with the practices that have built the Louvre and N.Y.U. And there is still time for Mr. Gehry to counter the ugly implications of Zaha Hadid's remarks after the deaths in Qatar, where she designed Al Wakrah stadium. "I have nothing to do with the workers," she said. "It's not my duty as an architect to look at it."

The U.A.E. is hardly alone in its dependence on tragically underpaid and ill-treated migrant workers. Every developed, and fast-developing, country has its own record of shame. But in the Persian Gulf States, the lavish lifestyle of a minority composed of citizens and corporate expats is maintained by a vast majority that functions as a servant class.

If liberal cultural and educational institutions are to operate with any integrity in that environment, they must insist on a change of the rules: abolish the recruitment debt system, pay a living wage, allow workers to change employers at will and legalize the right to collective bargaining. Otherwise, their Gulf paymasters will go on cherry-picking from the globalization menu— Lamborghinis, credit default swaps, liberal arts degrees, blockbuster exhibitions—while spurning the social contract that protects basic human rights.

49

"World Cities: Global? Postcolonial? Postimperial? Or just the result of happenstance? Some cultural comments"

Anthony D. King

EDITORS' INTRODUCTION

Anthony King is Professor Emeritus of Sociology and Art History at the State University of New York, Binghamton. In his work on global cities, King has been particularly interested in the social production of the building form, the relationships of colonialism and urbanism, social and spatial theory, postcolonial theory and criticism, as well as transnational cultures. In this chapter, King powerfully summarizes and integrates these various facets of his impressive body of work on global cities. King begins with a brief discussion of the concept of a postcolonial critique, which he defines as being "essentially concerned with questions of agency, representation, and especially, the representation of cultures under the asymmetry of global political and social conditions." On this basis, King points to the neglect in global cities research of historical and cultural considerations. After criticizing the present-oriented economism of this literature, King also calls into question the assumption that quantitative representations can adequately capture the lives of migrant populations in world cities. In the next section, King discusses the ongoing race among global cities to construct the tallest skyscraper. King contrasts this struggle for recognition, centrality, and symbolic power in the sphere of downtown architecture with the reality of global suburbanization, which he views as an expression of the new forms of sociospatial exclusion that are associated with global city formation. Consequently, in the shadow of the official quest for cultural recognition, King detects a global culture of the *banlieusards*, the inhabitants of the suburbs. Finally, King emphasizes the colonial legacies that are embedded in the spatial structures of many postcolonial cities and suggests that such legacies have important effects upon contemporary power structures.

INTRODUCTION

Postcolonial theory and criticism has been slow in penetrating the social sciences, not least, the literature of urban studies. At the basis of the postcolonial critique are questions of knowledge and power. A postcolonial critique is essentially concerned with questions of agency, representation and especially, the representation of cultures under the asymmetry of global political and social conditions. The critique of "Eurocentrism" is its basic task.

Postcolonial criticism most obviously assumes a knowledge of colonial histories in the contemporary world, not only those of major European imperial powers (France, Britain, Spain, Portugal, the Netherlands) but also of the USA, Russia and Japan.

Recent interpretations of globalization have done much to correct the initial Eurocentric and ahistorical focus of that process, both by examining globalization from the viewpoint of countries (and religions, including Islam) outside the west, but also tracing earlier, Islamic phases of globalization, before European hegemony, as well as later ones, including, since 1950, that of "post-colonial globalization" (Hopkins 2001).

Despite these theoretical developments, work on the historical origins of what have come to be called "global cities," either individually or in the form of a network or system (Taylor 2004), is still woefully neglected. A glance at any good 1900 atlas would, however, show that the majority, if not all, of the 82 cities "cited in world city research" were at that date (and earlier) either capitals or major cities of European or Asian empires or, alternatively, cities established in colonial or one-time colonial territories. (Here, I treat the "world cities" of north and south America, South Africa, Asia as well as Australia as resulting essentially from European and Asian colonial expansion. And in relation to their indigenous peoples, as colonizing or imperial.) The networks linking these cities to their real or one-time metropolitan core, whether by sea, over land and eventually by air, telegraph, telephone, the press, by economic and social processes of migration, were synchronized most powerfully through various economic, social, political and cultural institutions, adapted and transformed to various degrees by local and indigenous agency. Of these institutions, language is crucially important.

Other powerful networks, sustained by the formal institutions of particular belief systems (Catholicism in South America and also elsewhere; varieties of Protestantism) which were significant influences in some locations but in other, more secularized societies, were less powerful, have nonetheless left sedimented values. The earlier networks of Islam, before European hegemony, were equally powerful in the Asian and middle eastern world, known for radically different conceptions of space and time, and also in relation to the cosmology of Hindu, Buddhist and other belief systems.

In the context of language, following the onset of European hegemony from the sixteenth century onwards, anglophone, francophone, hispanophone and lusaphone networks, transformed by national, transnational, and regional identities, continue today to provide the basis for transnational processes in important spheres of public and commercial life. The knowledge regimes established through these

languages, both in the one-time metropole and its one-time colonies, and especially resulting from the expansion of the historically Anglophone, EuroAmerican cultural sphere (i.e. US hegemony), are primarily accountable for the fact that, in regard to the geographical sciences, for example (but also in many other areas of knowledge), what is frequently referred to as "international" knowledge (Gutierrez and Lopez-Nieva 2001) is, in fact, postcolonial (King 2004). How else can we explain why this book is published, and also being read, in ("international") English and not in Arabic, Mandarin Chinese or Hindi? And while religion has, with some minor exceptions effectively been ignored in the debate on global cities, it has taken the events of 9/11 to remind us that the vast majority of these 82 cities were in territories historically part of the Judeo-Christian ecumene (where, in Europe, political parties still describe themselves as "Christian Democrat" and in the Americas, religion is a significant factor in politics). Where this is not the case, as with Mumbai, Jakarta, Hong Kong, Kuala Lumpur, it is worth examining whether their putative world city credentials rest on institutions and practices developed, in conjunction with indigenous agency and knowledge, in their colonial past. It is also worth asking where the networks of Chinese diasporic capitalism fit into this picture. As for Cairo (another of the 82 "world cities" mentioned above), this was already described as the "mother (city) of the world," "the glory of Islam and the center of the world's commerce" as far back as the tenth century, well before the era of European hegemony (Abu-Lughod 1989: 149, 225). In brief, research with a cultural focus might provide an answer to one of the still unanswered questions in "global city" research: why so many major cities in the world are not, according to the criteria currently used to identify them, "global cities."

This does not, of course, deny the significance of the empirical findings in regard to which the present day concept of the "world" or "global city" has been (largely) defined, nor the discourses that circulate around them: the increasing concentration of producer services such as accountancy, advertising, banking, law, insurance and the rest, whether introduced from "the West" or emerging locally. What seems conceptually problematic, however, is the appropriation of the term "global city" to categorize a class of city based primarily on these criteria (Sassen 1991). It implies that as the *presence* of these producer services becomes increasingly universal in major cities worldwide the *effects* of these services will somehow be similar. Or to put it differently, that focusing on an apparent similarity manifested by the growth of these services – which have effects in relation to the realm of the world economy – is more important than focusing on the differences which may result from them in relation to the realm of global culture. The separation of both these realms is, of course, purely methodological. What I wish to argue here is that different questions not only demand different kinds of research but also different positions from which to start.

My argument about the imperial and colonial origins of many so-called world or global cities is not just a matter of "getting the record straight" or to add a missing history, even though these are legitimate tasks. The aim is rather to draw attention to the overly economistic nature of the criteria driving the "world city paradigm" and its framing within a narrowly restrictive framework of urban political economy. It is also to highlight the ahistorical and analytically feeble nature of the category, "*global* city." As others have pointed out, assumptions in much of the literature, including some of that on global cities, that "global urbanism can be regarded as a uniform or homogenous outgrowth from Europe and America, belatedly affecting Africa, Asia and South America" or that "cities in Africa, Asia or South America can be understood on the model of cities in Europe, Australia or the USA" (Bishop et al. 2003: 2) are seriously flawed. While the largely quantitative data which characterizes much of recent global city research may tell us something about the organization of the contemporary world economy and the worldwide growth of contemporary capitalism, it fails to address the distinctive *cultural* forms of that economy and also the cultural characteristics of all cities, including postcolonial cities, not least, those affecting, and in particular cases determining, the nature of contemporary economic and political activity. Here, I use the term cultural in its widest sense both to refer to questions of meaning, identity and representation as well as language, religion and ways of life.

Of major importance here has been the rapid growth in recent years of information technology enabled service employment between different parts of the one-time anglophone empire, particularly between the USA, Canada, the UK and India, resulting in the outsourcing of tens of thousands of jobs from high cost to low cost states. The rapid growth of call center employment in India, which has created 336 call centers between 1997 and 2002 employing over 100,000 operatives and generating a \$1.4 billion industry (King 2004: 152), is based on the existence of a highly articulate, *English-speaking*, under-employed graduate population, educated in an essentially postcolonial, Western-oriented university system. It is rarely, if ever, acknowledged that, after the USA, India has the largest English-speaking population in the world. The development of this language and technology based sector has not only created extensive employment opportunities but is also related to the massive construction boom on the outskirts of cities such as Delhi, Bangalore and Hyderabad.

Recognizing these postcolonial cultural links is equally vital for addressing economic developments in "world cities" in the West. At one level in the economic and social hierarchy, virtually half the cabs of New York City are driven by English-speaking migrants from India, Pakistan and Bangladesh; at another level, some 10,000 highly educated graduates of India's six elite and highly selective Institutes of Technology occupy some of the top ranks of business, banking, and IT development in the USA. We also need to recognize that the cultural politics of India include anti-English protests by "Hindutva" supporting nationalists.

MIGRATION AND ITS CULTURAL EFFECTS

In the discourse about the multiculturality of global cities, attention is often drawn to the presence of "other" nationalities and language groups. That one city has representatives from 102 nations or language groups and another from 192, is somehow taken as a "sufficient and necessary" condition of its cosmopolitan character. This issue needs addressing both in more detail and with more sophistication.

Viewed from a demographic viewpoint, it might be useful here to imagine what an "ideal" or "utopian" global city would be, for example, one whose population was composed of representatives (say 1,000) of each nation-state, and in proportion to its total population, from each of the almost two hundred nation-states in the world. Irrespective of the artificiality implied by this notion of the "nationally constructed subject," such an "international" city would in no way get close to the tens of thousands of ethnic and linguistic groups which make up these nation-states.

The relevance of suggesting this ideal – or perhaps absurd – model, however, is to demonstrate that, from a demographic viewpoint, no so-called world city can, or ever will, approximate towards it. This is because in all world cities (at the time of writing, at least), there is a numerically dominant population from the host society and in each, the proportion who are "foreign born" comes from a diverse range of cultures, ethnicities, religions and regions, and also as a result of very different historical circumstances. Moreover, the monolithic (and also xenophobic) category, "foreign born" can, for all except legal purposes (though this is obviously a highly significant exception) be dismissed. From specific points of view, to assume that "foreign birth" provides commonality amongst a vast range of peoples, in either eighty or even thirty "world cities" in five continents, is as absurd as suggesting that "domestic birth" can be used to characterize, along cultural or ethnic lines, those born in the host society. Similarly, to suggest that the 15 per cent of the Paris population who are "foreign born" and coming primarily from North Africa (Algeria, Morocco, Tunisia), Armenia or Mauritius; the 28 per cent in New York, over half of whom are from the Caribbean and Central America, with significant proportions from Europe, South America, East Asia and South and Southeast Asia; and some 20 per cent "foreign born" in London, from South and Southeast Asia, Ireland, continental Europe, East, West and South Africa, the Caribbean, North America and Australasia, have somehow more in common than they have differences is equally problematic.

Clearly, the historical, cultural and political status and power (or lack of it) possessed by migrants from different countries, when relocated in the cities of another society, are highly variable and differentiated. Given that a large proportion, both in Europe and North America, are from "Third World," postcolonial societies, their colonial histories – as I discuss below – place different kinds of migrants in very different situations of power and lack of it, irrespective of their relation to the (economic) labor market.

Plate 37 Shopping malls, Bangalore

Source: Xuefei Ren

Quite apart from their influence in different sectors of the economy, however, the influence of postcolonial subjects on the culture and politics (as well as cultural politics) of the dominant society can clearly be substantial. There is no better example than the significant impact which postcolonial criticism (largely developed by scholars such as Arjun Appadurai, Homi Bhabha, Dipesh Chakrabarty, Edward Said, Gayatri Spivak and others) has had on the epistemology of the Western academy. In terms of cultural politics and theoretical critique, numbers are irrelevant. Salman Rushdie, after all, is only one person.

REPRESENTING AND SYMBOLIZING THE GLOBAL AND POSTCOLONIAL

How are the global and postcolonial represented and symbolized in the spaces of the world, global or postcolonial city? Prior to September 11, 2001, a partial answer to that question might have been found in the global discourses prompted by the construction, at increasingly frequent intervals, of what is represented as "the world's tallest building." Roger Keil has suggested that the gigantic tower has become the most important symbolic product of the world economy (King 2004: 4), a sign used by states and cities both to challenge the existing economic order and to make their own claims to contemporary modernity. The twin-towered Petronas Building in Kuala Lumpur, for instance, built in 1996, was some meters taller than the previous claimant, Chicago's Sears Tower. The event generated global publicity, including predictions about "the decline of the West" and "the coming of the Asian century." The logic behind the construction of the "world's tallest building," rests on the acceptance of a conceptualization of globalization as a process by which "the world becomes a single place" (Robertson 1992). The logic also assumes that creating a similitude, or joining a contest, with other states and cities whose conceptions of modernity it wishes to share will, ipso facto, place that state or city in the same reference group. Other cities assume that it will attract inward investment.

Two recent instances of the "world's tallest building" phenomenon include the completion of Taiwan's "Taipei 101" Building in October 2003, the aim of which was said to be "to put Taipei on the global map" as well as "bring in foreign companies." This is also the case with the (elite) sponsors of other mega-projects in Asia which, like Taipei 101, aim to provide "world class" accommodation to attract financial and other producer services (Marshall 2002). Indonesia's proposed Jakarta Tower is "five meters taller than the Canadian National Tower in Toronto," currently the "world's tallest *tower*." This would be "like other famous structures in big countries" and, from the city governor's perspective but not necessarily that of the inhabitants, would "enhance the image of Jakarta as a metropolitan city."

What these few examples illustrate, however, is that, for both the producers as well as the public, there is not just one but rather many worlds and they do not necessarily coincide. Malaysia's Petronas Tower was not only a postcolonial gesture from Prime Minister Mahathir Mohammed but, in incorporating Islamic motifs in its design, was both a signifier, as well as signified, of the world of Islam. The mammoth multistorey tower designed for London by British architect, Richard Rogers, to celebrate the Millennium (though never built) prompted a leading paper, *The Guardian*, to suggest that, if constructed, it "would confirm the status of Britain as a Third World country." According to the paper's architectural correspondent, super high-rise towers are "increasingly symptoms of 'second city syndrome'." After September 2001, the discourse in the United States about the "world's tallest building" has been decidedly subdued, and the contest, both metaphorically and literally, has been Shanghaied by particular representatives of the Chinese elite.

If the city center high-rise tower is, for some at least, the prototypical sign of the global city and its postcolonial aspirants, equally important is the space and culture of the suburbs. Simultaneously both global and postcolonial, their transformation in many "world cities" is best expressed in Hopkins' term as postcolonial globalization (2001). In Paris, most of the postcolonial minorities (from Algeria, Tunisia, Morocco) are in the *banlieues* which, "in the 1990s, have become a byword for socially disadvantaged peripheral areas of French cities" (Hargreaves and McKinney 1997: 12). Structurally equivalent to British and American inner city areas, and often referred to as ghettoes, the *banlieues* provide a natural space in which to develop "a separarist cultural agenda marked by graffiti, music, dancing, and dress codes" with which the *banlieusards* (suburb-dwellers) reterritorialize the "anonymous housing projects" (Ibid). In Britain, the urban landscapes of the eastern, western, and southern suburbs of the postcolonial/postimperial global city of London are regenerated and transformed by South Asian Muslims, amongst others, from other networks, who, though "united by belief, are nonetheless divided along national, ethnic and sectarian lines" (Nasser 2003: 9). In the terrain and terraced housing of Britain's "second city" of Birmingham, the largest group of the 80,000 South Asian Muslims are from Pakistan, and comprise 7 per cent of the city's population (Nasser 2003: 9). Though not sufficient in themselves, postcolonial histories are nonetheless central to any analysis of the multicultural nature of both suburbs as well as central areas of global cities in Canada, the USA, Australia, the Netherlands, France, Spain and many other countries in Europe as well as elsewhere in the world.

HERITAGE, CULTURAL IDENTITY AND SOCIO-SPATIAL EQUITY

In other postcolonial states around the world, what was once the "Western" space of the colonial urban settlement – in Delhi, Karachi, Singapore, Jakarta, Shanghai, Accra, Cape Town, Kolkata and elsewhere – has, in the half century following independence, frequently become the preferred residential area for the indigenous elite, or more recently, under neo-liberal political regimes, been redeveloped to provide luxury housing in gated communities, transnational tourist hotels, shopping malls and other spaces of contemporary capitalism. In these real or potential "global cities," one of the most urgent questions to address is whether the social and spatial polarization, frequently seen as the key characteristic of the contemporary "world" or "global city," is simply a continuation of the social, spatial and racial divisions occupied by colonized and colonizer in the earlier layout of the city. Whether in their nomenclature, design, financing or use, new luxury housing developments – apartments, villas, high rise towers – from Delhi to Jakarta, reveal both the persistence of postcolonial cultural connections to the metropole as well as localized versions of transnational forms of consumption (King 2004).

In recent years, an increasing number of scholars, institutions and authorities have argued for the preservation of parts, sometimes all, of the architectural, urban design and planning forms of one-time colonial cities. In addition to reports by UNESCO's International Committee on Monuments and Sites (ICOMOS) there are studies on cities ranging from New Delhi to Jakarta, Singapore to Karachi. According to these commentators, fifty years after independence, the symbolic significance of colonial buildings has lost its old political meaning. Younger citizens of some republics (and their postcolonial capitals) see colonial architecture and urban design as more important for generating revenue from tourism than for summoning memories of colonial oppression.

Yet few of such reports address buildings and spaces as social and political as well as aesthetic objects. The elite Dutch colonial suburb of Menteng, outside Jakarta, with its Art Deco houses and spacious tree-lined boulevards, continues, as in colonial days, to house the rich and powerful (including ex-President Suharto). In New Delhi, government ministers fight for the privilege of living in spacious colonial bungalows set in three and four acre compounds. Thousands of low paid clerical workers cram into over-packed buses to traverse the colonial wastes or either walk or cycle the interminable avenues in 110 degree summer temperatures.

Despite the many studies that have been made on the "divided city" of colonial times, there has not yet been a study which puts together the views of both sides. That is, the urgent need for a "decolonization" of "symbolic space" and its more equitable redevelopment, along with a sensitive, socially critical acknowledgement of its value as cultural heritage. Such studies would provide different perspectives on the real and potential postcolonial global city.

The more challenging question, however, is how to view what are called "world" and "global cities" not only from a world outside the West, such as Asia, but also from a time, such as the future, that is beyond the present.

REFERENCES FROM THE READING

Abu-Lughod, J. (1989) *Before European Hegemony*, New York: Oxford University Press.

Bishop, R., Phillips, J., and Yeo, W.W. (eds.) (2003) *Postcolonial Urbanism*, New York, London: Routledge.

Gutierrez, J. and Lopez-Nieva, P. (2001) Are international journals of geography really international? *Progress in Human Geography*, 25, 1: 53–70.

Hargreaves, A.G. and McKinney, M. (1997) *Postcolonial Cultures in France*, London and New York: Routledge.

Hopkins, A. G. (ed.) (2001) *Globalization in World History*, London: Pimlico.

King, A. (2004) *Spaces of Global Cultures*, London and New York: Routledge.

Marshall, R. (2002) *Emerging Urbanity: Global Urban Projects in the Pacific Rim*, London: Spon.

Nasser, N. (2003) The space of displacement: Making Muslim South Asian place in British neighborhoods, *Traditional Dwellings and Settlements Review*, 15, 1: 7–21.

Robertson, R. (1992) *Globalization*, London and Thousand Oaks, CA, and New Delhi: Sage.

Sassen, S. (1991) *The Global City: New York, London, Tokyo*, Princeton, NJ: Princeton University Press.

Taylor, P. J. (2004) *World City Network*, London and New York: Routledge.

50
"'Global media cities': major nodes of globalising culture and media industries"

Stefan Krätke

EDITORS' INTRODUCTION

Stefan Krätke is the Chair of Economic and Social Geography at the Europa-Universität Viadrina in Frankfurt (Oder). He has been one of the most productive and creative European global city researchers and has collaborated with the GaWC research group to explore the role of media industries in the global cities network. Krätke's work on European urban regions combines detailed economic analysis with sophisticated modes of sociospatial explanation. His interdisciplinary approach to urban and regional studies is evident in the chapter below, in which he examines the institutional and geographical dimensions of the global media industry, emphasizing in particular the production of cultural commodities. According to Krätke, the products of cultural industries in global cities serve corporate demand in the centers of the global economy and are widely consumed throughout the world economy. However, as Krätke points out, the global map of media cities is only partly congruent with the global map of financial centers. Krätke's contribution also critically engages with debates on the role of the creative class in urban economic development. His book, *The creative capital of cities* (2011), challenges concepts such as the creative class and creative industries (Florida 2002) from a critical urban theory perspective. With empirical data analyses on Hanover and Berlin, Krätke demonstrates that the cultural industry exacerbates existing inequalities in the neoliberal labour market.

This article examines the link between cities and culture from the point of view of the *production* of cultural goods and media products. The culture and media industry is a prime mover for globalisation processes in the urban system, in which cultural production clusters act as local nodes in the global networks of the large media groups. The analysis of "global media cities" enables those locations to be identified, from which globalisation in the spheres of culture and the media proceeds. Global city research has predominantly emphasized the role of advanced producer services in the development of the contemporary world city network. This article emphasizes that for the process of globalisation the globally operating media firms are at least as influential as the global providers of corporate services, because they create a *cultural* market space of global dimensions, on the basis of which the specialised global service providers can ensure the practical management of global production and market networks.

MEDIA CITIES AND THE INSTITUTIONAL ORDER OF A GLOBALISING CULTURE INDUSTRY

Media city is a term currently used to describe culture and media centers operating at very different geographical levels. They range from small-scale local urban clusters in the media industry to the cultural metropolises of the global urban system. An *up-to-date* examination of culture and cities ought to have the "commodification of culture" as a central theme, i.e. the worldwide assertion of the market economy in the form of the market-focused production of cultural commodities and the market-related self-stylisation of individuals competing·for positions in societies characterised by the all-embracing mediatisation of social communication, consumption patterns and lifestyles. The culture and media industry embraces those branches of social activity that are determined to a large extent by creative work and the production and communication of symbolic meanings and images.

A main characteristic of the culture and media industry's geographical organisation is the selective concentration of culture and media producers in a limited number of large cities and metropolises within the *global* urban system. The other characteristic is the formation of clusters *within* the boundaries of large cities, i.e. the local concentration of cultural production in particular urban districts, preferably in the inner city area (Krätke 2002a). The locational patterns of the culture and media industry in selected global cities such as Los Angeles and London reveal that cultural production tends to the formation of local agglomerations of specialised firms (Scott 2000). The second, most important feature of the present-day cultural industry's institutional order is the *globalisation* of large cultural enterprises, which enables global media firms with their worldwide network of subsidiaries and branch offices to forge links between the urban clusters of cultural production. This supra-regional linkage of local media industry clusters lies at the heart of an emerging system of *global media cities* within the worldwide urban network.

Today's culture industry is a highly differentiated business incorporating diverse sectors that range from traditional artistic production to technology-intensive branches of the media industry. The products of these activities are of the utmost cultural importance in that they function as agents of information, influence and persuasion or as vehicles of entertainment or social self-portrayal. They include primarily the diverse branches of the entertainment and media industries, e.g. theatres and orchestras, music production, film production, television and radio productions, the printing and publishing trade, as well as design agencies and the advertising industry.

The "image production" activities of the cultural economy (Scott 2000) in today's marketing society include not merely the product images created by advertising and design agencies, but also the lifestyle images communicated via the programme formats of the entertainment and media industries. There is considerable overlapping between the culture industry and the media sector, since a large part of cultural production is organised directly or indirectly as a special value chain within the media industry. The media industry acts as a focus for the commercialisation of cultural production and is also to be found at the heart of the "culturalization of the economy," given that its market success is founded on the construction of images and extensive marketing activities that are supported by the media industry.

New communications technologies and the emergence of large multinational groups within the culture and media industry contribute to a global flow of cultural forms and products, whose reach, intensity, speed and diversity far exceed the cultural globalisation processes of previous eras (cf. Held et al. 1999). The emphasis below will be on the *corporate* infrastructure that produces and distributes the content and products for a globalized cultural market. The present-day flows of cultural globalisation proceed primarily from the industrialised nations of the West and their market-focused culture and media industry. However, the global firms in the culture and media industry are obliged to take account of specific tastes and cultural preferences in other countries. The market strategists employed by global media firms are well aware of the cultural variety and differentiation of their global audiences and customers, and have long given their products and programmes a "regional touch" with a view to stabilising or enhancing their global market success. In other words, they have adapted their products and programmes to specific regional or national tastes and cultural preferences. This trend towards cultural market differentiation is at the same time a driving force for the organisation of global production *networks* in the culture and media industries with "local" anchoring points in different regions and nation-states.

Today's cultural economy is characterised by a marked trend towards the globalisation of corporate *organisation*. The formation of huge media groups is accompanied by the creation of an increasingly *global network* of branch offices and subsidiaries. This global network of firms linked under the roof of a media group has its *local anchoring points* in those centers of the worldwide urban system that function as "cultural metropolises," i.e. as centers of cultural production. The global media groups are organising the worldwide spread of media content and formats which are generated in the production centers of the global media industry, in particular in Los Angeles, New York, Paris, London, Munich and Berlin. The culture and media industry is a prime mover for globalisation processes in the urban system, in which cultural production clusters act as local nodes in the global networks of large media groups.

The globalisation *strategy* pursued by media firms is not geared, as is the case in many industrial groups, to the use of "cheap" labor and the like, but primarily to market development and extension through the establishment of a presence in all the major international centers of the media industry. Secondly, the strategy of media firms reveals a strong trend towards using creativity resources on a global scale. A presence in the leading centers of cultural production offers global media firms the chance to incorporate the latest fashion trends in the cultural industry as quickly as possible.

In a "global" media city there is an overlapping between the location networks of *several* global media firms. The *local and the global* firms in the culture and media industry are linked here in a development context that fosters the formation of an urban media cluster, whose international business relations are handled primarily via the global media firms that are present. The local media cluster in Potsdam/Babelsberg on the outskirts of Berlin might be taken as an example (Krätke 2002b) to show that cluster firms are not only closely networked within the local business area, but are also integrated into the supra-regional location networks of *global* media firms: In the case of Babelsberg the *local* cluster firms are directly linked with the resident establishments of global media firms from Paris, London and New York. The establishment of a global network of business units that are integrated at the same time into the local clusters of the culture industry enables the large media groups to tap place-specific creativity resources on a global scale.

GLOBAL MEDIA CITIES IN A WORLDWIDE URBAN NETWORK

Most of the studies on global cities and the economic and functional structures of the international urban system reveal a tendency to reduce the "high-ranking" global cities to their function as financial centers and centers of specialised corporate services. This chapter highlights the particular role of the culture and media industry in the formation of a world city network, thus emphasizing the *variety* of economic activities which are involved in globalization processes.

There has been a severe data deficiency in the study of world cities, particularly for measuring inter-city relations. One solution, that pioneered through GaWC researches (Taylor 2003), has been to conceptualise the world city network as an *interlocking network* which then allows relations between cities to be measured through data collected on firms. An interlocking network has three levels: as well as the usual two levels comprising the nodal level (the cities) and the network level (all nodes and links, cities connected), there is a subnodal level comprising firms (corporate service firms, media firms, etc.). The latter "interlock" the cities to create a city network through their inter-city locational policies (Taylor 2003). Intra-firm flows of information, knowledge, instruction, ideas, plans and other business between offices/enterprise units are creating a world city network based upon the organizational patterns of global firms.

The requirement for an interlocking network analysis is a matrix of cities and firms showing which firms have establishments in which cities and the relative importance of the cities within a firm's organisational network. The research on the media industry's world geography started by identifying the most important urban nodes of the global media firms' locational network (Krätke 2003). The analysis covers the location networks of 33 global media firms with a total of 2,766 establishments. To qualify as "global" a media firm had to have a presence in at least three different national economic areas and at least two continents or "world regions." For the media firms included, the locations of all branch offices, subsidiaries and holdings were ascertained and entered in a list of 284 cities (distributed all over the world). The result is a relational data matrix of global media firms and cities, with the matrix cells indicating the number of establishments of a particular global firm in a particular city. These data can be used for a ranking of the cities based on the number of establishments of global media firms that are located in the city. By selecting certain threshold values it was possible to present a set of "global media cities" in the form of readily distinguishable groups (Krätke 2003). However, this relational data matrix can be used for more

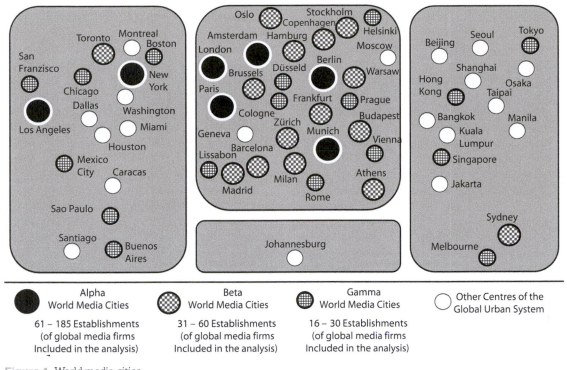

Figure 1 World media cities

Source: Stefan Krätke

thorough analyses, e.g. for measuring the *connectivity* of media cities and identifying distinct *geographical configurations* of global media firms' locational strategies (Krätke and Taylor 2004). In the basic analysis, global media cities were divided into three groups: *alpha*, *beta* and *gamma* world media cities. An *alpha world media city* had to have a presence of more than 50 per cent of the global players analysed here, and more than 60 *business units* from among the global media firms included had to be present. The characteristic feature of world media cities is that they can point to an *overlapping* of the location networks of a *significant number* of global (plus local) media firms. The results are presented here by a synthetic "world map" of the global media business network *nodes* (see Figure 1).

Global media cities as central nodes of global media firms' locational networks

The analysis shows, first of all, a *markedly unequal distribution* of the establishments of global media firms over no more than a handful of cities. The seven cities in the alpha group account for as much as 30 per cent of the 2,766 establishments of global media firms that were recorded, while the 15 cities in the beta group provide the location for a further 23 per cent of the registered establishments. Thus, over 50 per cent of the branch offices and subsidiary firms of global media groups are concentrated in 22 locations within the global urban system. If the gamma group is included, the share of the global media cities increases to 70 per cent of the establishments registered. The organisational units of the globalized media industry reveal a highly *selective* geographic concentration on a global scale.

Prominent among the *alpha world media cities* are New York, London, Paris and Los Angeles, which are ranked as "genuine" global cities in virtually every analysis of the global urban system. This research again stresses that global cities are to be characterized not only as centers of global corporate services, but also major centers of cultural production and the media industry. However, among other cities that qualify as global media cities there are interesting deviations from the widely employed global city system: The alpha group of global media cities also includes *Munich*, *Berlin* and *Amsterdam*, three cities that in global city research which focuses on corporate services were ranked as ("third-rate") gamma world cities (Beaverstock et al. 1999). In the system of global *media* cities, by contrast, Munich, Berlin and Amsterdam are included in the top group. These cities have achieved a degree of integration into the location networks of global media firms that qualifies them as internationally outstanding centers of the culture and media industry. Thus we can conclude that the diversity of globalized activities leads to *multiple globalizations* within world city network formation.

Whereas the global city network constituted by advanced producer services has major nodes which are relatively evenly represented *in all the major regions of the world* (with the exception of Africa), the network of global *media* cities reveals a strongly uneven distribution in favour of the *European* economic area. Of a total of 39 world media cities 25 are in Europe, nine in the USA, Canada and Latin America, and five in Asia and Australia. What is remarkable is the polarity in the group of US global media cities: integration into global location networks of the media industry is concentrated on just *two outstanding centers* – New York and Los Angeles. *Europe*, on the other hand, has the largest number of media cities with a high global connectivity. The reason for this is cultural *diversity*, since the European economic area has a large number of different nation-states compared with the USA and a multitude of distinct "regional" cultures. This cultural market differentiation is the driving force for the global media firms' strategy to establish local anchoring points in different nation-states. The media cities network as a whole is a reflection of the locational system run by the *western* media industry, which concentrates primarily on North America and Europe. Large media groups with a trans-national impact also exist in Asia (particularly in Bombay and Hong Kong), but the cities of Asia are not incorporated to the same extent as cities in Europe and North America in the western-style globalized media industry.

Global media cities as centers of creativity and the production of lifestyle images

To return to the point of departure – the comprehensive *merging of culture and market* – the relationship between urban development and the media industry shall be outlined. Global media cities are functioning as "lifestyle producers" which includes the *production of lifestyle images*. The current lifestyle producers in the culture and media industry are concentrated in leading media cities, from which they spread lifestyle images in the global urban network. In conjunction with the increasing mediatization of social communication and entertainment, the culture and media industry functions as a "trend machine" that picks up on the trends developing primarily in the leading media cities, exploits them commercially in the form of a packaging and re-packaging of lifestyle elements and transmits them worldwide as part of the phenomenon of globalisation.

The production locations of lifestyle images are *urban clusters* of cultural production. The local concentration of culture and media activity in specific "districts," which tend to be situated in inner-city areas, is not determined solely by economic driving forces of cluster building. In cities such as New York, London, Berlin, etc., culture and media firms prefer "sexy" inner-city locations in which living and working environments merge with leisure-time culture. The specific quality of urban life clearly becomes an attraction factor here. For corporate operators and employees in the media industry the local connection between working, living and leisure time activities is an attraction factor that is in harmony with their lifestyle. These people deliberately seek out locations in a "sub-cultural" urban district that they can use as an extended stage for self-portrayal during working hours and in their leisure time. In the local media clusters there is thus a direct link between certain lifestyles and urban forms of creative production activity, and thus a specific overlapping of the geographies of production and consumption.

The growth of culture and media industry clusters in selected urban areas is related to the fact that such cities have the socio-cultural properties to become a prime location of the "creative class" in terms of Florida's concept (2002). A concentration of the creative class attracts the music industry as well as other branches of cultural production and the media industry, and also a whole range of other knowledge-intensive activities (like the software industry, the life sciences sector, etc.). Florida (2002) emphasizes the socio-cultural properties which make a city like London, New York or Berlin particularly attractive as a place of living and working for the creative class: "Creative people (. . .) don't just cluster where the jobs are. They cluster in places that are centers of creativity and also where they like to live" (Florida 2002: 7). Thus lifestyle attributes of the creative class and a supportive socio-cultural milieu are at the center of a city's attractiveness to the creative economy. Florida highlights the role of a

> social milieu that is open to all forms of creativity – artistic and cultural as well as technological and economic. This milieu provides the underlying eco-system or habitat in which the multidimensional forms of creativity take root and flourish. By supporting lifestyle and cultural institutions like a cutting-edge music scene or vibrant artistic community, for instance, it helps to attract and stimulate those who create in business and technology. It also facilitates cross-fertilization between and among these forms, as is evident through history in the rise of creative-content industries from publishing and music to film and video games. The social and cultural milieu also provides a mechanism for attracting new and different kinds of people and facilitating the rapid transmission of knowledge and ideas.
>
> (Florida 2002: 55).

Moreover, the city *as a whole* can become an attraction factor for the media business in that the symbolic quality of the specific location is being incorporated into the products of the culture and media industry (Scott 2000). Hence production locations such as New York, Paris and Berlin are perceived in the sphere of the media as being "brand names" that draw attention to the attractive social and cultural qualities of the cities concerned. This includes, in particular, the perception of the respective city as a social space in which there is a *pronounced variety* of different social and cultural milieus. As regards the content and "design" of their products, media firms have to contend with rapidly changing trends. For that reason the media firms wish to be near the source of new trends that develop in certain metropoles such as New York, Paris and Berlin. A marked social and cultural variety and openness, therefore, represents a specific "cultural capital" of a city, which is highly attractive for the actors of the creative economy. On a local level, this cultural capital of a city might also be characterized as a specific "subcultural" capital of particular districts within the city. These thoughts support Florida's thesis that the metropoles' economic growth "is driven by the location choices of creative people – the holders of creative capital – who prefer places that are diverse, tolerant and open to new ideas" (Florida 2002: 223).

A flourishing creative and knowledge economy is based on place-specific socio-cultural milieus which positively combine with the dynamics of cluster formation within the urban economic space. Creativity and talent thus depend on the dynamic interplay of economic, socio-cultural and spatial factors, and might become a central basis of successful urban development in the future. However, with regard to the specific socio-cultural base of the creative economy, the concentration of knowledge-intensive activity and creative forces within the urban system is *highly selective*, so that only a certain number of particular cities and metropoles (i.e. those with "attractive" socio-cultural properties) can draw on the creative economy as a focus of their development strategy.

CONCLUSION

The present-day culture and media industry is characterised by the globalisation of large media groups. Within urban media clusters, these global players interact with specialised local media firms and form at the same time a global network of branch offices and subsidiary firms, by means of which urban centers of cultural production are being connected with each other on a global scale. This locational strategy enables the global players to make use of worldwide distributed creativity resources. An analysis of the global media cities leads to the identification of the prime locational centers of the contemporary culture and media industry, from which globalisation in the spheres of culture and the media proceeds. At the same time, major global media cities are locational centers of the "creative class," since they contain the specific socio-cultural properties and lifestyle attributes which are most important for a city's attractiveness to the creative economy.

REFERENCES FROM THE READING

Beaverstock, J. V., Smith, R. G., and Taylor, P. J. (1999) A roster of world cities, *Cities*, 16, 6: 445–58.

Florida, R. (2002) *The Rise of the Creative Class*, New York: Basic Books.

Held, D., McGrew, A., Goldblatt, D., and Perraton, J. (1999) *Global Transformations*, Cambridge: Polity Press.

Krätke, S. (2002a) *Medienstadt*, Opladen: Leske & Budrich.

Krätke, S. (2002b) Network analysis of production clusters: The Potsdam/Babelsberg film industry as an example, *European Planning Studies*, 10, 1: 27–54.

Krätke, S. (2003) Global media cities in a worldwide urban network, *European Planning Studies*, 11, 6: 605–28.

Krätke, S. and Taylor, P. J. (2004) A world geography of global media cities, *European Planning Studies*, 12, 4: 459–77.

Krätke, S. (2011) *The Creative Capital of Cities: Interactive Knowledge Creation and the Urbanization Economies of Innovation*, Oxford: Wiley-Blackwell.

Scott, A. J. (2000) *The Cultural Economy of Cities*, London: Sage.

Taylor, P. J. (2003) *World City Network*, London and New York: Routledge.

51

"Willing the global city: Berlin's cultural strategies of interurban competition after 1989"

Ute Lehrer

EDITORS' INTRODUCTION

Ute Lehrer is Associate Professor of Environmental Studies at York University, Canada, and has written extensively on architectural history and urban planning in Western Europe and North America. The chapter below stems from her larger research project on the role of the built environment, the culture of construction, and the "spectacularization" of the building process in the formation of global cities or "wannabe world cities." Lehrer's trademark style combines visual, planning, and geographical analyses to forge a cultural ecology of the city. Lehrer mobilizes this combination of methodological choices in order to explain the political and cultural strategies through which, following reunification, Berlin's local growth coalition embarked upon a deliberately aesthetics- and form-oriented debate on urban regeneration. By mobilizing approaches to redevelopment that were grounded upon stylized notions of history, geography, and place, the local growth machine attempted to reposition the city (once again) as a global command center. While competition with Germany's other global cities, most notably Hamburg and Frankfurt, was sidelined in this discourse, comparisons with the classic metropolises of Europe—such as Paris and London in the West and Budapest, Vienna and Moscow in the East—served as backdrops to a localized *Kulturkampf* (cultural struggle) around the meaning of architecture, built form, symbolic place, and historicity.

After the fall of the Berlin Wall in November 1989, the largest city in central Europe had suddenly lost its significance as the major switching station and place of confrontation of the Cold War. Previously assured of its national and international importance through the rivalry between the showcase of the West in the East and the capital of the richest East bloc country, Berlin was now devoid of its mission both sides of the crumbling dividing line, which had physically and militarily separated the city's communities and by extension – symbolically – the world. Not surprisingly, the first reaction of decision- and image-makers in the newly unified city banked on globalization and Berlin's potential role in it as a process through which the city could grow together. When leaders within the government and the business community of Berlin promoted images of the future, their shared vision was that Berlin would become a major player within the global economy, a world city, a service metropolis, a bridge between East and West, and the old/new capital city of the reunified Germany. The social construction of these images that would lead to a discursive formation over Berlin's world significance, represents a strategic attempt to position the city within the accelerated global interurban competition.

Fantasies of global city status were rampant in the early 1990s, when demographic growth projections predicted to turn the city of 3.46 million into 6 million inhabitants within two generations, when investments from global corporations would bring European-scale headquarters and general "buzz" would attract global cultural attention to Berlin, potentially reviving the capital's glory days of the Roaring Twenties. Pink Floyd's *The Wall* performed at The Wall in 1990, U2's *Zooropa* album and tour from 1993, a gigantic Love Parade techno fest in 1995 with rapidly increasing numbers of dancing bodies in the central city to over a million in the following years, as well as Christo/Jeanne-Claude's packaging of the Reichstag, which as visited by 5 million spectators, before its make-over were some indication that cultural attraction, at least at the scale of global mega-events, was working as planned. In addition, investors and developers at the central Potsdamer Platz construction site organized operatic performances and philharmonic concerts. Fireworks illuminated the sky over the construction site on a number of occasions, and arts projects at the Daimler-Benz site elevated the banal everyday place of a construction site to a temporarily spectacular space. The employment of star architects and the labelling of the mega-project at Potsdamer Platz as "Europe's largest construction site" all were part of Berlin's strategy to become recognized as a significant player on a global stage.

The dominant forces behind the redevelopment schemes of Berlin believed that a transformation of the built environment was one of the means to this end; hence, architecture was supposed to work as the catalyst for Berlin's quest for a new identity as global city. By the mid 1990s, the inner city had turned into a major construction site, with Potsdamer Platz as the most spectacular of Berlin's projects.

With these practices, Berlin had put itself on the map of global cultural spectacle once again. However, much to the disappointment of local boosters and investors, economic development and demographic expansion did not quite come along in the same way. And this despite concerted efforts by many among the city's decision-makers to "will" the global city and "channel" the global. Blueprints of the global city were scaled down to the realities of the capital city, which had become Berlin's official function from 1998 on.

In this text, I will sketch this process of "willing" the global city through a range of explicit cultural strategies by Berlin's growth regime. The underlying argument is that the building culture of a specific locality is changed by the dynamics of globalization where the building *process* – and not only the outcome

– becomes part of a concerted marketing strategy (Lehrer 2002). I will specifically look at three aspects of this development in the context of redeveloping Potsdamer Platz: first, the phenomenon of the spectacularization of the building process; second, the projection of Berlin's globality as a place of centrality and culture via a selective and manipulative presentation of local history; and third, a brief discussion of Berlin's architecture debate as "*Kulturkampf*" (cultural struggle).

THE SPECTACULARIZATION OF THE BUILDING PROCESS

Individual buildings and whole building complexes are being used increasingly as a means of establishing a city on the map of world locations and destinations. In this spatial transformation of cities, three distinct expressions of the city's "symbolic politics" stand out: the trophy building, the mega-event, and the large-scale project (Lehrer 2003). The trophy building usually results from hiring a world-renowned architect as a certain guarantee to get recognition on a world scale. The second strategy for putting a city on the map of world locations is the mega-event. These temporally limited events serve as impetuses and legitimizing forces in the structural and physical redevelopment schemes of major parts of cities and their regions. The third symbolic strategy in the transformation of cities is a straightforward attempt to compete with other cities for symbolic preeminence in a global environment by emphasizing scale. Large-scale projects, however, play a dominant role in image production of cities not only because of their sheer size but also because of the impact they have on the urban, and therefore social, fabric.

The importance of the built environment's symbolic value in an advanced service economy has been the subject of extensive discussion since at least the mid-1980s (Debord 1994; Harvey 1989; Fainstein 1994; King 1996; Zukin 1991). What is new is that the process is being sold. In other words, the building process no longer involves just the conception of an idea and its realization; it is equally about the production of images between the project's inception and completion. In underscoring the significance of these developments, I call this new strategy for producing global images of place the "spectacularization of the building process." In this process, not only the scales

Plate 38 Construction of Potzdamer Platz

Source: Ute Lehrer

and scopes of projects, but also the speed and the assertiveness with which images are produced, leave a physical imprint in the global production of images, long before the building is able to "speak" for itself.

Potsdamer Platz in Berlin is a prime example of such a spectacularization of the building process, one in which the full range of strategies have been deployed to draw attention to the construction site and its future. Each of the three modes of architectural image production – trophy building, large-scale project and mega-event – found its way into the redevelopment scheme of Potsdamer Platz. Promoted as Europe's largest construction site, Potsdamer Platz is certainly a large-scale project. It can be identified as a series of trophy buildings since world-renowned architects participated in its design. By turning the construction process into a major spectacle, the redevelopment of Potsdamer Platz had become also a mega-event.

The spectacularization strategy was based on the usage of superlatives, the creation of tangible objects, and the production of hands-on events on construction site. From the beginning, investors at Potsdamer Platz labelled the mega-project "Europe's largest construction site." Politicians, bureaucrats and the mainstream press were quick to adopt this affinity for superlatives. While one could easily have believed these hyperbolic claims – after all the project's scale and scope were huge – it was unclear on what grounds such a statement was based. Did it refer to the footprint of the construction site, to the scale of the construction activities or to its budget? Many additional superlatives

were exploited during the course of construction to describe and celebrate the site's remarkable scale: the number of cranes being used, the number of construction workers employed and the ethnic and cultural diversity of the people engaged in the project. The investors declared even the overcoming of logistical hurdles and the unique means of removing debris as a "documentation" of the success of their "unusual" and "spectacular" building techniques.

Playing a crucial role in the spectacularization process, the so-called *Info Box* was the most successful object of image production at Potsdamer Platz. Shaped like an oversized version of those containers that usually appear next to construction sites, this temporary building alluded directly to its immediate surroundings. With its bright red metal façade and its elevated position, it worked simultaneously as a billboard and as a place of orientation within the messiness of the construction site. In fact it represented a microcosm of the building activities at Potsdamer Platz, since it was home to an exhibition that explained the scale and scope of the construction project as well as the site's historical significance. As a marketing strategy, the Info Box exhibition/building had two main objectives. On the one hand, Info Box created a concrete "place" in the middle of a wide-open space in Berlin's geographic center, one that became a point of attraction for tourists and Berliners alike. On the other hand, Info Box drew attention away from the nuisances associated with such a large-scale project. Moreover, it turned the site and its building process into the happening place for a New Berlin.

Plate 39 Info Box and Berlin Wall

Source: Ute Lehrer

Another temporary but hugely successful strategy to produce images during the construction period was a personalized experience of construction sites at Potsdamer Platz and other places around the city. Introduced for the first time in the summer of 1996, *Schaustelle Berlin* (Show-site Berlin – a play on words implying both stage and construction site) allowed Berliners and tourists alike to participate in guided tours and other innovative ways of sightseeing of buildings under construction during the summer months of June, July and August. Daimler-Benz announced that during *Schaustelle Berlin* as many as 50,000 people per day visited its construction site. The driving force behind this summer event was *Partner für Berlin*, a public-private agency between the Berlin Senate and about 120 national and international corporations, created in fall of 1994 with the mandate to promote Berlin as a location for good investment.

THE PROJECTION OF GLOBALITY: THE CULTURE OF THE IMAGE AS GLOBALIZATION

Despite some initial scepticism, within a few years of its erection the Potsdamer Platz redevelopment found wide acceptance among tourists and Berliners alike. In 2001, Alexander Smoltczyk argued, "No place in today's Berlin is as history-less as Potsdamer Platz. Perhaps this is why it is so successful" (Smoltczyk 2001: 60–1). To become "history-less," however, was a process that seems to have been produced intentionally. The paradox is that it was in fact history that was exploited as an argument by the supporters of Berlin's attempt going global, however, a very selective history that to a good part avoided drawing attention to the problematic past of Berlin. Therefore, the treatment of history played a central role in the debate over redevelopment schemes in Berlin. Because Potsdamer Platz had become a non-place in most Berliners' mental maps after the erection of the Wall, the challenge was to reclaim the space and fill it with new meaning, meaning that would contribute to the projection of Berlin's globality as a place of centrality and culture.

Image production understands the built environment not only in physical and aesthetic terms but also as an outcome of socio-economic relations and as discursive practices. This means that along with the production process of the "real-material" built environment, there is a production process of the imaginary, a social construction of a particular image and meaning. Images, as they are understood here, include three overlapping and communicating levels of visual, symbolic and metaphorical products and processes. The production of these images has to be understood as processes through which members of society make sense of their individual worlds and of each other's discursive and visual contributions to the general process of communication in society (Deutsche 1996). Therefore images should be treated as substantial elements in the three-pronged spatiality people encounter in cities (Lefebvre 1991).

Ridding Berlin of certain symbolic elements, while manipulating others and constructing new ones, was the preferred method of re-envisioning history in Berlin. While the new Potsdamer Platz can be seen as amnesic in terms of its treatment of history, certain historical details and allusions were used to enhance aspects of the project that were considered key to the reappropriation of space; others were crucially elided. The promotion of a new version of Potsdamer Platz's history accompanied its physical renovation, and took the form of a visual deconstruction and reconstruction of Berlin's material markers.

In the process of rewriting its history, Potsdamer Platz was described as the heart of Berlin. It is true that in its earlier incarnation, Potsdamer Platz was a major traffic intersection with five large streets, several streetcar lines and the site for the first traffic light in Europe, and with two nearby train stations constantly feeding people and goods into the area. However, when investors promoted Potsdamer Platz's cosmopolitanism during the Roaring Twenties, they assiduously avoided any mention of elements that would have linked the site to the Nazi period of the Thirties and Forties, or to its own masters of image production – in particular Albert Speer with his master plan for Germania. The proximity of the Hitler Bunker, the Gestapo prison and the Ministry of Propaganda to the North of the redevelopment of Potsdamer Platz was hardly mentioned. Omitted was any inkling that today's Daimler-Benz site also housed the *Volksgerichtshof*, where about 13,000 death sentences were issued beginning in 1935 (Winter 2001: 24). There was little reference to the expropriation of Jewish-owned businesses in the area, such as the Wertheim department store on Leipziger Platz, during the Nazi period. The same silence prevailed about Daimler-Benz's Second World War role in the production of weaponry and its forced labor camps. The official story of Potsdamer Platz even fails to mention

the fact that on 2 May 1945, a delegation of the Nazi leadership initiated Germany's capitulation to the Allies at Potsdamer Brücke (Winter 2001: 24). This selective deployment of historical detail was not only part of the marketing strategy of the investors when Potsdamer Platz was under construction, but has continued ever since.

In the first years after reunification, the area behind the Info Box became a safe haven for Berlin's historical relics, a bizarre open-air museum and dumping place for the remnants of the GDR period. In the end, however, those historical markers had to make room for the prospect of new development. The displacement of watchtowers and wall sections was also a symbolic act of historical annihilation. Berlin's officials consciously sought to free the city of most reminders of its partitioned past.

The dismantling of Info Box in 2000 can be interpreted as a similar kind of historical deconstruction. Because of its peculiar shape, its distinctive position on the construction site, and its function as the foremost symbol of the rebuilding of Berlin, the persistence of Info Box would have been a constant reminder of a transitional phase in Berlin's new history and identity as global city. Ironically, the space that Info Box occupied is replaced by the perfect octagon of Leipziger Platz – as if there never had been a Second World War or two distinct German states.

While certain images were suppressed, others were manipulated. One of the very few buildings that had survived the bombing in 1944/5 as well as the bulldozers of the 1960s and 1970s was the *Kaisersaal* (Emperor's Hall), which used to belong to the Grand Hotel Esplanade. Built in 1907 as a speculative investment, this hotel became one of Berlin's most exquisite places to be and to be seen. During the Second World War about three-quarters of the hotel was destroyed, and after the city was partitioned, the surviving halls, staircases and bathrooms were turned into a dance hall and theater. When construction began at the site that Sony had purchased for its European headquarters, this remnant of a formerly grandiose Berlin was in the way of the architect's plan. With great fanfare, and superior media coverage, the Kaisersaal was relocated from its original site to about 70 meters to the West in March 1996. It was the most spectacular event on Potsdamer Platz and it demonstrated the literal deconstruction of a historic relic. This relocation represents a plain instance of the manipulation of Potsdamer Platz's

Plate 40 Potsdamer Platz, Berlin

Source: Roger Keil

material history: After being surgically separated and partly relocated, the remainder of the Kaisersaal became an enshrined part of the Sony Corporation where pieces of it were enclosed into a glass box.

In spite of a strong local building culture that favoured clear height limitations and stone facades, images that fit into the rhetoric of global city formation were strengthened and elaborated: highrise buildings, glass and metal facades, an enclosed inner city shopping center all fit this agenda. They all demonstrate the urge to create the image of a global city that is in sync with other global cities by using an architectural language as well as building materials that are global in their uniformity.

THE ARCHITECTURE DEBATE: *KULTURKAMPF* IN THE GLOBAL ARENA

Architecture plays a significant role within the public discourse on urban development and local politics in Germany. Understood as more than merely aesthetically or economically motivated, the discourse about architecture draws on previous time periods and their ideologies about architecture, including the time when architecture became one of the various strategies used by the Nazi regime to claim cultural superiority over other "races" (Frank 1985). Hence, the relevance that architecture plays within public discourse on city-building processes in Germany in general and in Berlin in particular often bewilders outsiders.

The roots of Berlin's current discourse on architecture and urban planning go back to the 1970s and 1980s, when locally specific strategies for urban redevelopment met with internationally celebrated architectural approaches. With the International Building Exhibition and at the height of a postmodern style in architecture, Berlin had become an open-air museum for buildings designed by world renowned architects. Yet, at the same time, the city also could demonstrate the value of a more ecological approach to urban planning (the so-called *Behutsame Stadterneuerung*). With a lively squatter scene, the discursive realm of architecture and urban redevelopment was not left to the self-acclaimed experts, but instead became part of a general public discourse. Not surprisingly, therefore, all of the proposed visions of a reunified Berlin were heavily contested, both within the expert world as well as in the general public sphere.

While the discourse was about creating an identity through the built environment, and about the role of marketing this identity, it was also about a localized *Kulturkampf* (cultural battle) over the meaning of architecture. The dispute was articulated as one between the American city and the European city. There were those who wanted to use new projects in Berlin in order to connect with the global language of office towers made out of glass and steel. Opposing them were those that wished to save the European city by demanding various types of restrictions on building style and form. Whereas the former position was shared among the investors and some of the architects, the latter was favored by a local alliance composed of various city planners, politicians and architects. They saw the redevelopment of Potsdamer Platz as a defence of the European city; it was about streetscape, density, scale and urban pattern. This clash of ideologies appeared, at first glance, to be about stylistic considerations; in effect, however, it concerned the relationship between the construction of built environment and larger social, cultural and economic transformations. The dispute led the investors in Potsdamer Platz to urge the municipal government to reject the image of the European city as backwards and provincial. Of course, this was a slap in the face to a city that was trying hard to strengthen its position within the global economy through visual statements in the built environment. In the end, the redevelopment of Potsdamer Platz presents a compromise between the European and the American city: height limitations that were in reference to Milan throughout the development, but skyscrapers right at the tip of Potsdamer Platz as well as on the opposite side; predominant use of stone facades, but glass and steel for the Sony complex; a street plan that is relatively fine grained, but an entire section of a street is roofed off and turned into a shopping mall.

CONCLUSION

Long-established forms of boosterism have now evolved into coordinated efforts to turn cities into spectacles and the urban experience into image consumption. This is particularly true for large-scale projects, where it is difficult – for both the specialist and the non-specialist – to imagine the future shape of new built environments and to anticipate their impact on the urban fabric. What is new is that the building culture has changed in such a way that it is no longer about the end result but also about what images are created and disseminated during the building *process*. Therefore, not only buildings and the advertisement of their potential success contribute to locational competitiveness in a global marketplace but also the marketing of the building process, which draws attention to the real physical as well as messy and procedural aspect of global city formation.

REFERENCES FROM THE READING

Debord, G. (1994) *The Society of the Spectacle*, New York: Zone.

Deutsche, R. (1996) *Evictions*, Cambridge: MIT Press.

Fainstein, S. (1994) *The City Builders*, Cambridge: Blackwell.

Frank, H. (1985) *Faschistische Architekturen: Planen und Bauen in Europa. 1930 bis 1945*, Hamburg: Christians.

Harvey, D. (1989) *The Condition of Postmodernity*, Cambridge: Basil Blackwell.

King, A. (1996) Worlds in the city: Manhattan transfer and the ascendance of spectacular space, *Planning Perspectives*, 11: 97–114.

Lefebvre, H. (1991) *The Production of Space*, Oxford: Basil Blackwell.

Lehrer, U. (2002) Image production and globalization: City-building processes at Potsdamer Platz, Berlin, UCLA, Dissertation.

Lehrer, U. (2003) The spectacularization of the building process: Berlin, Potsdamer Platz, *Genre: Forms of Discourse and Culture*, Fall/Winter: 383–404.

Smoltczyk, A. (2001) Auf neutralem Boden, *Der Spiegel*: 60–1.

Winter, F. (2001) Verpatzter Potsdamer Platz: Geschichtsverleugnung im Europäischen Städtebau, *Die Wochenzeitung*, 14 June: 24.

Zukin, S. (1991) *Landscapes of Power*, Berkeley: University of California Press.

52

"The transnational capitalist class and contemporary architecture in globalizing cities"

from *International Journal of Urban and Regional Research* (2005)

Leslie Sklair

EDITORS' INTRODUCTION

In this essay, Leslie Sklair, Professor Emeritus at the London School of Economics, analyzes the impact of globalized architectural practices on our cities. The 1990s and 2000s marked the emergence of "starchitects"—a small group of elite architects very much sought after by both private and public clients to design signature buildings. Starchitects and their designs can have tremendous branding effect in marketing properties. One well-known example is the Guggenheim Museum in Bilbao, designed by Frank Gehry in the late 1990s, which has helped revitalize the city and which continues to draw millions of tourists annually. These well-connected, highly mobile elite architects may be small in number, but they have delivered a significant number of high-profile buildings around the world. Drawing upon his previous work on the transnational capitalist class, Sklair examines the four fractions of the transnational class in architectural practice: the corporate, the state, the professional, and the consumerist. He describes how the four groups interact with one another and how the built environment is increasingly shaped by the rise of the global architectural design network of prestigious architects, firms, and globalizing local patrons.

The focus of this article is on the specific role of each of the four fractions of the transnational capitalist class (TCC) in and around architecture in globalizing or world cities (see Taylor 2004). The TCC consists of people who typically have globalizing as well as (rather than in opposition to) localizing agendas. These are people from many countries who operate trans-nationally as a normal part of their working lives and who more often than not have more than one place that they can call home. This reflects their relationships to transnational social spaces and the new forms of cosmopolitanism of what I have conceptualized as generic globalization (Sklair 2005). These forms

encourage both local rootedness and transnational (globalizing) vision. New modes of rapid and comfortable long-distance transportation and electronic communications make this possible in a historically unprecedented fashion. It is for this reason that the new concept of globalization is most appropriately reserved for the new economic, political and cultural conditions that began to develop in the middle of the twentieth century and have rapidly accelerated since then.

I have conceptualized the transnational capitalist class in terms of the following four fractions (Sklair 2001):

1 Those who own and/or control the major transnational corporations and their local affiliates (corporate fraction). In architecture these are the major architectural, architecture–engineering and architecture–developer–real estate firms, listed in the magazine *World Architecture*. In comparison with the major global consumer goods, energy and financial corporations, the revenues of the biggest firms in the architecture industry are quite small. However, their importance for the built environment and their cultural importance, especially in cities, far outweigh their relative lack of financial and corporate muscle.

2 Globalizing politicians and bureaucrats (state fraction). These are the politicians and bureaucrats at all levels of administrative power and responsibility who actually decide what gets built where, and how changes to the built environment are regulated.

3 Globalizing professionals (technical fraction). The members of this fraction range from the leading technicians centrally involved in the structural features and services (including financial services) of new buildings to those responsible for the education of students and the public in architecture. (There is obviously some overlap between the technical and corporate fractions.)

4 Merchants and media (consumerist fraction). These are the people who are responsible for the marketing and consumption of architecture in all its manifestations.

Those who lead the TCC see its mission as organizing the conditions under which its interests and the interests of the capitalist system can be furthered in the global and local context. The concept implies that there is one central TCC that makes system-wide decisions, whose members also connect with transnational capitalist class fractions in each locality, region and country, as well as globally. While the four fractions are distinguishable analytic categories with different functions for the global capitalist system, the people in them often move from one category to another. A common form of this intra-class mobility is the 'revolving door' between business, education and government, and vice versa. Many leading architects are simultaneously practitioners, professors and agents of the state at various levels.

The transnational capitalist class is transnational in the following respects, in architecture as in any other sphere:

1 The economic interests of its members are increasingly globally linked rather than exclusively local and national in origin.

2 The TCC seeks to exert economic control in the workplace, political control in domestic and international politics and culture-ideology control in everyday life through specific forms of global competitive and consumerist rhetoric and practice.

3 Members of the TCC have outward-oriented globalizing rather than inward-oriented localizing perspectives on various issues.

4 Members of the TCC tend to share similar lifestyles, particularly patterns of higher education and consumption of luxury goods and services.

5 Finally, members of the TCC seek to project images of themselves as citizens of the world as well as of their places of birth.

In his pioneering study of the sociology of architecture, Gutman (1988) elaborates on three types of contemporary architectural firms, an analysis that is of relevance for the relationship between the transnational capitalist class and architecture. The first type consists of the 'strong-idea firms' (his examples are those led by Frank Gehry, Michael Graves, Richard Meier and Robert Venturi). These tend to be practice-centred businesses. Second are 'strong-service firms' (notably Skidmore, Owings & Merrill, hereafter SOM), and they tend to be business-centred practices. Third are 'strong-delivery firms', commercial firms that rarely win awards but build a great deal. Gutman argued that personal inspiration was being replaced by more conventional marketing for architectural services and that it was getting more difficult to ascertain how much and exactly what the architect contributed to big projects. Michael Graves, for example, was careful to ensure that he was not held responsible for the aspects of his Portland municipal building that he did not personally design, and the architect of the Getty Center overlooking Los Angeles makes it clear that not everything that appeared on the site was to his liking (Meier 1999). Gutman (1988: 59) makes the point that except for a few special cases, 'the trade press, but even more so magazines such as *Architectural Digest* or the *New York Times* — whose editors regard architecture as if it were furniture, fashionable clothing, or gourmet cooking — ignore the complex relations among the cast of characters who now participate in a major building project.' By the late 1980s, 25–30 museums in the US had

Plate 41 Office buildings designed by international architects, Beijing

Source: Xuefei Ren

important architectural collections, major publishers like MIT Press, Rizzoli and Princeton had expanding architectural programmes, and there were other manifestations of the commodification of architecture and architects. Gutman reprints the advertisement for the expensive Dexter shoe, featuring the afore-mentioned postmodernist architect Michael Graves, whose Public Services Building in Portland, Oregon, had created such a stir in the early 1980s, as a sign of the growing celebrity status of architects. As Gutman concludes: 'There has been a tremendous expansion in opportunities to consume architectural culture over the last few decades' (Ewen 1988: 93). This trend has intensified since the 1980s and the production and marketing of architectural icons and architects as icons (the so-called signature architects or starchitects) have been at the centre of it, and I shall discuss this in more detail below in the context of the impact of the culture-ideology of consumerism on architecture.

CORPORATE FRACTION OF THE TCC IN ARCHITECTURE

The corporate fraction of the TCC in architecture consists of the major globalizing architectural, architect–engineering and architect–developer–real estate (mixed) practices, in terms of fees and fee-earning architects employed. These are the 'strong-delivery firms' who build many buildings but do not produce much contemporary iconic architecture. In addition, a few large 'strong-service firms' compete with these majors and are distinguished by the fact that they are responsible for a few buildings that have achieved some measure of iconicity. Most iconic buildings in the global era, however, have not been built by the biggest firms but by a relatively small number of firms identified with individual architects, often with substantial reputations based more on publication than on actual buildings. These are the 'strong ideas firms' and to understand them we have to understand the central role of celebrity in the culture industries and how it operates in the specific culture-ideology of consumerism for iconic architecture.

Most contemporary professional icons that have also attracted a measure of public icon status, however, have been produced by architects whose practices are relatively small in terms of revenues but whose symbolic capital in the industry is high. A good proxy for such architects of iconic buildings is the list of winners of the Pritzker Prize (widely regarded as the equivalent of the Nobel Prize for architects). The Pritzker was introduced in 1979. Of the twentieth century architects who were dead by then, Wright, Le Corbusier and Mies certainly have both professional and public iconicity in the senses used here (and most enthusiasts will wish to add to this list—my own additions would include Gaudi, Aalto and Louis Kahn). It is noteworthy that while the first three are still influential for architects and historians today, none has left an enduring corporate presence, in comparison with SOM, founded in 1936–9 and still going strong—though this is possibly the only example. Wright, Corbusier and Mies all have institutional legacies and plenty of enthusiasts, but with the partial exception of the Mies-related firms of Lohan Associates (Dirk Lohan is the grandson of Mies) and Murphy/Jahn, there are no firms to carry on their names and work. Furthermore, while new buildings are now on post-cards and trinkets everywhere, until recently, with few exceptions—notably Wright's Guggenheim Museum in New York—buildings were not iconized in this way.

STATE FRACTION OF TCC IN ARCHITECTURE

The state fraction of the TCC in architecture comprises globalizing bureaucrats and politicians who promote and award contracts for important subnational (usually urban), national and sometimes transnational projects in global competitions. Where the sites of these projects are in or near the most notable globalizing cities (New York, London, Tokyo, Paris and perhaps a few others) they may achieve global significance. Some cities that would not normally be considered global cities clearly set out deliberately to establish global credentials through promotion of iconic architecture. The best current example is Barcelona, to which we can add Bilbao and Glasgow, Los Angeles, Berlin and many others. And in China, cities are competing against each other for icons and are using international architects to drum up that 'something different'. In Chongqing (China), city authorities are racing to create the necessary public buildings. Rather in the manner of a shopping spree, they say they want 10 and have decided half should go to foreign architects.

Many of the buildings intended by urban boosters to be global icons that will put their city 'on the map' start off with high-profile competitions, often open only to a restricted group of already famous architects who are invited to submit entries, and are often paid to do so. The topic of architectural competitions has attracted a great deal of attention within the industry and, indeed, has its own journal *Competitions*. The topic occasionally spills over into the mass media, most notably in the case of the rebuilding of the Twin

Plate 42 Mumbai International Airport, designed by SOM

Source: Xuefei Ren

Towers site in New York post 9/11. The competition system varies from country to country, but cases where national governments or local authorities restrict entry to or specifically invite entries from architects who are co-nationals are much less common now than in the past. Where competition juries for 'public' buildings have foreign members this may be taken as evidence of a globalizing tendency among state bureaucrats and politicians. For example, the juror from Britain for the Brasilia competition in 1956, William Holford, is credited with having had the imagination and independence to pick the rather schematic design by Lucio Costa (Holston 1989). Some projects are marketed as sites and buildings with genuinely transnational (globalizing) significance. Notable examples of these are the original building and subsequent rebuilding of the United Nations HQ in New York, new buildings associated with major sporting events like the Olympics and the football World Cup, repositories of world heritage like the proposed Cairo Museum of Antiquities and the Acropolis Museum in Athens and perhaps major theme parks of the Disney type. These and similar sites are examples of the transnational social spaces from above that capitalist globalization has brought.

TECHNICAL FRACTION OF THE TCC IN ARCHITECTURE

The globalizing professionals of the transnational capitalist class in architecture are a very mixed bunch, ranging from those who work with (or for) those who own and control the major architectural firms (some of which overlap with the corporate fraction) to those engaged in the education of architects, from designers in general to professional architectural historians and critics. What unites them all is their globalizing agenda within, more or less, the confines of capitalist globalization. However, there are frequent debates between them and other professionals who oppose their agenda of capitalist globalization while pursuing their own, sometimes more generic alternative globalization agendas as, for example, engineers and consultants working with cheap and sustainable local materials and building methods, and teachers, historians and critics who give them theoretical and practical support. A little speculatively, I suggest that radical developments of movements like Critical Regionalism and New Urbanism could provide some hints about what

alternative globalizations in architecture could look like. The long tradition of architecture without architects connects with these more recent theories and practices in ways that entirely undermine the pretensions of much iconic architecture and the more general issue of the role of celebrity in the culture-ideology of consumerism.

CONSUMERIST FRACTION OF THE TCC IN ARCHITECTURE

The consumerist fraction of the TCC in architecture consists of those who use their control of and/or access to the commercial sector and the media to promote the idea of contemporary architecture as a transnational practice in the realm of culture-ideology. In this fraction we find retailers with an interest in architecture and signature architects as a means of globalizing the appeal of their own businesses as well as those who control those parts of the media who see commercial opportunities in the promotion of signature architects and the use of iconic buildings.

This operates in a variety of spheres. The connection between architecture and shopping is well illustrated by the relationship between the boutique Prada and its architects of choice, notably Koolhaas, Herzog & de Meuron, and Kazua Sejima, all of whom have deliberately designed iconic stores for them in globalizing cities. As Speaks (2002) reports, there are many other retailers who see the advantages of such connections. In England, a long-established central London store, Selfridges, had a new store designed for them in Birmingham by Future Systems. This was instantly dubbed iconic and the architects were asked if some of their visuals could be used on the store credit card. Recognition of the outline of a building or a skyline is one of the great signifiers of iconicity.

The extent to which and the ways in which buildings, spaces and architects are represented as iconic in the mass media are, obviously, crucial for my thesis. The use of iconic buildings in movies and television is there for all to see, but representatives of the architecture industry regularly complain that, generally speaking, architecture, like most serious culture, is virtually ignored by mainstream TV.

Finally, a few words about the significance of new museums for the relationship between iconic architecture and capitalist consumerism are in order. Reference has already been made to the almost instant iconicity of Frank Lloyd Wright's Guggenheim Museum in New York, and 40 years later Frank Gehry's Guggenheim Bilbao repeated the feat. Three reasons are commonly given to explain why some museums become iconic. First, the two Guggenheims and many other successful museums have unusual sculptural qualities and people visit them to see the buildings as much as—if not more than—for the art inside. Second, museums like all cultural institutions have become much more commercialized in the global era.

Third, museums often become iconic when they can be seen to successfully regenerate rundown areas. This is certainly true for the Guggenheim in Bilbao, and in London, the 'Tate Modern effect' helps to explain how a converted disused power station has transformed a grimy area south of the Thames and, by connecting it via the new Millennium Bridge with St Paul's Cathedral on the north side of the river, has created a new urban pole of attraction. There is, therefore, a good deal of evidence to suggest that architecture has a higher public profile than ever before and that this is connected to the culture-ideology of consumerism and the agency of the consumerist fraction of the TCC in architecture.

CONCLUSION

This article has attempted to probe the agents most responsible for this transformation, namely the transnational capitalist class, and to suggest that it is becoming a global phenomenon, specifically a central urban manifestation of the culture-ideology of consumerism. While this article has concentrated on the four fractions of the TCC and their specific functions for the class as a whole in promoting the project of capitalist globalization, further research might usefully focus on how members in each fraction work together in the production, marketing and consumption of iconic architecture, which clearly is the case for contemporary globalizing cities. The study of the production of architectural icons might prove to be not only a fruitful site for research on these issues but may also suggest alternative modes of iconicity beyond capitalism. It is now also appropriate to think about how some new iconic architecture in neighbourhoods and cities might meet the needs of all those who live there without simply pandering to the culture-ideology of consumerism. But this would imply the end of capitalist globalization as we know it.

REFERENCES FROM THE READING

Ewen, S. (1988) *All Consuming Images*, Basic Books: New York.

Gutman, R. (1988) *Architectural Practice: A Critical View*, New York: Princeton Architectural Press.

Holston, J. (1989) *The Modernist City: An Anthropological Critique of Brasilia*, Chicago: University of Chicago Press.

Meier, R. (1999) *Building the Getty*, Berkeley and Los Angeles: University of California Press.

Sklair, L. (2001) *The Transnational Capitalist Class*, Oxford: Blackwell.

Sklair, L. (2005) Generic globalization, capitalist globalization, and beyond: A framework for critical globalization studies, in R. Applebaum and W. Robinson (eds.), *Critical Globalization Studies*, Routledge: New York: 55–63.

Speaks, M. (2002) Design intelligence and the new economy, *Architectural Record*, January: 72–6.

Taylor, P. (2004) *World City Network: A Global Urban Analysis*, London and New York: Routledge.

53

"Shanghai nightscapes and ethnosexual contact zones"

from *Shanghai Nightscapes: A Nocturnal Biography of a Global City* (2015)

James Farrer and Andrew Field

EDITORS' INTRODUCTION

James Farrer, Professor of Sociology at Sophia University, Tokyo, is the author of *Opening Up: Youth Sex Culture and Market Reform in Shanghai* (2002), and Andrew Field, Associate Dean of Undergraduate Studies at Duke Kunshan University, is the author of *Shanghai's Dancing World: Cabaret Culture and Urban Politics, 1919–1954* (2010). In this essay, the two veteran ethnographers of Shanghai examine the reemergence of the city's lurid nightlife scene that had thrived as far back as the Jazz Age of the 1920s. Following a decades-long lull during the socialist era, the authors explain, Shanghai's clubbing scene has returned because of larger socioeconomic processes–the rising economic power of local Chinese women, more tolerance toward premarital sex and interracial marriages, as well as a large foreign population, which comprises roughly 20 percent of the total foreign population in all of China. The authors point out the dearth of research on leisure and play in global cities, and by vividly depicting Shanghai's nightclubs as "ethno-sexual transzones," they argue that global cities are nodal points where sexuality is performed and sexual subjectivity is negotiated.

Our focus is what we call "cosmopolitan nightlife," referring to the transnational, ethnically mixed, and culturally hybrid nature of nightlife practices found in large cities around the world. Developing in the age of steamships, mass journalism, and phonographs, the first truly transnational nightlife cultures spread through networks of world cities, producing widely shared styles of music, dance, and sociability, though all with their local variations. The term cosmopolitan also refers to spaces of cultural hybridity and social mixing, or "contact zones" between people of different social backgrounds, often with vastly different resources and contrary cultural outlooks. For much of its recent history, Shanghai nightlife has been cosmopolitan in the sense that all sorts of people living and traveling in the city—whether of Asian, African, or European heritage—met on its bars and dance floors. Another key feature of cosmopolitan nightlife is that it involves both men and women as paying customers. Unlike the domestic spaces of the courtesan houses they eventually displaced, the commercial dance halls featured women as both employees and paying guests.

This cosmopolitan nightlife culture is historically associated with the rise of global cities, or world cities. There are many definitions of the global city, but all emphasize transnational flows of money, people, and goods as well as the concentration of cross-border administrative and financial functions in these cities. Few researchers focus on leisure in the definition of global cities, though arguably urban leisure cultures have fueled urban development throughout human history. For our purposes, a global nightlife city is one whose urban nightscape is shaped as much by transnational cultural flows as it is by local, regional, and national influences. Shanghai, except for the Mao years, has been a global nightlife city, receiving and localizing worldwide cultural trends and transmitting them to other parts of China.

THE NOCTURNAL BIOGRAPHY OF A GLOBAL CITY

Like a person, a city has a biography, which encompasses its formative periods, its crises, rivalries, glories, and incorrigible habits, or in the words of urban sociologist Gerry Suttles its "cumulative texture" (Suttles 1984). And, just as people have a "nocturnal self," a drunken swagger or ruby-lipped smirk that might deviate wildly from a polite daytime persona, modern cities have lurid neon visages that differ from their grey-tone daywear, warmly colored by players' sensuous laughter, and echoing with the drunken dithyrambs of celebrants. In short, the city at night is not the city by day, though it is shaped by some of the same forces.

The nocturnal history of global Shanghai can be summed up in three distinct phases. Phase one occurred during the 1920s–30s. In this era, Shanghai became internationally famous as a sinful city of Jazz Age nightlife. Hundreds of cabarets or dance halls operated within the "semi-colonial" environment of the city—either inside or on the outskirts of the International Settlement and French Concession. From the famed Paramount Ballroom to Ciro's Nightclub and dozens of others, these legendary establishments

Plate 43 A Shanghai nightclub

Source: Roger Keil

featured jazz orchestras, ballroom dancing, and an endless variety of shows for their customers. Many featured hostesses, whose job was to accompany male customers both on and off the dance floor. These were the dance hostesses or *wunü*—akin to taxi-dancers in American culture—and thousands of them hit the city's dance floors on a nightly basis. Most of them were Chinese, though foreign women—most famously the Russians—also served as dance hostesses in Shanghai. These women featured heavily in the city's collective identity and mythology, both as a metropolis of nighttime pleasures and as a magical space of transformation, where countless folk from rural China—both men and women—learned modern city ways.

Phase two occurred during the wartime and immediate postwar Revolutionary Era. The Japanese invasion of 1937 brought the city's nightlife to new heights of decadence, but the eight-year war between China and Japan took a heavy toll on the city and its world-famed nightlife. In the late 1940s, the Nationalist government under Chiang Kai-shek attempted to shut down the dance industry in Shanghai, but ultimately failed. By the mid-1950s, however, the new government of the Chinese Communist Party had succeeded in closing the cabarets, condemning them as vestiges of a bourgeois, vice-ridden society.

The third phase took place between the 1980s and 1990s, when the city reemerged as a nocturnal metropolis. By the late 1980s, Shanghai boasted hundreds of commercial dance establishments catering to Shanghai's broad working classes. In the 1990s a new class order was clearly emerging, and these pretensions of socialist equality fell way. Palatial nightclubs emerged as glittering stages for competing social elites (including foreigners, overseas Chinese, and "returnee" Chinese) to display their wealth and status. By the mid-1990s, when we first arrived in Shanghai to conduct research, we found the city sprouting up a new nightscape of bars, discos, and underground dance clubs. As more and more people streamed into the city from overseas to work and live, and as local Shanghainese embraced new forms of international culture, Shanghai regained its identity as a global center of nightlife. Nightspots became hip and happening places for men and women from many different cultural, social, and ethnic backgrounds to get together and enjoy new styles of music, dancing, and play imported from the West or from other Asian metropolises such as Hong Kong, Tokyo, or Taipei.

ETHNO-SEXUAL CONTACT ZONES

Urban nightlife has served as a space of interracial sexual encounter ever since its advent. Indeed, "slumming" by middle-class urbanites in the working-class underworlds of major Western cities is one of the origins of modern nightlife practices. This was the case for Chicago, Paris, and New York in the nineteenth century, and most famously for Harlem in the 1920s, when affluent whites went on "safaris" to the clubs of this largely African American district in New York City. A vogue for "negrophilia" enraptured Parisian club goers in the same period. Shanghai nightscapes from the 1920s onward were also sites for the exploration of "urban exoticism" in the form of racialized sexual spectacles. Kevin Mumford uses the term "interzone" to describe these spaces of asymmetric social exchanges across racial boundaries (Mumford 1997). Although the interzones of nightlife racial mixing were spaces of great cultural creativity and social mixing, Mumford writes, they ultimately did little to challenge existing racial and gender hierarchies in the U.S., and they may even have served to reinforce or clarify the boundaries between blacks and whites.

Even now, the nightscapes of global cities can be conceptualised as "ethnosexual contact zones," spaces in which boundary-crossing sexual encounters are facilitated by commercial nightlife entrepreneurs catering to a transient and diverse urban population (Nagel 2003; Farrer 2011). For the contemporary period we employ the concept of "transzone" as a space in which people move more rapidly between cities, more fluidly cross racial boundaries, transgress gendered norms of sexual behavior, and more easily acquire forms of transnational cultural capital in the context of a globalized nightscape. At the same time, we develop an ethnographically grounded description of the "eddies" and "pools" in this transnational "space of flows" (Castells 2005), describing the interracial sexual scenes and the gendered norms of sociability that shape the self-presentations—or nocturnal selves—of participants in these zones.

REFORM ERA NIGHTSCAPES AS TRANSZONES

The "reform and opening" policies, beginning in 1978, reestablished Shanghai as a cosmopolitan social and sexual mixing ground. This happened gradually in the

1980s, then far more rapidly during the 1990s and beyond. In the early 1980s, Shanghai's internationalization lagged behind southern cities such as Shenzhen and Guangzhou. However, with the designation of its Pudong district in 1990 as a "Special Economic Zone," Shanghai reassumed its position as the "dragon head" of the Yangzte River region. This remaking of Shanghai as one of China's neoliberal "zones of exception" funneled transnational capital into the city, and with it, flows of new people. The resident foreign community in the late 1990s still numbered less than ten thousand persons, but with the entry of China into the World Trade Organization in 2001, it ballooned. By 2013, the number of foreign residents in Shanghai reached 173,000, about a quarter of China's entire foreign population, while the number of foreign visitors entering yearly through Shanghai rose to over eight million, also nearly 20 percent of China's total. With much denser and speedier connections to the larger world than in the 1930s, Shanghai's nightscape also became a transzone, a space of human flows in which Chinese and foreigners mingled freely, though not without frictions.

This did not happen at once. In a country only gradually emerging from the anti-Western fervor of the Mao years, many Chinese in the early 1980s were afraid to socialize openly with foreign visitors. Chinese citizens who engaged in sexual relationships with a foreigner could be disciplined by their *danwei* (work unit), jailed for "hooliganism," or face social ostracism, especially if the relationship did not quickly result in marriage. Some foreign companies investing in China responded to Chinese strictures by imposing informal policies of "no fraternization" with locals. By the latter half of the 1980s, however, the social and political climate was changing. The rebirth of commercial nightlife was part of this change. Although it was still very risky to bring a Chinese lover back to a heavily monitored international hotel or foreign apartment complex, the small bar scene in Shanghai had emerged as an ethnosexual contact zone in which foreign men and Chinese women (usually "bar girls") could meet and hook up. The relationships formed there were complicated by the huge gaps in income between foreigners and Chinese, the hopes many Chinese held for help with emigration visas, the mutual lack of cultural understanding, and the illegal and irregular nature of both casual sex and prostitution.

The 1990s was the key transitional period when nightlife "transzones" emerged in Shanghai. Clubs like

D.D.'s and Y.Y. were spaces in which Westerners and Chinese mingled promiscuously and on an increasingly equal social footing. Three key social changes—largely external to nightlife itself—were behind this new phenomenon.

The first was the growing presence and power of female Chinese customers. In the earlier decades of the twentieth century, Chinese women had enjoyed much less access to leisure than men. Those who did often worked there or else were members of the new bourgeoisie. In the new socialist China from the 1950s onward, women's participation in leisure became more equal, but there was little of it. As dance halls returned to Shanghai in the 1980s, however, young women and men, now earning comparable wages, hit the dance floors in roughly equal numbers. By the 1990s young, white-collar Chinese women had become important and visible customers in elite establishments, able to pay and play by their own rules.

The second important change was the weakening taboo on premarital sexual relations among Chinese youth. Sex among unmarried young people remained stigmatized and socially disapproved among most Shanghai youth in the 1980s, even though it sometimes occurred. During the "sexual opening" of the 1990s, which partly took place in nightlife itself, young men and women in Shanghai increasingly understood premarital sex as a normal life experience, trends which continued through the 2000s. As a result of these broader changes, "fooling around" (*baixiang*) in a nightclub or bar, resulting in a "one-night stand" or casual sexual relationship with a Chinese or foreign partner, was becoming much more acceptable among younger Chinese in Shanghai.

Finally, there was the greater acceptance of intermarriage and interracial relationships. Under Maoist socialism intimate relations between foreigners and Chinese had become politically taboo, with formal and informal restrictions persisting into the 1980s. Only in the 1990s did it become increasingly common to see mixed couples (usually a foreign man and a Chinese woman) walking hand-in-hand on Shanghai's streets. And although such relationships still met some societal resistance, by the early 2000s marriages between Chinese and foreigners had risen to roughly 3 percent of all marriages registered in the city. In tandem with these trends, flirting and socializing across national and racial lines became part of the performance of a worldly, cosmopolitan lifestyle within Shanghai nightlife spaces.

As the restrictions and taboos on dating, casual sex, same-sex relations, and dating and marriage between Chinese and foreigners disappeared one by one, Shanghai nightlife spaces were increasingly experienced as sexually permissive and tolerant of difference. Although there were occasionally rude snubs by staff or conflicts with a bouncer or other clubbers that seemed racially motivated, no one was systematically excluded from a Shanghai bar or club on the basis of gender, race, or nationality. Still, white and Asian, Chinese and foreign, men and women, experienced the sexual scenes of Shanghai nightlife quite differently. The transzone of Shanghai nightlife was indeed a space of transnational flows, but one with racialized "pools" and gendered "eddies" in which some felt included while others floated on by or were marginalized in the still waters of social interaction.

The general sociological point is that people learn whom to like—at least in part—by learning who they are liked by. This could be described as an erotic version of the "looking glass self," the symbolic interactionist notion that the self is constructed in a reflexive interaction with others. Western men saw Shanghai nightclubs as a "sexual paradise" because they felt sexually appreciated (or "spoiled") by Chinese women. Some described dozens of fleeting sexual encounters with Chinese women they met in dance clubs (not including those they paid for). Of course, not all foreign men and Chinese women responded to this scene positively, but those who did returned for more, ensuring that the scene would persist despite rapid changes in the larger social world. By the 2010s many Chinese women were earning more money than Western men their own age in Shanghai, and rich Chinese men were the new hot marriage tickets, but the interracial scene did not disappear. If anything, more income gave women more opportunities to explore their sexual interests without a concern for a long-term "result" such as marriage. People came into such a sexual scene by chance or by vague inclination, but some stayed because such spaces provided definition and confirmation of a specialized nocturnal self. Initial investments, such as learning to flirt in a foreign language, appreciate foreign music, and play foreign drinking games, became parts of a nocturnal self-image that players of both sexes adopted.

When we look at the ethnographic realities of particular nightlife scenes, however, the picture is more complicated than the sociological analysis above might suggest. For example, one typical club in

which foreign men and Chinese women pursued one another in the mid-2000s was a club called Velvet Lounge. Near the corner of Changshu Road and Julu Road, Velvet was a raging ethnosexual contact zone in its heyday from 2007 to 2010. With its dark red velvet curtained walls, it was popular with an older mixed crowd of Western and overseas Asian men and women, and local Chinese and foreign women who mostly seemed interested in meeting foreign men. A constant flow of revelers hungrily traversed the four salon-like rooms, making Velvet an easy space to encounter someone new. The music was funky but not overpowering and the rooms were full of dancers, jostling for position on the small dance floor, while other customers lounged on sofas underneath oil paintings in imitation of Toulouse-Lautrec's famed artworks showcasing the brothels and cabarets of fin-de-siècle Montmartre. Hooking up with an attractive stranger was clearly on the minds of many patrons.

On one August night in Velvet in 2010, when both co-authors of this book were visiting the club, we met Ella, a thirty-year-old Chinese employee of a prestigious U.S. multinational. She was enjoying a night out after work with two female friends. She was looking to enjoy herself with men as well. While dancing, she kept an eye on the other women on the dance floor. Looking at a svelte, long-legged, and curvy African woman in a tight print dress that revealed a great deal of ebony skin, Ella pointed her out, and said admiringly: "That dress looks like it was photo-shopped onto her." As Ella appraised the foreign sex worker, she remarked: "She must be in a different line of business than me."

Glancing at her friend in a squirmy embrace with a drunken man, Ella herself grew eager to dance, and as we left the club past 1:00 a.m., she was eagerly chatting with an American businessman. Outside on the narrow street, in the glare of the streetlamps, taxis, beggars, cigarette vendors, and streetwalking Chinese prostitutes waited for customers, while some men and women who "got lucky" left on each other's arms.

Our many evenings in Velvet revealed it as a transzone, a complex space of human flows, in which several transnational sexual fields intersected and overlapped. At a minimum, we could distinguish three: the interracial sexual scene between thirty-something foreign men living in Shanghai and the thirty-something Shanghainese women interested in these men, a commercial sex scene between middle-aged men (mostly married business travelers) and much younger sex workers from around the world, and finally, a somewhat separate transnational sexual scene among short-term sojourners, nearly all foreigners, who found themselves temporarily in Shanghai, sometimes only for a few days. In the end, however, not everyone fit into any of these scenes, and such people could drift off into other nightlife spaces in the city.

CONCLUSION

Overall, the twenty-first century transzone of Shanghai nightlife was most certainly not a space of social or even sexual equality, but it was a space in which racial and national differences were re-coded in individualized or market-oriented terms—as differential access to symbolic, cultural, economic, and sexual capital. In this neoliberal transzone, transfers of resources, ideas, and even sentiments were accelerated by increased cultural familiarity, linguistic competence, and a shared orientation toward consumerism, cosmopolitanism, and individual choice. One was not excluded from participation on racial or gender grounds, but might be for lacking money, style, linguistic skills, or sex appeal.

For many, of course, the promised fluidity of this cosmopolitan space was illusory. Crudely put, people with money, beauty, and international backgrounds had more fun, while those who lacked these traits struggled to find a footing in the scene. Nightlife encounters could produce painful shocks of marginalization or exclusion. Nonetheless, the nightlife was also a place to acquire social, cultural, and sexual capital. Both men and women treated nightlife as a school in the carnal arts of attraction and seduction that made up the specialized sexual capital of a scene. And nightlife was a space to pick up forms of cosmopolitan cultural capital—especially language ability—that had value outside nightlife. In short, play is not inconsequential. Just as sandboxes and jungle gyms

nurture young children into the ethics and etiquette of group play, the playgrounds of urban nightlife were one of the important ways in which millions of young adults—from around the world—came to terms with the mores and codes of social and sexual life in the big global city.

Shanghai's global nightscape now served as an unpredictable and productive contact zone for Chinese and foreigners living in the city. These were busy beehives of creative social and sexual energy that linked people from many parts of China and the world together in ever-shifting combinations, like molecules bumping and combining, and producing new chains of social life.

REFERENCES FROM THE READING

Castells, M. (2005) Space of flows, space of places: Materials for a theory of urbanism in the information age, in B. Sanyal (ed.) *Comparative Planning Cultures*, London: Routledge, 45–63.

Farrer, J. (2002) *Opening Up: Youth Sex Culture and Market Reform in Shanghai*, Chicago: University of Chicago Press.

Farrer, J. (2011) Global nightscapes in Shanghai as ethnosexual contact zones, *Journal of Ethnic and Migration Studies*, 37, 5: 747–64.

Field, A. (2010) *Shanghai's Dancing World: Cabaret Culture and Urban Politics, 1919–1954*, Hong Kong: The Chinese University Press.

Mumford, K. (1997) *Interzones*, New York: Columbia University Press.

Nagel, J. (2003) *Race, Ethnicity and Sexuality: Intimate Intersections, Forbidden Fruits*, New York: Oxford University Press.

Suttles, G. D. (1984) The cumulative texture of local urban culture, *The American Journal of Sociology*, 90, 2: 283–304.

54

"Graffiti or street art? Negotiating the moral geographies of the creative city"

from *Journal of Urban Affairs* (2012)

Cameron McAuliffe

EDITORS' INTRODUCTION

Cameron McAuliffe is Senior Lecturer of Human Geography at the University of West Sydney. This essay focuses on the most visible visual aspect of global cities–graffiti. From Paris, Berlin, Milan to New York and São Paulo, many of today's global cities are playgrounds for street artists who use the city as canvas– sometimes invited but more often not. City authorities and residents often react ambivalently toward graffiti and, as McAuliffe notes, it is hard to draw the line between what is (legitimate) street art and what is (illegitimate) graffiti. Using metropolitan Sydney as an example, McAuliffe compares the different approaches that local governments adopt to deal with graffiti, oscillating between removal and engage- ment. He traces two types of cultural flows that shaped the graffiti scene in Sydney: first, the transmission of New York hip-hop culture to a global audience in the 1970s, and second, the travelling of the "creative city" discourse from the US to other parts of the world in the early 2000s (Florida 2002). McAuliffe points out that the creative city discourse has paved the ground for the partial normalization of graffiti as legitimate street art in Sydney. This essay represents the scholarship that empirically examines how globalizing cities are connected with one another through cultural flows.

Graffiti in its various forms has become a perennial feature of life at the edges of the contemporary city. Implicitly set as a challenge to urban relations, as the transgressive act of property crime, graffiti has become an emotionally charged public order issue. Attempts by urban managers to eradicate graffiti have resulted in spiraling costs as increasingly more sophisticated methods are deployed in the various urban "wars on graffiti" (Dickenson 2008; Iveson 2009, 2010). Despite the increased mobilization of anti- graffiti technologies, backed by explicit anti-graffiti policies and laws, graffiti and other forms of unsanc- tioned "public art" persist. The persistence of graffiti can in part be attributed to subcultural responses to the urban wars on graffiti. By framing graffiti as out of place, urban authorities have ensured that succes- sive generations of predominantly young men have taken up graffiti as a risk-laden behavior, as fame and respect are accrued among peers through the brazen transgression of laws and social norms (Ferrell 1996; MacDonald 2001). As the wars on graffiti have esca- lated, so too have the subcultural rewards for those willing to engage in graffiti (Iveson 2010). But graffiti is also finding a place in the city via the presence of discourses that challenge an indiscriminate crimi- nalization of graffiti. Chief among these discourses is the promise of the creative city. Beyond the framing of tensions between urban authorities and graffiti writers over the presence/absence of graffiti, the rise of creative cities discourses (Florida 2002; Landry 2000) has afforded the opportunity to rethink the way the creative practices of graffiti writers and street artists are valued. A reevaluation of graffiti in the light of the importance of creativity to the postindustrial economy aids in the understanding of the continued presence of graffiti and the appearance of new forms of public art derived from graffiti.

For those involved in graffiti subcultures, or in the post-graffiti worlds of street art, "lowbrow" contemporary art, reverse graffiti, cup-rocking, scratchiti, stickers, paste-ups, and posters (Dickens 2008; Manco 2004), the creative city offers new paths to recognition. The commodification of graffiti and street art in advertising, on t-shirts, or through successful crossover into the contemporary art marketplace, has raised the profile of individual artists and the genres of graffiti and street art more generally (Dickens 2010; Snyder 2009). Following the success of Jean-Michel Basquiat and Keith Haring, contemporary artists like Shepherd Fairly and Banksy command a notoriety that extends well beyond the boundaries of their artistic peers, challenging assumptions that graffiti and street art are fueled purely by subcultural recognition (MacDonald 2001; Snyder 2009). Beyond recognition, the creative city promises more substantive rewards, as jobs blossom in the creative sector, aided by strategic investment in creative hubs, quarters, clusters, and precincts. These more discrete, predominantly state-run initiatives are augmented by cultural-planning processes at the local scale producing cultural plans and public art policies.

Yet the promise of the creative city continues to be plagued by questions concerning the sustainability of creative lives. Recent work has focused on the precarious nature of creative workforce employment at a time when the policy landscape is embracing the cultural economy. Despite the fervor with which the creative city has been received by urban managers in many cities, there remain serious questions about how creativity will live up to its promise of being the salve for the wounds of the postindustrial city. One aim of this article is to present some of the ways the revaluing

Plate 44 Creative office space, Sydney

Source: Roger Keil

of graffiti and street art in the creative city offers possibilities for recognition and a measure of vocationalism for young artists. At the same time, the endurance of normative views of graffiti as transgressive disorder continues. For those whose creative practice stems from the transgressive worlds of graffiti, the promise of recognition may not be enough, as artistic practices continue to be framed by dominant discourses of criminalization. New efforts to criminalize graffiti and graffiti writers intersect with processes of recognition producing profound ambivalences. The graffiti writer/ street artist has always had to negotiate the ambivalences that exist at the edges of legality (Lachmann 1988). However, now, with the opportunities afforded by the valorization of creativity, the inconsistencies may be more keenly felt.

While there is no strict and consistent differentiation between graffiti and street art, it is worth noting some of their commonly held features. Graffiti typically relates to a range of practices from tagging through to elaborate "pieces" (from masterpiece) with a focus on stylized words and text, usually including the tag name of the artists/writers and their associated crew. The complexity of letter styles, which often renders graffiti illegible to the wider public, supports a position that graffiti is an egocentric form of private communication among writers—an appropriation of public space. In contrast, street art as a wider set of artistic practices often presents a more public address, less tied to the subcultural practices and conventions associated with graffiti. The differentiation of graffiti and street art is often arbitrary. Street art, like graffiti, is more often than not done without permission (i.e., illegal) and often practiced by people who are or have been involved in the graffiti subculture. As this article will endeavor to show, the sometimes arbitrary separation of graffiti from street art by metropolitan agencies has allowed an embrace and even valorization of the power of "street art" to activate space, at a time of increasing criminalization of "graffiti."

NORMALIZING GRAFFITI AS OUT OF PLACE

Modern graffiti traces its lineage to Philadelphia and New York in the late 1960s and early 1970s, developing into a subculture in the 1970s that soon became enmeshed with the emerging rap and breakdance scene, with the dance-rap-graffiti assemblage

becoming recognized collectively as "hip-hop" culture. For Australian graffiti writers a key moment in the transmission of the underground movement beyond New York was the screening on Australian television of the video clip for Buffalo Gals by Malcolm McLaren, which brought the New York hip-hop culture to a global audience. The film clip itself featured the Rock Steady Crew breakdancing and young men using spray paints to produce graffiti as the visual setting for the soundtrack.

With this specialization came the first systematic attempts to mitigate the presence of graffiti. Wilson and Healy's 1986 report for the New South Wales State Rail on behalf of the Australian Institute of Criminology marks the commencement of Sydney's version of the war on graffiti, part of a "new military urbanism" (Iveson 2010) where "war has ... become virtually indistinguishable from police activity" (Hardt and Negri 2004: 14, in Iveson 2010: 118). In cities like Sydney, political statements and anti-graffiti laws have reflected the political will to continue to fight this urban "war without end." Bespoke anti-graffiti laws have been developed to support statements made by state and local government politicians that continually reaffirm the commitment to eradicate graffiti from the urban environment.

MANAGING GRAFFITI AND STREET ART IN CREATIVE SYDNEY

While they continue to criminalize graffiti, state and local governments have bought into creative city discourses. The promise of the creative economy has led to investment in cultural planning mechanisms and public art policies. In the wars on graffiti the moral battleground of the city has moved on, with the embrace of the transformative (and economically significant) capacity of creativity heralding a new era of zones of toleration, of legal graffiti walls, and street art galleries. One of the main issues of contention has been the continual pronouncements that it is good to have public art but bad to have graffiti. The rise of street art has gone some way to unsettling this dichotomy, opening possibilities to reappraise the moral geographies of creative practice in public space.

The three primary practices in contemporary graffiti management are enforcement, removal, and engagement. The first two are the means through which property relations are protected from graffiti,

Plate 45 Graffiti and art, Sydney

Source: Roger Keil

involving coordination with police, reporting and recording mechanisms, and efforts to physically remove graffiti from public and private property. The third of these programmatic responses to graffiti involves attempts to engage with graffiti writers and young people at risk in order to limit their involvement with destructive graffiti practices, and alleviate the slide into risky behaviors and criminal activity. One form of engagement is the implementation of legal graffiti walls, which have been used in Sydney to delimit the impact of graffiti and as sites that facilitate engagement and diversionary programs. The presence of "legal walls" unsettles the normative criminalization of graffiti, supplying demarcated sites where this illegal activity is legal.

CITY OF SYDNEY

The City of Sydney, as the wealthiest LGA in the Sydney metropolitan area and the symbolic and financial center of the city, has actively pursued policies and practices that promote Sydney as a global city. Attempts by Melbourne to celebrate street art, including graffiti, through the implementation of zones of tolerance in several central Melbourne laneways, including the declaration by the CEO of Melbourne City Council that Melbourne is "the street art capital of Australia," has led to emulation in

Sydney. In 2010, the City of Sydney implemented an annual Laneway Art project and Streetware program designed to transform Sydney's laneways (City of Sydney 2011). More generally, the City of Sydney has recently implemented a cultural plan, Sydney 2030: A Cultural and Creative City, and a Public Art Policy. Each year the City celebrates public art through the month-long Art and About public art event. Regarding unsanctioned public art, the City of Sydney maintains both a Graffiti Management Plan (2004) designed to manage the presence of illegal graffiti, and an (Interim) Aerosol Arts Guidelines (2006) that have been implemented in part to recognize the "artistic and social value" of legal aerosol art. However, even here the separation of policies on graffiti management is not so clear, as the Aerosol Arts Guidelines validate existing illegal "murals" "which have been accepted by the City as contributing to the local character of a precinct" (City of Sydney 2006). Despite being "illegal," these murals have been evaluated as "public art" and as such are not subject to removal. Further, new aerosol art murals require an extensive vetting process, involving development applications (minimum cost \$110 with a 13-page application), and review by the council's Public Art Panel against a long list of criteria including the relevance and appropriateness of the work to the City of Sydney, and consistency with the vision of the City of Sydney. Despite the good intentions to include "good" graffiti practices within the policy purview of the Council, these costs and restrictions appear to have been an effective barrier to successful incorporation of legal aerosol art practiced by graffiti writers in this LGA.

WRITING WAYS IN THE CREATIVE CITY

Some graffiti writers/street artists, those who have refined their skills enough, and have persisted with graffiti and street art practice in the face of public resistance, are now part of a contemporary arts scene that has begun to include street art. In 2010, two events signified the wider acceptance of "street art" in the public sphere. First, Exit Through the Gift Shop, featuring Banksy, was selected as an opening night film for the 2010 Sydney Film Festival. The success of the film ensured it received a limited release in Sydney, showing in a cinema in the graffiti-friendly suburb of Newtown bridging the Marrickville and Sydney LGAs. Second, at the national level, the acceptance of street art in the wider arts community was signaled by its inclusion in the premier national arts institution, the National Gallery of Australia. The exhibition, Space Invaders, ran from October 2010 to March 2011 and included 40 street artists with varied associations to graffiti and street art.

CONCLUSION

What I have presented here is the diversity at the edges of debates about graffiti. The overwhelming majority of discussions continue to assume that all graffiti is out of place, supporting the implementation of more innovative and technologically mediated ways of getting rid of it. As a testament to this, the recent NSW Inquiry into Graffiti and Public Infrastructure (NSW Legislative Assembly 2010) included sixteen submissions from local governments in Sydney. Most simply stated how much money they spent trying to eradicate graffiti. Even Councils that have more complex approaches, as noted in this paper, neglected to discuss their engagement practices, pushing instead the costs of graffiti to the local community, conforming to the normative valuation of graffiti despite the complexity present in the local governance context.

Norms change with time. Time continues to play an important role in graffiti as ephemeral artworks are increasingly captured, stored, and circulated on the Internet. The Internet has become central to the prolongation of presence beyond the physical removal of the aerosol art piece. With regard to the destruction of the artwork in the MCA, you can watch it again and again on Vimeo, or try to track down one of the limited edition prints of the completed piece sold online to hang on your private wall. The artworks themselves are also on the move as they travel to regional art galleries in New South Wales, the Australian Capital Territory, South Australia, and Queensland. Attempts to control space are subverted and the moral codification of urban space is challenged. Declarations of a moral geography that renders graffiti as uniformly out of place, as in the case in contemporary Parramatta, appear too rigid. In fact the shift to recognize and formalize graffiti in public space, in the form of street art galleries, or legal walls, or even through recognition of the value of Banksy, reflects the way moral codes have been unsettled by the presence of creative city discourses. Rather than seeing these street artists and graffiti writers as purely transgressive actors whose morality of behaviors exists outside and relative to the moral codification of urban space, it is possible to see them as generative of new ways of conceiving of spaces in the city as creative. By paying attention to the moral geographies of creative practice we can reveal the discontinuities between the way we are seeking to regulate creativity in the city and the informal creative practices that are an integral part of the creative city. With the embrace of the transformative character of creativity and the postindustrial promise of the creative industries, the unsanctioned presence of graffiti and street art in the city becomes not merely subject to a moral framing from without, but also contributes to the continuous production of new urban moral codes.

REFERENCES FROM THE READING

City of Sydney (2006) *City of Sydney Interim Aerosol Art Guidelines: For the Creation and Management of Murals in the Public Domain*, Sydney: City of Sydney.

City of Sydney (2011). Streetware, *City of Sydney website*, http://www.cityofsydney.nsw.gov.au/cityart/special/Streetware.asp.

Dickens, L. (2008) Placing post-graffiti: The journey of the Peckham Rock, *Cultural Geographies*, 15: 471–96.

Dickens, L. (2010) Pictures of walls? Producing, pricing and collecting the street art screen print, *City*, 14: 63–81.

Dickenson, M. (2008) The making of space, race and place: New York city's war on graffiti, 1970– the present, *Critique of Anthropology*, 28, 1: 27–45.

Ferrell, J. (1996) *Crimes of Style: Urban Graffiti and the Politics of Criminality*, Boston: Northeastern University Press.

Florida, R. (2002) *The Rise of the Creative Class*, New York: Basic Books.

Hardt, M. and Negri, A. (2004) *Multitude: War and Democracy in the Age of Empire*, New York: Penguin.

Iveson, K. (2009) War is over (if you want it): Rethinking the graffiti problem, *Australian Planner*, 46, 1: 24–34.

Iveson, K. (2010) The wars on graffiti and the new military urbanism, *City*, 14: 115–34.

Lachmann, R. (1988) Graffiti as career and ideology, *American Journal of Sociology*, 94: 229–50.

Landry, C. (2000) *The Creative City: A Toolkit for Urban Innovators*, London: Earthscan Publications.

MacDonald, N. (2001) *The Graffiti Subculture: Youth, Masculinity and Identity in London and New York*, Basingstoke: Palgrave.

Manco, T. (2004) *Street Logos*, London: Thames and Hudson.

NSW Legislative Assembly (2010) *Report on Graffiti and Public Infrastructure*, Report No. 6/54, Sydney: Standing Committee on Public Works.

Snyder, G. J. (2009) *Graffiti Lives: Beyond the Tag in New York's Urban Underground*, New York: New York University Press.

SIX

55
"Spaces and networks of musical creativity in the city"

from *Geography Compass* (2009)

Allan Watson, Michael Hoyler, and Christoph Mager

EDITORS' INTRODUCTION

The geography of creative and media industries has attracted much research scrutiny for global city researchers. In this essay, Allan Watson (Senior Lecturer in Human Geography at Staffordshire University), Michael Hoyler (Senior Lecturer in Human Geography at Loughborough University), and Christoph Mager (Department of Geography, University of Heidelberg) review the scholarly literature on the relational geography of music production. They point out that the role of urban space in music creativity and production has often been neglected, and that particular physical places within cities–bars, cafes, recording studios, and ethnically diverse neighborhoods–are magnets for creative talents. They also highlight the circulation of music knowledge between cities, through the movement of musicians, DJs, producers, and music industry executives, and the sharing of pre-recordings across borders so that songs can be mixed by specific engineers. They urge a spatial turn in the study of musical creativity, and large and diverse global cities such as London and New York offer privileged sites to examine how musicians interact with other actors, urban physical environments, and technology in their creative endeavors. Along with the previous chapter on graffiti, this essay provides another example of how globalizing cities are closely connected with one another with flows of cultural production.

Certain cities have a privileged history of creativity. Peter Hall's (1998) historical account of creativity in cities suggests a link between cities of large size and episodes of extraordinary creativity. Hall demonstrates how, throughout history, the most creative cities have been the true global cities of their time. However, musical creativity can spark in any city at any given time. Whether one thinks of classical music in 18th Century Vienna, New York's Tin Pan Alley, Nashville's Music Row, Motown in Detroit in the 1960s, or the guitar music of Liverpool and Manchester, specific types of music are associated inextricably with particular cities. More generally, the city provides the concrete places which offer spaces for musical

creativity. Certain spaces, such as recording studios, are specifically organised for this purpose, although music is produced in many spaces, from the bedroom, garage or home studio, to community and youth centres, to street corners and clubs. However, music is not only made in urban spaces, but also for urban spaces. Specific sites link the production and consumption of music, for example night clubs and concert halls, but also abandoned and reclaimed spaces such as empty warehouses and former factories and public spaces like the street. Urban geography, both material and imagined, is then a crucial mediating factor in the production and consumption of music, although we concede that musical creativity is not exclusively an urban phenomenon.

Cities also sustain networks that foster and support musical creativity. These networks may persist over time, or exist only for a short creative episode. Thus some cities are associated with one particular musical style, while others provide a constant stream of musical creativity. These networks come together in locales of creativity and production, for example live music venues, cafes, and bars allowing networking, along with music industry infrastructure, and therefore find 'fixity' in the concrete spaces of the city. However, networks of musical creativity are at the same time fluid. While mobility within musical creative networks has undoubtedly been enhanced by new internet technologies, allowing for the increased sharing of knowledge and for the wider distribution of musical products, there is a materiality to this mobility that stretches further back than the widespread introduction of the internet. Musical knowledge has always moved within and between cities through mobile creatives, including musicians and DJs, producers, and music industry executives. (Pre-)recordings have also always been mobile, having been sent and continuing to be sent throughout the world to be mastered and mixed in different studios by specific engineers.

Individuals with unique skills and creativity are thus the main prerequisite for the maintenance and renewal of these creative networks, with certain cities acting as magnets for talented individuals from across the globe (Scott 1999). City diversity is seen to be a significant factor in encouraging skilled labour to locate to a particular city, contributing to an open, dynamic, and cool 'people climate' valued by creatives. Nowhere is this more marked than in the buzzing, heterogeneous, ethnically diverse, and tolerant neighbourhoods of cities.

MUSICAL CREATIVITY IN THE CITY: SPACES AND NETWORKS

As simultaneously structure and event, creativity finds newness in both space and time through the mixing, encounters, and contacts between people and cultures, across multiple spatial scales. Therefore, while the imaginative capacities of an individual are indispensable to the process of creativity, creativity does not reside exclusively within isolated individuals. To understand creativity there is then a need to understand the social and existential conditions that are its foundations.

Furthermore, as Scott (1999) contends, even at their most intimate moments of birth, creative moments and episodes connect with concrete social conditions. Therefore, it is important to give attention to the social and physical environments in which creativity happens. Notions of space and place are inherently tied to culture, with cultures not only forming within a certain place, but also being active in producing the identity of places.

Cities however do not only provide places and spaces for creativity. Certain cities act as leading 'cultural metropolises' in a global urban network (Krätke and Taylor 2004), and channel and articulate creativity from different (urban and non-urban) places to consumers in other cities across the world. In a global media industry that is concentrated in and around the key cities of global capitalism, if musical output is to be recognised as *creative* it must go through the cultural contexts and distinct spaces of specific cities. Musicians and other artists have a historical tendency to concentrate in the creative and bohemian enclaves of particular cities in search of inspiration and experience; see for example Lloyd (2006) on the Wicker Park neighbourhood of Chicago, and Foord (1999) on the Hackney area of London.

While certain cities have developed an intimate relationship with music, and are celebrated as distinctive sites of productions for particular forms of music, cities are not however single homogeneous entities. Certain neighbourhoods and spaces within these cities are identifiable places of musical creativity, containing specific spaces of musical production and consumption. This creativity will be influenced by the physical landscapes and cultural diversity of particular neighbourhoods. It will also be influenced by the presence of supporting networks of musicians, other creatives, audiences, and music industry players, and by a presence of a cultural and economic infrastructure.

Diverse neighbourhoods provide the opportunity for the mutual exchange of musical styles and practices amongst different cultural groups, increasing wider exposure to a set of *atonal ensembles* of diverse musical cultural expressions. Musical creativity from cultural fusion in and across such neighbourhoods has produced some of the most successful and influential genres of music. Hip hop, for example, finds its roots in the Caribbean but materialised as a distinct genre when mixed with urban musical cultures in

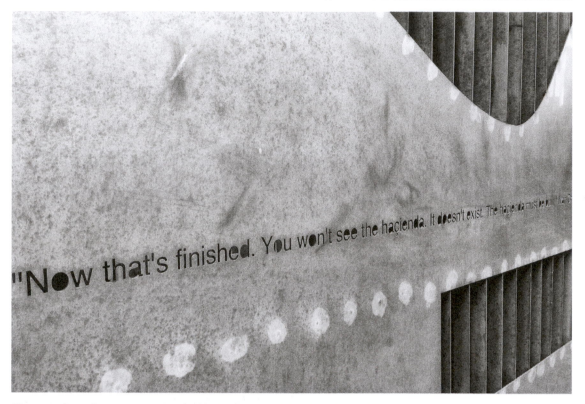

Plate 46 Artwork commemorating the Hacienda music club in Manchester, UK

Source: Roger Keil

Western cities. Emerging in the deprived inner-city neighbourhoods of US cities, in particular the Harlem and South Bronx neighbourhoods of New York, hip hop was and remains to be intense in its territoriality, and in particular in its focus on the *ghetto* as both a real and imagined space. Jazeel's (2005) examination of British-Asian soundscapes emanating from the UK highlights the new soundscapes which develop when musical creatives draw on fluid, transnational cultural and technological influences in both their work and life.

In examining local musical creativity and music industries, we must also recognise the role of supply and demand in the local economy. For local music industries and infrastructure to be economically successful or even viable, there must be a sufficient number of customers, and density of human capital and resources to economise on production costs, to make production profitable. Recording studios, for example, can have large fixed costs from continuous investment in new technologies, and must be able to attract a sufficient number of musicians and

producers to use the studio in order to cover these costs. Therefore the largest and most successful studios are predominantly located in cities, where the density of musicians will be highest. Furthermore, while new technologies may empower musicians within studios, it cannot guarantee commercial success. Almost all music that is commercially successful has to pass through urban spaces, in which cultural innovators practice their vocations on products for both localised consumption and also distribution to more remote places. Musicians may find it difficult to sell music without using the supporting industry infrastructure of such places. This infrastructure includes live music venues where consumers and record companies can see the music performed in a concrete space as opposed to the fluid space of the internet, serving to distinguish authentic products and give value to local music in an increasingly global market. Pubs and clubs remain the main sites for engagement with live music, and are central to the development of local music scenes. Live music performance, for example, is recognised as a key source of revenue in the music industry. However, large live music venues have high fixed costs, and they must be able to attract a sufficient number of consumers within a distance that allows ease of travel to the venue. Larger music venues are therefore almost exclusively located in cities with considerable population density. Demand within local markets is then crucial to the economic viability of the music industry.

RECORDING STUDIOS: FORMALIZED SPACES OF MUSICAL CREATIVITY

Recording studios are the most formal of all spaces of musical creativity in cities. Largely acting as an independent service within the contemporary recorded music industry, they form the direct link between the record companies and artists and the creation of the final recorded musical product. Many are owned and operated by entrepreneurial producers and engineers, whilst record companies maintain control over a number of larger studios. Recording studios are privileged to the most intimate moments of musical creativity and emotive performance. Viewed from Gibson's (2005) relational perspective, these creative moments are produced not by the musician alone, but through relations between musicians, producers, and engineers.

Although recording studios are often regarded in the popular imagination as a closed and guarded environment, it should be recognised that it is not only the relationships operating *inside* the studio that affect creative moments. Recording studios are at once insulated spaces of creativity, isolated from the city outside, and spaces influenced directly by the wider contexts in which the studios operate. Studios exist in neither a musical nor cultural vacuum, and music scenes, local aesthetics, musicians, and skilled labour play an important role in the development of approaches to recording and an influence on the resulting sounds. For Scott (1999) the recording studio is a sort of microcosm of a much more extensive domain of activities in the creative field. The attraction of creative workers to a city supports a different infrastructure, which in turn may correspond to concomitant developments in musical life in those same places. The location of studios within large cities thus reflects the locational preferences of musicians and skilled workers from throughout the music industry including the producers and sound engineers critical to the studios. This creative talent is

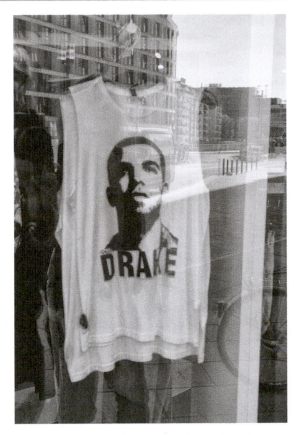

Plate 47 Toronto's hiphop artist Drake in a Frankfurt storefront

Source: Roger Keil

crucial to the performance of the recording studios, being required to know how to operate technical complex equipment, but also to have the tacit knowledge and *craft skills*, gained from experience, which are indispensable to artistic creativity within the studio. In this sense the studio is a unique place of learning and knowledge transfer that may cut across artists, genres, and styles. Here lie the roots of the current artist-producers in popular music.

OUTSIDE THE STUDIO: CREATIVITY AND PERFORMANCE IN THE URBAN ENVIRONMENT

While recording studios are amongst the most conspicuous spaces of musical creativity, urban creative spaces may take a variety of forms, from the bedroom and garages to street corners, clubs, and dancehalls. In the 1920s and 1930s in North America, big band swing music developed into a distinct genre through being played in large urban dance halls. As southern blacks migrated to northern cities, their dance traditions fused with European traditional formal ballroom dancing, resulting in a new dance form known as the Lindy Hop. As the best bands followed the dancers, and developed swing music suited to the dancers' needs, the dancers in turn followed the best bands and created dance moves that emphasised the musicality of the emerging 'swing' music. Harlem, with its dense concentration of speakeasies and dance halls, nightclubs and ballrooms, became the epicentre for this new dance form. In particular, it was the Savoy Ballroom, spiritual home to some of the most famous personalities of the time, such as the 'swing master' Chick Webb and dancers Norma Miller and Frankie Manning, that acted as the key space for musical creativity and consumption. Similarly, in the 1950s and 1960s in the Pacific Northwest Region of North America, a distinctive musical style called the Northwest Sound developed around the key social institution of teenage dance, as both a social event and an opportunity to hear live music. The music at these dances captured and created the excitement, power, and illicitness of the events, as well as reflecting the physicality of the worklife of the Northwest. The urban spaces at the centre of these local social activities were 'big band' era ballrooms, where promoters presented local groups. These groups developed local dance hall alternatives to the rock and roll being written in a factory style in New York, which were produced in a layered and artificial style that made the music difficult to reproduce in a live setting. The creative process was driven by the need to produce music which young people could dance to when played live in the dance halls. The sound they created was necessarily elemental and energetic, loud and hard-edged with a driving dance beat. These examples show how dance-beat oriented people, through their preferences and demands from the dance floors of a specific set of urban spaces, directly influenced the development of the musical styles of big band swing and the Northwest Sound.

The advent of club cultures, raves, and other forms of dance music, has predicated certain urban spaces being symbolically transformed by music. This is due to the ways in which dance music producers have traditionally been quick to embrace new technologies and modes of production. Dance music focuses on DJs using and mixing pre-recorded material in a live environment, mediating fragments of other texts from diverse geographical contexts in re-combined forms. Using available technologies to compose new sounds, dance music creativity links directly to the spontaneous moments of live performance, and spaces of performance are at once spaces of production and consumption of dance music. Unlike more commercial forms of dance music performance which have permeated more widely into many diverse spaces of production and consumption, rave and 'acid house' performances deliberately took place in large abandoned spaces, often previously used for industrial and manufacturing production such as old warehouses and factories, turning the cracks in urban landscapes into temporary *lived* spaces and *imaginative* landscapes.

The process of creativity in space is of course not limited to urban environments. In Goa, India, psychedelic rave music has agency in *natural* spaces where very different people come together as audiences and dancers, enabling the connection of bodies to the physical conditions of the environment (Saldanha 2005). On Goa's beaches, music develops meaning through its spatial-temporal connection to the natural environment - the sun, the moon, the smells and noises of the beach – arranging and transforming the physical setting and taking bodies 'elsewhere.' In the mid-1980s, as electronic house and techno music were developing in North American cities, DJs were able to play and create new forms of music in Goa, which became known as 'Goa trance', enabling the audience to participate in drug-fuelled dancing and making Goa the 'rave capital of the third world.' As young people travel to hear and dance to the music in Goa, so they bring new music with them, participating directly in the development of the rave scene.

CONCLUSION

Urban creative spaces may take a variety of forms, from the formal creative space of the recording studio, to the informal spaces of bedrooms, garages, community centres, clubs, and street corners. In the case of recording studios, we have argued that creative moments happen through the relations between skilled creative technologists and artists. The location of the most successful studios within large cities therefore not only reflects the locational preferences of musicians, but also those of the skilled workers (producers, sound engineers) who are crucial to the performance of the studios. As tools and techniques for networking studios in geographically distant locations continue to become more sophisticated, studios are increasingly able to service highly mobile musical creatives, enabling them to co-ordinate musical recordings on a global scale. Thus we are seeing the development of new relational geographies of music creativity across multiple spatial scales. In the case of rave cultures, urban spaces are shown to be important symbolic resources in the creative process. This creative process, we argue, is a material and embodied practice that links directly to the spontaneous moments of simultaneous live performance and consumption. Large abandoned urban industrial spaces, such as old warehouses and factories, are transformed symbolically in imaginative landscapes through the material practices of musical creativity. This, we suggest, clearly highlights the need to situate creativity more squarely in its material and embodied contexts of production and consumption.

REFERENCES FROM THE READING

Foord, J. (1999) Reflections on 'hidden art', *Rising East*, 3, 2: 38–66.

Gibson, C. (2005) Recording studios: Relational spaces of creativity and the city, *Built Environment*, 31, 3: 192–207.

Hall, P. (1998) *Cities in Civilization*, London: Phoenix.

Jazeel, T. (2005) The world is sound? Geography, musicology and British-Asian soundscapes, *Area*, 37, 3: 233–41.

Krätke, S. and Taylor, P. J. (2004) A world geography of global media cities, *European Planning Studies*, 12, 4: 459–77.

Lloyd, R. D. (2006) *Neo-Bohemia: Art and Commerce in the Post-industrial City*, New York: Routledge.

Saldanha, A. (2005) Trance and visibility at dawn: Racial dynamics in Goa's rave scene, *Social and Cultural Geography*, 6, 5: 707–21.

Scott, A. J. (1999) The cultural economy: Geography and the creative field, *Media, Culture & Society*, 21, 6: 807–17.

SIX

56
"Provincializing the global city: from Bombay to Mumbai"

from *Social Text* (2004)

Rashimi Varma

EDITORS' INTRODUCTION

Mumbai, India's primary commercial capital, is a familiar setting for film and literature on India. Rashimi Varma (Associate Professor of English and Comparative Literary Studies at Warwick University, UK) compares urban representations of Mumbai in literature and film in the 1950s and 1980s, and she contrasts them with Mumbai after 2000. The depiction of Bombay in those early Bollywood films from the 1950s, post-independent India is often a city divided by class rather than by ethnicity or religion. It is a cosmopolitan place for both nationalist ambitions and global aspirations, a magnet attracting migrants from across the country. The political climate changed drastically in the 1980s with the rise of right-wing Hindu nationalism, which sparked anti-Muslim riots in the city in 1992. Mumbai in the 1980s as represented in the Indian literature is a city with "provincializing" tendencies. Fast-forward to the present: Mumbai today is a center of finance, real estate, film industries, aspiring to be India's global city. Varma questions whether the Western-conceptualized global city narrative can be applied to understand the contradictions and realities in postcolonial Mumbai, and she urges us to revisit the cultural politics of urban representation, to uncover provincializing currents in the process of global city formation.

One key paradox that this article explores is that the moment in which Bombay begins to be embraced by theorists and practitioners of global cities and globalization in the North is also paradoxically the moment of Bombay's provincialization. I use *provincialization* as a primarily relational term. Though it has historically sedimented notions of particularisms (regional or local), homogeneity, and anticosmopolitanism, I want to suggest that it gathers all these meanings only in relation to how globality and cosmopolitanism are constructed. The following discussion entails an account of how provincialization has been central to the project of the Hindu Right in Bombay and how that project is integrally articulated with the material transformations engendered by global capitalism. I track the contradictory articulation of the global and the provincial in Bombay through the circulation of the two texts—*Shri 420* (1955 film, by Raj Kapoor) and *The Satanic Verses* (1988 novel, by Salman Rushdie) as marking different histories of the representational space of Bombay. The 1950s and the 1980s are thus read as key moments in which narratives of citizenship were constituted, contested, and rearticulated with transformations in both the national and the global economy as represented in the city of Bombay.

Much of recent scholarship on Bombay has emerged in the wake of the perceived dissolution in the 1990s of Bombay's iconic status as the nation's cosmopolitan center. The dissolution was marked violently by the riots of late 1992 and early 1993, in which thousands of Muslims were massacred in Bombay, and about a quarter million of them fled the city. The origins of what Arjun Appadurai has called the "decosmopolitanization" of Bombay (Appadurai 2000), or what I call its provincialization, are most often situated in the rise of the right-wing party Shiv Sena that was formed in 1966 but really gained political hegemony in the 1980s and 1990s. The party's ideology promotes regional chauvinism (in which Bombay belongs to Maharashtra, the state in which it is located, and thus to Maharashtrians), and Hindutva, or Hindu supremacism (in which Bombay is part of the sacred geography of a Hindu nation and Muslims are "outsiders"). Accounts of the rise of the right-wing party usually link the growth of ethnic chauvinism to struggles over economic resources in the city. They trace a trajectory of Bombay's transformations from its status as a cosmopolitan center of trade, commerce, and bourgeois nationalism to a postcolonial dream city of the possibilities of enterprise and industry. The city's promise of civic order and efficiency had attracted the nation's poor, the disenfranchised, and "unproductive" members of the educated class who arrived there to become productive citizens of the nation, making it India's most populous city in the immediate decades around and after independence. But that promise was violently transmuted into the present nightmare of an ethnically cleansed city in which Maharashtrians and Hindus are Bombay's only legitimate citizens. This has caused a serious rupture in the image of Bombay as India's capital of hope both for many of its own residents and for those living in less cosmopolitan parts of the country. This thoroughly divided city, far from being a model for the secular Indian imagination, now seems to be emulating a different model of nationalism rooted in provincial and exclusionary identifications.

This article seeks to open another front—the ground of representation—as a way to think about how Bombay's realities have been imagined and how its imaginations have been transmuted into its realities. Attentiveness to the traffic between imagination and reality as constituting the different modalities of being in Bombay might perhaps lead us to examine the political stakes in remembering Bombay's cosmopolitan past—an insistent trope in all critiques of Bombay's decline—and to imagine new realities for the future of this embattled postcolonial city.

SHRI 420

In 1955 Raj Kapoor, the maker and hero of *Shri 420*, was already an established star in Bombay. On the eve of Prime Minister Nehru's visit to the USSR that year, *Shri 420* was shown in Moscow, and Kapoor, or rather his screen persona, the innocent, Chaplinesque Raju, became the symbol of a globalizing socialism in the wake of third world decolonization. *Shri 420* was an incredibly powerful narrative in 1950s India. Its plot commented on the hopes and desires of countless migrants who flocked to the city looking for both economic opportunity and social justice. Films from this period were often variations on that theme, and the story of an innocent tramp surviving and questioning the soulless frenzy of a capitalist city became a staple of Kapoor's cinema of social commitment. This was also the period of Nehruvian effervescence about the possibilities of a modern, socialist, and secular nation as embodied in the space of a well-planned city. This optimism has now come under attack for its blindness to social difference and to the limits of imagining universal citizenship in the nation.

Key to *Shri 420*'s thematization of citizenship in postcolonial Bombay is a radical opening up of the processes by which enfranchisement is achieved. The city is the space where citizens are made and where they make themselves. Housing as a mode of being at home in the city seems to materialize the question of citizenship in the concrete space of Bombay, subject as it is to the vagaries of capitalist development in the city. As a complex of individual, communal, and state desires, it is, after all, the concentrated site of the experience of inequality in the city. The footpath where Raju spends his first night in Bombay belongs to "mai-baap" (mother- father), referring to the paternalistic state, even while goons extract rent from the homeless for a spot on it. Nevertheless, it is here that Raju finds a community of the homeless that is morally superior to the world of the rich in which hearts are sold for gold and silver (to paraphrase another song from the film). A memorable shot in the film has the camera panning from the footpath upward to the mansion of the rich Seth, directly dramatizing the spatial inequities of the city and its housing, a theme that has been central in both political and artistic imaginings of Bombay. But despite homelessness, the poor have claims on the city, and the search for dwelling becomes a focal point of being in the city.

S I X

THE SATANIC VERSES (1988)

The Satanic Verses offers a narrative that produces Bombay as cinema. Clamoring to be renamed Mumbai (signifying the cynical use of its precolonial history by the Hindu Right), which Rushdie calls Bombay's "stage name," the city of Rushdie's fiction blurs the gap between reality and image. Straddling both East and West, Rushdie writes of Bombay as "a culture of remakes. Its architecture mimicked the skyscraper, its cinema endlessly re-invented" Hollywood. Breaking off from the fabular mode of the rest of the narrative, Rushdie paints Bombay through a pastiche of newspaper headlines, magazine stories, advertisement hoardings, film sequences, and television serials. As the narrative disintegrates into fragments, it breaks up the city into discontinuous realities. In contrast with the deep and Manichaean evocation of space in *Shri 420* that was rendered in black-and-white tones, in *The Satanic Verses* Bombay is flat like a cinema screen, a city of lurid and tawdry surfaces.

The Satanic Verses uses the idea of culture as surface and fragment to mark the shift from the emphasis on modern production as the ground for the consolidation of the state and national identity (as in the film) to the postmodernizing emphasis on consumption and fragmentation in the economy. Both are evidenced in the proliferation and insinuation of various media in domestic and public spaces. The shift from manufacturing to service and finance industries and a growing informalization of labor in 1980s Bombay was another crucial feature of how capital was restructuring the very experience of being in the city. The latter shift came on the heels of the historic 1982 strike of textile workers and effectively broke the back of union power in Bombay. Amid the backdrop of the global dominance of neoliberalism (as depicted in the figure of Thatcherism in the London sections of the novel), the historic defeat of the textile workers paved the way for the smoother entry of transnational capital into the city, and the rampant growth in real estate value that catapulted Bombay into the company of "global cities" such as New York, London, and Tokyo. Transnational capital also had another side— it unleashed the forces of the underworld in virtually all spheres of the city's life—encompassing business (film production, real estate, gold smuggling), administration, and politics, establishing a shadowy counterpoint to the supposedly aboveground entry of global capital.

The impact of a neoliberalizing economy concentrated in the city's topography. Capital's endless capacity to first invest in built environment and then destroy it in order to create newer opportunities for capital accumulation has been of course key to Bombay's capitalist development. As the state cedes more and more space to private capital, Bombay becomes increasingly vulnerable to the vagaries of private investment and the global economy that produce vast inequities in the labor markets and often fictitious scarcities of space in the city. By the 1980s the state had already reneged on its role as guarantor of housing to the public and handed the task to corrupt private developers and builders. In the novel, on his return "home" to Bombay, Saladin finds his ancestral property is worth millions of rupees and is up for sale to developers. Its owners (his two stepmothers) were characteristically "wholly unsentimental about real estate." For them the decision to convert history and sentiment, as embodied in family property, into capital merely indicated the changes the city was going through, even as the agents of this change were rendered simultaneously omnipotent and invisible: "One more high-rise, one less piece of old Bombay. . . What's the difference? Cities change." Looking out the window of this "home" in Bombay, Saladin muses how "the view from this was no more than an old and sentimental echo." Now he can only exclaim: "To the devil with it! Let the bulldozers come. If the old refused to die, the new could not be born." He turns away from the view, in stark contrast to the vista of public housing that Raju looks onto at the end of *Shri 420*, imagining a socialist Bombay that would house every being in Bombay. What stands out in the narrative of this latter moment of modernity, looking back onto the ruins of the first, is the representation of that prior moment as a sentimental impossibility.

These major shifts in capital's topography are linked to seismic shifts in the political formations in Bombay. The novel narrates the transformation of Bombay's history from universalism to particularism. The 1950s euphoria of the potential for postcolonial universal citizenship and commitment to a global, socialist ideal had paradoxically rested on nationalism and a strong idea of the state. It is transmuted now into communal and provincial identifications emerging on the heels of the failures of the state and the exhaustion of its potential for providing a universal narrative. This disavowal of the state's role in articulating visions of social justice has been key to neoliberalism's denial of society itself, as expressed in Margaret Thatcher's famous pronouncement about the demise of society. In a marked departure from the ethical imperatives of the state in Kapoor's vision, the novel depicts the Indian state as implicated in global corruption. The shadows of the "armaments scandal" (involving the purchase of Swedish guns in which Indian politicians were exposed for having received kickbacks) fall across the Bombay of Rushdie's narrative. The ethic of free enterprise that was the moral equivalent of theft in *Shri 420* is rearticulated as patriotic duty in 1980s India, and Bombay is heralded as the engine of capitalist growth.

CONCLUSION

In Mumbai, we see that the globalization of capital has unleashed contradictory forces of provincialization. Globalization is no longer the necessary precondition for cosmopolitanism. Far from "unbundling" nation-state and territory, we can see how the nation-state is being defined in more exclusivist and territorial terms. While contemporary postmodern and Left analyses of the city maintain a binary opposition between the city and nation, and maintain as well an enduring faith in the progress narrative (the idea that things are in general becoming better as they become more global), we see how the state and nation are articulated with urban form in both provincial and transnational networks. Paradoxically, IBM's slogan "Solutions for a small planet," with its emphasis on the increasing contraction of places, represents the degree to which the unevenness of economic development, as well as the national site on which political struggles are waged, is glossed over within the commodified celebration of global cities. Indeed, global capital has been all too eager to embrace the idea of the global city.

All this goes to show that the "case" of Mumbai does not exactly fit the global city narrative. But is it enough to suggest that Bombay as a global city of the South challenges many of the assumptions of global city theory? Is that not a tacit acquiescence to the very hierarchical global grid constructed in the global city theory with at least two tiers of global cities (those in the North and those in the South) that has been key to theorizations of global cities?

Postcolonial cities, because of their multiple legacies, can occupy only a tenuous position in this particular story of globalization. The contradictory ways in which capital is reconstituting both citizenship

and urban form in the global South is starkly visible in Bombay's mutation into Mumbai. The act of renaming Bombay as Mumbai embodies the interaction between multiple spatialities, and it is in the transactions between the provincial and the global that the constitution of locality, globalism, and the simultaneously mythic and very real idea of the homeland are played out. Thus the global city of the South is perhaps better understood not only by noting its position in the global economy but also by analyzing complex processes of provincialization and nativization that are unfolding even as globalization proceeds apace. In Mumbai we see in many ways how it is possible to displace the global both into an earlier historical moment (as in the discussion of the 1950s moment) as well as beyond the terrain of the city region as capital moves into ever newer sites.

Contemporary forms of global capital and citizenship—articulated in the space of the city—are offering different challenges to our very being in the city. Capital's visible, tangible, material forms and its invisible, speculative, mysterious flows through the space of the city (crisscrossing between the multiple layers of the official and the unofficial economies) exist in productive tension with each other and with citizenship in terms of who belongs, moves into, and through the city, in both official and unofficial capacities. Crucially, both capital and citizenship are dependent on intersecting economies of representation. The materialization of value in bodies, property, money, on the one hand, and the anxious gap between reality and image, legal status and non-status, use and speculation, on the other, produces contradictions that have become key to the constitution of the global cities of the South, releasing great uncertainties about the direction this particular political moment is going to take. What I have attempted to foreground in this article is the role of the cultural politics of texts, enmeshed in the messy terrain of popular imaginings of Bombay in cinema and poetry, in journalism and activism. What we see in the transmutation of Bombay into Mumbai is an example of such a contradictory articulation in which the globalization of capital confronts the provincialization of citizens within the postcolonial state.

REFERENCES FROM THE READING

Appadurai, A. (2000) Spectral housing and urban cleansing: Notes on Millennial Mumbai, *Public Culture*, 12, 3: 627–51.

PART SEVEN

Frontiers

Plate 48 Financial center, Singapore

Source: Roger Keil

INTRODUCTION TO PART SEVEN

The chapters in this final section are more recent contributions to the debate on globalization and urbanization. These essays, perhaps more than others, are abridged and distilled down to the core arguments in order to illustrate the ongoing debates. All chapters are, in some manner, critical of the traditional global/world city research agenda and all are equally wary of the possibly contradictory political valences that may be associated with it, both in scholarly and policy contexts. Some of these chapters do, however, represent a sincere and thorough defense of the adaptive methodology used in global city research by the founders of the field. In general terms, the contributions below may be divided into two groups: those advocating for conceptual and methodological reformulations within the current parameters of global city theory, and those adopting a fundamentally critical stance and proposing an incisive change of direction in the field. Concomitantly, the critiques differ according to whether they are primarily social-scientific, and thus target the "research" element of Friedmann's original agenda, or whether they are more explicitly political in orientation, and thus strive to address the question of "action" in globalizing city-regions. More recently, critiques of global city research have addressed more fundamental questions of urban inquiries, such as epistemology (the way we see things), ontology (the way we imagine ourselves in the world), and methodology (the ways we find out about things and ourselves). These questions are at the heart of most chapters below (for another comprehensive collection of articles dealing with the challenges of theorizing globalizing urbanization, see the debate curated by Jennifer Robinson and Ananya Roy in the *International Journal of Urban and Regional Research* in 2015).

The prologue by Doreen Massey is recognition of the late geographer's tremendous influence on the field of global urban studies. The essay is an excerpt from her acclaimed and in many ways prophetic book *World City*, a love-hate story of London as a global city. Massey did not live to see either the inauguration of the British capital's new mayor Sadik Khan, the first British Muslim to hold this highest office in the country's largest city since his election in early 2016, or the Brexit vote of June 23, 2016, that left the city as an island of ostensive internationalists in a sea of majority anti-Europeans. Yet she would have surely appreciated the contradictions, opportunities, and ironies that such developments entail. As she notes in *World City*, "the question 'where does London (or any city) end?' must at least address the issue of those recruited into the dynamics of the urban economy and society by the long lines of connections of all sorts that stretch out to the rest of the country and on around the planet" (2007: 216). Referencing Paul Gilroy's notion of "the challenge of being in the same present", she speaks about "the challenge of space, the challenge of the full recognition of coeval others" (Ibid.).

The opening chapter (Ch. 58) in the section is explicitly critical of the methodology used in many studies on global cities. Excerpted from Michael Peter Smith's work, *Transnational Urbanism* (2001), the chapter critiques the structuralist bias of much global cities research. M. P. Smith argues that, despite the extensive empirical research on global cities, there can be no positive determination of a "real place" that could be legitimately described as a "global city." As he explains, "the global city is best thought of as a historical construct, not a place or 'object' consisting of essential properties that can be readily measured outside the process of meaning-making." For M. P. Smith, the structuralist bias of global city theory blinds researchers to the transnational forms of human agency that, in practice, actually construct and sustain the global city.

Richard G. Smith (Ch. 60) takes an even more radically critical approach and questions the very foundational thinking of global city research, especially its ostensive obsession with "command and control" capacities.

Chapters 59, 61, and 62 in this section, however, extend the original "mission" of global city research further into a more elaborate agenda for research on particular centralities, such as command and control functions in cities under a regime of financialization. While Taylor (Ch. 59) charges ahead by building an even more solid (if not uncontested) framework for his research on "interlocking networks" of world cities, a newer generation of GaWC trained urban scholars put up a spirited defense of the primary assumptions and methodologies in political economy based on research Taylor and his group have been doing for the last 20 years (see also Ch. 20 and Ch. 22). Taken together, these chapters both continue and adapt many of the original conceptual and methodological frameworks in the study of power structures of global urban networks.

The rest of the chapters in this section mark significant departures from the traditional research agenda of global cities. While each of these contributions comes with a different set of assumptions on globalizing urbanization, they share the notion that the debate has to move away from any fixed notions of the "city." Focusing on suburbanization, Keil (Ch. 63) examines the intersection of globalization and urbanization from the perspective of the growing periphery of urban regions. Roy (Ch. 64), Robinson (Ch. 66), and Ren (Ch. 67) raise important issues of boundaries and comparative methods in a postcolonial urban world. Brenner and Schmid (Ch. 65) introduce the concept of "planetary urbanization" as a critique of "city-ist" ideology.

This section, and the Reader, ends with an excerpt from an author who was arguably one of the most important voices in 20th century urban thought, the French sociospatial theorist Henri Lefebvre. Lefebvre never wrote explicitly about global cities in the sense that they have been examined in this book. Nonetheless, in one of his major works, *The Urban Revolution* (2003 [originally published in 1969]), Lefebvre alludes to globalized urbanization in at least two ways that, in our view, have considerable significance for future work in this field. First, Lefebvre contends that we have now collectively entered into a phase of world history in which questions of human survival are tied, intrinsically, to processes of urban development. For Lefebvre, then, the world city has emerged not because certain types of places have become command and control centers for the global economy, but rather because urbanization has become a generalized process unfolding worldwide. Lefebvre referred to this process as the arrival of "urban society." Second, in a prescient critique of Maoist revolutionary strategy, Lefebvre argues against the notion that there could, under contemporary conditions, be a rural alternative to global urbanization. Like other authors after him (Friedmann 2002; Magnusson 1996), Lefebvre insists that there is no "outside" to the urban world. We all reside in it; therefore, one of the basic political questions of the current period is: what kind of urban world do we want to live in?

Beyond its significance to urban specialists, does research on globalizing cities make a more general contribution to our understanding of contemporary social life? According to the late John Berger (2003: 13), we live in a globalized tyranny grounded upon a decentralized power structure "ranging from the 200 largest multinational corporations to the Pentagon"; for him, this imperial system rules in a style which is "interlocking yet diffuse, dictatorial yet anonymous" and its "aim is to delocalize the entire world." Research on globalizing cities, in our view, offers us some bearings, some intellectual and political grounding, as we attempt to orient ourselves within this disjointed, yet profoundly authoritarian, new world order. Whether or not this intellectual perspective can help open up possibilities for radical or progressive social change is ultimately a political question that can only be decided through ongoing social mobilizations and struggles. For Friedmann and Wolff (Ch. 3), the different worlds of the "citadel" and the "ghetto" within the global city co-constituted one another; yet the inhabitants of the ghetto were said to be "isolated like a virus" from the hegemonic power structures. By contrast, as we learn from Doreen Massey in the prologue, "the constraints are undeniable (from the global movements of capital to the corsets imposed by national policy), but there are possibilities for responses that question and even rework and undermine those constraints." There are political options, and there are different ways of doing research on global cities. None of them is hardwired to support power, corporations, and the structures of global finance, and all of them can be shaped to point towards alternatives in the "urban commonwealth" (Kohn 2016). Staying with the image of the "virus" for another moment, we also now know that the global city's rigid, fortress-like boundaries are dissolved into fluid flows of actor-networks, where viruses seem to co-exist quite easily alongside corporate leaders and bankers within the global urban system (Ali and Keil Ch. 24; Smith 2001; Swyngedouw 2004). The politics that spring from such radical openings in the world city fortress may generate unexpectedly transformative outcomes—for instance, new types of citizenship claims, new modes of political struggle, and a new globalized urban political ecology. The contributions to this section begin to lead us down this path of possibilities.

REFERENCES

Berger, J. (2003) Where are we? *Harper's Magazine*, March 13.
Friedmann, J. (2002) *The Prospect of Cities*, Minneapolis and London: University of Minnesota Press.
Kohn, M. (2016) *The Death and Life of the Urban Commonwealth*, Oxford: Oxford University Press.
Lefebvre, H. (2003) [1969] *The Urban Revolution*, Minneapolis: University of Minnesota Press.
Magnusson, W. (1996) *In Search of Political Space*, Toronto: University of Toronto Press.
Massey, D. (2007) *World City*, Cambridge: Polity.
Smith, M. P. (2001) *Transnational Urbanism*, Cambridge, MA: Blackwell.
Swyngedouw, E. (2004) *Social Power and the Urbanization of Water*, Oxford: Oxford University Press.

57 Prologue "World city"

from *World City* (2007)

Doreen Massey

Cities are central to neoliberal globalisation. The increasing concentration of humanity within them is in part a product of it. Their internal forms reflect its market dynamics (the shining spectacular projects, the juxtaposition of greed and need). The competition between them is both product and support of the neoliberal agenda. And in certain cities (those we call world cities, or global cities) is concentrated the institutional and cultural infrastructure that is key to all of this. Cities, then, are crucial to neoliberal globalization but they figure in very diverse ways within it. London as a centre of command and orchestration and as, indeed, a focus of migration and a home to an astonishing multiplicity of ethnicities and cultures is a part, and a powerful part, of the same dynamics that produce, elsewhere in other cities, [Mike] Davis's 'planet of slums.'

'What does this place stand for?' is a question that can and should be asked of any place. Its import and urgency will vary between places (global cities may have more possibility in the sense of room for manoeuvre, and more responsibility in the sense of the magnitude of their efforts), but it is a question that makes each and every place a potential arena for political contest about its answer. The constraints are undeniable (from the global movements of capital to the corsets imposed by national policy), but there are possibilities for responses that question and even rework and undermine those constraints. Conceptually, it is important to recognize that the global is as much locally produced as vice versa, that an imaginary of big binaries of us and them (often aligned with local and global) is both politically disabling and exonerating of our own (and our own local place's) implication,

and that the very fact of specificity (that places vary) both opens up the space for debate and enjoins us to invent. Moreover, not only is it politically possible, it is also a political responsibility, to find some way of addressing that question. It is a challenge not only for the local state, but for the grass roots of the city, too, indeed for all those who in one way or another take a part in their identity from the fact that they are there.

In this more complex picture, then, London's character as an image of a future world is at least ambiguous. Internally, too, like most cities, it is both enormously pleasurable and a site of serious deprivation and despair. The account presented in *World City* attempts to weave a course between on the one hand the dystopian visions and apocalyptic urban accounts, generalized perhaps from experience in the USA (and further generalized through the power-geometries of academy and the publishing industry), and on the other hand those over-easy skateboarding celebrations of cities as poor fun. In part, this is a general position (few cities if any are solely one or the other); moreover an insistence on complexity leaves open more opportunities for politics. . . . [Global cities] have become crucial bargaining chips, vital components in the struggle to assert neoliberalism politically. Within urban policy discourse around the world a global-city rhetoric has emerged. . . . It becomes necessary to ask: *whose* city is at issue here? The very characterization of cities as 'global' is a strategy whereby the part stands in for the whole, where the city is defined by its elite and the rest are consigned to invisibility. . . . In his consideration of the possibility of spreading outwards the implications of the demotic cosmopolitanism that exists within the city, [Paul] Gilroy writes of 'the challenge of being in the same present'. This is precisely the challenge of space, the challenge of the full recognition of coeval others. What are at issue here are the responsibilities of place. These may concern the politics within the city, the question of the city within the country, or the question of the city in the wider world. But in any case, this is a 'local' politics that asserts and actively politicizes both the fact of multiplicity within and the essential openness of place to the beyond.

58

"The global cities discourse:
a return to the master narrative?"

from *Transnational Urbanism* (2001)

Michael Peter Smith

EDITORS' INTRODUCTION

Michael Peter Smith is Distinguished Professor Emeritus in Community Studies and Development at the University of California, Davis. Smith has written, edited or co-edited more than twenty books on urban development and globalization, including *City, State and Market* (1988), *Transnational Urbanism* (2001), and *Citizenship Across Borders* (2008). Smith's earlier work examined the political economy of US cities during the post-1970s period and the changing role of state policy in the restructuring of urban space. While Smith's writings of the 1980s contributed significantly to the advancement of global cities research, he has more recently developed a provocatively critical relationship to this literature, in part through his ongoing ethnographic research on transnational migration flows, social movements, and the rise of "extra-territorial" citizenship. Smith's "agent-centered" critique of global cities scholarship is developed at length in his 2001 book, *Transnational Urbanism*, from which the chapter below is excerpted.

For Smith, global cities research has focused excessively on the construction of objectivist, structural typologies that bracket the everyday experience of life and struggle within globalizing city-regions. According to Smith, it is essential to reconceptualize global cities as spaces of intense sociocultural interaction in which transnational social networks – composed of diverse, interconnected people and institutions – converge and interact. Smith argues that this perspective can transcend the methodological limitations of earlier approaches to global cities research by underscoring the role of social mobilization "from below" in the production of globalized urban spaces. In the selection excerpted here, Smith concretizes his methodological agenda through a brief case study of grassroots urban protests against the Suharto regime in Jakarta, Indonesia, following the Asian financial crisis. Smith reinterprets the process of "globalization" in this context as a multifaceted localization of diverse transnational social networks to produce contextually specific conditions of sociopolitical contestation. Smith's critique of global cities research raises a number of key questions regarding the limitations of political-economic approaches to the study of urbanization. While Smith's theoretical position remains controversial, it is evident that his alternative, agent-centered approach opens up an illuminating and highly original perspective on the social life of globalizing urban spaces. The debate about transnational urbanism has been one of the critical departures in the conversations with and beyond the original global city literature. Smith's work in particular has been influential for a new generation of scholars who have engaged with the concept. Among the exemplary work in this tradition has been an edited book by Stefan Krätke, Katrin Wildner, and Stephan Lanz, *Transnationalism and Urbanism* (Routledge 2012).

Debates about "the global city" have taken on a recognizable, if not formulaic, character, poised somewhere on a conceptual and epistemological borderland where positivism, structuralism, and essentialism meet. The tendency to focus this debate around positivist taxonomies, urban hierarchies constructed on the basis of these taxonomies, and empirical efforts to map or even to formally model the "real" causes and consequences of global cities, leads participants in the debate to overlook the fact that the global cities discourse takes place within a wider public discourse on "globalization," which is itself a contested political project advanced by powerful social forces, not some "thing" to be observed by scientific tools. The global cities discourse constitutes an effort to define the global city as an objective reality operating outside the social construction of meaning. The participants in this debate argue about which set of material conditions are attributes of global cities and which cities possess these attributes. The debate generates alternative positivist taxonomies said to be occurring entirely outside our processes of meaning-making.

My own position on this debate is framed within the wider epistemological and ontological debate about social constructionism and the critique of ideology. The basic starting point of my argument is this: there is no solid object known as the "global city" appropriate for grounding urban research, only an endless interplay of differently articulated *networks, practices,* and *power relations* best deciphered by studying the agency of local, regional, national, and transnational actors that discursively and historically construct understandings of "locality," "transnationality," and "globalization" in different urban settings. The global city is best thought of as a historical construct, not a place or "object" consisting of essential properties that can be readily measured outside the process of meaning-making.

The global cities literature varies in its specification of the financial, informational, and migratory flows that intersect to constitute a global city. Representations of global cities nonetheless share a common conceptual strategy in which these global flows are envisaged as "coming together" within the jurisdictional boundaries of single cities like New York,

Plate 49 Mexican American store, fashion district, Los Angeles

Source: Roger Keil

London, Paris, Tokyo, or Los Angeles. This strategy thus localizes within the boundaries of particular cities highly mobile, transnational processes of capital investment, manufacturing, commodity circulation, labor migration, and cultural production. In so doing, the strategy sharply demarcates an "inside" from an "outside" and largely highlights what goes on inside global cities. When the global cities framework addresses relations among global cities it imposes a hierarchical ordering on the economic functions and production complexes assumed to be "integrating" the global cities across space. By hierarchically nesting criss-crossing transnational connections and imposing an economic ordering mechanism on global cities, the effort to construct a global urban hierarchy belies the often marked disarticulation among the financial flows, political alliances, mediascapes, and everyday sociocultural networks now transgressing borders and constituting what I term transnational urbanism.

RECONSIDERING THE GLOBAL CITY THESIS

Viewed from our current vantage point, global city assumptions about the systemic coherence of the urban hierarchy, the transterritorial economic convergence of global command and control functions, and the declining significance of the nation-state, are more difficult to maintain than they once were. For better or worse, cities have different histories, cultural mixes, national experiences, and modes of political regulation of urban space. These must be taken into account in any nuanced analysis of the localization of global processes. Their absence weakens the usefulness of the global cities thesis.

The quest for a fixed urban hierarchy should be abandoned, in my view not only because hierarchical economic taxonomies are too static a formulation in the face of the volatility *of* capital investment flows, but also because of the multiple and often contradictory compositions of these flows (e.g. speculative vs. direct fixed investment). More importantly, this quest for a hierarchical ordering mechanism is fruitless because the local cultural spaces that are sites of transnational urbanism are also far less static than global city theorists assume. Far from reflecting a static ontology of "being" or "community," localities are dynamic constructions "in the making." They are

multiply inflected sites of cultural and political as well as economic flows, projects, and practices. The cultural and political processes that go into their constant making and remaking are simply too dynamic to predict the course of urban development in advance of the actual political struggles through which contending spatial practices and cultural conflicts that constitute transnational urbanism are played out.

The global cities thesis centrally depends on the assumption that global economic restructuring precedes and determines urban spatial and sociocultural restructuring, inexorably transforming localities by disconnecting them from their ties to nation-states, national legal systems, local political cultures, and everyday place-making practices. In the past decade extensive research questioning this core assumption on empirical grounds has opened up the discourse on global cities and globalization more generally (e.g., Cox 1997). I will extend the argument by advancing a social constructionism analysis which exposes the entire discourse of globalization as a "tightly scripted narrative of differential power" (Gibson-Graham 1996/1997: 1) that actually creates the powerlessness that it projects by contributing to the hegemony of prevailing globalization metaphors of capitalism's global reach, local penetration, and placeless logic.

Attention to the analysis of transnational social networks is one way out of this impasse. The practices of transnational political networks exemplify the ways in which situated actors socially construct historically specific projects that become localized within particular cities throughout the world, thereby shaping their urban politics and social life. This type of urban politics is constituted as people connected across borders in transnational networks interact with more local institutional structures and actors, as well as with such putatively more global actors as multinational corporations and international agencies, to produce urban change. The local and the global are thus mutually constitutive in particular places at particular times.

THE GLOBAL GOVERNANCE AGENDA

Today's grand narrative of economic globalization has been advanced most forcefully not by academics but by an emergent international monetarist regime, a set of institutional actors who have instituted a political offensive against developmental states and institutions. The globally oriented institutions spearheading this offensive were established under the auspices of the 1980s debt crisis to advance the monetarist agenda of global efficiency and financial credibility, against the nationally oriented institutions of developing countries. This globalist political project has, in turn, produced a series of struggles over the meaning of "the local," as cities and other localities, via their political, economic, and cultural actors and institutions, seek either to find a niche within the new global public philosophy, or resist pressures to "globalize," i.e. to practice fiscal austerity and conform to monetarist principles and policies.

The origins of the contemporary ideology of globalization are thus historically specific. They constitute efforts by powerful social forces to replace the developmentalist institutional framework of the 1960s to 1980s, premised on modernization theory, with a new mode of economic integration of cities and states to world market principles. Abandoning the argument that a convergent path of development will follow from the spread of Western-oriented "modernization," it is now argued that locational specialization or "niche formation" is an inevitable by-product of globalization. The principles legitimating economic globalization have been posited by their advocates as inevitable byproducts, or ruling logics, of the material condition of globalization "on the ground," rather than, as they are, the social constructs of historically specific social interests, including transnational corporate and financial elites, heads of international agencies, state managers that have embraced neoliberalism, various academic ideologues, and the managers of the IMF, the World Bank, and the World Trade Organization. This regime seeks to transform the discourse of development to one in which shrinking the state is a social virtue and structural adjustment an unavoidable imperative. This neoliberal regime of "global governance" is viewed as an incipient global ruling class whose efforts to achieve global economic management can only be thwarted when globalization itself is recognized as a historically specific and hotly contested project of social actors and agents rather than an inevitable condition of contemporary existence.

Many academics overestimate the coherence of this global project and underestimate the potential effectiveness of the oppositional forces they acknowledge. As grassroots social-movement scholars have shown, forceful national, local, and transnational political identities have sprung up to resist the hegemonic ideology and austerity policies imposed by the global neoliberal regime. Thus it is not possible to agree entirely with this pessimistic assessment of the global future. It is surprising, nonetheless, that the global cities scholars seem not to have given much thought to the question of whether their research agenda and its "objective" findings implicitly naturalize that very project by legitimating the "reality" of global cities as part and parcel of an unstoppable process of economic globalization. Unintentionally, their epistemology thus becomes the ontology of global cities.

Once globalization is unmasked as the most recent historical version of the free-market ideology, a more fruitful way forward is suggested by the trenchant critique of the globalization discourse offered by two leading post-Marxist feminist theorists writing under the pen name J. K. Gibson-Graham (1996/1997). Gibson-Graham argue that the globalization thesis, both in terms of the economy of global production and finance and in terms of the culture of global consumerism fueled by the assumed capitalist domination of global telecommunications, suffers from theoretical and empirical overreach, but has telling political consequences. These consequences – a politics of fear and subjection by workers, communities, and other potentially oppositional forces – are most apparent when the two threads of the globalization thesis are combined into a single masculinist grand narrative of the penetration of capitalist social relations not only of production and consumption but also of meaning-making and the constitution of subjectivity.

The global penetration metaphor may be challenged on theoretical and empirical grounds. Theoretically, Gibson-Graham argue that by relying on the sexual metaphor of penetration to invoke the power of capital, but failing to acknowledge that sexuality is a process of interpenetration, not a one-way flow, the globalization discourse refuses a contingent countervision of the capitalist economy "as penetrable by non-capitalist economic forms" and social relations (Gibson-Graham, 1996/1997: 17–18). Relatedly, the very logic of the metaphor of the global reach of capitalist command and control arrangements can be reinterpreted as a project inherently subject to over-extension, a set of tasks often frustrated by global overreach.

The potential efficacy of active political resistance to global neoliberalism is insightfully illustrated in

Gibson-Graham's narration of a transnational political campaign conducted during the early 1990s against a TNC, the Ravenwood Aluminum Company (RAC). The campaign was led by the United Steelworkers of America (USWA), working through the channels of several international labor organizations. It combined cross-border political lobbying by labor unions in 28 countries, on five continents, with a US-based consumer boycott. By going global, the unions and transnational grassroots activists involved in this campaign succeeded in obtaining a favorable labor contract in the USA and imposed restrictions on the key global trader who had organized the purchase of RAC in a leveraged buyout. Gibson-Graham (ibid.: 9–10) conclude their narrative with a metaphor of dogged determination, noting that "terrier-like, the USWA pursued the company relentlessly around the globe yanking and pulling at it until it capitulated. Using their own globalized networks, workers met internationalism with internationalism and eventually won."

The moral of the story is clear. Far from being the exclusive preserve of global capital, global space is a discursive arena and very much a contested terrain. One key dimension of globalization, the telecommunications revolution, is often assumed to be a straightforward tool of global capital in organizing production, directing financial flows, and orchestrating consumer desire. While it may facilitate these processes to some extent, it also has become a viable channel for spatially extending the contested terrain of globalization, enabling oppositional forces to jump scale and go global. New technologies of communication such as e-mail, fax, and the internet have become viable mechanisms facilitating the transnationalization of culture and politics, enabling cross-border information exchange, transnational political networking, and the sociopolitical organization of a wide variety of new forms of "transnational grassroots politics." Such modes of transnational political networking include, but are not limited to, transnational labor organizing, international human rights campaigns, indigenous peoples' movements, multinational feminist projects, and global environmental activism. For better or worse, the transnationalization of communications provides access as well to such transnational networks of social action "from below" as Islamic fundamentalism, neo-Nazism, and the militia movement, a development that progressive scholars and social activists doubtless find less salutary.

TRANSNATIONAL URBANISM: BEYOND REIFICATION

In light of these dynamic political and cultural developments, it is time to move beyond the boundaries of the global cities discourse. Instead of pursuing the quest for a hierarchy of nested cities arranged neatly in terms of their internal functions to do the bidding of international capital, it is more fruitful to assume a less easily ordered urban world of localized articulations, where sociocultural as well as political-economic relations criss-cross and obliterate sharp distinctions between inside and outside, local and global. These partially overlapping and often contested networks of meaning are relations of power that link people, places, and processes to each other transnationally in overlapping and often contested social relations rather than in hierarchical patterns of interaction.

Numerous examples of these criss-crossing linkages that literally "re-place" the urban, reconfiguring "the city" from a global epiphenomenon to a fluid site of contested social relations of meaning and power, could be investigated. To bring this chapter to a close, however, a historically specific example of what I have in mind may be helpful. I offer the case of the political crisis in Indonesia in which grassroots student protests and urban unrest in Jakarta and other Indonesian cities brought down the Suharto regime. These street-level forms of political practice have been represented as "local" responses to a perceived national crisis. Nevertheless, the protests, occurring in various cities throughout Indonesia, were occasioned by the transnational IMF austerity policies imposed on Indonesia from a putative "outside" as a condition for an IMF bailout of the Suharto regime in the face of the Asian financial crisis. Moreover, global media representations of the interplay between Suharto and the IMF reinforced the outside – inside duality by featuring a widely circulated photograph of an IMF official standing behind Suharto with arms folded as he signed an IMF austerity agreement. While the bodily gesture was meaningless to the IMF official, it was pregnant with meaning within Indonesian cultural practice, symbolizing personal domination of one body by another and creating an impression of Suharto's declining strength as a symbol of national development and modernization. This impression was reinforced when the austerity policies were at first avoided and then implemented selectively "on the

Plate 50 Indian textile store, Singapore

Source: Roger Keil

inside" by Suharto, who imposed sharp price increases on daily necessities such as food and fuel for ordinary citizens, while striving to protect his family's vast fortune in the face of a declining currency market and maintain the collusive business practices with ethnic Chinese business elites that had been the source of the family fortune.

A key target of the popular urban protests that eventually led to Suharto's ouster were the wealthy Chinese business elites of the Chinese diaspora, a transnational entrepreneurial class tied to networks of Chinese business elites operating in other cities in Asia and throughout the world. This group, comprising less than 3 per cent of the Indonesian population, dominates the nation's economy and has nearly 70 per cent of the nation's wealth, but has been absent from the military and the public political arena. Thus, a "Chinese" ethnic background became an easy initial target of nationalist and racist discourses and practices during the early stages of the financial crisis. Indeed, as criticism of Suharto and his inner circle mounted, the government tried to deflect the blame onto a more generalized target, the entire ethnic Chinese population.

This effort was initially partially successful, but when the student component of the urban protests emerged in Jakarta and other Indonesian university cities, the various student groups focused like a laser beam on political democratization as well as economic grievances. Their frequent demonstrations charging corruption and calling for Suharto's ouster captured as much, if not more, global media attention than the initial round of anti-Chinese violence. The growing student protests in various cities throughout Indonesia also further emboldened the discontented urban masses, who lashed out in a new round of urban rioting, looting, and burning directed largely against banks, shops, supermarkets, electronics stores, and the home of a prominent Chinese billionaire. These developments, in turn, refocused discontent on the vast wealth of Suharto and his family. In this changed and highly charged political context, Suharto was unable to defend his interests by commanding the army to repress the students or the urban masses. Instead the army largely stood by during the demonstrations and street rioting, as various military leaders maneuvered behind the scenes to oust Suharto and jockeyed for position in the new regime.

Suharto's fall from grace and power in Indonesia was accompanied elsewhere in Asia by the emergence of anti-IMF protests in cities like Seoul, Korea, where in late May 1998 tens of thousands of workers walked off their jobs and participated in street demonstrations in central Seoul directed against a state law making it easier to lay off workers. The law was passed by the South Korean government under pressure from the IMF, which had made increased "labor flexibility" a condition of granting Korea a $58 billion bailout package. While the workers railed against their government and the Korean corporate conglomerates, or *chaebol,* whose elites have escaped the hardships experienced by ordinary Koreans during the financial crisis, they left little doubt that they perceived the IMF as responsible for their plight. Standing in front of Seoul Station, the city's transportation hub, the workers raised their fists and chanted slogans like: "Let's destroy restructuring! Let's fight to secure our jobs!" The Korean Confederation of Trade Unions representing over 500,000 members followed up the demonstration by demanding that the state renegotiate the "labor flexibility" issue with the IMF.

In the face of increasingly visible popular opposition against their policies, both the IMF and the World Bank have been forced to reconsider the rigidity with which their agents have pursued the neoliberal agenda. For example, in July 1998 the World Bank resumed aid to Indonesia by granting a $1 billion loan under far less stringent terms than had previously been expected of the Suharto regime. The new accord allowed for more social spending than previous agreements and permitted the state to run a sizable budget deficit, an item explicitly prohibited in previous agreements.

In the Indonesian case, a complex web of social relations and conflicting discourses of "globalization" came together and were localized in confrontations in Jakarta and other Indonesian cities. In the historically specific context of a general economic crisis in Asia, accompanied by urban protests in Indonesia and transnationally, a state-centered power structure that had colluded for three decades with a transnational economic network to regulate and appropriate the lion's share of Indonesia's economic wealth and expansion was significantly challenged. This alliance

Plate 51 Hairdresser, Istanbul

Source: Roger Keil

SEVEN

was caught in a collision course. They were squeezed between transnational regulatory bodies such as the IMF and the World Bank seeking to legitimate neoliberalism by ending "crony capitalism," and a series of geographically "local" irruptions of discontent in Indonesian cities by students who were connected to transnational communication circuits by internet information flows, and perceived themselves as central agents of democratization operating on a "global" stage. The student protests were accompanied by other episodes of urban violence by popular classes venting frustration at Chinese ethnic scapegoats or simply seeking survival in the face of austerity-induced economic hardships. These local and global crises of confidence were mediated by national cultural understandings, played out in the global media, and had effects beyond the boundaries of the nation-state. Though differently inflected, a related global-national-local discourse on economic "globalization" was being played out in Seoul. This is the stuff of the new urban politics of transnational urbanism.

The future course of transnational urbanism in Jakarta and Seoul are open-ended. Yet even a brief examination of these cases clearly illustrates the fruitfulness of viewing cities as sites where national and transnational practices become localized; local social actions reverberate transnationally, if not globally; and the networks connecting these social relations and practices, intersect in time and space and can be discerned, studied, and understood.

A further advantage of this approach to urban studies is that a wide variety of cities, rather than a handful of sites of producer service functions, or a score of interwoven, mainly Western, centers of global command and control, become appropriate sites for comparative research. Viewing cities as contested meeting grounds of transnational urbanism invites their comparative analysis as sites of (a) the localization of transnational economic, sociocultural, and political flows; (b) the transnationalization of local socioeconomic, political, and cultural forces; and (c) the practices of the networks of social action connecting these flows and forces in time and space.

These emergent transnational cities are human creations best understood as sites of multicentered, if not decentered, agency, in all of their overlapping untidiness. This is a project in which studying mediated differences in the patterns of intersecting global, transnational, national, and local flows and practices, in particular cities, becomes more important than cataloging the economic similarities of hierarchically organized financial, economic, or ideological command and control centers viewed as constructions of a single agent – multinational capital.

The "new urban politics" uncovered by this move is a disjointed terrain of global media flows, transnational migrant networks, state-centered actors that side with and oppose global actors, local and global growth machines and green movements, multilocational entrepreneurs, and multilateral political institutions, all colluding and colliding with each other, ad infinitum. The urban future following from this contested process of "place-making" is far less predictable but far more interesting than the grand narrative of global capital steam-rolling and swallowing local political elites and pushing powerless people around that inevitably seems to follow from the global cities model. Rather than viewing global cities as central expressions of the global accumulation of capital, all cities can then be viewed in the fullness of their particular linkages with the worlds outside their boundaries.

REFERENCES FROM THE READING

Cox, K. (ed.) (1997) *Spaces of Globalization*, New York: Guilford.

Gibson-Graham, J. K. (1996/1997) Querying globalization, *Rethinking Marxism*, 9, 1: 1–27.

Smith, M. P. (1988) *City, State and Market*, Cambridge, MA: Blackwell.

Smith, M. P. (2001) *Transnational Urbanism*, Cambridge, MA: Blackwell.

Smith, M. P. and Bakker, M. (eds.) (2008) *Citizenship Across Borders: The Political Transnationalism of El Migrante*, Ithaca: Cornell University Press.

59

"External urban relational processes: introducing central flow theory to complement central place theory"

from *Urban Studies* (2010)

Peter J. Taylor, Michael Hoyler, and Raf Verbruggen

EDITORS' INTRODUCTION

This chapter is another contribution from the larger GaWC group of scholars. Taylor, Hoyler, and Verbruggen take on the task of theorizing external relations between cities. While this type of analysis has been at the core of global city research for decades, this contribution is well placed in this section on research frontiers as it opens up new ways of theorizing the ubiquitous urban hierarchies and networks. Note how Taylor and his co-authors, rather than borrowing from technology-influenced thinking on rapid societal change (see Graham Ch. 23 and Castells Ch. 21), refer instead back to a critical re-reading of classic theoretical work on central places and urban economies. The authors take up the fundamental concerns introduced by Castells (Ch. 21) but elaborate on the types of vertical and horizontal relationships cities and towns are embedded in. They introduce key terms such as "town-ness" and "city-ness" to refer to variously scaled hinterland and hinterworld relationships. By introducing "central flow theory" to our vocabulary, Taylor and his co-authors break new ground and extend the reach of the conventional global city research. This work has received extended treatment in the second edition of Taylor's co-authored book *World City Network* (Taylor and Derudder 2016).

INTRODUCTION

In this chapter we propose to rethink how we study the external relations between urban places. Traditionally, conceptualisation of these relations has been satisfied by central place theory with its depiction of a spatial-hierarchical arrangement of settlements. However this theory has had a curious recent history within contemporary urban scholarship. On the one hand its formal spatial modeling has become unfashionable in geography so that it has all but disappeared from urban geography research agendas (see Taylor and Derudder 2016: Ch. 2). We would argue, though,

that 'central place thinking' is represented in the 'new economic geography/geographical economics' through the ubiquitous assumption that towns and cities are ordered in hierarchies. The 'rethinking' that we engage in here is to argue that hierarchical relations between urban places constitute only a partial understanding of inter-city relations. As well as the 'vertical' relations emanating out of 'central place thinking', it is necessary to treat distinctive and separate 'horizontal' relations that define city networks.

Although we are also critical of central place theory, it is undeniable that the basic processes identified in it continue unabated whether or not social scientists choose to study them. We interpret central place theory as describing a generic urban process, one of relations between an urban place and its hinterland. Thus we have no interest in jettisoning central place theory. However, the consensus that current urban external relations require more than this one theory has led us to identify a different external urban process that can be theorised to produce a complementary set of conceptual tools. As well as the hierarchical structure postulated by central place theory, we argue that there is a network structure between cities. Whereas the former is a vertical spatial structure linking local scales of interactions (hinterlands), the latter is primarily a horizontal spatial structure linking non-local interactions. We treat both as generic urban processes and therefore both are required to adequately describe external urban relations now and in the past.

There is, of course, nothing new in understanding that cities have relations with other cities beyond their hinterlands and that this is an important process: this is the basis of the traditional research field of macro transport geography, especially port geography. What we offer here is a more formal approach to such interactions that we set alongside central place theory: the latter is modeled as interlocking hierarchies, we introduce a central flow theory modeled as interlocking networks. Clearly this differentiation of external urban relations into two theories draws upon Castells' (1996/2001) classic identification of two distinct social spaces: spaces of places and spaces of flows. Following Arrighi (1994) in eschewing Castells' particular use of the terms just for characterizing his interpretation of contemporary society, we argue that both social spaces exist in all societies and that they need to be understood in tandem. This is what we attempt below for external urban relations.

ON THE NEED FOR A SECOND THEORY OF EXTERNAL URBAN PROCESS

Central place theory has been developed in two directions: settlement geography and retail geography; in the latter this includes intra-urban relations. Here we focus on limitations of the former as a means of understanding the spacing and hierarchy of towns and cities. A common criticism of the theory that we set aside straightaway is dislike of its normative nature. Berry (1967: vii) treats the theory as a 'deductive base'

through which to explore urban settlement patterns and we use it in this manner here: without fixation on the formal theory, and following Berry and Pred (1965: 10), we "view central place studies broadly" in order to understand the "functioning of cities as retail and service centers for surrounding areas". We focus upon two critical elements of central place studies that highlight the need for a second theory of external urban process: the inherent hierarchical relations and the varying scales at which they operate.

CRITIQUES OF INTER-URBAN HIERARCHICAL RELATIONS

Hierarchies imply competitive and cooperative inter-city relations. This argument has been used also to underpin the concept of a world city network where it is argued that mutuality is necessary for the operation of the network. Complementarity, mutuality, cooperation, of course there is nothing new in such arguments: it was just such 'non-hierarchical links' that Pred (1977) illustrated so clearly in his classic study of the space-economy three decades ago. The conclusion is that non-hierarchical inter-city links require an alternative theorizing to central place theory. The

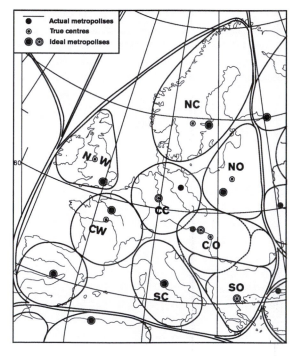

Figure 1 Christaller's European spatial structure

Source: Christaller, 1950, extract from map 1

complex pattern of inter-city relations 'à la Pred' – (1977) are a starting point for our discussion here.

This hierarchy-competition versus network-cooperation contrast goes beyond inter-city relations for it is a basic distinction in social organization in general: hierarchy and network are fundamentally different and should never be confused or used interchangeably. And yet this is often the case in urban studies, for instance in Sassen's (1991/2001) description of the inter-city relations of global cities.

We are not arguing here that cities are not organised into hierarchies, but we are arguing that there is more to inter-city relations than such hierarchies. Central place theory takes reasonable care of hierarchical relations, but is deficient for understanding complex non-hierarchical relations. Unfortunately several scholars have attempted to stretch central place theory beyond its competences and this is well illustrated in debates on the hierarchical scales to be found in central place studies.

LIMITS TO HIERARCHICAL SCALES?

Central place theory incorporates a model of interlocking hierarchies for which no upper limit is specified. Although Christaller (1933/1966) initially limited his studies to the regional scale within Germany; in relaying his ideas to an Anglophone audience the world became the limit. Berry and Pred's (1965: 7) hierarchy, for instance, "culminates in a world economy, [serviced] by 'world cities'" and Mayer (1969: 19) invokes Hall's (1966) classic *The World Cities* to claim that "at the top of the hierarchy is the 'world city', whose service area for some functions may be intercontinental". But this was not really a presage for the contemporary interest in global/world cities since in practice central place thinking was limited to two main scales of analysis: rural-regional and national.

The second major scale of analysis in central place thinking has been the national, encompassing 'national urban systems' with 'national urban hierarchies'. As their name suggests, these concepts were treated as bounded systems enabling further concepts such as entropy to provide new tools of measurement for development (Berry 1961) and modernization (Gould 1970) of states. This state-centric thinking was generally translated into spatial planning tools for national policy studies.

It has been this thinking that has been transferred to the global/world cities literature via Friedmann

(1986, 1995). His world city hierarchy with London, New York and Tokyo at the top appears to be etched into globalization consciousness. This is a case of concepts from the national scale being 'up-scaled' to the global level (Taylor 2004). Hall (2002) has attempted to convert this broad thinking into a more strict Christaller framework by extending his hierarchy to the global scale. Maybe the world city hierarchy is best interpreted as illustrating the limits of central place ideas for understanding city development. Certainly London, New York and Tokyo are very important central places, but their economic power is based upon much more than their respective central place prowess. It is time to look again at network in relation to hierarchy.

INTRODUCING TWO PROCESSES: TOWN-NESS AND CITY-NESS

The common denominator in our misgivings about the salience of the central place process for understanding cities is the neglect of non-local and non-hierarchical relations. The starting point is to name two distinct processes: the external relations that link an urban place to its hinterland we term 'town-ness'. We argue that since all urban places have hinterlands they are products of town-ness but the importance of this process will vary across urban places. Generally the larger urban places are less constituted by town-ness and more by the second urban external relations process: city-ness. This process represents inter-city relations that are broadly horizontal and beyond the hinterland. Town-ness is described by central place theory (more specifically by Christaller's marketing principle) and is modelled as urban hierarchies, whereas city-ness is described by central flow theory and is modelled as urban networks. Two key points arise from this formulation.

First, there is the important move from seeing the urban as process rather than place. Cities have been interpreted in this way by both Jacobs (1969) and Castells (1996/2001) and we extend this form of thinking to towns. The basic reason for treating towns and cities as processes rather than places is to overcome the spatial exclusivity of the latter in which an urban place is deemed either a town or a city. As processes, town-ness and city-ness can and do occur simultaneously in urban places. Every urban place, therefore, is constituted through both town-ness processes and city-ness processes. The interesting question is the balance between the two processes for any given urban example.

Second, we need to briefly explain our particular terminology. City and town are English language terms that are sometimes used interchangeably; in dictionaries cities are commonly just defined as 'large towns'. Treating them as distinct processes is therefore a new conceptual departure: this chapter is about arguing for the utility of this lexicon departure. The use of the terms town-ness and city-ness can be interpreted as both an opportunistic application of the dual urban ascription in English and a conceptual clarification of the confusion caused by their inherent inter-changeability in common usage. The remainder of this chapter provides a restatement of central place theory as town-ness and an introduction to central flow theory as city-ness.

We claim central place theory to be a formal description of a generic urban process. This is clearly suggested by the wide range of its applications across both time and space. We concur with Berry and Pred's (1965: 11) assessment that central place functions are "universal" in the make up of urban places (despite its positivistic suggestion); that is to say, they are generic mechanisms of what it is to be urban.

THE INTERLOCKING HIERARCHY MODEL

Central place process can be interpreted as the local dimension in urban external relations. In the formal specification of the model this is explicitly designated as bounded hinterlands (hexagons) around each urban settlement. But note that the concept of what is 'local', and therefore particular to a central place, is variable: the scale of hinterlands increases with the level of the central place. This is a model that is premised upon multiple inside-outside definitions that describe a hierarchical space of places. Each central place is located on the border between two higher-level places so that its hinterland is divided in half when allocated to servicing at the higher level. This produces an interlocking hierarchical pattern rather than a simple linear tree ordering. It is theoretically shown to be most efficient in bringing buyers to sellers, the agents of the town-ness process. It is a model that is premised upon multiple between-ness relations that define an interlocking space of places. Our basic argument is that the central place process is essentially simple in comparison to central flow theory.

TOWN-NESS AS SIMPLE URBAN EXTERNAL RELATIONS

Town-ness is a local affair and as such is inherently non-dynamic as an economic process. Following Jacobs (1969), economic expansion does not occur as a result of servicing a hinterland, however large. Therefore no small central place ever grew to become a metropolitan economy through external relations limited to its own hinterland. The town-hinterland relation is a relatively stable relation, not prone to rapid economic changes. In a rapidly changing economy, urban-hinterland relations will certainly change but they will never be at the cutting edge of economic development.

The basic reason why central place processes do not create economic development is because they include no local mechanism for expanding economic activity. Urban places grow by economic expansion deriving from the introduction of 'new work' creating a more complex division of labour (Jacobs 1969). Such dynamism will require inter-urban relations beyond servicing the local, whatever the hierarchical level. Thus economic change is something that occurs through a different process (city-ness) that does not restrict (simplify) inter-urban relations through a hierarchical structure. In contrast, town-ness is a process that inherently generates dependence through hierarchy rather than opportunity through more complete and complex inter-urban relations.

Town-ness may be inherently simple as an economic process but this does not mean that central place process is not important as the spatial organization through which society reproduces itself (distribution and consumption). In other words, town-ness is a generic process that is a necessary composite of urbanized societies but there is more to urban external relations than the central place process. '

INTRODUCING CENTRAL FLOW THEORY AS GENERIC NON-LOCAL CITY-NESS

Central flow theory is about bringing the non-local into an urban place to create a cosmopolitan mix of peoples, commodities and ideas. City-ness incorporates an inter-urban process, a network process that links together cities across different regions: this defines a broad hinterworld (Taylor 2004), beyond the hinterland. The result is to make cities special places,

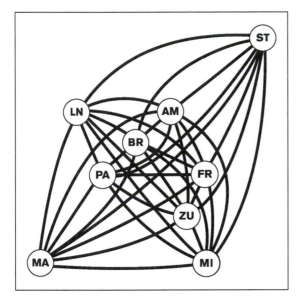

Figure 2 Inter-city links between nine European cities as practised by advanced producer services

Source: Peter J. Taylor, Michael Hoyler, and Raf Verbruggen

unique settlements within which economic expansion occurs (Jacobs 1969). We argue that city-ness as a generic feature of being urban differs from town-ness through its inherent complexity.

CITY-NESS AS COMPLEX URBAN EXTERNAL RELATIONS

Cities are dynamic and complex and this derives from the city-ness network process. According to Jacobs (1969) city networks are central to economic expansion through the mechanism of import replacement. This is how 'new work' in a city is created: local production replaces imports from other cities. This contrasts with economic growth by expanding current old work, which only increases the size of an economy but not its complexity. Adding new work makes the division of labour broader and more varied. This is expansion of economic life based upon an increasingly complex economy. Jacobs (1969) argues that such import replacement tends to occur in economic spurts and in this way can convert an 'ordinary town' into an 'extraordinary city'. In the argument developed here an urban place dominated by town-ness process may be quickly changed by enhanced city-ness process: its economy will change from being simple

and local (hinterland-based) to being complex with important non-local-links (hinterworld-based).

Because of city-ness, larger urban places are the locus of economic expansion: 'dynamic cities' are central to economic development. In addition, because they are complex economic units they are resilient to adverse change. And because they are a network process, their relations define mutuality: all cities in a network need each other in both good times and bad. This is why cities never exist alone; they come in assemblages, ordered as networks. Thus the spatial organization of economic development (production of commodities as goods and services) is a space of flows: a network of dynamic cities. We claim city-ness is a generic feature of all urbanized societies. It has certainly been a feature of the modern world-system.

Following Sassen (1991/2001), we interpret advanced producer services (professional, creative and financial) as a critical cutting edge sector in economic globalization. They service global capital through solving the problems of operating in a large transnational economy. These services have massively expanded in the last few decades and have contributed greatly to the new work that has created the dynamic and complex urban places that are called global or world cities. A key feature of these cities is that the import replacement mechanism has operated on a worldwide scale to produce a world city network. City-ness is a process and therefore there have to be agents who operationalize the process: cities do not replace imports, firms in cities do. In the case of the world city network the agents are the advanced producer service firms with global clientele. To service the latter, they operate through extensive office networks in cities across all world regions. It is the amalgam of these firms' office networks that constitutes the world city network. Inter-city relations are the flows of ideas, knowledge, information, plans, instructions, personnel, etc. that are made in the everyday business of carrying out advanced producer service projects (e.g. inter-jurisdictional contracts, global advertising campaigns). Thus it is that the service firms 'interlock' the cities and this can be formally specified as an interlocking network model. The world city network has three levels: sub-nodal (the firms), nodal (the cities) and network (city network). And it is in the extra sub-nodal level that we locate the agents, the service firms that are the network makers: world cities do not make the world city network, advanced producer service firms do.

CONCLUSION

Although the interlocking network model was devised to study contemporary cities in globalization we now contend that it constitutes a generic model for cityness, for describing city networks beyond current globalization. Vibrant, dynamic cities have always been interlocked by 'foreign' commerce – this has been what has made them cosmopolitan. The point is that city networks are constituted by the interlocking of cities by commercial agents in the everyday course of their business practice – ergo, the interlocking network model is a generic central flow theory.

We will conclude by illustrating this critical point by comparing recent findings from the interlocking network model with Christaller's (1950) European spatial structure (Figure 1). Christaller is describing a relatively simple urban space of places. A roughly equivalent urban space of flows can be created by computing the relations between the leading nine cities that dominate the European section of the world city network. Abstracting just the leading nine European cities from these data, the interlocking network model can be used to produce estimates of the business connections between the cities. Note that our units of interest are not cities per se, but rather city dyads. These are depicted in Figure 2. This model describes a very complex European space of flows. This is just a glimpse of the complexity of central flow theory in relation to central place theory.

REFERENCES FROM THE READING

Arrighi, G. (1994) *The Long Twentieth Century*, London: Verso.

Berry, B. J. L. (1961) City size distributions and economic development, *Economic Development and Cultural Change*, 9: 573–88.

Berry, B. J. L. (1967) *Geography of Market Centers and Retail Distribution*, Englewood Cliffs, NJ: Prentice-Hall.

Berry, B. J. L. and Pred, A. (1965) *Central Place Studies: A Bibliography of Theory and Applications*, Philadelphia: Regional Science Research Institute.

Castells, M. (1996/2001) *The Rise of the Network Society*, Oxford: Blackwell.

Christaller, W. (1933/1966) *Die zentralen Orte in Süddeutschland*, Jena: Gustav Fischer Verlag; translated by C. W. Baskin as *Central Places in Southern Germany*, Englewood Cliffs, NJ: Prentice-Hall.

Christaller, W. (1950) *Das Grundgerüst der räumlichen Ordnung in Europa: Die Systeme der europäischen zentralen Orte*, Frankfurt am Main: Kramer (Frankfurter Geographische Hefte, 24, 1).

Friedmann, J. (1986) The world city hypothesis, *Development and Change*, 17: 69–83.

Friedmann, J. (1995) Where we stand: a decade of world city research, in P. J. Taylor and P. L. Knox (eds.) *World Cities in a World-System*, Cambridge: Cambridge University Press, 21–47.

Gould, P. (1970) Tanzania 1920–63: The spatial impress of the modernization process, *World Politics*, 22: 149–70.

Hall, P. (1966) *The World Cities*, London: Heinemann.

Hall, P. (2002) Christaller for a global age: redrawing the urban hierarchy, in A. Mayr, M. Meurer, and J. Vogt (eds.) *Stadt und Region: Dynamik von Lebenswelten*, Leipzig: Deutsche Gesellschaft für Geographie, 110–28.

Jacobs, J. (1969) *The Economy of Cities*, New York: Vintage.

Mayer, H. M. (1969) Cities and urban geography, *Journal of Geography*, 27: 6–19.

Pred, A. (1977) *City-Systems in Advanced Economies*, London: Hutchinson.

Sassen, S. (1991/2001) *The Global City*, Princeton, NJ: Princeton University Press.

Taylor, P. J. (2004) *World City Network: A Global Urban Analysis*, London: Routledge.

Taylor, P. J. and Derudder, B. (2016) *World City Network: A Global Urban Analysis* (2nd Edition), London and New York: Routledge.

60
"Beyond the global city concept and the myth of 'command and control'"

from *International Journal of Urban and Regional Research* (2014)

Richard G. Smith

INTRODUCTION

Why the global city concept caught on and met with wide support as a seemingly valid way to understand contemporary globalization is an interesting and fundamentally important question for those engaged in the study of cities in globalization. The idea of the world economy being subject to control and coordination through major cities emerged as a core concept for neo-Marxist urban studies in the context of the political economy of the 1970s and 1980s. Then, in Britain and the US, Thatcher and Reagan respectively forced a replacement of the Keynesian postwar democratic settlement with a neoliberal formula(tion) founded on the 'free-market fundamentalism' of Milton Friedman and an obeisance to global capitalism—'legitimized' through such venal ideologies as 'wealth creation', selfish hyper-individualism, privatization, financialization, computerization, liberalization, denationalization, deregulation, free trade and the 'invisible hand' of market forces—that skewed their national economies both sectorally, towards finance and servicing globalization, and geographically, with those sectors becoming ever more concentrated in the urban centres/city-regions of London and New York. In other words, the Götterdämmerung of neoliberalism came to town and the idea of world or global cities chimed with the need of neo-Marxist critique for the world's financial centres—exemplified at the time by New York, London and Tokyo—to be understood as centres of control for an economic and financial globalization that represents just the latest stage in the development of an uneven and polarizing capitalism.

In this chapter I argue that this way of thinking about cities in globalization, about the role and function of a certain type of city as the *nec plus ultra* locations for subjecting a world economy to control, coordination and 'command', is a mistake. To make the argument, the article proceeds in three parts. Firstly, the neo-Marxist origins of the global city concept, and that concept itself, are discussed in a way that specifies how the widely acknowledged lack of evidential proof for the global city concept's claims for 'command' as a produced functionality is a consequence of its theoretical foundation in a neo-Marxist envisioning of the world economy as a structured totality to assume a need for strategic cities of a certain type to be in control. Secondly, there is a sorites paradox at the root of Taylor's construction of a 'world city network' via an interlocking network model (INM). A classic telling of the sorites paradox concerns the difficulty of answering the seemingly simple question: how many pebbles make a heap? Two or three are not a heap; 1,000 clearly are. There is no agreed threshold for 'heapness', a 'heap' is intrinsically vague, but we think we know one when we see one. The sorites paradox (from the Greek *soros*, meaning a 'heap') highlights how little-by-little arguments are a flawed form of reasoning. Discussing this with regards to Taylor's methodology discloses a further, more fundamental problem. This is because, whilst purporting to be an extension (Taylor 2001: 183) of Sassen's work, Taylor's model not only fundamentally contradicts the global city concept in its technical minutiae, but also contains no empirical

proof for centres of authority (i.e. global 'command' centres) in the network. Indeed, it is demonstrated, for the first time, exactly how Taylor's widely adopted INM represents the end-game for neo-Marxist work on cities in globalization because as the apogee of structuralism the city vanishes all together. This is because its emphasis on relations is to the detriment of the terms (the cities or 'nodes'), so that cities are reduced to no more than place-less 'containers' of unrelated autonomous corporate service firms. Finally, the article moves beyond the rubrics of neo-Marxism, structural-functionalism, structuralism and structural networks through a juxtaposition of empirical results from the social studies of finance (SSF) literature with the 'command' assumptions of Sassen (1991)—i.e. global cities as 'highly concentrated' sites where 'command and control' is produced—and Taylor (2004)—i.e. an interlocking 'world city network' where 'command and control' is a structural effect of the network—to advance the relevance of SSF literatures for understanding financial centres as socio-technical assemblages and eventful multiplicities.

THE GLOBAL CITY CONCEPT DE NOVO

The idea that certain cities act as centres of 'highly concentrated command' (Sassen 1991: 3) for the globalization of distanciated economic activity has a particular heritage, emerging from Hymer (1972), several years before most commentators currently assume. An understanding of the protracted neo-Marxist background to the global city concept (Sassen 1991) is essential for understanding its assumption of 'command', and why it is diametrically opposed to Taylor's (2001; 2004) INM and its assumptions about 'command centres' in a 'world city network', even though it appears to be similar given its common focus on advanced producer services.

The specific idea of certain cities having a 'strategic role' in controlling the world's economy is relatively new and can be traced back to a handful of authors writing in the 1970s and 1980s who focused on multinational corporations (MNCs) in cities as the locus of control for the world economy (Hymer 1972; Friedmann 1986; Friedmann and Wolff 1982). However, it was Cohen who most directly foreshadowed Sassen's (1991) and Taylor's (2004) theses, because he not only considered cities as centres of corporate headquarters but also as centres of international banking and strategic corporate services

(Cohen 1981: 302). Cohen extends Hymer's focus on MNCs in cities to consider corporate demand for advanced producer services, and he 'links these new demands to . . . the emergence of a series of global cities' (1981: 287–88) that 'serve as international centers for business decision making and corporate strategy formulation' (ibid.: 300). Thus, Cohen lays the path for Sassen's (1991) subsequent emphasis on advanced producer services, rather than MNCs, in her global city concept, because he identifies the agglomeration of corporate services and other key international functions in a few cities.

'[B]y looking at the system as a whole', Friedmann and Wolff (1982: 319) explicitly set out to establish an agenda for research and action on the premise that the world's dominant cities (i.e. those few cities 'in which most of the world's active capital is concentrated' (ibid.: 309) are increasingly integral to the formation of a worldwide economic system which must be 'viewed as a totality' (ibid.: 314). Transnational capital allocates control to just a select few of those cities at any one time; and that without these 'world' cities 'the world-spanning system of economic relations would be unthinkable' (ibid.: 312). These axioms of the world city 'approach' allowed Friedmann and Wolff to speculate as to the possibility of a hierarchy of world cities, and of global influence and control, in the capitalist world.

It was by working from this established neo-Marxist tradition that Sassen (1991: 3) could propose her global city concept and a 'new strategic role for major cities'. She envisioned a poly-nodal world economic system with New York, London and Tokyo as the leading global cities in 'command' of a process of globalization (not urbanization) predominantly spread across North America, Western Europe and Asia. For Sassen, global cities are those that now function 'as centers of finance and as centers for global servicing and management' (ibid.: 324, original emphasis), and that are afforded a centrality as 'highly concentrated command points' (ibid.: 3) in the world economy's organization because they are crucial sites for finance and for specialized producer services. It is in this post-Weberian sense (i.e. focusing on production, not power) that Sassen differs from Friedmann, and consequently this is why Taylor's (2001; 2004) work directly contradicts Sassen's (1991), because Taylor draws explicitly (see Taylor 2004: 87–8) on Friedmann's work on the spatial organization of power in urban systems as a given, to frame his

conceptualization of the power of 'global command centres' in the interlocking 'world city network'.

The city vanishes into structure (a.k.a. the interlocking 'world city network' impasse).

It is to Taylor's work that the chapter now turns, debunking the INM by exposing how its bedevilment by a sorites paradox serves to disclose a further more fundamental flaw which means that a 'world city network' of global 'command' is a non sequitur, not, in fact, logically following on from its underpinning INM, because through that methodology the 'place-ness' of all the cities in the network vanishes. A number of critiques of Taylor's interlocking 'world city network' have been made (e.g. Nordlund 2004; Robinson 2005; Neal 2011). As this chapter is about the myth of 'command and control', the central tenet of the literature on both 'world' and 'global cities', let us now examine that specific aspect in Taylor's (2004) book.

To attempt to address the paucity of data on inter-city relations across the world—rather than the dearth of intra-city data about relations between advanced producer firms within the global cities that Sassen identifies—Taylor (2004) has worked with numerous co-researchers to identify 315 cities in which just 100 independent rival firms, predominantly specializing in one of just six business services (accountancy, advertising, banking/finance, insurance, law, management consultancy), have located what are assumed to be their headquarter and branch offices. He has done this because he believes that by computing that data he can meaningfully determine the 'global network connectivity'—'where the direct instrumental power lies within the world city network' (Taylor 2004: 89)— of those cities with headquarter offices so as to locate 'command and control' in the global economy. But this approach is fatally compromised.

The empirical basis of Taylor's numerous writings is that the 'sub-nodal' data he has collected concern headquarter and branch 'nodal' locations of the intra-firm office networks of globalized business-service firms (not the 'sub-nodal' inter-firm networks). Thus, independent, unrelated service-firm headquarter offices are amalgamated to score and define the importance of any assumed global 'command' centres in a network of intercity relations, when logically that manoeuvre is meaningless. Taylor expresses no interest in, and has no data on, the relations, connections and networks between the 100 service firms—ironically, it would be evidence of those inter-firm relations within 'nodes' (i.e. cities) that might constitute some partial empirical grounding for Sassen's assumption of 'concentrated command' according to which corporate service firms do work together ('joint production')—to make his 'command centres' more than just placeless 'containers' of unconnected rival advanced producer service firms, with the assumed function of 'connectivity-through-dominance' and 'connectivity-through-subordination' within a 'world city network' (Taylor 2004: 88).

Taylor is undone by his need to identify and classify some world cities as global 'command' centres. Taylor is caught out by his ambition to want to tell a big neo-Marxist story about the global economy as a 'world network', to visualize some of the world's cities as ranked types ('Mega', 'Major', 'Medium' and 'Minor') in order to see a hierarchy of 'command and control' in a so-called 'world city network', because, if cities (the 'nodes') are mere 'containers' of independent in vacuo commercial service firms (i.e. they are unrelated/unconnected to one another), any attempt to classify cities as 'global command centres' is inevitably bedeviled by a sorites paradox that serves to disclose a much more fundamental flaw in the INM and consequent 'world city network'. A classic telling of the sorites paradox concerns the difficulty of answering the seemingly simple question: how many pebbles make a heap? Two or three are not a heap; 1,000 clearly are. There is no agreed threshold for 'heapness', a 'heap' is intrinsically vague, but we think we know one when we see one. The sorites paradox (from the Greek *soros*, meaning a 'heap') highlights how little-by-little arguments are a flawed form of reasoning.

The inconvenient truth Taylor's approach needs to confront is the fact that no hierarchy of global 'command' centres within the 'world city network' can be claimed because any such classification is only ever predicated on the 'nodal' headquarter location of one firm's overall office network. One headquarters can never logically have the effect of defining a 'global command centre' because it is completely autonomous in its operations and relations to all the other different firms' headquarter offices. Taylor's data are only intra-firm. I refer to the sorites paradox because it enables the disclosure of an underlying problem: that the fatal flaw in the INM and so 'world city network' is that the data need to be inter-firm, not intra-firm, if the city is not to be reduced to a place-less 'container' and vanish into structure. Not to be misunderstood, my point is that it is because Taylor's INM is only based on intra-firm data (i.e. data on the assumed head-quarter and branch offices in the network of offices of

autonomous individual corporate service firms) that his typology of global 'command' is sorcery. And this is crucial because all of Taylor's work on the 'world city network' from 2001 onwards is conjured through his notion of the INM, which serves as the pinion on which the very existence of his meta-geographical 'world city network' subject to 'command and control' depends. *In nuce*, unlike in Sassen's global city concept whereby the city takes 'place' as a 'factory' for the production of 'highly concentrated command', the city in Taylor's research has no 'place': 'world city network'.

BEYOND THE GLOBAL CITY CONCEPT

So, how might one further critique, and definitively move beyond, the neo-Marxist tradition, the global city concept, the interlocking 'world city network' and consequently a whole field of endeavour that is now moribund through its assumption that the global economy is subject to 'command and control'? I would propose taking heed of Thrift's (1993) sign-posting of a way out of studying cities through such a false assumption by engaging with those literatures that are, in a certain sense, descended from the 'new economic sociology' (i.e. the work of Mark Granovetter, Paul DiMaggio, and so on) that Thrift (1993: 232) first indicated to be a 'rather convincing' challenge to the taken for granted rhetoric of 'command and control'. Indeed, to develop Thrift's purpose of refuting the notion of 'concentrated command' by arguing for the social and cultural performance of cities 'as sites of social contact and narrative innovation, as places where this new world presents and represents itself, as places for story telling rather than strategy' (ibid.: 233), I will now draw out some of the implications of the findings of the social studies of finance (SSF) research for challenging both the global city concept's (Sassen 1991: 3) core assumption of 'highly concen-trated command', and the 'world city network's' (Taylor 2004) contention that the global economy is now under distributed, yet still centralized, 'command and control' across a worldwide network of cities as a functional whole. Inspired by actor-network theory and its insistence that purportedly 'economic' prac-tices are always already socio-technical, ethnographic SSF are attentive to the everyday practices, perfor-mances and geographies of financial firms. So, what are the ramifications of this SSF research if it is explicitly juxtaposed to both Sassen's (1991) global city concept and Taylor's (2004) interlocking 'world city network' model?

Buenza and Stark (2003) conducted ethnographic research in an international investment bank based in New York's World Financial Center before and after the terror attacks of 9/11 to detail how '[a] trader is not an isolated and contemplative thinker, but is engaged in cognition that is socially distributed across persons and things' (Buenza and Stark 2003: 141). However, what is interesting for the purposes of this chapter, is what can be drawn out from their study of how the trading strategy of arbitrage is achieved through the formation of 'socio-technical networks' (Latour 1991) and 'communities of interpretation'—how investment banks actually perform their business practices—in order to question the purchase of extant explanations for how such advanced producer services are organized and function as an explanation for where they are located.

Wall Street has, over several decades, become more and more of a metonym for the financial sector as a whole, because through a 'veritable quantitative revolution, based on three legs: high-speed network connectivity, high-powered computation and the development of mathematical finance' (Buenza and Stark 2003: 154), finance firms have dispersed their operations across Manhattan, and beyond it, so as not to be 'highly concentrated' (Sassen 1991: 3) either in Lower Manhattan or elsewhere. However, that dispersal cannot be explained solely through recourse to a technological determinism, the contention that urban geographies of finance can be 'read off' from the changes in the technologies of trading. Indeed, a location is chosen by an investment bank for more than the, nevertheless undoubtedly highly important, technological reason of winning a 'speed contest' against its rivals for competitive advantage. It is by investigating the role of locality through an understanding of technology as always a socio-technology—'Technology is society made durable' (Latour 1991: 103)—that Buenza and Stark (2003) have made numerous important observations that run contrary to extant assumptions about the global economy being subject to 'command and control' from major cities.

First, they note how in investment banking '[p]roximity has become crucial for some companies and obsolete for others, a source of profits for some departments and a threat to the existence of others' (ibid.: 157). Second, Buenza and Stark (2003: 158, my emphasis) note how '[T]he real locus of modern finance is not the Exchange but *the trading rooms*' of individual investment banks. The practice of trading is *within* individual firms, whose socio-spatial-technical organization is designed to enable innovation through producing new configurations (not by adding new resources). Third, because any individual firm's trading rooms are 'knowledge intensive' they need to be designed so as to foster a 'community of interpretation' and so are necessarily organized in a heterarchical fashion as independent centres of creative association, internally organized so as to distribute intelligence, organize diversity, forge trust and foster lateral ties. Taylor's assumptions about the modern organization having a command chain—with strategy and decisions flowing down from atop a hierarchical ladder of authority—is an understanding of organizations from the mid-twentieth century according to the notion of dependence, not an understanding of the contemporary firm where it is interdependence that is the essential in situ ingredient for the success of an investment bank's actual business practice. Finally, as 'communities of interpretation' and 'distributed intelligence' there is a 'place-ness' to the trading rooms of financial centres that is highly complex and simply cannot be captured by assuming either in-place 'joint production', or out-of-place 'command and control' through the transnational headquarter offices of advanced producer firms.

Investment banking now resides primarily with the machines, programmes, algorithms, software and trading methods—'black-box trading', 'algo trading', 'robo trading', 'high-frequency trading'—that make financial centres 'fly by wire', and markets much more volatile. But it is important to understand that the volatility is, to a significant extent, the consequence of the computerization of trading through algorithms by 'Quants' not being founded on value (worth) (Zaloom 2010: 23). Thus, it is what does not belong in the algorithmic models, what is not included in moving money around to 'make nothing but money' (to invoke Bear Stearns' infamous self-description), that can force the destabilization and unbinding of the networks of financial capitalism, and that is precisely why the global economy is beyond 'command and control'.

In sum, the emphasis on communities of interpretation, and socio-technical networks in SSF, is an important advance because it forces an understanding that, just because major cities host the headquarters of many of the world's major MNCs, contain identifiable 'clusters' of corporate and financial services who might work together, or are the chosen 'nodes' for the headquarter offices of transnational advanced producer service firms office networks, that does not mean that, ipso facto, those major cities are 'organizing nodes' for a world economy that is subject to 'command and control'. In short, by focusing on the practices of financial firms in major cities, and by considering these practices as performed (as events) rather than preformed (as functions), SSF scholars have challenged the modus operandi, and consequently the locus operandi, of both Sassen's and Taylor's research. Financial centres are multiplicities that are assembled, fixed and lent consistency through their performance, practice and enactment, to be sure, but the financial centre as a socio-technical assemblage is only able to be a performance because it is based on an element that does not belong to that performance, i.e. that which is inconsistent and incalculable. In 'algo trading', this element was revealed to be the baselessness of value, a fact that means that a financial centre cannot be simply reduced to its causes or conditions to assume it is in 'command and control' of the global economy.

REFERENCES FROM THE READING

Buenza, D. and Stark, D. (2003) The organization of responsiveness: Innovation and recovery in the trading rooms of Lower Manhattan, *Socio-Economic Review*, 1, 2: 135–64.

Cohen, R. B. (1981) The new international division of labour, multinational corporations and urban hierarchy, in M. Dear and A. J. Scott (eds.), *Urbanization and Urban Planning in Capitalist Society*, New York and London: Methuen: 287–318.

Friedmann, J. (1986) The world city hypothesis, *Development and Change*, 17, 1: 69–83.

Friedmann, J. (1995) Where we stand: A decade of world city research, in P. Knox and P. Taylor (eds.) *World Cities in a World-System*, Cambridge: Cambridge University Press: 21–48.

Friedmann, J. and Wolff, G. (1982) World city formation: An agenda for research and action, *International Journal of Urban and Regional Research*, 6, 3: 309–44.

Hymer, S. (1972) The multinational corporation and the law of uneven development, in J. Bhagwati (ed.) *Economics and World Order*, London: Macmillan.

Knox, P. and Taylor, P. (eds.) (1995) *World Cities in a World-System*, Cambridge: Cambridge University Press.

Latour, B. (1991) Technology is society made durable, in J. Law (ed.) *A Sociology of Monsters: Essays on Power, Technology, and Domination*, London: Routledge: 103–31.

Neal, Z. (2011) Structural determinism in the interlocking world city network, *Geographical Analysis*, 44, 2: 162–70.

Nordlund, C. (2004) A critical comment on the Taylor approach for measuring world city interlock linkages, *Geographical Analysis*, 36, 3: 290–96.

Robinson, J. (2005) Urban geography: World cities, or a world of cities, *Progress in Human Geography*, 29, 6: 757–76.

Sassen, S. (1991) *The Global City: New York, London, Tokyo*, Princeton, NJ: Princeton University Press.

Smith, R. G. (2003) World city topologies, *Progress in Human Geography*, 27, 5: 561–82.

Smith R. G. (2013a) The ordinary city trap, *Environment and Planning A*, 45: 2290–304.

Smith, R. G. (2013b) The ordinary city trap snaps back, *Environment and Planning A*, 45: 2318–22.

Smith, R. G. and Doel, M. A. (2011) Questioning the theoretical basis of current global-city research: Structures, networks and actor-networks, *International Journal of Urban and Regional Research*, 35: 24–39.

Taylor, P. J. (2001) Specification of the world city network, *Geographical Analysis*, 33, 2: 181–94.

Taylor, P. J. (2004) *World City Network: A Global Urban Analysis*, London: Routledge.

Thrift, N. (1993) An urban impasse? *Theory, Culture and Society*, 10, 2: 229–38.

Zaloom, C. (2010) The derivative world, *The Hedgehog Review*, 12, 2: 20–27.

61

"World cities under conditions of financialized globalization: towards an augmented world city hypothesis"

from *Progress in Human Geography* (2015)

David Bassens and Michiel van Meeteren

EDITORS' INTRODUCTION

David Bassens is Assistant Professor of Economic Geography at the Free University of Brussels and Associate Director of Cosmopolis, an interdisciplinary urban research team on city, culture, and society. He has also been a decade-long contributor to GaWC research and publication. Michiel van Meeteren is a postdoctoral researcher at the Free University of Brussels and his primary research focus is on economic and financial geography, although he is also interested in methodology, history of geography, and political geography. This chapter is the first of two interventions that can be read in direct contradistinction to the previous chapters by M. P. Smith and R. G. Smith. While M. P. Smith and R. G. Smith considered the notion of 'command and control' a myth of world city research, Bassens and van Meeteren argue the opposite. By incorporating financialization processes in Friedmann's world city hypothesis, they emphasize that world cities, as home to advanced producer services, remain obligatory passage points for the realization of surplus value under conditions of financialized globalization. Geographically, Bassens and van Meeteren point out that the world city archipelago extends far beyond international financial centers, as the advanced producer service sector has expanded its global reach.

WORLD CITIES: CONTINUITY AND CHANGE

The networked perspective in global city research (see Taylor's prologue in Part 3 and Taylor et al., Ch. 59) invoked a lively empirical debate about the actual geographical articulation or 'footprint' of the world city archipelago (WCA) and its backbone of International Financial Centers (IFCs) in the past decade. The overall consensus is that capitalist 'command and control' is exercised from a limited set of cities which function as nodes for transnational flows of capital, goods, people, and information from which actors operating from these places draw their power. Accepting the more or less agreed-upon geographies of the WCA, this chapter asks what goes on in WCA space.

Friedmann (1986) conceptualized 'world cities' as key basing points of capital from where the escape of the labor market rigidities of the 1970s crisis were orchestrated, while retaining capital accumulation in the world's economic core areas. The related work on 'global cities' (Sassen 1991 [2001]) theorized places where global control functions are produced by Advanced Producer Services (APS) in accountancy, advertising, finance, law, and management consultancy. Expanding these pioneering works, an augmented world city hypothesis has to address the different nature of the current crisis compared to the one of the 1970s. Both crises can be considered as different expressions of a general crisis of overaccumulation, even though the barriers to capital realization are currently very different. We expect the increasing importance of financial capital circuits – i.e. financialization – and rent-seeking behavior to change the character of command and control functions exercised from world cities.

While still functioning as a space from where global production is coordinated to an important degree, we argue that these practices related to command and control have been subsumed in a logic of financialization, making the world city archipelago an obligatory passage point for the relatively assured realization of capital under conditions of financialized globalization (cf. Allen 2010). The rise of global financial markets has increased the global level of uncertainty about capital realization to such an extent that APS in world cities – as providers of the necessary and seemingly sufficient authority and expertise – collectively hold a fee-earning 'class monopoly' that yields 'class monopoly rents' (Harvey 1974). Such an augmentation blurs differences between world cities research and financial geography since the social processes and associated geographies overlap significantly. Yet, at the same time, it is overly reductionist to collapse the more extensive geography of the WCA onto a concise network of IFCs. Typologically, whereas all IFCs are world cities, the WCA does greatly extend beyond that shortlist because, as we shall extensively argue below, APS practices relating to the worlds of finance and production have increasingly become interdependent.

THE CHANGING MODE OF COMMAND AND CONTROL IN WORLD CITIES

The defining feature of world cities – at least from its political-economy perspective – is their role as 'basing points' for global capital from which 'command and control' is exercised over capitalist accumulation (Friedmann 1986; Sassen 1991[2001]). We argue current modes of command and control seem to come to a crucial extent with the ability of agents in

world cities to produce 'strategic narratives' which rely to a significant degree on storytelling to make the crisis-prone global financial system legible, but which also involves circumscribed expertise of the specific APS trade. For instance, as is evident in offshoring, law firms and investment banks have a set of standard or customized solutions on offer that allows clients to circumvent tax regulations. The enduring concentration of command and control in world cities and its qualitative changes under financialization then become a crucial conundrum to disentangle.

To understand this shifting emphasis in the modality of command and control it is important to remember the changes in the political-economic context from the crisis of the 1970s and 1980s onwards. We identify three epochs in particular, which exemplify distinct dominant practices of command and control that morphed into one another: the introduction of a New International Division of Labor (NIDL) in the 1970s, 'the rise of a service economy' and 'the coming of postindustrial society' (Bell 1973) in the 1980s as well as new postfordist growth where services were interpreted to lubricate the process of flexible specialization in a globalizing economy that became increasingly interconnected through information technology. Here we hypothesize that financialization dynamics require continued and accelerated agglomeration in world cities beyond the IFC shortlist as clients are in dire need for services that allow indirect control over financial assets. While financialization entails a wide set of processes, it can generally be understood as the growing relevance of financial motives, financial markets, and financial institutions in the operation of domestic and international economies (Epstein 2005).

WORLD CITIES AS OBLIGATORY PASSAGE POINTS UNDER FINANCIALIZED GLOBALIZATION

The emergence of an APS class monopoly due to overaccumulation

In order to understand the enduring relevance of world cities for command and control over capital we have to delve into the specific role that the APS complex plays in the valorization and realization of capital in the context of financialization. We start our inquiry with the central axiom that contemporary

Plate 52 Frankfurt Messetower and its Westend neighbors

Source: Roger Keil

capitalism is increasingly characterized by a state of overaccumulation of capital that is the root cause of the increasing prevalence of economic crises. 'Capital' is a process of value in motion: money capital is brought into a circulation process consisting of a combination of constant and variable capital that eventually allows for the realization of its initial value and a surplus. This necessarily implies the consumption of a use value of some sorts (Marx 1885 [1992]). The lag between the initial capital layout and the moment of final consumption, and therefore the ultimate realization of surplus value, can be stretched across space and time. Until this moment of realization, capital has to be regarded as 'fictitious' since it is unknown whether or not it will finally realize itself: an investment can fail (Marx 1894 [1991]: 594–606; cf. Harvey 1982 [2006]: 267). In order for capital to reproduce itself and expand there have to be enough investment opportunities to realize surplus value. And these opportunities have to be orchestrated by socially

constructing material circuits of value (Lee 2006, 2011). Switching capital between these various circuits involves significant transaction costs which function as barriers to capital and surplus realization and a source of crises (Harvey 1982 [2006]).

The alleviation of these transaction costs is generally considered one of the important functions of the financial system, which collects a fee in return. In absence of overaccumulation, one could argue that the APS complex serves to alleviate transaction costs and thereby make markets more efficient: it is through the labor of the APS complex that valorizations of capital come into existence that otherwise would not have occurred. However, in a period of overaccumulation there is 'a situation where there is excess capital relative to the opportunities to use capital profitably' (Harvey 2010: 45). Inevitably, some capital will not realize itself and will hence be devalued or destroyed. This generates a fundamental uncertainty about the profitability of capital. Consequently, capitalists who are better informed about which segments of capital will eventually realize themselves with a surplus, will incur a lower uncertainty than others. If this knowledge about lower uncertainty is derived from a structural dominant position in circuits of value, it allows the actor to sell 'relative certainty of capital realization' for a fee, offloading the risk of the actual investment to the client. This fee is essentially a form of rent, since it is incurred regardless of capital realization. These surplus- or superprofits will not dissolve as long as there is overaccumulation combined with knowledge asymmetries about potential surplus value realization.

The 'structural dominant position in circuits of value' from which these knowledge asymmetries are derived is not the exclusive property of one single actor, but emerges from the combined activities of a class of actors (Swyngedouw 1992). As an abstract category, the APS complex holds exactly such a position under the current conditions of overaccumulation. Collectively, the APS complex forms a 'class monopoly' as their collective knowledge resources allow the appropriation of class-monopoly-rent (Harvey 1985: 65). The APS class monopoly involves a strong spatial expression, since its underlying structural dominant position in circuits of value draws on a network of localization economies where the right combination of people and information can be converted into knowledge about potential surplus generation. The effects of this localization economy reverberate throughout the cities that form its main nodes and provide it with structured coherence, as urban land and labor markets, politics, and growth regimes tend to co-evolve with the prevalent accumulation logic. The system of interconnected localization economies reveals itself as the particular geography of the WCA, while the salience of this structured coherence can be observed in its main nodes.

Class monopoly exploitation and reproduction through APS practices

Having reasoned the emergence of the APS class monopoly in an abstract way, this section theorizes the associated command and control position from the practices within the APS complex in the WCA that exploit and reproduce that position. In our view, the APS complex class monopoly emerges from the interplay of three distinct processes or 'elements' at

the juncture of conceptual spaces of production and finance (Figure 1). These conceptual spaces can be understood through the various moments of capital in the accumulation process, i.e. the classical M-C-M' formula, Marx (1867 [1976], 1885 [1992]) that can be disaggregated and spatialized (Lee 1989). Four distinctive moments in such a circuit of value can be identified that elucidate the functions of command and control: (i) the origination/pooling of capital (M), (ii) the transformation of respectively capital into commodities (M-C) and (iii) commodities into expanded capital (C-M'), and (iv) the optimization and 'multiplication' of the surplus (M').

This formula can be read in terms of two ideal-type spaces: production space typifies the industrial and commercial circuits of capital, which are driven by the maximization of capital accumulation through the process of finding the optimal combination of labor and capital in production on the one hand, and allocation to the most profitable markets on the other. The modus operandi of this space has been greatly influenced by the challenges and opportunities of globalization in terms of coordinating global production (Henderson et al. 2002). Financial space, on the other hand, explains capital accumulation through the maximization of returns on credit that – even if only in the very last instance – has been extended to the production circuit. The workings of financial space itself have profoundly changed by the globalization and virtualization of financial markets and the growing proliferation and trading of debt products and derivatives. Both spaces ultimately reflect perspectives, or 'windows', on the same process of capital circulation. Both spaces also crucially depend on knowledge-producing practices by the APS complex that drives a clear-cut tendency towards the further agglomeration and urbanization of capital in world cities (Krätke 2014; Sassen 1991 [2001]).

The crucial and self-reinforcing role that the APS complex plays draws on the three elements that intertwine financial and production space under financialized globalization. The first element emerges from the aggregation of service interventions in the production space. The second element is related to the social construction of material circuits of value (Lee 2006). The third element relates to enabling capital switching. These elements are not a-priori connected to specific actors and these elements can be found in endless different configurations of roles divided among several actors within the APS complex. However, each

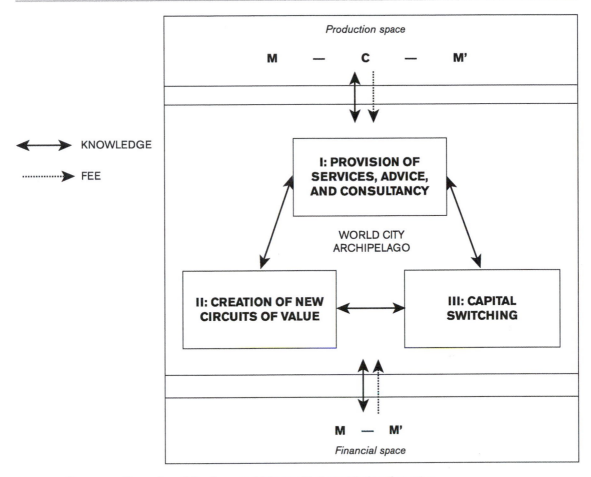

Figure 1 A conceptual overview of the elements driving world city archipelago formation

Source: David Bassens and Michiel van Meeteren

element thrives on the two others. Let us take each element in turn.

First, APS attain a favorable knowledge asymmetry vis-à-vis the production space through consecutive rounds of service provision in the context of financialization. The APS complex still provides knowledge about how to organize production (the M-C transformation) and help orchestrate demand for the produced commodities and support their actual sale (the C-M' transformation). Knowledge about consumer preferences, sensitivities, and susceptibilities is provided by advertising agencies, while a range of legal, accountancy, and financial services tackle effective demand problems by providing consumer credit. But the APS complex is also strongly implicated in the two other moments of capital, M and M'. To secure the inflow of capital into the circuit, firms

will hire investment bankers and lawyers operating from world cities to do an initial public offering or to structure a financial product in over-the-counter markets directly to institutional investors which results in cheaper credit access. Resultantly, APS firms are argued to be 'important influencers' of corporate decision-making, especially in the field of legal practices, financial markets, and real estate markets (Parnreiter 2010: 45). In sum, the APS complex acts as a gatekeeper of an enormous stock of sought-after knowledge that ultimately rests on the monopolization of knowledge out of previous fee-based interactions.

Second, the fruits of the APS monopoly position are not only direct service fees, but their global and up-to-date stock of knowledge constitutes an important ingredient in the process of the construction of (seemingly) competitive and market-beating new investment opportunities. Folkman et al. (2007) attach two roles to market intermediaries like APS. In the first instance APS can be considered what they call 'responsive functionaries' that do routine market intermediation. The by now routine work at the M and M' moment in the previous section can be categorized as such. However, in the second instance, APS can also be considered 'proactive initiators' whose highly remunerated fee-based work entails the innovation of new products. The continuous routine interaction with the productive space generates knowledge on the materialities – read economic geographies – that are a necessary element for the social construction of new circuits of value that together constitute financial geographies. Such new circuits of value can be produced via the 'innovation' of new financial products and legal contracts (e.g., 'Islamic' financial products, see Bassens et al. 2013), often in tandem with the construction of new markets for investment altogether.

'Emerging markets' are arguably the best-documented example showing the growth of dedicated funds or similar in-house activities at investment banks since the 1990s at least. Knowledge on markets is not constructed from the boardroom of these corporations in a top down fashion. Rather, reaching the boardrooms' strategic aims for accumulation depends on decentered and more spatially dispersed knowledge production systems. The 'imagineering' of emerging markets entails drawing on a variety of personal and impersonal data sources on companies and economies to enable investment (Lai 2006). Project ecologies are often stretched over space; for example, London-based fund managers depend on Singaporean brokers and analysts who hold knowledge on Asian economic geographies (cf. Faulconbridge 2007). Nevertheless, the obligatory passage of APS project ecologies through WCA nodes is still fundamental in the 'emergence' of new markets.

Third, the gatekeeping role with regard to access to profitable investment channels is not a static affair, but hinges crucially on the permanent search for the most profitable channel offering the best return-on-investment for end-investors. Under conditions of overaccumulation, competition between investors hence drives the process of constant capital switching between various existing and newly constructed circuits of value. Harvey defines switching as macro processes whereby capital shifts between production, the built environment, and reproductive functions. Switching, however, can also be understood

as the practices on the micro- or meso-scale that enable the non-frictionless pouring of money in or out of a range of investment channels. The use of APS becomes obligatory since increased competition for scarce 'safe and profitable investments' renders knowledge on where and how financial capital can realize itself most profitably necessary. APS are likely to hold knowledge of the existing circuits of value they co-produced (i.e. the financial products and markets) and the underlying production space they have advised (i.e. the economic and financial operations of firms). Preferential access to knowledge about political or financial movements in emerging markets will offer a premium.

Capital switching is a fee-based business we most directly associate with the activities of asset management services or the (hedge) fund industry, but it can more generally refer to all legal, financial, accountancy, and other professional services catering to capital that is channeled through financial markets in search for the highest profit. Sovereign wealth funds (SWFs), for instance, are moving out of plain government bond investments and increasingly invest in equities, commodities, and real estate. APS firms have responded by establishing offices to maintain good face-to-face relations with SWFs in their home financial center. This also generates business in a relational way: SWF investments in London and New York have increased, while the need for expertise available in mainstay financial centers like London has received a considerable boost because of overseas deals.

Consuming the monopoly: Finance feeding upon itself

It might be argued that the notion of the APS complex as a class monopoly falls apart in the face of the global financial crisis of 2008. In our view, the recurrence of such major crises is indicative of the ability of a group of actors that is structurally dominant in circuits of value to engage in practices that involve high-risk taking. Pre-crisis, but also post-crisis practices in many world cities indicate indeed that APS are far more than just the neutral operator of a now financialized global economy, but rather a consumer proper of that preferential knowledge as part of an accumulation strategy during intense global competition within the APS complex. This is most evident in the financial services industry where retail and investment banks alike have continuously reinvented themselves as financial service providers over the past few decades, focusing on new activities that increasingly depend on fee income (Erturk and Solari 2007). Competition within the financial services section of the APS complex has urged institutions to participate in the very casino practices they supposedly controlled (Froud et al. 2012; Strange 1988 [1994]).

REFERENCES FROM THE READING

Allen, J. (2010) Powerful city networks: More than connections, less than domination and control, *Urban Studies*, 47: 2895–911.

Bassens, D., Engelen, E., Derudder, B., and Witlox, F. (2013) Securitization across borders: Organizational mimicry in Islamic finance, *Journal of Economic Geography*, 13: 85–106.

Bell, D. (1973) *The Coming of Post-Industrial Society: A Venture in Social Forecasting*, New York: Basic Books.

Epstein, G. A. (ed.) (2005) *Financialization and the World Economy*, Cheltenham: Edward Elgar.

Erturk, I. and Solari, S. (2007) Banks as continuous reinvention, *New Political Economy*, 12: 369–88.

Faulconbridge, J. R. (2007) Relational knowledge networks in transnational law firms, *Geoforum*, 38: 925–40.

Folkman, P., Froud, J., Johal, S., and Williams, K. (2007) Working for themselves? Capital market intermediaries and present-day capitalism, *Business History*, 49: 552–72.

Friedmann, J. (1986) The world city hypothesis, *Development and Change*, 17: 69–83.

Froud, J., Nilsson, A., Moran, M., and Williams, K. (2012) Stories and interests in finance: Agendas of governance before and after the financial crisis, *Governance*, 25: 35–59.

Harvey, D. (1974) Class-monopoly rent, finance capital and the urban revolution, *Regional Studies*, 8: 239–55.

Harvey, D. (1982 [2006]) *The Limits to Capital* (2nd Edition), London and New York: Verso.

Harvey, D. (1985) *The Urbanization of Capital*, Oxford: Basil Blackwell.

Harvey, D. (2010) *The Enigma of Capital and the Crises of Capitalism*, London: Profile Books.

Henderson, J., Dicken, P., Hess, M., et al. (2002) Global production networks and the analysis of economic development, *Review of International Political Economy*, 9: 436–64.

Krätke, S. (2014) Cities in contemporary capitalism, *International Journal of Urban and Regional Research*, 38, 5: 1660–77.

Lai, K. (2006) "Imagineering" Asian emerging markets: Financial knowledge networks in the fund management industry, *Geoforum*, 37: 627–42.

Lee, R. (1989) Social relations and the geography of material life, in Gregory, D. and Walford, R. (eds.) *Horizons in Human Geography*, Basingstoke: Macmillan: 152–69.

Lee, R. (2006) The ordinary economy: Tangled up in values and geography, *Transactions of the Institute of British Geographers*, 31: 413–32.

Lee, R. (2011) Spaces of hegemony? Circuits of value, finance capital and places of financial knowledge, in Agnew, J. and Livingstone, D. N. (eds.) *The Sage Handbook of Geographical Knowledge*, London, Thousand Oaks & New Delhi: Sage: 183–99.

Marx, K. (1867 [1976]) *Capital*, Volume 1, London: Penguin Books.

Marx, K. (1885 [1992]) *Capital*, Volume 2, London: Penguin Books.

Marx, K. (1894 [1991]) *Capital*, Volume 3, London: Penguin Books.

Parnreiter, C. (2010) Global cities in global commodity chains: Exploring the role of Mexico City in the geography of global economic governance, *Global Networks*, 10: 35–53.

Sassen, S. (1991 [2001]) *The Global City*, Princeton, NJ: Princeton University Press.

Strange, S. (1988 [1994]) *States and Markets* (2nd Edition), London and New York: Continuum.

Swyngedouw, E. A. (1992) Territorial organization and the space/technology nexus, *Transactions of the Institute of British Geographers*, 17: 417–33.

62

"Can the straw man speak? An engagement with postcolonial critiques of 'global cities research'"

from *Dialogues in Human Geography* (2016)

Michiel van Meeteren, Ben Derudder, and David Bassens

EDITORS' INTRODUCTION

In this second contribution to this section by Bassens and van Meeteren, co-authored with Ben Derudder, the authors defend the work on global urban networks as an effective research strategy to understand processes of globalized urbanization. This essay is the most direct and coherent response to poststructuralist and postcolonial critiques as leveled in particular by Jennifer Robinson, Ananya Roy, and others. In these critiques, the research on global urban networks, with a focus on advanced producer services firms and top-tier global cities, is perceived as flawed by economism, ignoring cities that are not command and control centers, and serving the boosterist city marketing strategies. In their defense, the authors argue that researchers studying global urban networks have been paying increasing attention to cities "off the map"; that their focus on political economy is not a fallacy of economism, but a deliberate strategy for better understanding the articulation of global economic processes, and that city marketing by use of global city rankings predates the research on global urban networks, and the causes for uneven development have to be found elsewhere. Overall, instead of imagining a "mainstream geography" that does not exist and dropping the studies on global urban networks "off the map," this essay advocates for open conversations among scholars with different theoretical and methodological orientations. Only by embracing pluralism can we better understand the complex, intensified urbanization on a planetary scale.

INTRODUCTION

An unfortunate one-sided representation of the 'global cities research' (GCR) literature is prevalent in many postcolonial and 'ordinary city' writings on globalized urbanization. In our reading, the core purposes of GCR tend to be misrepresented, while its sizable ontological, epistemological, and methodological diversity is de facto ignored or deemed irrelevant in the broad-brush stroke of criticism. Moreover, with each iteration of the critique, GCR is further compressed into a homogenous and allegedly fundamentally flawed 'Other'. Having all the rhetorical bearings of a classical straw man fallacy – committed when one – willingly or not – misrepresents a (perceived) opponent's position by imputing it implausible commitments, and then refutes the misrepresentation instead of the (perceived) opponent's actual view (Talisse and Aikin 2006). The GCR Other is only summoned in order to casually reject it. It merely serves to distinguish one's own position from an imagined mainstream in human geography or urban studies more widely.

Witnessing this trend, our key concern is not to deplore the silencing of GCR voices per se: in spite of our title's playful reference to Spivak (1988), GCR can speak for itself, as it constitutes an active circuit of knowledge for critical urban scholars. What we regret is the under- or misrepresentation of these critical voices in many postcolonial writings, which we interpret as a way to provincialize GCR and stress its limits for comprehending globalized urbanization. The common misrepresentation of GCR generates a barrier, an unproductive silence, for intellectual exchange across polemically accentuated global city/ordinary city divides, which ultimately hinders the development of the broader literature on globalized urbanization. This paper is therefore not intended as an exegetic excursion into what GCR is 'really' about, but rather as a step toward a more open and engaged field of global urban studies that recognizes and respects theoretical diversity.

Jennifer Robinson's postcolonial critique of GCR has clearly been a catalyst for a part of the field to distance itself from GCR (Robinson Ch. 9 and Ch. 66). Robinson has raised a number of legitimate concerns with respect to GCR as she cautioned against particular tendencies in the urban studies literature of that time. However, when reading the plethora of postcolonial writings following Robinson, we observe that these legitimate concerns have gradually morphed – through iteration – into a set of apparent truisms that fall short as an interpretation of GCR. We interpret this discrepancy to be part of a gradually routinized straw man rhetoric that emerged as a rallying signifier for postcolonial urban scholars. Given this rhetoric's potentially debilitating effects on the field, we conclude by advocating the idea of 'engaged pluralism' as outlined in Barnes and Sheppard (2010) as a possible way out of this problematic 'othering'.

THE SUBSTANCE OF POSTCOLONIAL CRITIQUES

Postcolonial criticisms of GCR have increased in number following the work of Jennifer Robinson. In our interpretation, her critique of GCR revolves around three inter-related concerns that can be paraphrased as follows: (i) the subjugation of alternative accounts of globalized urbanization and/or research on allegedly non-global cities and the consequent need for the parochialization of GCR; (ii) GCR's

economism as visible in its narrow focus on specific sets of economic processes as well as an associated focus on hierarchies of 'performance'; and (iii) the tendency for (research on) 'the global city' to travel as an aspirational model and standard for urban economic dynamism around the world. We interpret these critiques to be legitimate concerns that, when properly addressed, can push the field forward.

STRAW MAN RISING

Curiously, Robinson's critique has provoked limited dialogue with GCR itself. This lack of dialogue has probably contributed to the gradual morphing of Robinson's legitimate concerns into overstretched truisms that are not debated. The procedure of these truism-based studies tends to (re)present, and subsequently often discard, GCR as a unified paradigm or theory that de facto (1) silences or misrepresents processes in 'other' cities, (2) does so because it is flawed by economism and/or structuralism, as visible in its goal to devise 'city hierarchies', and (3) usefully serves the boosterist planning practices of political elites around the world. Although the (re)production of these apparent truisms is obviously not a feature of all postcolonial writings on globalized urbanization, nor are all of these truisms (equally) (re)presented, we observe that this has become a common way in which GCR is portrayed. While it is of course impossible to provide a detailed or comprehensive account of GCR's (re)presentation across the literature, a number of concrete examples should suffice to illustrate our point.

First, GCR as a whole is casually condensed into 'being about' devising hierarchical rankings, which are held reflective of the literature's economism and structuralism, given the focus on the location strategies of multinational firms. A favorite target is Beaverstock et al.'s (1999) identification of different 'levels' of global cities (alpha/beta/gamma), which is deemed to be the core preoccupation of GCR at large. A second example is Roy's (2009) account of GCR in her call for 'new geographies of theory' for the 21st century metropolis. In her paper, GCR is presented as a body of 'authoritative knowledge from the North . . . [mapping] a hierarchy of city-regions', which translates into a 'Darwinian logic' of 'the survival of the fittest in the keen competition of network capitalism'. A third and more recent example is Koch's (2013: 111) analysis of urban policy-making in Astana. The paper interestingly shows how Astana's policy makers are primarily oriented towards mimicking Ankara, hence making the point that aspirational models are not necessarily grounded in paradigmatic global cities in the region (e.g. Dubai) or beyond (e.g. London). While these insights could serve to pluralize the politics of global city formation, they are instead utilized to distance her research from GCR (Koch 2013, 111). In a fourth example, a recent extended editorial of several financial geographers in the British newspaper *The Guardian* (Engelen et al. 2014), researchers who have spent their careers as critical scholars of globalized capitalist urbanization and its uneven consequences occurring in and through cities, seem to be suggestively recast as the masterminds of global city formation themselves. But even when postcolonial scholars do acknowledge nuance, there seems to be a solid but empty box left that can be criticized relentlessly. The construction of this empty box is visible in the fifth and final example, which is a paper by McCann (2004: 2316–2317) who usefully and convincingly shows that the globalization-urbanization nexus can be studied in and through a diverse range of cities, yet his introduction contains the decontextualized assessment by Robinson (2002) that "a limited range of cities still end up categorized in boxes or in diagrammatic maps, and assigned a place in relation to a priori analytical hierarchies". Let us ask the rhetorical question of what the often alluded to 'majority of global cities research' actually is? Does that majority exist, or is it a phantom that merely functions as an imaginary yardstick for one's own argument? A brief appraisal of GCR's subject matter and internal diversity is required to rectify the picture being painted here.

THE PLURALIST SUBSTANCE OF CONTEMPORARY GLOBAL CITIES RESEARCH

While Robinson herself was pretty clear as to whom she targeted in her critique and why she targeted them, this has become increasingly unclear over time. The indiscriminate critique of GCR is aimed at an invisible, yet apparently comprehensive, literature. Speaking from 'the inside', we first wish to question Acuto's (2014) characterization of the GCR field as a unified 'paradigm' or even 'theory.' It is true that John

Friedmann (1995) casually characterized GCR as a paradigm, but – following Saey (2007) – we distinguish political-economic, cultural, critical, and 'politically naive' perspectives coming together in GCR, which do not add up to a paradigm in the Kuhnian (Kuhn 1962) sense. It may be better to think of (and therefore address) GCR as an 'invisible college', as has been proposed earlier by Acuto (2011, drawing on Friedmann 1995: 28). An invisible college refers to the sociological formation of a group of authors in a particular research field who constitute a social circle, but have varying degrees of involvement on the basis of diverging research interests. This circle has an (informal) stratification, and is characterized by internal disagreements, debate, and openness to internal mavericks and criticism. To a large extent, then, GCR is simply what global city researchers do: the praxis of doing research defines the evolving research subject and prevails over a rigid definition or conceptualization of what a world or global city is, how it should be researched, and how one should interpret the results. Even one of the arguably most prominent emblems of GCR – the model-based mappings and measurements of the world city network by a specific team within GaWC – can best be read as one particular building block within an international division of labor among those interested in understanding and studying global city formation. Ironically, reducing GCR to global city rankings or league tables – besides often resting on a misrepresentation of that exercise – is to take one particular line of research to stand for the whole. The result is a narrow essentialization of GCR by the rhetorical strategy of the pars pro toto, which resultantly, to paraphrase Robinson, drops numerous other GCR contributions 'off the map' of urban studies.

THE STRAW MAN SPEAKS

Now that we have addressed the 'truisms' that are stifling debate, we can return to the question of how actually existing GCR ought to deal with the legitimate concerns that are at the root of postcolonial critiques. En route, we take the opportunity to further elaborate on some features of GCR that are in our view commonly misrepresented. We will reflect on three elements that were originally raised by Robinson, namely the tendency towards economism, the subjugation of alternative geographies, and the performativity of GCR, to structure our response.

Plate 53 Occupy Sydney

Source: Roger Keil

First, we evaluate the risk of economism. The actually existing varieties of GCR do not intend to provide a comprehensive account of globalized urbanization or construct a universally valid urban theory. Instead, GCR has a long tradition of focusing on the changing role of cities in the global capitalist system. The very diversity within the invisible college is the best guarantee that the assumptions of GCR are critically evaluated and that a reduction to economism is averted. GCR is not a collective exercise in ignoring agency as Ley (2004) argues: the structure of the world city network is not simply treated as the 'outcome' of globalization, but as the collective product of the agency of – amongst others – world city network makers in producer services. Admittedly, the focus in this emerging literature continues to be almost exclusively on political-economic processes and agents, especially when global city formation is linked to wider processes of rescaling. This emphasis on political-economy is, however, not simply a fallacy of economism but the result of a deliberate focus on the articulation of global economic processes in cities, which are a key force of neoliberal social change. Moreover, the urban-economic geographies under scrutiny stretch well beyond the global city shortlists as they explicitly probe the contours of the world city network by studying world-city formation beyond the 'old core'.

Second, we come to address the legitimate concerns about GCR's tendency to subjugate alternative geographies of globalized urbanization, especially when considering cities in the Global South. Inexorably, a critical focus on the how, why, and where of capitalist accumulation as organized from global cities makes particular modes of accumulation seem inevitable and alternatives unreachable (Gibson-Graham 2006 [1996]). We acknowledge this epistemological problem, but see no other way out of it than a division of labor between scholars who critically research capitalism-as-practiced on the one hand and people who investigate alternatives on the other. In hindsight, it is clear that the GCR agenda has emerged from issues located in very specific geo-historical circumstances (cf. Friedmann and Wolff 1982). The starting point of 'deindustrializing cities' in the Global North as the 'place' of origin of GCR has influenced the research questions that were asked (cf. Barnes 2004). More recently, GCR has been reflexive of, and has acted upon, the acclaimed 'parochialism'. The most obvious indications for this are: i) the reduction in the number of places and processes that are effectively 'off the map' in new rounds of quantitative data gathering (Taylor and Derudder 2015: chapters 4–5), ii) the application of analytical moves to decenter the economic geographies through which cities are enrolled in global urban networks, and iii) the deliberate effort to include more and more cities, actors, and geographies in empirical case studies of the world city network.

Third, postcolonial critiques have heightened our awareness of processes pertaining to the performativity of knowledge. The problems anticipated by Michael Peter Smith (Ch. 58) and Robinson have partially materialized. Particular practices associated with GCR did end up reinforcing neoliberalism despite its critical origins, as is evident from the popularity of 'league tables' amongst policy-makers and global firms alike. The genealogy of how this vulgarized conception of 'global city rankings' exactly ended up in the toolbox of 'business consultant urbanism' (Amin 2013) remains to be written, although we stress that scientific city rankings and typologies certainly predate the emergence of the GCR, as does their use in consultancy circles. Yet, despite these material effects, it is useful to recall Sayer (2000: 11), who reminds us not to put too much weight on the performativity of our knowledge in the present. For him, social scientists are mostly in the business of 'construing' rather than 'constructing' the social world. That we call something an 'alpha city' probably has limited effects in the short run on neoliberal urban development, and to easily attribute causes of neoliberal urbanism to GCR, can equally be diverting attention from the fact that there are far more important and powerful explanations out there that cause uneven development.

Nevertheless, we agree with the problem identified by Jazeel and McFarlane (2007) that a critical researcher needs to take the (unintended) effects of research into account. Relatedly, we have to acknowledge the power-effects that, even critical, accounts have had in our historical understanding of capitalism by muting voices emerging from the Global South (Slater 1992). But this should not hold us back from developing critical concepts and analyses that aspire to say something about capitalist urbanization worldwide. It should be clear that (predatory) accumulation organized from global cities, including its influence on cities that are regarded to be 'off the map', will not stop if we stop researching it. We will simply understand it less well.

PROSPECTIVE REMARKS: TOWARD ENGAGED PLURALISM

Our discussion of the variety and critical character of GCR as practiced contrasts remarkably with the strong claims made by postcolonial critics. How can we explain the emergence of this gap between actual research and perception? We have argued above that critics of GCR have gradually erected a straw man. But how did it emerge, and more interestingly, why has so little response been voiced by researchers engaged in GCR? The absence of a comprehensive response is in fact the logical outcome of the straw man fallacy itself since a straw man, by its very nature of being a misrepresentation, 'cannot speak' – it has no real existing interlocutors. Everybody who does respond is therefore rendered not to be part of the Other.

The muteness of the straw man is reinforced by the deafness of postcolonial critics to revisions and refinements in the field of GCR over the years. These nuances may or may not be convincing (and thus be subject to scientific debate), but the point here is that they hardly get taken into account when postcolonial critiques are repeated in each cycle of academic publishing. Surprisingly often an argument is made in opposition to some alleged and often unnamed 'mainstream' urban theory (Brenner and Schmid Ch. 65; Sheppard et al. 2013; Jazeel 2014), but if it is not the invisible college of GCR described above, we cannot locate that mainstream anywhere in either the urban studies journals we read or the citation indexes we observe.

We contend that the examples presented above point to a rhetorical strategy that Barnes and Sheppard (2010: 194) have called 'polemical pluralism', in which an invocation of other viewpoints in academic debates only serves as an ideological weapon to advance one's own orientation. This ideological weapon is most clearly expressed by Gibson-Graham (2006 [1996]: 5–11) where they make the case that erecting a straw man, 'a non-existing hyperbolic monster', can be productive to elucidate alternative perspectives. However, comparisons between real world cases on the one hand and such hyperreal representations on the other, carry methodological risks. We interpret Roy's (2009) call for 'strategic essentialisms' as a similar kind of polemic rhetorical strategy: as the putative 'process geographies' of the former are contrasted with the putative 'trait geographies' of GCR, research findings are (mis)represented as merely pertaining to major-cities-in the-First-World theoretical ideal types to be falsified elsewhere.

This rhetorical strategy is counterproductive to the research agenda of globalizing urbanization for three important reasons. First, it might lead scholars to draw wrong conclusions about the value of the work of colleagues. Second, it leads to a toxic culture of dialogue. Speaking counterfactually, if postcolonial critics would have truly engaged with 'real life' GCR researchers – much like the above-mentioned GCR mavericks, who would probably not position themselves in the 'mainstream' of the field – they would have crossed the diffuse line of GCR's invisible college and likely participated in the collective process of setting research agendas from the inside. While we do not wish to argue that this makes GCR the only viable candidate to have such critical encounters, it does illustrate how that community is open and receptive to critique on its central assumptions. And third, as global capitalism relentlessly expands and induces intensified urbanization and crisis on a planetary scale, we need each other's insights in order to comprehend these processes and hopefully change them for the better. But foremost, we need to strive for an intellectual atmosphere where that exchange can be productive.

By means of conclusion, let us reiterate unambiguously that we believe that at the heart of the postcolonial critique are a number of legitimate concerns that may help GCR to retain its critical edge. Scientific fields can only flourish when theories are challenged and when different research methods alternate in probing different assumptions and validities of a model. The epistemological and onto-logical differences that characterize urban studies have been effective in sharpening debate, providing room for difference, and safeguarding valid knowledge from the risk of being unduly subjugated. Divisions of labor between the critic and the criticized – as long as the roles tend to alternate – are important in any healthy research field. Hence our call to continue the substantial debates on GCR as sharp as it can be; but

in the spirit of engaged pluralism. As a virtue when encountering difference, engaged pluralism encourages us to 'however much we are committed to our styles of thinking, we are willing to listen to others without denying or suppressing the otherness of the other' (Barnes and Sheppard 2010: 194).

REFERENCES FROM THE READING

Acuto, M. (2011) Finding the global city: An analytical journey through the 'invisible college', *Urban Studies*, 48, 14: 2953–73.

Acuto, M. (2014) Dubai in the 'middle', *International Journal of Urban and Regional Research*, 38, 5: 1732–48.

Amin, A. (2013) Telescopic urbanism and the poor, *City: Analysis of Urban Trends, Culture, Theory, Policy, Action*, 17, 4: 476–92.

Barnes, T. J. (2004) Placing ideas: Genius loci, hetero-topia and geography's quantitative revolution, *Progress in Human Geography*, 28, 5: 565–95.

Barnes, T. J. and Sheppard, E. (2010) 'Nothing includes everything': Towards engaged pluralism in Anglo-phone economic geography, *Progress in Human Geography*, 34, 2: 193–214.

Beaverstock, J. V., Smith, R. G., and Taylor, P. J. (1999) A roster of world cities, *Cities*, 16, 6: 445–58.

Engelen, E., Johal, S., Salento, A., and Williams, K. (2014) How to build a fairer city, *The Guardian*, September 24.

Friedmann, J. (1995) Where we stand: A decade of world cities research, in Knox, P. L. and Taylor, P. J. (eds.) *World Cities in a World-System*, Cambridge: Cambridge University Press, 21–48.

Friedmann, J. and Wolff, G. (1982) World city formation: An agenda for research and action, *International Journal of Urban and Regional Research*, 6, 3: 309–44.

Gibson-Graham, J. K. (2006 [1996]) *The End of Capitalism (As We Knew It)* (2nd Edition), Minneapolis and London: University of Minnesota Press.

Jazeel, T. and McFarlane, C. (2007) Responsible learning: Cultures of knowledge production and the North-South divide, *Antipode*, 39, 5: 781–9.

Jazeel, T. (2014) Subaltern geographies: Geographical knowledge and postcolonial strategy, *Singapore Journal of Tropical Geography*, 35, 1: 88–103.

Koch, N. (2013) Why not a world city? Astana, Ankara, and geopolitical scripts in urban networks, *Urban Geography*, 34, 1: 109–30.

Kuhn, T. S. (1962) *The Structure of Scientific Revolutions*, Chicago: University of Chicago Press.

Ley, D. (2004) Transnational spaces and everyday lives, *Transactions of the Institute of British Geographers*, 29, 2: 151–64.

McCann, E. J. (2004) Urban political economy beyond the 'global city', *Urban Studies*, 41, 12: 2315–33.

Robinson, J. (2002) Global and world cities: A view from off the map, *International Journal of Urban and Regional Research*, 26, 3: 531–54.

Roy, A. (2009) The 21st-century metropolis: New geographies of theory, *Regional Studies*, 43, 6: 819–30.

Saey, P. (2007) How cities scientifically (do not) exist: Methodological appraisal of research on globalizing processes of intercity networking, in Taylor, P. J., Derudder, B., Saey, P., and Witlox, F. (eds.) *Cities in Globalization, Practices, Policies and Theories*, London and New York: Routledge: 286–300.

Sayer, A. (2000) *Realism and Social Science*, London, Thousand Oaks, and New Delhi: Sage.

Sheppard, E., Leitner, H. and Maringanti, A. (2013) Provincializing global urbanism: A manifesto, *Urban Geography*, 34, 7: 893–900.

Slater, D. (1992) On the borders of social theory: Learning from other regions, *Environment and Planning D: Society and Space*, 10, 3: 307–27.

Spivak, G. C. (1988) Can the subaltern speak? in Nelson, C. and Grossberg, L. (eds.) *Marxism and the Interpretation of Culture*, Urbana, IL: University of Illinois Press: 271–313.

Talisse, R. and Aikin, S. F. (2006) Two forms of the straw man, *Argumentation*, 20, 3: 345–52.

Taylor, P. J. and Derudder, B. (2015) *World City Network: A Global Urban Analysis* (2nd Edition), London: Routledge.

63
"Global suburbanization"

Roger Keil

EDITORS' INTRODUCTION

In this chapter, Roger Keil, one of the co-editors of this book, makes the argument that a theoretical approach to globalizing cities needs to acknowledge the relevance of suburbanization to urban theory in the 21st century. The chapter summarizes work of a broader research project on global suburbanization (www.yorku.ca/suburbs) and Keil's own work on the subject (Keil 2013, 2017). It makes the observation that global city formation in the classical form as depicted by Friedmann and Sassen relies not just on a set of high level financial service firms in central business districts, but on a suburbanizing economy in global city-regions (see Keil 2011). Suburbanization—as initially a linear process of population and economic growth at the periphery—and postsuburbanization—as the complex interplay of multiple processes of suburban maturation and emerging suburban ways of life (i.e. 'suburbanisms') characterized by a proliferation of differences (such as through immigration)—play a central role in the spatial and social relationships of the globalizing city.

INTRODUCTION

It is time to argue for an intervention into urban theory on the basis of the suburban explosion. An aspect of what Henri Lefebvre refers to as the "explosion" of the city, suburbanization has now become a pervasively planetary phenomenon. Often, suburbs have rightly been discussed in association with the Anglo-Saxon societal model. Private land ownership, consumerist capitalism, and the ideology of freedom prevalent in the United Kingdom and British settler societies have made Australia, Canada, the United States, and to some degree Britain itself, ideal places for the prototypical single-family home residential suburb to thrive. But Anglo-Saxon suburbia needs to be re-evaluated in a global context. We depart from the common wisdom on suburbanization as a chiefly American domain in two important ways. First, through its dominance the Anglo-Saxon model obscures historical parallels and alternatives in suburbanization.

There have always been different pathways to peripheral urban development. Secondly, and more importantly, in recent years, newer forms of suburbanization that give rise to the need for rethinking urban theory overall have sprung up around the world. This is, as we will see below, of particular significance in the context of distinct spatial agglomerations that have been seen as global or world cities. The focus on intensification of centrality and core economies has eclipsed the extensive emergence of the suburban as a basis for the global city.

GLOBAL CITY FORMATION: THE ROLE OF THE METROPOLITAN FRINGE

Global city formation has been predominantly a narrative of centrality. The very notion of a global city conjures up central places, central functions, and concentrated urbanity. Much of this is justified as agglomeration economies in globalized cities have created easily recognizable built and social environments of centrality: extensive and ever taller skylines of gleaming office and bank towers in this most recent period of financial-industry based urban accumulation and now increasingly also recentralization of residential populations predominantly in privatized, often gentrified, and secured condominiums (see Smith Ch. 35 and Lehrer Ch. 51). The focus of global city research has been predominantly on advanced producer services, often located in the central business districts, that support the burgeoning, fast-changing, and pace-setting financial industries at the core of the globalized urban region (see Taylor and other GaWC contributions in this volume). This has pinned the

attention of researchers and the general public on the inner cities of large financial centres that were widely associated with the production of services that are indispensable for the running of a globalized financial economy.

The functional emphasis on those core services brought with it a geographical emphasis on the downtown. Early global city researchers decried the "dual city" characteristics of polarized labour markets where bankers and fast food workers, lawyers and dry cleaners shared space and joined in a division of labour geared towards financial profit (although at opposite ends of the pay scale). And while there was sometimes talk about back offices in suburban office parks that took up the more routine aspects of this

Plate 54 Suburban town centre, Mississauga

Source: Roger Keil

production process, those were rarely seen in relation to what was going on in the prime network spaces downtown. Policies of expansion and renewal were recorded as in step with the rising spatial and service appetites of those industries and their high-flying white-collar workforce (at the expense of dramatic social polarization and socio-economic segregation). Likewise, cities' push for global relevance has often been accompanied by spectacular urban development through megaprojects and iconic cultural and architectural objects (museums, etc.); more recently the rise of creative class discourse has been used as the hegemonic frame for global city expansion. Whether it is transportation terminals or talent, the geographical and functional focus of any such strategy has usually been the central city. Even geographically peripheral developments such as expanded airports or ports were tightly linked in their functionality and logistical connectivity to the needs and logics of the central city economy. This narrative, sketched here in the simplest terms and with a lot of omissions, overlooks an important trend: *that global city formation is dependent on the accelerated suburbanization of the global city region*. This entails the story of back-offices, regional infrastructures, residential dispersal, etc. as well as the proliferation of ethnoburbs as typical world city phenomena.

If centrality is what the world city is about, this chapter discusses the significance of the peripheral in the 'dialectic of centrality' which, following Stanek (2008: 74) "consists not only of the contradictory interdependence between the objects but of the opposition between center and periphery, gathering and dispersion, inclusion (to center) and exclusion (to periphery)". Accordingly, globalizing suburbs and global city cores constitute themselves mutually in a dialectical fashion. In turn, then, it can be argued that suburbanization is a necessary and integral, constitutive element of the global city's centrality.

STUDYING GLOBAL SUBURBANIZATION

One can define *suburbanization* as the combination of non-central population and economic growth with urban spatial expansion. Although suburbanization is not uniformly applicable to all parts of the world, one can carefully deploy this generic term to mean all manner of peripheral growth: from the traditional single-family home subdivisions of the United States, to the high rise-dominated old suburbs of Europe and

Canada, the new towns outside Chinese cities, and the slums and squatter settlements in African and Latin American cities. Spaces of suburbanization are not just residential, they contain crucial infrastructural, logistical, and employment functions that support the globalizing regional economy (see Addie Ch. 25 and Negrey et al. Ch. 26). A new typology and global nomenclature of suburbanization is taking shape as we begin to see the entire spectrum of peripheral, non-central development under one conceptual umbrella (Harris 2010; Harris and Vorms 2017). Emerging forms of suburbanism across the world come into relief as do the challenges of aging modernist suburbia and we are facing new realities of globalized urbanization, where central city and fringe are remixed. In this period, new forms blend together on the periphery; the result has been a different suburbia and a different city. In simple terms and following the definition of historian Jon Teaford (2011: 15), we can now speak of global *post*suburbanization which refers to myriad forms of "suburbanization carried to the extreme, the end product of two centuries of continuous deconcentration of metropolitan population" (and one might add of economic activities).

Suburbanism can be broadly defined as the growing prevalence of qualitatively distinct 'suburban ways of life'. Unique land use patterns of suburbs, relative to the central city (although there are hybrid forms of mixed patterns), have engendered differing social and cultural norms of suburban life. Among the causes of such variation and dynamics is density as it relates to transportation and socioeconomic distinctions. Infrastructure layout determines not just the larger strategic opportunities that come with suburbanism, it also works as a significant constraint in the long term as its fixed pattern discourages alternative uses (Filion and Keil 2017). Importantly, while in the past, "suburban lifestyles" were considered stereotypical and predictable, they have now become the subject to much interest in urban studies as populations and morphologies in suburbia have changed.

The distinctions between central city and suburbs differ widely across countries and regions. While there is often less density in suburbs, this is by no means the case everywhere: density is not just a feature in American edge cities but also in European *banlieues*, Asian new towns, and Canadian suburbs. For instance, there are important socioeconomic distinctions between suburbs and central city that may construct diametrically opposed value systems affecting democracy, justice, and sustainability, although suburbs are now recognized as important arenas for struggles around issues of collective consumption, culture, and citizenship as well as labour issues.

THE GLOBAL CITY REGION

Global city formation can be seen, to some extent, as a form of place making in a space of flows (see Castells Ch. 21 and Sassen Ch. 4). Meanwhile, it has become increasingly clear that the dynamics of the space of flows are being harnessed precisely in globalized urban cores because a regional, suburbanized network of relations supports the process. Let us quickly look at one European and two North American cities to illustrate this point:

Los Angeles

The Los Angeles region has been restructured since the 1970s to serve global needs through recalibrated and globally oriented infrastructure projects, most of which span the suburban expanse of Southern California (Erie 2004). As a global city, Los Angeles has built on the fragmented and increasingly privatized structure created in earlier decades. Largely suburban local states have been integrated into the spatial and political logic of the region on the basis of distinct politics and policies which have granted them special status in the spatial division of labour. Dispersed suburban local governments have pursued the project of globalization by weaving the rhetoric of world city *grandesse* into a new political economy of place. This led to a decentralized and diversified tapestry into which global city functions were woven: At one end was West Hollywood, an enclave wedged between Los Angeles and Beverly Hills incorporated into a separate municipality in 1984, which marketed itself as an 'avant-garde, culturally sophisticated, creative city'. The director of its Marketing Corporation said in the 1980s: 'Much of what people associate with Los Angeles – which is this newly self-conscious world capital – emanates from West Hollywood'. At the

Plate 55 Birthing a new Compton

Source: Roger Keil

other end is Carson, a suburban municipality in the South of the Los Angeles Basin; the city capitalizes on the opportunities that have sprung up through the northward expansion of the harbor economy along the Long Beach and Harbor Freeways. Carson's explicit development strategy was to provide a safe ground for international investment. Carson's development, thus, has been largely a product of the entangled interests of (inter-)national capital and of local real estate capital. Global city formation in the Los Angeles urban region provided a new arena to this relationship: Carson became a turnstile for the integration of global economic tendencies into the local urban structure. In addition, Carson is also a residential suburb with one of the most ethnically diverse populations in the United States. Its population is almost evenly distributed among the region's major ethnic groups that count for about one quarter each.

The insertion of the region's suburban communities into the global city was taken to a certain extreme in Compton, south of Los Angeles, one of the poorest suburbs in the United States. Once a center of gangs fighting for the control of territory and markets in the 1980s, Compton became, for a while, the unrivaled world headquarters of hiphop music. When the rap group N.W.A. sold half a million copies of its album 'Straight Outta Compton' in the summer of 1988, a flood of defiant self-stylization was triggered in the world-forsaken community. The creation of communal identity was achieved through a decidedly anti-establishmentarian attitude that expressed itself in blunt realist style and with an unapologetically capitalist sales strategy fueling an entire industry of Compton marketing. In contrast to the conscious simulacra of the marketing strategies of West Hollywood, the rappers of Compton subscribed to an ostensibly unmediated representation of reality and truth expressing the identity of place in Compton. The 'realist' strategy aims at the production of visibility in a situation of complete marginalization and criminalization. It is the inversion of Compton's southern neighbor Carson's internationalization strategy that targets global business investment with images of a peaceful multinational community. While other municipalities marketed themselves as localities of innovation, stability, and success, Compton's global music market handlers sold the city globally as the home of the drive-by-shooting: 'It's the Compton Thang'.

A generation after N.W.A., a new artist from Compton, Kendrick Lamar, reflects on this legacy in a song, co-performed with N.W.A.'s Dr. Dre with the title "Compton" (2013). Featuring the refrain "Compton, Compton, Ain't no city quite like mine", Lamar (and Dr.Dre) lament(s):

> "Now we can all celebrate, we can all harvest the rap artist of NWA
>
> . . .
>
> This was brought to you by Dre
> Now every muthafucka in here say
>
> Look who responsible for taking Compton international
> I make 'em holla oooooooh
>
> . . .
>
> Ain't no city quite like mine."

In the new Compton of Kendrick Lamar, the suburb has become just as much a fictional reality that contains all of the (now sampled) dreams and nightmares of the art form of hiphop and a new municipal reality in which a majority African American suburb takes control of "heritage", the wind in the back of consecutive Obama administrations and a set of new infrastructural connectivities between the Alameda corridor and the Blue Line light rail transit that transect the southern suburbs. It is in this postsuburban, globalized environment of thickly layered urbanities that Los Angeles has to be read now as it matures into a post-auto city of mature suburbs, racial diffusion, and post-crisis recovery.

Frankfurt up the country

The German global city of Frankfurt presents a markedly different but related story. Frankfurt's suburbs have also established their visibility in the global city region. The periphery has become the staging ground for important sectors of the global city economy. The concentration of office towers in the downtown often eclipses the tendency of the "real" economic center of Frankfurt to shift away from the downtown into the forest on the fringe of the city: Frankfurt airport, the second largest in Europe after London Heathrow, has developed into the crucial modem of the global flows of people, goods, and information as air traffic related infrastructures along with first class office, conference, and hotel space have been continuously (re)developed there. A regional ring of small and

Plate 56 Suburban town centre, Eschborn

Source: Roger Keil

semi-autonomous suburban communities has become a significant target of foreign investments and the terrain of the peripheralization of globalization; one such place is Eschborn, in the northwestern periphery of Frankfurt. A town of 21,000, Eschborn hosts 31,000 mostly service sector jobs, 14 percent of which are in the financial service industry. Most spectacularly, perhaps, since 2008 Eschborn has been the seat of the Deutsche Börse stock exchange corporation that left Frankfurt to take advantage of lower business taxes (Belina and Lehrer 2017).

The spatial form of Frankfurt's periphery is being produced by a series of often contradictory dynamics. The periphery has not just been colonized by the core but appears as the product of economic exchange both between core and periphery and of among various subcenters on the fringe. Ambitious large-scale housing projects near old village cores reflect this urbanization trend as much as the often mannerist and pastiche 'reurbanization' of such cores and the intensification of historic settlements. Even in smaller communities, social housing has

now become a ubiquitous issue while high tech companies in newly designated industrial areas and cultural centers in medieval city halls have become part of the image we have of former country towns. Thus, uniform 'urban sprawl' is, in fact, increasingly replaced by a multiplication of centralization effects. We are starting to find, in the periphery, a socio-spatial structure equally complex as (or even more complex than) in the centre of the Rhein-Main global city region.

Toronto: Creating the Greater Golden Horseshoe

The Toronto case confirms the overall narrative we have presented here. The urban region has been a veritable testing ground for centre-periphery re-regulation for at least a generation. Prompted by the urban region's dramatic demographic and economic growth after World War II, suburban sprawl of residences and employment has been a major factor in Toronto's formation into "Canada's global city". While the urban economy is often seen as synonymous with the strong presence of a financial services and creative cluster in the centre, its strengths have also been regional manufacturing (such as automobile assembly and parts production), producers services, transportation, cultural production, media, education, and tourism. Pearson International Airport, in the west of the city, has become a major – suburban – centre of employment and economic activity. Around the airport as well as other suburban employment hubs, suburban residential expansion has occurred.

While the periphery of Toronto has internationalized power centres such as the city of Markham with 400 corporate head offices, the important story in Toronto is that immigration has become a suburban process. Long thought as dependent on the dense institutional and cultural structures of the inner city, immigrant settlement has increasingly shifted directly to areas peripheral to the urban region. In Markham, more than 70 percent of the population of 300,000 are non-white; almost 60 percent were born outside Canada. One-third of the population speaks Mandarin or Cantonese (Belina and Lehrer 2017). And Markham is not alone in the region. Suburban immigrant power houses Brampton and Mississauga mirror Markham's profile. Demographically, culturally, and increasingly also economically, the world city spreads into the global suburb.

CONCLUSION

Extended urbanization is now a global phenomenon. Involving ongoing processes of traditional suburbanization and emerging forms of postsuburban diversity, it marks the moment of our shared experience in cities around the world. While often referred to as an "urban revolution" (Lefebvre 2003; Brugmann 2009), most urban growth worldwide now takes the form of peripheral or *suburban* development. Global and creative city boosters and scholars of global urban economies have often displayed a bias towards the (gentrified) central and dense urban neighbourhoods and production spaces that have triggered yearnings for widespread re-urbanization. Yet, suburbanization remains the dominant mode in which cities are built. The universality of the suburban trend and the boundless divergence in its real processes and outcomes pose an encompassing yet particularized set of scenarios.

Although the case studies sketched in this brief essay are all from cities of the global North, the growing presence of the global suburb in the metropolitan areas and small towns of the global South has equally been documented (Roy 2015; McGee 2015). Similarly, the intellectual domain of the global city cannot longer be ignored. This is particularly so as the suburbs of the global city contribute more visibly and thoroughly to the galloping modernities and diversities one finds throughout the urban world that defy the common patterns of western urbanization. Looking "from the expanding edge of the global city" into the dynamic process that drives the urban revolution today (Brugmann 2009), students of globalizing cities must now address salient (post)suburban challenges. By studying the suburbs of the global city, or to put it differently, the globalizing suburbs of the world's cities, we can better understand emergent forms of urbanization and of urbanism around the world more generally, but also specifically bring the suburban experience into view for research on globalizing cities. This includes looking at *suburbanism* as the growing prevalence of qualitatively distinct 'suburban ways of life' that are indispensable to the formation of the global city region. These ways of life are now themselves entirely globalized and not the "provinces" to the central city's global economic "metropolis".

Tim Bunnell and Anant Maringanti (2010) bemoan the tendency to see the most important financial metropoles, particularly those in the West, placed in the centre of research. This, they argue, happens at the expense of other urban experiences that are left in the conceptual and empirical dark. The authors emphasize that it is necessary to see "diversity as constitutive of contested global processes" and that we need "to pay attention to the criss-crossing messy pathways through which ideas circulate, connecting cities in ways that can neither be ignored nor reduced to one-way traffic. The process of arriving at this idea conceptually and confronting the anxieties of dealing with the unpredictability of actual research requires building shared agendas and collaborative work practices among scholars at multiple locations" (2010: 418).

In the future, we need to address both the difference between central city and suburbs as well as the diversity within the latter, including their hybrid, in-between, and postsuburban forms. Suburbanisms – as distinct ways of life – include elements such as centrality/peripherality, scale, mix of land uses, population characteristics, power/control, governance, mobility modes, services, and amenities. While one must consider the general tendency towards global suburbanisms through the lens of localized manifestations, one needs to envision these as an integral part of the global city.

REFERENCES FROM THE READING

Belina, B. and Lehrer, U. (2017) The global city region: Constantly emerging scalar fix, terrain of inter-municipal competition and corporate profit strategy, in: Boudreau, J., P. Hamel, R. Keil and S. Kipfer (eds.), *Governing Cities Through Regions: Transatlantic Perspectives*, Waterloo: Wilfrid Laurier Press: 83–97.

Brugmann, J. (2009) *Welcome to the Urban Revolution: How Cities are Changing the World*, Toronto: Bloomsbury Press.

Bunnell, T. and Maringanti, A. (2010) Practising urban and regional research beyond metrocentricity, *International Journal of urban and Regional Research*, 34, 2: 415–20.

Erie, S. (2004) *Globalizing L.A.: Trade, Infrastructure, and Regional Development*, Stanford: Stanford University Press.

Filion, P. and Keil, R. (2017) Contested infrastructures: Tension, inequity and innovation in the global suburb, *Urban Policy and Research*, 35,1: 7–19.

Harris, R. (2010) Meaningful types in a world of suburbs, in M. Clapson and R. Hutchison (eds.) *Suburbanisation in Global Society. Research in Urban Sociology*, 10, Bingley: Emerald Group Publishing.

Harris, R. and Vorms, C. (eds.) (2017) *What's in a Name? Talking about Urban Peripheries*, Toronto: University of Toronto Press.

Keil, R. (2011) Suburbanization and global cities, in B. Derudder, M. Hoyler, P. J. Taylor, and F. Witlox (eds.) *International Handbook of Globalization and World Cities*, London: Edward Elgar.

Keil, R. (ed.) (2013) *Suburban Constellations*, Berlin: Jovis Verlag.

Keil, R. (2017) *Suburban Planet: Making the World Urban from the Outside In*, Cambridge, UK: Polity.

Lefebvre, H. (2003) *The Urban Revolution*, Minneapolis: University of Minnesota Press.

McGee, T. (2015) Deconstructing the decentralized urban spaces of the mega-urban regions in the global south, in P. Hamel and R. Keil (eds.) *Suburban Governance: A Global View*, Toronto: University of Toronto Press: 325–36.

Roy, A. (2015) Governing the postcolonial suburbs, in P. Hamel and R. Keil (eds.) *Suburban Governance: A Global View*, Toronto: University of Toronto Press: 337–48.

Stanek, L. (2008) Space as concrete abstraction: Hegel, Marx, and Modern urbanism in Henri Lefebvre, in K. Goonewardena, S. Kipfer, R. Milgrom, and C. Schmid (eds.) *Space, Difference, Everyday Life: Reading Henri Lefebvre*, London and New York: Routledge: 62–79.

Teaford, J. (2011) Suburbia and post-suburbia, in N. A. Phelps and F. Wu (eds.) *International Perspectives on Suburbanization: A Post-Suburban World?* Basingstoke: Palgrave Macmillan: 15–34.

64

"What is urban about critical urban theory?"

from *Urban Geography* (2015)

Ananya Roy

EDITORS' INTRODUCTION

Ananya Roy is Professor of Urban Planning and Social Welfare and inaugural Director of the Institute on Inequality and Democracy at UCLA Luskin. She holds the Meyer and Renee Luskin Chair in Inequality and Democracy. Trained as an urban planner, Roy was previously based at UC Berkeley, where she held the Distinguished Chair in Global Poverty and Practice. Roy has published widely on urban transformations in the global South, specifically on the dispossessions and displacement wrought by the making of "world-class" cities. Amongst her books are *City Requiem, Calcutta: Gender and the Politics of Poverty* (2003) and *Worlding Cities: Asian Experiments and the Art of Being Global* (co-edited with Aihwa Ong, 2011).

Ananya Roy has been one of the drivers of a thorough rethinking of urban theory from developments in the global South (Roy 2009). Her work has been hugely influential in studies on postcolonial urbanization, global poverty, and planning. While it may be possible to view urban theory as the common denominator of Roy's far-reaching contributions to various debates, her work is particularly effective, as we will see in the excerpt below, in elegantly weaving ethnographic fieldwork into sophisticated theoretical and methodological arguments. Roy's interventions, therefore, find broad audiences in a variety of arenas of urban theory and practice. In the chapter below, Roy engages an ongoing debate about globalizing cities regarding the nature of the urban, based on empirical fieldwork conducted in West Bengal, India. In a playful interpretation of Nancy Fraser's and Neil Brenner's previous reading of critical theory, Roy argues that the urban has to be understood as a historical construct, as a governmental technology, and also in relationship with the rural. Today's urban question, according to Roy, is a land question which "encompasses regulations, registers, and rights that are not urban and that are not simply making way for the urban." By emphasizing the incompleteness of the urbanization process, she puts herself in direct contradistinction to the idea of an "urban society" discussed in the two essays in this section by Brenner and Schmid (Ch. 65) and Henri Lefebvre (Ch. 68).

BEYOND THE FAMILIAR

For several years now, the Government of India has sought to implement an ambitious policy paradigm: inclusive growth. Remaking its existing welfare programs and launching new ones, the effort to include

the poor is busily underway. In 2015, I spent a few months tracing the life of a particular anti-poverty program in India. From the ministries of Delhi, imposing in their rituals of centralization and standardization, I made my way to the municipalities that ring the Kolkata metropolitan region. Beyond the zone of familiarity, beyond the city, beyond what I had once called the rural-urban interface, was an urban India I barely knew. Along one-lane roads that claimed to be highways, trucks precariously piled with onions and potatoes barreled alongside ubiquitous but snail-speed bicycles. Tractors lumbered to paddy fields while incessantly honking cars tried to claim a right of way, as if the tolls they had paid to enter these highways were an entitlement of sorts. These highways, if we are to call them that, did not lead to municipalities. To get to them, one had to turn onto even narrower roads, pitch-dark at night, roads that snaked through ponds lush with greenery, bustling markets spilling out onto that sliver of asphalt, a montage of village huts and pucca houses, tiny stores offering services, from hairdressing to xeroxing. Often the road would pass a railway station and then one would know one had arrived. At the urban. At the municipality. At the administrative unit designated in 1992 by India's 74th Constitutional Amendment as an "urban local body."

It is thus that I arrived one morning at Dankuni, a new municipality that had been constituted a few years ago out of three and a half panchayats (the administrative unit in India for villages). The Dankuni municipality like other urban local bodies in India has been "empowered" by the 74th Amendment to undertake various functions, ranging from town planning to the provision of water and sanitation infrastructure to urban forestry to slum improvement to the keeping of vital statistics, including registration of births and deaths. Alongside the bustle of these vital statistics – the long lines of men and women waiting to register a birth or death or obtain a trade license – is the work of planning, conducted by municipal-level urban planners and engineers. This is the first location from which I want to pose the question, What is Urban about Critical Urban Theory? A newly painted teal colored building that houses the offices of the Dankuni municipality, depicted here with my interlocutors – the urban planner, the engineer, the assistant engineer (Plate 57). It is from this location that I launched my research tours of the slums of Dankuni, the familiar tangle of improvised shelter and upgraded houses rendered unfamiliar outside the grid of urban agglomeration and human density. Here fields stretched to the horizons, houses could conceivably stretch their elbows, infrastructure could be laid through empty swaths of land. In what way were they slums? Did the arrival of the governmental category, "slum," mark the ineluctably urban character of these places? Was

Plate 57   Dankuni Municipality, West Bengal, India

Source: Ananya Roy

the arrival of the slum the arrival of the urban in these zones beyond the familiar city?

But newness does not help us analyze these social topographies. In Bhatpara, north of Dankuni, as I pored over the plans, maps, and documents of the municipality, I confronted an extraordinary piece of historical geography – much of the land-use of the municipality was designated as slum (Plate 58). This pixelated map displayed on the screen of the municipal engineer's computer is the second location from which I want to pose the question, What is Urban about Critical Urban Theory? Marked in blue, slum boundaries dominate the map, leaving us to wonder what, in Bhatpara, is not a slum. But this is not a manifestation of the apocalyptic story of a planet of slums as Davis (2006) would have us consider; there is no inexorable hand of global neoliberalism invisible behind this GIS map. These are old slums, necessary companions to the coolie lines of the jute and paper mills once built by the British in this valuable agro-industrial hinterland of the empire. I grew up, I am told, in one such paper mill where my father served as director of a British managing agency house. Needless to say, I do not remember the slums.

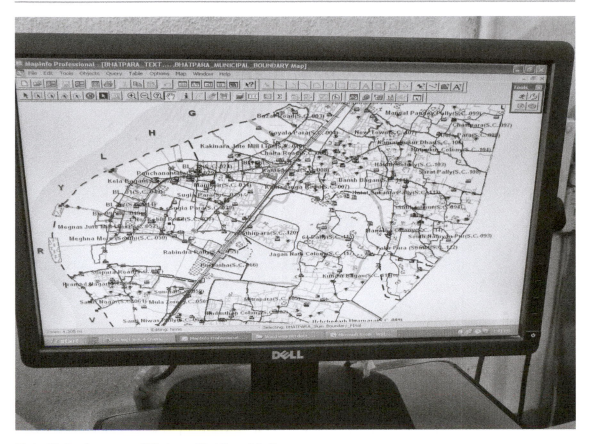

Plate 58 Land-use map of Bhatpara, West Bengal, India

Source: Ananya Roy

I was in Dankuni and Bhatpara studying a program by the Government of India to regularize and upgrade all slums in the country. The program signaled a dramatic shift in the relationship between government and slum, a relationship that until recently had been framed as that between sovereign and encroacher. With the new program, the relationship was recast as state and beneficiary. Inscribed on the walls of slum houses in Dankuni are the signs of different iterations of these programs of government, each preceded by a comprehensive survey that numbers and marks homes of the beneficiaries. In the offices of Bhatpara municipality are the DPRs – the detailed project reports that must be sanctioned by the Delhi bureaucracy – with lists, photographs, thumbprints, and signatures of slum-dwellers recast as beneficiaries. These lists of state-designated and self-styled beneficiaries are the third location from which I want to pose the question, What is Urban about Critical Urban Theory? Here then is the third location recast: the

fossil of an old DFID-funded program, KUSP (Kolkata Urban Services for the Poor), along with the recent stamp of the ruling political party in West Bengal, the TMC, as it prepared for the 2015 municipal elections (Plate 59).

It is from these three locations that I want to arrive at three analytical points about the urban. First, in contrast to conceptual frameworks that emphasize the urbanization of everything, I insist on paying attention to the "constitutive outside" (Mouffe 2000) of the urban and to the always incomplete processes of becoming urban. I am particularly interested in the "rural" as a constitutive outside of the urban. By no means is the rural the only or even a privileged constitutive outside. But given my ongoing research in India, the rural is foremost on my mind. Analytically and empirically, this means understanding the entanglement of the agrarian and urban questions. Methodologically, this means understanding feminist and poststructuralist practices of deconstruction. Second,

Plate 59 Hooghly-Chinsurah pilot slum, West Bengal, India

Source: Ananya Roy

I foreground the urban as a historical geography, indeed as a historical category. In contrast to arguments, often drawn from Lefebvre, that we live in the age of an urban revolution, I suggest that we pinpoint the conjunctures at which the urban is made and unmade, often in highly uneven fashion across national and global territories. Third, I argue that the urban and rural are governmental categories. Whether or not they are faithful depictions of actually existing socio-spatial processes is one set of analytical considerations. But an equally important analytical task is to understand why these categories matter and how they are deployed in the repetitious work of government. Another way of framing this is to ask if there is a distinctive "politics of the governed" – Partha Chatterjee's (2004) felicitous phrase – that is at stake in the urban. Indeed, recent debates in urban studies suggest that this is the case, be it assertions of the right to the city or insurgent citizenship by city-dwellers. But in municipalities like Dankuni, Hooghly-Chinsurah,

and Bhatpara, beyond the familiar zones of street politics and rights-based movements, I return to the category of beneficiary. It is through the collection of beneficiaries, the recipients of those functions designated to urban local bodies by the 74th Constitutional Amendment of India, that the urban comes to have meaning as a governmental category. Rural as a similar governmental category is not the antonym of urban; it is its necessary supplement, marking populations where the government of poverty is different, where the relationship between state and beneficiary is of a different socio-spatial character.

TODAY'S URBAN QUESTION

My question, What is Urban about Critical Urban Theory?, is inspired by the essay written by feminist philosopher, Nancy Fraser (1985), "What's Critical about Critical Theory?" In urban studies, Neil Brenner

(2009) has asked, "What is critical urban theory?" For him, critical urban theory is not only a critique of existing social relations but also a search for emancipatory alternatives. Such a critique and search, Brenner notes, are always historically specific. With this in mind, he calls for a critical urban theory appropriate for the early 21st century, a historical moment that he argues is characterized by "nothing less than an urbanization of the world" (Brenner 2009: 200, emphasis in the original). In other words, for Brenner, now, critical theory is necessarily critical urban theory. But if this is indeed the case, what is urban about critical urban theory?

The effort to conceptualize the urban is of course an integral part of the practice – and existential crisis – of urban studies. Let me return to my three analytical points – the constitutive outside of the urban, the urban as an uneven historical geography, and the urban as a governmental category – to explore what might be the contours of today's urban question. I start with the conceptual framework of planetary urbanization (see Brenner and Schmid Ch. 65). Relatedly, Brenner and Schmid (2014: 750) note about the relationship between the rural and the urban: "There is, in short, no longer any outside to the urban world; the non-urban has been largely internalized within an uneven yet planetary process of urbanization." They continue, "the urban/rural binarism is an increasingly obfuscatory basis for deciphering the morphologies, contours and dynamics of sociospatial restructuring under early twenty-first century capitalism."

Thinking from the locations I have introduced, I want to argue that the rural is much more than the non-urban, that it is in fact a "constitutive outside." It is not the only or even the privileged constitutive outside, but it is a vitally important one. This means that even if we are to concede the urbanization of everything, everywhere, we have to analytically and empirically explain the processes through which the urban is made, lived, and contested – as a circuit of capital accumulation, as a governmental category, as a historical conjuncture. I want to consider why the pathway that lies through small towns in the global South, places like Dankuni and Bhatpara, might matter and how it is here that the necessarily incomplete and uneven process of being urban is laid bare.

My own interest is in how a postcolonial government initiates urban land reforms, in a process that both mirrors and veers away from agrarian land reforms, and in doing so seeks to convert complex informal regimes of inhabitation, occupation, and tenancy into property, cadastral property of course but most crucially property that is legible to the state and its projects of urban development. What I am studying in India is a simultaneous effort to convert informal regimes of urban inhabitation into secure tenure all under the watchful eye of the state. This process is a key way in which the urban is being reconstituted as a governmental category, with populations that must be governed. Put bluntly, I argue that today's urban question is a land question but that this land question very much encompasses regulations, registers, and rights that are not urban and that are not simply making way for the urban. The social relations of production as well as the political identities and struggles evident in such territories cannot be encompassed by the urban. It is this persistence of historical difference that is of concern to me and that I wish to pose as an analytical challenge to current conceptualizations of the urban.

The processes of becoming urban, of making urban, are both old and incomplete. The rural is not the antonym of urban. It is not not-urban. Needless to say, the rural, like the urban, is not a morphological description but rather an inscription of specific regulations and logics of territory, land, and property. It is only in this rural-urban matrix, at least in this corner of the world, that what we globally conceptualize as urbanization is underway. This is at once today's urban question and today's agrarian question.

The urban question, as a project of critical theory, has always been concerned with historical alternatives. In fact, urban studies today is overflowing with arguments about urban citizens, residents, occupants, movements, and experiences as the new political subject. In particular, I am interested in the relationship between the "complete urbanization society" – Merrifield's (2014: 2) borrowing of Lefebvre's idea – and a new political subject that is ineluctably urban. Here is how Merrifield (2014: 69) phrases this relationship: "the new urban question is about creating. . .a movement that can loosen the neo-Haussmannite grip on our society. . .an urban political movement [not social] that struggles for generalized democracy." Who can resist being inspired? Who can resist, on the path of ethnography, searching for such movements? Perhaps less so in Merrifield, but quite definitely in Warren Magnusson (2011: 34), the urban becomes a political ontology, "a particular way of being political."

But in Dankuni and Bhatpara, I could not gather together the claims pressed by newly urban beneficiaries as a struggle for generalized democracy. There, in the old slums adjoining the coolie lines of jute mills and paper mills, on land settled by the East India Trading Company, I could find neither a neo-Haussmannite grip nor an urban political movement to resist such predations. It is not my intention to use a single empirical case to challenge a conceptual framework and to mark an exception. Instead, I want to share the deep ambiguities and ambivalences that attend the urban, always accompanied by the not-urban. That mix of constitutionally empowered urban local bodies geared up for the cycles of municipal elections and project sanctions with intimate transactions of rural patronage and governance requires our attention. For this may very well be the grassroots of the complete urbanization society. And if this is the case, then I must strongly disagree with Magnusson (2011: 5) when he notes that "the fundamental question is of the city, not the state." The urban, I am arguing, is a particular way of being governmental – my rephrasing of Magnusson's line. The urban, as I encountered it, beyond the zones of the familiar city in India, is a state designation, an administrative category that creates distinctive governed populations, including self-styled beneficiaries for housing programs and land reforms. That designation is often contested. I therefore suggest that we take up the question of whether the urban is a particular way of being political as precisely this, a question, rather than an ontological truth.

The places whose stories I am trying to tell cannot be understood as geographies of urbanization. Nor can the politics of space present in these locations be read as urban in the ways in which critical urban theory would lead us to believe. They are urban because statecraft has decided that they are so. Yet, such municipalities are more than simply a formal administrative designation. I have also emphasized that they mark a distinctive form of government, the urban as government. They are examples of urban government without geographies of urbanization or without urban politics. With this in mind, I argue that such places allow us to think about the urban as an incomplete and contingent process as well as an undecidable category. Perhaps such contingency and undecidability is especially visible in India where both the administrative designation of urban and forms of urban government are fluid and contested. If rural

market towns or agro-industrial hinterlands have been designated as municipalities, then a related category is that of "census towns," a fast growing designation in India. Meeting Indian census definitions of the urban (size, density, structure of nonfarm employment), these are governed by rural administration, notably gram panchayats. But as Guin and Das (2015: 68) note, in some parts of India, including West Bengal, "because of the huge increase of agricultural labourers (in 2011), many new census towns might be reclassified as villages for the next census in 2021." This is not simply the case of arbitrary classifications and reclassifications. Instead, this is a glimpse of complex forms of rural-urban differentiation that exceed our analysis of urban political economy and its patterns of accumulation and dispossession.

"FROM THE STANDPOINT OF AN ABSENCE"

What is at stake is not only what is urban about critical urban theory but also from where on the map we produce the body of authoritative knowledge that we are willing to acknowledge as theory. In his recent essay in *Regional Studies*, Jamie Peck (2015) acknowledges that a new moment is underfoot in urban studies. He notes the "opening up of new spaces in and for urban theory, and new ways of thinking about urban theory" (Peck 2015: 162). But Peck charges such "post-millennial reinvention" as prone to "difference-finding and deconstructive manoeuvres" rather than being "projects of urban-theoretical renewal." Peck seems ambivalent about what he interprets as the "embrace of particularism and polycentrism." He concludes, "uniqueness and particularity are back (again) and finding exceptions to – as well as taking exception to – general urban-theoretical rules have become significant currents in the literature."

I argue instead that laments about the erosion of coherent urban theory (often couched as a confrontation between political economy and postcolonial theory) misread historical difference as empirical variation. To find difference is not to sidestep general processes for particularities. It is to theorize historical difference as a fundamental constituent of global urban transformation. This changes, as Spivak (2014) has noted, the "required reading" that we must undertake. For to analyze the sheer extent of slum lands in Bhatpara today, to do that "rethink" of the

Plate 60 Gram panchayat plaque, Dankuni municipality

Source: Ananya Roy

categories of urban and the urban political that Merrifield calls for in the context of the complete urbanization society, I must turn to the challenge that Dipesh Chakrabarty (1989) presents to Marxist working-class history and European modernity. It was a challenge from those very same jute mills of Bengal. To think via historical difference is not to avoid generalization but it is to insist that general processes (in this case a rural-urban matrix of property and power) are not necessarily universal, that the jute mills of colonial and postcolonial Bengal might yield a different working-class politics, a different urban transformation, a different way of being political.

I conclude then with the invitation to read the urban from the standpoint of absence, absence not as negation or even antonym but as the undecidable. I conclude too with the provocation that theory, including a theory of the urban, can be made from the teal colored building at the edge of the world that is the Dankuni municipality, a panchayat office repurposed for urban government. But in a gesture befitting the task of provincializing the urban, I note that the dedication plaque for the panchayat building references a fin de siècle poet, Jibanananda Das and his writings on "rupasi bangla," or beautiful Bengal, envisioned as rural and verdant. But Das is also the first urban poet of Bengal, with a set of starkly neo-urban poems that are now etched into the region's self-imagination of urban modernity. The plaque can thus be read as a serendipitous anticipation and premonition of the urban yet to come but its rurality cannot be effaced or erased (Plate 60). The sign of a constitutionally demarcated urban local body it is the undecidability of the urban.

REFERENCES FROM THE READING

Brenner, N. (2009) What is critical urban theory? *City*, 13, 2–3: 198–207.

Brenner, N. and Schmid, C. (2014) The 'urban age' in question,' *International Journal of Urban and Regional Research*, 38, 3: 731–55.

Chakrabarty, D. (1989) *Rethinking Working-Class History: Bengal, 1890–1940*, Princeton, NJ: Princeton University Press.

Chatterjee, P. (2004) *The Politics of the Governed: Reflections on Popular Politics in Most of the World*, New York: Columbia University Press.

Davis, M. (2006) *Planet of Slums*, New York: Verso.

Fraser, N. (1985) What's critical about critical theory? The case of Habermas and gender, *New German Critique*, 35: 97–131.

Guin, D. and Das, D. N. (2015) New census towns in West Bengal: 'Census activism' or sectoral diversi-fication? *Economic and Political Weekly L*, 14: 68–72.

Magnusson, W. (2011) *Politics of Urbanism: Seeing Like a City*, New York: Routledge.

Merrifield, A. (2014) *The New Urban Question*, New York: Pluto Press.

Mouffe, C. (2000) *The Democratic Paradox*, New York: Verso Books.

Peck, J. (2015) Cities beyond Compare? *Regional Studies*, 49, 1: 160–82.

Roy, A. (2009) The 21st-century metropolis: New geographies of theory, *Regional Studies*, 43, 6: 819–30.

Spivak, G. C. (2014) Postcolonial theory and the specter of capital, *Cambridge Review of International Affairs*, 27, 1: 184–98.

65
"Planetary urbanization"
from *Urban Constellations* (2011)

Neil Brenner and Christian Schmid

EDITORS' INTRODUCTION

This chapter was originally published in *Urban Constellations*, edited by Matthew Gandy in 2011. Drawing upon works by Henri Lefebvre (see Ch. 68), Neil Brenner and Christian Schmid make a bold statement that we have arrived at the age of "planetary urbanization" in the early 21st century. By "planetary urbanization," they refer to a worldwide condition in which political economy, infrastructure, and landscapes are all integrated into the urban fabric. To put in other words, the urban can no longer be understood just as the city (and suburbs), and places outside metropolitan areas have also been integrated into the urban tissue. Thus, Brenner and Schmid challenge us to think of the urban as broader processes of (capitalist) production relations, instead of as concrete settlement types. They recognize that this will necessarily entail an epistemological shift in how we conceive the urban world, but such a shift will open up new possibilities for theoretical and conceptual innovation. The thesis on "planetary urbanization" is further elaborated in several recent publications by Brenner and coauthors, including *Implosions/Explosions* (Brenner ed. 2014) and *Critique of Urbanization* (Brenner 2016).

During the last several decades, the field of urban studies has been animated by an extraordinary outpouring of new ideas regarding the role of cities, urbanism and urbanization processes in ongoing global transformations (see Sassen 2000; Soja 2000; Roy 2009). Yet, despite these advances, the field continues to be grounded upon a mapping of human settlement space that was more plausible in the late nineteenth and early twentieth centuries than it is today.

The early twentieth century was a period in which large-scale industrial city-regions and suburbanizing zones were being rapidly consolidated around the world in close conjunction with major demographic, socioeconomic and environmental shifts in the erstwhile "countryside." Consequently, across diverse national contexts and linguistic traditions, the field of twentieth-century urban studies defined its theoretical categories and research object through a series of explicit or implied geographical contrasts. Even as debates raged regarding how best to define the specificity of urban life, the latter was universally demarcated in opposition to a purportedly "non-urban" zone, generally classified as "rural." As paradigms for theory and research evolved, labels changed for each term of this supposed urban-rural continuum, and so too did scholars' understandings of how best to conceptualize its basic elements and the nature of their articulation. For instance, the Anglo-American concept of the "suburb" and the French concept of *la banlieue* were introduced and popularized to demarcate further sociospatial differentiations that were occurring inside a rapidly urbanizing field (Fishman 1989; Forsyth 2012). Nonetheless, the bulk of twentieth-century urban studies rested on the assumption that cities—or, later, "conurbations," "city-regions," "urban regions," "metropolitan regions," "megacities," and "global city-regions"—represented a particular *type* of territory that was qualitatively specific, and thus different from the putatively non-urban spaces that lay beyond their boundaries.

The demarcations separating urban, suburban and rural zones were recognized to shift historically, but the spaces themselves were assumed to remain discreet, distinct and universal. While paradigmatic disagreements have raged regarding the precise nature of the city and the urban, the entire field has long presupposed the existence of a relatively stable, putatively non-urban realm as a "constitutive outside" for its epistemological and empirical operations. In short, across divergent theoretical and political perspectives, from the Chicago School's interventions in the 1920s and the rise of the neo-marxist "new urban sociology" and "radical geography" in the 1970s, to the debates on world cities and global cities in the 1980s and 1990s, the major traditions of twentieth-century urban studies embraced shared, largely uninterrogated geographical assumptions that were rooted in the late nineteenth and early twentieth century geohistorical conditions in which this field of study was first established.

During the last thirty years, however, the form of urbanization has been radically reconfigured, a process that has seriously called into question the inherited metageographical assumptions that have long underpinned urban theory and research. Aside from the dramatic spatial and demographic expansion of major megacity regions, which has been widely discussed (Champion and Hugo 2005; Scott 2001; Burdett and Sudjic 2006), the last thirty years have also witnessed several equally far-reaching implosions and explosions of the urban at all spatial scales. These include:

■ *The creation of new scales of urbanization.* Extensively urbanized interdependencies are being consolidated within extremely large, rapidly expanding, polynucleated metropolitan regions around the world to create sprawling urban galaxies that stretch beyond any single metropolitan region and often traverse multiple national boundaries. Such mega-scaled urban constellations have been conceptualized in diverse ways, and the representation of their contours and boundaries remains a focus of considerable research and debate (Hall and Pain 2006; Nelson and Lang 2011; Florida, Gulden and Mellander 2008). Their most prominent exemplars include, among others, the classic

Gottmannian megalopolis of "BosWash" (Boston-Washington DC) and the "blue banana" encompassing the major urbanized regions in western Europe, but also emergent formations such as "San-San" (San Francisco-San Diego) in California, the Pearl River Delta in south China, the Lagos littoral conurbation in West Africa, as well as several incipient mega-urban regions in Latin America and South Asia.

■ *The blurring and rearticulation of urban territories.* Urbanization processes are being regionalized and reterritorialized. Increasingly, former central functions, such as shopping facilities, corporate headquarters, multimodal logistics hubs, research institutions, cultural venues, as well as spectacular architectural forms, dense settlement patterns and other major infrastructural arrangements, are being dispersed outwards from historic central city cores, into erstwhile suburbanized spaces, among expansive catchments of small- and medium-sized towns, and along major transportation corridors such as superhighways and rail lines.

■ *The disintegration of the hinterland.* Around the world, the hinterlands of major cities, metropolitan regions and urban-industrial corridors are being reconfigured as they are operationalized, infrastructuralized and enclosed—whether as back office and warehousing locations, global sweatshops, agro-industrial land-use systems, data storage facilities, energy generation grids, resource extraction zones, fuel depots, waste disposal areas, recreational areas or corridors of connectivity—to facilitate the metabolism of industrial urbanization and its associated planetary urban networks.

■ *The end of the wilderness.* In every region of the globe, erstwhile "wilderness" spaces are being transformed and often degraded through the cumulative socio-ecological consequences of unfettered worldwide urbanization, or are otherwise being converted into bio-enclaves offering "ecosystem services" to offset destructive environmental impacts generated elsewhere. In this way, the world's oceans, alpine regions, the equatorial rainforests, major deserts, the arctic and polar zones, and even the earth's atmosphere itself, are increasingly being interconnected with the rhythms of planetary urbanization at every geographical scale, from the local to the global.

In our view, these geohistorical developments pose a fundamental challenge to the entire field of urban studies as we have inherited it from the twentieth century: its basic epistemological assumptions, categories of analysis and sites of investigation require a foundational reconceptualization in order to remain relevant to the massive transformations of worldwide sociospatial and environmental organization we are witnessing today. Under contemporary conditions, therefore, the urban can no longer be understood with reference to a particular "type" of settlement space, whether defined as a city, a city-region, a metropolis, a metropolitan region, a megalopolis, an edge city or otherwise. Consequently, despite its continued pervasiveness in scholarly and political discourse, the category of the "city" has today become thoroughly problematic as an analytical tool. Correspondingly, it is no longer plausible to rely upon the inherited urban/rural (or urban/non-urban) distinction to characterize the variegated differences that obtain between densely agglomerated zones and the less densely settled zones of a region, a territory, a continent or the globe. Today, the urban represents an increasingly worldwide condition in which all political-economic relations, infrastructural geographies and socio-environmental landscapes are enmeshed.

This situation of *planetary urbanization* means, paradoxically, that even spaces that lie well beyond the traditional city cores and suburban peripheries—from territories of agro-industrial production, zones of industrialized resource extraction and energy generation, "drosscapes" and waste dumps, transoceanic shipping lanes, transcontinental highway and railway networks, and worldwide communications infrastructures to alpine and coastal tourist enclaves, "nature" parks and erstwhile "wilderness" spaces such as the world's oceans, deserts, jungles, mountain ranges, tundra and atmosphere—have become integral parts of a worldwide urban fabric. While the process of agglomeration remains essential to the production of this new worldwide topography, political-economic spaces can no longer be treated as if they were composed of discrete, distinct, bounded and universal types of settlement. In short, in an epoch in which inherited notions of the rural and the countryside appear increasingly to be ideological projections derived from a largely superseded, preindustrial geohistorical formation, our image of the urban must likewise be fundamentally reinvented.

Already four decades ago, Henri Lefebvre put forward the radical hypothesis of the complete urbanization of society, a transformation that required,

he argued, a radical epistemological shift from the analysis of urban form to the investigation of *processes* of urbanization (Lefebvre 2003). However, a systematic application of this fundamental thesis has yet to be undertaken. Perhaps, in the early twenty-first century, the moment is now ripe for such an undertaking. In our view, the epistemological foundations of urban studies must today be fundamentally transformed (Brenner 2014; Brenner and Schmid 2015). The epistemological shift towards the analysis of planetary urbanization requires new strategies of concrete research and comparative analysis that transcend long-entrenched, city-centric assumptions regarding the appropriate object and parameters for urban research. In close conjunction with such new research strategies, the investigation of planetary urbanization will require major theoretical and conceptual innovations. We need, first of all, new theoretical categories through which to decipher emergent transformations of sociospatial organization, infrastructural configurations, political regulation, social mobilization and everyday life across places, scales, territories and landscapes. To this end, a new conceptual lexicon must be created for identifying the wide variety of urbanization processes that are currently reshaping the planet, and their intense contestation through diverse political strategies and social forces. Lastly, we require adventurous, experimental and boundary-exploding methodological strategies to facilitate the empirical investigation and visualization of these processes. Whether or not a distinct field of "urban" studies can persist amidst such theoretical, conceptual and methodological innovations is a question that remains to be explored in the years and decades ahead.

REFERENCES FROM THE READING

Brenner, N. (ed.) (2014) *Implosions/Explosions: Towards a Study of Planetary Urbanization*, Berlin: Jovis.

Brenner, N. and Schmid, C. (2015) Towards a new epistemology of the urban, *City*, 19, 2–3: 151–82.

Brenner, N. (2016) *Critique of Urbanization: Selected Essays*, Berlin and Basel: Birkhauser Verlag.

Burdett, R. and Sudjic, D. (eds.) (2006) *The Endless City*, London: Phaidon.

Champion, T. and Hugo, G. (eds.) (2005) *New Forms of Urbanization*, London: Ashgate.

Fishman, R. (1989) *Bourgeois Utopias*, New York: Basic Books.

Florida, R. Gulden, T., and Mellander, C. (2008) The rise of the mega-region, *Cambridge Journal of Regions, Economy and Society*, 1, 3: 459–76.

Forsyth, A. (2012) Defining suburbs, *Journal of Planning Literature*, 27, 3: 270–81.

Hall, P. and Pain, K. (eds.) (2006) *The Polycentric Metropolis*, London: Earthscan.

Lefebvre, H. (2003) *The Urban Revolution*, trans. by R. Bononno, Minneapolis: University of Minnesota Press.

Nelson, R. and Lang, R.E. (2011) *Megapolitan America: A New Vision for Understanding America's Metropolitan Geography*, Chicago: APA Planners Press.

Roy, A. (2009) The 21st century metropolis: New geographies of theory, *Regional Studies*, 43, 6: 819–30.

Sassen, S. (2000) New frontiers facing urban sociology at the millennium, *British Journal of Sociology*, 51, 1: 143–59.

Scott, A. J. (ed.) (2001) *Global City-Regions*, London: Oxford University Press.

Soja, E. W. (2000) *Postmetropolis: Critical Studies of Cities and Regions*, Cambridge, MA: Blackwell.

66

"New geographies of theorizing the urban: putting comparison to work for global urban studies"

from *The Routledge Handbook on Cities of the Global South* (2014)

Jennifer Robinson

EDITORS' INTRODUCTION

In this following chapter, Jennifer Robinson, whom we already encountered in the first section of this book, argues that a comparative method is indispensable for theorizing in urban studies, especially in the investigation of globalizing urbanization. Robinson carefully advances her argument by reviewing extant practices of comparative urban thinking in light of her own empirical research and poststructuralist theoretical and methodological preferences. She argues that conventional, quasi-scientific comparisons restricted to similar cases have to be reinvented, and she offers a number of alternative comparative approaches: composing comparisons across different cases; tracing connections across cities; and launching analyses of specific urban contexts into wider conversations. These comparative strategies can lead to new concepts and perspectives that take into account the diverse urban experiences worldwide, and make a meaningful intervention to urban scholarship dominated by theoretical frameworks derived from the West. Robinson's call for urban comparisons appeals to a wide audience, and today we see an emerging comparative scholarship undertaken by the younger generation of scholars.

INTRODUCTION

The spatiality of theorizing the urban experience, of thinking cities in a world of cities, maps well onto the core elements of a comparative imagination – thinking across different 'cases' to produce conceptualizations which contribute to wider understandings of the processes being analysed, and which might then, in turn, be considered in relation to other contexts or cases. However, the comparative methods which have conventionally underpinned urban studies have not been well placed to serve the project for a more global urbanism. Theorizing cities can benefit from a comparative imagination, but comparative methods need to be refitted to support a more global urban analytical project, including a substantial reconfiguring of the ontological foundations of comparison. For example, what might be considered a 'case' needs to be redefined to avoid the restricting and territorializing trap of only comparing (relatively similar) 'cities': we might rather compare, for example, specific elements or processes in cities, or the circulations and connections which shape cities, thus rendering urban experiences comparable across a much wider range of contexts, and building research strategies which are adequate to the complex spatiality of urban forms

(Ward 2010). Comparators – the "third term" which establishes the grounds of comparability of cases (Jacobs 2012) – need to be selected so they are relevant to a diversity of urban contexts, rather than seeking relatively similar cities for comparison as is conventional. In this way, they will be able to stretch theoretical concepts to the breaking point required for the reinvention of urban studies for global analysis, and not simply reinforce parochial and limited understandings. More generally, the status of the case itself needs to be reimagined, in relation to both the wider empirical processes shaping particular outcomes, and the conceptualizations which are an important ambition of comparative strategies; this is essential to ground an adequate post-structuralist comparative method which moves beyond both quasi-scientific explanations and a view of the world in which wider structures are drawn on to explain complex specific outcomes.

SOME ANALYTICAL TACTICS OF GLOBAL URBANISM

It is widely agreed that urban studies needs to embrace a much wider repertoire of urban contexts in building understandings of contemporary urbanism and to revisit parochial interpretations of the past: the challenge of thinking cities in a world of cities is not somehow only newly arrived! But what exactly is to be done? Different scholars have set out on this project for a global urbanism in a variety of directions, establishing productive and interesting lines of analysis. Pulling them together here I would like readers to join me in being inspired by the extent and range of work which supports global urban studies, and to try out these different paths, while at the same time reflecting on how collectively they also bring into the open some of the emerging challenges of thinking across the diversity of urban experiences.

Some 'strategic essentialisms': Critiquing the parochial state of urban studies in the 2000s – and the numerous occasions previously when scholars have pointed out the limited resources shaping dominant understandings of the urban – some initial manoeuvres for a global urban studies involved locating and then dislocating the conceptual foundations of what appeared as 'Western' urban studies (see Robinson Ch. 9). Now, over a decade later, it seems this work of critique needs to be ongoing – for it turns out to be far too easy for scholars working in such contexts to simply name their location – northern cities, or regions such as US or EU cities – and to continue to ignore the rest of the world of cities while forming their analyses. Given that such well-resourced scholarship still dominates the key journals, not to mention the institutional resources dedicated to international urban research, again and again, I expect, those of us concerned for a more global perspective, or who live in, work on and think through different kinds of cities, will need to review, critique and extend such parochial conceptualizations. The challenge here is that such ideas, even if named as parochial, continue to be styled as universal arguments, and thus gain traction and set themselves on the move, becoming hard to avoid when cities everywhere are analysed. The claims to universalization which are embedded in authorial voice (confident, dominating, authorizing, unmarked) and enabled through the practicalities of unevenly resourced circuits of knowledge and publishing could be further dissipated through embedding expectations of quite different practices of theorization – not only a critical reflexivity on the part of writers, but the propagation of modes of theorizing which are actively open to being revised, much more modest in their voice and precise about locational co-ordinates (both physical and social). Some of the practices of comparative urbanism outlined in the next section could be helpful in framing different practices of theorizing.

In turning to articulate new dimensions of theorizing cities from elsewhere, authors have sought new geographies of authority and voice (see Roy Ch. 64). "The region" as a ground for theorization can be a substantively important and generative context in terms of scholarly networks, traditions and practices, but it is already open to a wider array of contexts, influences and theoretical conversations. Insofar, then, as cities are more than regionally determined, framed by globalizing processes, circulating policies and the numerous re-iterations of urban forms, the "region" as a basis for theoretical voice might be claimed, momentarily, but is quickly called into question through internal differentiation across the region, the wider processes of globalization and competing geographical frames of explanation (not least the nation, the city). Our analyses of cities cannot necessarily be circumscribed by the regions or even countries in which they are located. Cities are, as Mbembe and Nuttall have it 'embedded in multiple elsewheres' (2004: 348). This same challenge of deploying strategic geographical imaginations as grounds from which to speak new accounts of the urban faces the idea of a 'southern' urban theory. In this case the geography of the south is even more complicated to work with. It borrows the geopolitical metaphor linking poor nations beyond the capitalist heartland and western core to figure the broadly critical 'ex-centric' positionality of the scholar whose perspective does not derive from the mainstream (western, northern) academy. Both these moves, to evoke a 'southern' positionality, and to excavate regional distinctiveness, are important moments in the emergence of new approaches in urban studies. But they are both 'strategic' opportunities, interim moves, and more sustained formulations for building global urban analyses will be required. But it is intellectually important and analytically essential to be precise about these spatialities, not least because even these shared histories are unevenly experienced and strongly exteriorized, part of wider global processes (see Bunnell 2013). Such precision would open up the possibility for on-going conversations across diverse but interconnected urban outcomes across the globe, rather than render cities incommensurable or irrelevant to analyses elsewhere through the invocation of a blunt spatial analytics. While postcolonial moves are seeking to stretch the imagination of urban theory to draw on the resources of urban experiences across the world, debates within mainstream urban studies also offer some new openings for such efforts. Let me sketch two of them here briefly.

The disappearing city: A growing interest in the idea of 'planetary urbanization', drawing on Henri Lefebvre's (2003[1974]) suggestive hypothesis of the 'complete urbanization of society' has brought a concern for the fate of cities (literally) everywhere into the heart of urban theory (see Brenner and Schmid Ch. 65). Certainly the territorial referents of the term, 'city', do service for only a small portion of urban processes – perhaps the territorializing moments of political demarcations of administration and governance which frame inter-city competition (although these are supremely exteriorized (globalized) in the circulations of urban policy); or the constitutive nature of social relations of place (but again, urban social and cultural processes are also highly globalized). This search for new vocabularies of the urban to replace the increasingly unworkable territorial shadow contained in the notion of the city is a significant opening for building urban theorization which takes account of the world of cities. While planetary urbanization as a response to the challenge of the disappearing city

(since the city is now everywhere) is largely pursued through the lens of political-economic analyses, mainstream urban studies has also drawn a strong engagement with alternative post-structuralist traditions which offer competing vocabularies for more global conceptualizations of cities. I review one of these now, largely inspired by the work of the French philosopher, Gilles Deleuze.

Repeated instances: The second emergent approach to global urbanism which I will review here draws its inspiration from Gilles Deleuze's (1994) account of repetitious differentiations which suggests we attend to how we come to a conceptualization of singularities or specific outcomes through working with "difference". As memory and intuition bring objects to "thought", we enter the realm of the 'virtual', where objects can come to be conceptualized through actively exploring connections with other possible configurations and instances in a field, and in relation to an ongoing sensibility to matter. Thus (our concepts of) specific urban outcomes could be understood as intimately connected to many other possible (preceding and future) outcomes, through their shared conditions of conceptualization (as, for example, many instances of "gentrification" could be drawn on in trying to specify how a certain instance of urban regeneration might be understood), And in the case of urbanization (and certainly for "gentrification") many of the processes of production are shared across different instances (such as with policy circulation, or circuits of investment related to urban regeneration). In this view, then, each instance or repetition is only a step aside from other instances, or singularities, distinctive but intimately connected with other specific outcomes (Deleuze 1994). This philosophical intervention offers much food for thought in trying to imagine an interconnected conceptual project across a world of distinctive urban outcomes. In urban studies the most succinct example of this thinking is to be found in Jane Jacobs' (2006) analysis of the globalizing residential high-rise. Here the distinctive achievement of each repetition – almost-the-same – through globalizing circulations and specific assembling of diverse elements to produce each building provides an insight into what it might mean to think with the productivity of the virtual in the sphere of the urban (Farias 2010: 15). The achievement of urban modernity in the repetitive architecture of international modernism emerges from the relatively unpredictable multiplicity of circulations and manifold elements able to be assembled into each construction – buildings which are both repeated and yet produced as original objects, with an equally original yet partly repeated and interconnected set of meanings crafted locally, each time (King 2004): 'the making of repetition – or more precisely, repeated instances in many different contexts – requires variance, different assemblages of allies in different settings' (Jacobs 2006: 22). For Jacobs, each instance produces the global effect of international modernism in her comparative research on the residential high-rise: each case is a singularity, and not an example of an already given global process (Jacobs 2012).

Unfortunately the lines of opposition have been drawn between political economy approaches and the kinds of analyses which have been inspired by Deleuze and Actor Network Theory after Bruno Latour. Thus McFarlane (2011a) invokes the idea of 'assemblage' as a metaphor for interpreting the city as composed of relatively unpredictable but agentful combinations of objects, techniques, practices and human actions using the examples of the informal production of housing and mobilization of resistance amongst poorer urban dwellers. And as we saw above, Jane Jacobs focuses on how each instance of residential high-rise is produced through the "assemblage" of different human and non-human actors in specific places.

Insofar as we might seek to find some common ground between planetary urbanization's search for new conceptualizations of the urban, and the now extensive urban research in the idiom of actor network theory this could be found in the more Deleuzian-inspired opening to reformulating how the urban might be thought across the diversity of urban outcomes. This places the urban as a conceptualization profoundly open to reformulation in response to both differentiated emergent forms of urbanization and the nonetheless interconnected conceptualizations which we might be provoked to consider as this 'urban manifold' makes itself known to us (Simone 2011). This sets the scene, I would propose, for a new formulation of urban comparativism.

REFORMATTING URBAN COMPARATIVISM

Building critically on the practices of urban comparison we could reimagine comparisons as involving the broad practice of thinking cities/the urban through elsewhere (another case, a wider context, existing theoretical imaginations, connections to other places), in order to better understand outcomes and to contribute to broader theorizations and conversations about (aspects of) the urban. Thinking comparatively can highlight the differentiation of outcomes, it can bring into view the distinctive (or shared) processes shaping a certain urban outcome, it can put to work theoretical insights drawn from other instances or cases; it can insist on the incompleteness of analytical insights drawn from different contexts; moreover, it can suggest new objects of analysis by displacing ethnocentric assumptions which arise from the inevitable locatedness of all theory. In the case of cities, the opportunity to think comparatively is ubiquitous by virtue of the multiplicity of urban outcomes, used here to indicate some specific urban space, or process, or the simple fact of having to think cities in a 'world of cities': any act of urban theorization from somewhere is by necessity a comparative gesture, putting a perspective informed by one context or outcome into conversation with wider theorizations. Thus one of the most useful comparative tactics in urban studies is the case study, brought into creative conversation with a wider literature. In many ways this format, the case study – whether understood as a city, a specific urban phenomenon or form, or wider circulating urban processes – brought into conversation with theoretical debates and other cases, is well suited as a model for global urban studies. It insists on taking seriously the scholarly output of people working in different places, thinking through that work to inform one's own located analysis, and in turn, suggesting new lines of theorization based on the new case study. The call for a more global urban studies is in some ways well formulated as an insistence on more critical 'planetary' reading practices (see Jazeel 2012).

The intrinsic comparativism of urban studies can also be put to work more purposively, and here the repertoire of comparative strategies has been expanding through attention to relational comparisons and to the need to formulate comparative methods which are adequate to the specific spatialities of cities. Thus the project of 'composing' comparisons can be reconfigured to map better on to current understandings of the urban. Thus, the territorialized figuring of the individual 'case' as a city is clearly redundant and instead we would seek to put a comparative imagination to work to consider: the range of urbanization processes which stretch far beyond the physical form of cities; the diverse array of social and spatial forms which emerge in different urban settlements; the repeated instances and circulating phenomena (such as policies, forms, visions) which draw highly differentiated urban outcomes into the same frame of analysis. We might draw analytical insights by considering cities through the specific shared connections which shape each, highlighting the impact of different histories and contexts, as Hart (2003) pioneered in her consideration of the effects of rural dispossession on small industrializing towns in northern KwaZulu-Natal, South Africa, through tracing the largely Taiwanese industrialists who chose to locate there in response to a late-apartheid industrial incentive scheme. We might compare the webs of relations which creatively draw cities into practical engagements with circulating policies, economic networks, transnational political influences or direct engagements with actors from specific other cities, as Söderström (2013) does in his comparison of the two 'cities in relations', Ougadougou and Hanoi. And the proliferation of repeated instances across cities provides a basis for a locationally promiscuous research agenda to inform a global conversation about many aspects of contemporary urban life (on neoliberalism, see Goldfrank and Schrank 2009; on gentrification, see Harris 2008; on the residential high-rise, see Jacobs 2006).

Reformatting comparison to support a global urban studies therefore has many possible practical tactics for proceeding: building comparisons through reading strategies to put case study work into wider conversations; composing bespoke comparisons across diverse outcomes or repeated instances; tracing connections amongst cities to inform understandings of different outcomes or to compare the wider interconnections and extended urbanization processes themselves; and, as we discussed in the previous section, launching distinctive analyses from specific urban contexts or regions into wider conversations.

A new conceptualization of the meaning of comparison is also required, which amounts to a new geography of theorizing, or generating concepts. An account of the process of building concepts through engaging across the differentiated field of the urban needs to provide for a radical deterritorialization of the 'case'. Theorizing the urban 'now' provides one possible resource for re-imagining the spatiality of comparison, and a method for *theorizing cities in the midst of elsewhere*. Whether through the logistics of composing, tracing or launching, the comparative imagination calls for attention to the strong exteriorization of the

urban – i.e. the fact that cities are shaped by processes that stretch well beyond their physical extent. Thus we find here resources to insist on both the empirical multiplicity of the urban (a diversity of inter-related urban outcomes) and the multiplicity and interconnectedness of possible conceptualizations or analyses of the urban.

Drawing all cities into the conceptualization of urbanization would benefit from reimagining this relationship so that theorization of the urban can be informed by the widest range of urban experiences. This clearly requires considerable further specification but, as a prosaic example, empirical investigations into the many shared processes shaping different cities could both contribute to a more open theorization of the urban and draw inspiration from a wider range of urban contexts. Rather than being seen as abstract "structures", perhaps the 'rule regimes' of global capitalism (Peck et al. 2009), or extensive urbanization processes (Brenner and Schmid Ch. 65), alongside the circulations of urban forms (Söderström 2013) and the emergence of specific state capacities (McFarlane 2011b) might be empirical objects of investigation in their own right (i.e. not assumed as already conceptualized structures), and open towards conceptualizations of the urban after "virtuality" – in which understandings of distinctive outcomes emerge through exploring interconnected conceptualizations and the range of also interconnected relations, dynamics and processes which help to specify one among a multiplicity of possible urban outcomes. This would perhaps offer a way to appreciate the diversity of urban outcomes while supporting wider conversations and explanations with purchase on the significant political challenges associated with making and surviving urban life.

Thus the key comparative ambition to explain outcomes can benefit from re-framing the meaning of the 'case' in comparative analysis as not simply an example (perhaps hybridized) of singular overarching processes (Jacobs 2012), but as specific outcomes (singularities) which open opportunities to interrogate and conceptualize the wide range of dynamics constituting the urban. In this framing, we might propose the precarious nature of conceptualizations – their openness to revisability, and the necessary instability of their empirical referents. This prompts processes of conceptualization, presents new entities to our imaginations, and draws us in to revising theories, remaking interpretations, generating new concepts, or performing new iterations of an emergent urban. In this imagination, conceptualization is a dynamic and generative process, one subject to rules of experimentation and revisability (Deleuze and Guattari 1994).

Conceptualisations of the urban, if they are to respond to the interconnectedness and emergent unpredictability of the diverse forms of cityness in a world of cities need to be formulated as radically revisable: as Simone puts it, 'the urban is always 'slipping away' from us, always also somewhere else than where we expect it to be' (2011: 356). A reformatted comparative imagination thus draws us to proliferate the grounds for comparability across "cities" through: re-specifying the spatiality of the case in order to more adequately theorize cities in the midst of elsewhere; reconsidering the status of the case to be able to explain outcomes in such a way as to ensure that cases are not seen as simply exemplars of pregiven overarching processes; and insisting that the concepts generated through comparative analysis can be understood as revisable. This reformatted comparativism opens up the possibilities for new geographies of theorizing to support a more global urban studies.

REFERENCES FROM THE READING

Bunnell, T. (2013) City networks as alternative geographies of Southeast Asia, *TRaNS: Trans-Regional and -National Studies*, 1, 1: 27–43.

Deleuze, G. (1994) *Difference and Repetition*, New York: Columbia University Press.

Deleuze, G. and Guattari, F. (1994) *What is Philosophy?* London: Verso.

Farías, I. (2010) Introduction, in I. Farías and T. Bender (eds.) *Urban Assemblages: How Actor–Network Theory Changes Urban Studies*, London: Routledge, 1–24.

Goldfrank, B. and Schrank, A. (2009) Municipal neoliberalism and municipal socialism: Urban political economy in Latin America, *International Journal of Urban and Regional Research*, 33, 2: 443–62.

Harris, A. (2008) From London to Mumbai and back again: Gentrification and public policy in comparative perspective, *Urban Studies*, 45, 12: 2407–28.

Hart, G. (2003) *Disabling Globalisation: Places of Power in Post-Apartheid South Africa*, Berkeley: University of California Press.

Jacobs, J. (2006) A geography of big things, *Cultural Geographies*, 13, 1: 1–27.

Jacobs, J. (2012) Commentary: Comparing comparative urbanisms, *Urban Geography*, 33, 6: 904–14.

Jazeel, T. (2012) Spatializing difference beyond cosmopolitanism: Rethinking planetary futures, *Theory, Culture and Society*, 28, 5: 75–97.

King, A. D. (2004) *Spaces of Global Cultures: Architecture Urbanism Identity*, London: Routledge.

Lefebvre, H. (2003 [1974]) *The Urban Revolution*, Minneapolis: University of Minnesota Press.

McFarlane, C. (2011a) Assemblage and critical urbanism, *City*, 15, 2: 204–24.

McFarlane, C. (2011b) On context: Assemblage, political economy and structure, *City*, 15: 375–88.

Mbembe, A. and Nuttall, S. (2004) Writing the world from an African metropolis, *Public Culture*, 16, 3: 347–72.

Peck, J., Theodore, N., and Brenner, N. (2009) Neoliberal urbanism: Models, moments, mutations, *SAIS Review*, XXIX, 1: 49–66.

Simone, A. (2011) The surfacing of urban life, *City*, 15, 3–4: 355–64.

Söderström, O. (2013) *Cities in Relations: Trajectories of Urban Development in Hanoi and Ougadougou*, Oxford: Wiley-Blackwell.

Ward, K. (2010) Towards a relational comparative approach to the study of cities, *Progress in Human Geography*, 34: 471–87.

67

"Governing the informal in globalizing cities: comparing China, India and Brazil"

from *Housing Policy Debate* (2017)

Xuefei Ren

EDITORS' INTRODUCTION

This chapter by Xuefei Ren, one of the book's co-editors, presents a large-scale, comparative analysis of the politics of informal settlements in globalizing cities in China, India, and Brazil. Unlike the first generation of global city research in the 1990s that concentrated on major financial centers in the global North, in the past decade there has been an increasing body of scholarship focusing on the global South. More and more urban researchers themselves come from developing countries, and the diversity of urban experiences in the global South is no longer treated as marginal and irrelevant to urban theory. Scholars today pay particular attention to the ways in which cities in the global South are unevenly integrated into the world economy (see Wu Ch. 16, Dupont Ch. 17, and Shatkin Ch. 18), and how their global connectivity (or lack of) has altered socio-spatial relations and the built environment.

While significant progress has been made in explaining urban processes in the South from the perspective of global linkages, there has not been sufficient effort to situate globally induced urban social change within historical and institutional contexts. For example, from China to India, from Brazil to South Africa, diverse urban governance patterns are often indiscriminately labeled as "neoliberal" and "entrepreneurial," and the rich and divergent historical-institutional processes are lost in the transnationalized knowledge production. Ren believes that there is a pressing need for the scholarship on global South cities to move beyond the meta-narratives of "urban entrepreneurialism" and "neoliberalism," and a useful research strategy to achieve this is engaging urban comparisons. Country-specific area studies should open up to a bolder, comparative approach by studying globalizing cities across borders, historical eras, and stages of development.

Echoing Robinson's (Ch. 66) call for more comparative research, this chapter illustrates how large-scale comparisons can be carried out to explain globally induced urban restructuring in relation to local-specific, historical-institutional developments. The empirical focus of this chapter is informal settlements—China's "urban villages" (*chengzhongcun*), India's "slums," and Brazil's "favelas"—which house a large swath of the global subaltern class and have become targeted for a new round of "redevelopment." Based on fieldwork in Guangzhou, Mumbai, and Rio de Janeiro, Ren compares different redevelopment policies in China, India, and Brazil, centered on removal, rehabilitation, and upgrading. She traces these policy choices to the varying capacity of local states, which, in turn, is shaped by larger socio-political and institutional processes such as inter-governmental relations, electoral politics, municipal financing structures, and mobilization by the civil society.

China, India, and Brazil offer fertile ground to study informal settlements and policy responses in the global South in comparative perspective. In all three countries, an ever-increasing number of city dwellers can find places to live only in informal settlements with ambiguous land ownership, precarious tenure security, and substandard infrastructure. Based on the latest census in 2011, 41 percent of Mumbai's population lives in "slums," and 22 percent of Rio de Janeiro's population lives in "favelas." In Guangzhou, the largest city in the Pearl River Delta in south China, more than five million people, about 35 percent of the city's population, find cheap rental apartments in the hundreds of "urban villages," China's homegrown informal settlements developed on collectively owned rural land. The proliferation of informal settlements—described as "planet of slums" (Davis 2006), "shadow cities" (Neuwirth 2004), or "undercity" (Boo 2012)—is perhaps the most striking example of urban poverty and entrenched housing inequality at the global scale.

There has been substantial scholarship on informal settlements in individual cities and countries, but few studies to date have compared the development, policies and politics of informal settlements across countries and in the context of global urban restructuring. This chapter echoes Robinson's call for a comparative imagination in global urban studies (see Ch. 66), and it believes that the area scholarship on Chinese, Indian, and Brazilian cities can all benefit from a comparative sensibility. It is acknowledged that there is great variation in economic structures, political regimes, and development trajectories of informal settlements both within and across countries. But these differences do not necessarily weaken comparability. As Robinson (2011; 2016) points out, urban comparison need not be conducted only within similar cases following a quasi-scientific method of controlling for independent variables; rather, the conception of "cases" should be broadened to incorporate a wide range of elements, processes, circulations, and connections in a diversity of urban contexts. At a minimum, she argues, engaging in comparative analyses is about the ability to "think the urban through elsewhere" (Robinson 2016).

The comparators assembled here are major policy initiatives launched in Guangzhou, Mumbai, and Rio de Janeiro to address informal settlements, specifically, urban village removal, slum rehabilitation, and favela upgrading. This article shares Robinson's view that cases should be treated as singularities, i.e., "distinctive

outcomes on their own terms," (Robinson 2016: 14) instead of derivative instances of a wider process such as neoliberalism or capitalism. By juxtaposing different informal housing policies, the comparative analysis highlights the larger socio-political contexts that shape the policies and politics of informal settlements in these globalizing cities.

THE VARYING CAPACITY OF THE LOCAL STATE

Urban restructuring in the global South today is often interpreted through David Harvey's concept of "urban entrepreneurialism" (Harvey 1989). Harvey (1989) argues that what distinguishes entrepreneurial urban governance from conventional managerial modes of governance is a public-private partnership in which local state power combines private capital to promote projects oriented for competition and capital accumulation. Prior studies understood the redevelopment of informal settlements as examples of entrepreneurial governance, aimed to recapture land value by private and public elites (Banerjee-Guha 2002; Broudehoux 2007, 2013). This reading of the local state as becoming entrepreneurial in order to attract investment is largely correct, but the narrative overlooks significant variations in the entrepreneurial capacity of the local state, which in turn shapes policy choices targeting informal settlements. The variations especially come to sharper focus when cities and countries are compared.

The scholarship on Chinese cities has embraced urban entrepreneurialism as a dominant narrative of urban social change in China, but this narrative underestimates the heavy intervention of the local state on all fronts of policy-making and urban-affairs management. Local officials in Chinese cities no doubt act as entrepreneurs while they pursue sources of revenue and investment, but any characterization of the local state as simply "entrepreneurial" misses the expansive reach of Chinese local governments in urban affairs. Local governments in China enjoy far greater administrative autonomy and a much larger revenue base—heavily dependent on land financing—than cities elsewhere. They are not simply partners in growth coalitions, but also the architects of pro-growth agendas who determine the scope and terms of participation for non-state actors.

If China presents a case where "urban entrepreneurialism" fails to capture the deep reach of the local state in running urban affairs, then India presents the opposite case in which "urban entrepreneurialism" overstates the capacity of the weak local state. Although the 73rd and 74th Constitutional Amendments in 1993 mandated devolution of power from state to local governments, decentralization in India did not advance as much as in China and Brazil. Most urban policies regarding land and housing are still the domain of state governments, and municipal governments play only minor roles in policy design and implementation. Due to lack of strong municipal institutions and fragmentation of power at the local level, it is often unclear who is being entrepreneurial and about what. Public-private partnerships in India are often unable to undertake large-scale and complex development projects involving multiple stakeholders.

The framework of urban entrepreneurialism better captures urban restructuring in Brazil. In general, Brazilian cities exhibit strong tendencies of urban entrepreneurialism, with a powerful alliance of financial capital, property capital, political party leaders, and state bureaucracy (Rolnik 2011; Ribeiro 2014). In Brazil, municipal institutions have been significantly empowered since the country's transition from military dictatorship to democracy in the 1980s, to undertake major urban policies and projects. In Rio de Janeiro, for example, the local state actively formed partnerships with the private sector to pursue major infrastructure and redevelopment projects in preparation for the World Cup and the Olympics. Ribeiro and Santos Junior (2014) describe urban governance in Brazil as a hybrid "neoliberal Keynesianism." On one hand, the state strongly intervenes in the economy and implements national-level redistributive policies, but on the other hand, at the city and metropolitan scale entrepreneurial policies seem to be in full force, spurring greater inequality and uneven development.

INFORMAL SETTLEMENTS IN GLOBALIZING CITIES: AN ANALYTICAL FRAMEWORK

Local governments in China, India, and Brazil have experimented with different housing policies to deal with informal settlements. China's approach has been to demolish existing urban villages and encourage private developers to build high-rise apartment blocks, and in the process migrant tenants are displaced while

indigenous villagers (collective owners of land) are enriched through market-rate compensation. In India some slum dwellers have been "rehabilitated" and resettled for no charge in newly built apartment buildings (both on-site and off) but many fail to qualify. Brazil's policies toward informal housing are comparatively more progressive, with an emphasis on upgrading infrastructure, providing social services, and better integrating the favelas with the rest of the city. In all three countries, the policy choices for redeveloping informal settlements are intensely contested, especially as land value rises and cities prepare to host mega-events and build infrastructural projects, often on land occupied by informal settlements.

This section proposes an analytical framework to understand the divergent policy choices targeting informal settlements in globalizing cities. It argues that housing policy responses to informal settlements are largely shaped by four factors—central-state-local government relations, electoral politics, municipal finance, and the capacity of civil society. By no means are these the only factors shaping policy choices—in India and Brazil, for example, slum upgrading programs are also influenced by international donor agencies such as the World Bank and Inter-American Development Bank. Nor are they meant to offer a deterministic account of all policy choices. However, closely intertwined, these four factors comprise the institutional and politico-social context for steering housing policy choices across cities.

Inter-governmental relations

Inter-governmental relations define the parameters of the authority devolved to local governments to make housing policies. The degree of decentralization largely affects a city's autonomy to make its own housing policies, and for China, India, and Brazil, decentralization has unfolded under very different economic and political environments, and it has led to highly differentiated outcomes. China's decentralization took place under the single-party, non-democratic political system, and it is characterized with partial political devolution but substantial administrative and economic devolution of power to local governments (Lin, Tao, and Liu 2006). Decentralization in Brazil and India took place in democratic settings and it is characterized by comprehensive political devolution but partial and uneven economic and administrative devolution (Bardhan and Mookherjee 2006).

China and Brazil are much more fiscally and administratively decentralized than India, and local governments have more autonomy to devise their own housing policy and undertake entrepreneurial projects. Since China's market reform that began in 1978, decision-making authority and responsibility over urban affairs have been largely devolved to municipal governments, for example, housing, health, education, and social security are all domains of municipal governments. For urban villages, there are no national-level policies and municipal governments can devise their own policies and programs, and also adjust them frequently without much interference from provincial and central governments.

In Brazil, decentralization took place in the period of transition to post-dictatorship and it was mandated in the 1988 Constitution. Some of the major economic and administrative functions were devolved to local governments (Baiocchi 2006). Municipalities are considered equal members of the national federation as state governments, and they are given substantial discretion over land legislation and housing policies over squatter settlements. But as the Constitution did not specify which level of government is responsible for service provisions, so that in addition to municipal authorities, federal and state governments can also implement their own housing programs. Currently all three levels of government intervene in favela housing policies, which often leads to competition and lack of coordination. India is the least decentralized among the three countries, and local governments lack autonomy, authority, and resources to devise slum policies and carry out implementation.

Electoral politics

In democratic regimes, informal housing policies and their implementation are often highly politicized by competing political parties to gather votes. Electoral competition, especially at the municipal level, can lead to initiation of populist housing policies by contending candidates, and contestation and discrediting of existing housing programs especially if they are adopted under previous administration. In non-democratic regimes, inter-party competition is largely missing and there is little public input on housing policy making. Electoral politics matter most in Brazil and least in China.

In Brazil, favela housing policies have been largely shaped by mayoral elections and administrative transitions. Rio de Janeiro's favela upgrading program—Favela-Bairro—was established and expanded under mayor Cesar Maia's first administration (1993–1996), but during the mayoral campaign in the late 1990s, the program was highly politicized and used as an instrument by Maia and Luis Paulo Conde—the housing secretary under Maia's administration, to gain political capital and discredit one another (Becerril Miranda 2014). The administrative transition in the early 2000s significantly weakened and disrupted the program.

In India, slum housing programs are also strongly influenced by party politics and election cycles, but not in the medium of mayoral campaigns, as mayors in Indian cities are largely symbolic posts and appointed for short terms. Instead, state government elections are the central platforms affecting slum housing policy changes. Mumbai's Slum Rehabilitation Scheme can be traced back to the 1995 Maharashtra State elections. Shiv Sena, a Mumbai-based right-wing, anti-migrant party, won the state election on a set of ambitious campaign promises, including clearing slums and providing housing to four million slum dwellers. The free housing to four million slum dwellers was never delivered, but the campaign promise turned into the current Slum Rehabilitation Scheme which invites private capital to partake in slum rehabilitation. Due to their numerical strength, slum dwellers are often described as "vote banks" in India, and they are courted before elections and ignored after (Nijman 2008).

In China, inter-party competition and electoral politics are completely missing as a factor influencing housing policies. Under the single-party (Chinese Communist Party) rule, mayors are appointed from the level of government above, and once appointed, mayors and municipal party secretaries can have tremendous influence on policy making. Housing policies are less politicized in China, as the small circle of techno-bureaucrats and officials within municipal governments devise housing policies with little input from the public.

Municipal financing structures

Informal housing policies with different balances between public and private investment reflect different municipal financing structures at large. In China, local governments have retreated from contributing funding directly to redeveloping urban villages. The current informal housing policies in China heavily rely on the private sector. The retreat of the local state in providing funding can be traced to municipal finance structures. In China, local governments in general have a large revenue base but also are responsible for a wide range of expenditures, and to increase revenues, most local governments turned to land leasing. The pressure of raising revenues from land leasing is the key driver for Guangzhou to devise its 2009 policy of redeveloping urban villages based on removal. In other words, urban villages cannot stay because they are obstacles for the local government to raise revenues—once redevelopment sets in motion, the land where urban villages stand is acquired by the state and converted from rural to urban land. In India, municipal governments are weak in fiscal capacity to provide housing for the poor, which leads to private models of slum rehabilitation such as in the case of Mumbai. Local governments in Brazil have the fiscal capacity and are also obligated to provide funding for favela upgrading. Large cities such as Rio de Janeiro have a strong revenue base, as they can raise revenues through sales, vehicle, and services taxes. Since the 1980s, the municipal government of Rio de Janeiro has co-funded Favela-Bairro and Morar Carioca programs, together with international lending agencies, federal, and state government.

Capacity of the civil society

Lastly, informal housing policies are conditioned by the organizational capacity of the civil society. In all three countries, mobilization from the civil society sector has succeeded, albeit in different degrees, to halt demolitions and evictions. Housing rights movements in India and Brazil tend to be led by powerful NGOs with rich experience working with slum dwellers and in community organizing. Under pressure from the civil society, the state has gradually extended protection for housing rights among informal settlers: in the case of Mumbai, the cutoff date upon which eligibility of slum dwellers is determined has been extended from January 1, 1995, to January 1, 2000, and in Rio de Janeiro, the 1992 master plan officially recognized favelas as part of the city, after decades of favela demolition under the military dictatorship. By comparison, powerful housing rights NGOs are almost

non-existent in China and resistance against evictions often comes from individual property owners. Property owners in urban villages in Guangzhou have put the removal-based policies to almost a complete stop. Their mobilizations, not led by any housing NGOs, but building on family ties and clan networks, have forced local governments to recalibrate their policy choices (Al 2014).

CONCLUSION

In the global city literature, housing inequality in the global South is often read through the lens of urban entrepreneurialism and neoliberalism. This chapter has shown that the local state has varying entrepreneurial capacity to carry out market-oriented housing policies. With examples from Guangzhou, Mumbai, and Rio de Janeiro, this chapter argues that policy responses to informal settlements in these globalizing cities are largely shaped by four intertwining forces—inter-governmental relations, electoral politics, municipal finance, and capacity of the civil society. Land financing drove the removal-based, private sector-led urban village policy in Guangzhou; electoral politics and weak municipal revenue base gave rise to the private model of slum rehabilitation in Mumbai; and democratic transition and competitive mayoral elections shaped the integration-based favela upgrading programs in Rio de Janeiro. The analytical framework is not meant to be deterministic, but instead it is to map the large contours of varying local state capacity and political-social context that shape the policies and politics of informal settlements.

Currently in all three cities studied, informal housing policies are facing a crisis. Rio de Janeiro's favela upgrading program was shelved up as other Olympics-related infrastructural projects were given priority. Evictions and removal were carried out under the name of favela integration, which triggered a new wave of housing rights activism. Mumbai's slum rehabilitation program has been strongly contested by those excluded. Guangzhou's plan to redevelop urban villages is unraveled by property-owning villagers, who became a new interest group bargaining intensely with developers for better terms of compensation. The analytical framework presented here invites debate and future comparative research on divergent paths of urban restructuring in cities in the global South.

REFERENCES FROM THE READING

Al, S. (2014) *Villages in the City: A Guide to South China's Informal Settlements*, Hong Kong: Hong Kong University Press.

Baiocchi, G. (2006) Inequality and innovation: Decentralization as an opportunity structure in Brazil, in P. Bardhan and D. Mookherjee (eds.), *Decentralization and Local Governance in Developing Countries*, Cambridge, MA: MIT Press: 53–80.

Banerjee-Guha, S. (2002) Shifting cities: Urban restructuring in Mumbai, *Economic & Political Weekly*, 37: 121–8.

Bardhan, P. and Mookherjee, D. (2006) *Decentralization and Local Governance in Developing Countries*, Cambridge, MA: MIT Press.

Becerril Miranda, H. (2014) Slum upgrading's role for housing policy and governance transformations: From Favela-Bairro to Morar Carioca, investigating the case of Rio de Janeiro in Brazil. PhD dissertation, London: University College London.

Boo, C. (2012) *Behind the Beautiful Forever: Life, Death, Hope in a Mumbai Undercity*, New York: Random House.

Broudehoux, A. M. (2007) Spectacular Beijing: The conspicuous construction of an Olympic metropolis, *Journal of Urban Affairs*, 29: 383–99.

Broudehoux, A. M. (2013) Neo-liberal exceptionalism in Rio de Janeiro's Olympic port regeneration, in M. W. Leary and J. McCarthy (eds.), *The Routledge Companion to Urban Regeneration*, New York: Routledge: 558–68.

Davis, M. (2006) *Planet of Slums*, London: Verso.

Harvey, D. (1989) From managerialism to entrepreneurialism: The transformation in urban governance in late capitalism, *Geografiska Annaler, Series B, Human Geography*, 71: 3–17.

Lin, J., Tao, R., and Liu, M. (2006) Decentralization and local governance in China's economic transition, in P. Bardhan and D. Mookherjee (eds.), *Decentralization and Local Governance in Developing Countries*, Cambridge, MA: MIT Press: 305–27.

Neuwirth, R. (2004) *Shadow Cities: A Billion Squatters, A New Urban World*, New York: Routledge.

Nijman, J. (2008) Against the odds: Slum rehabilitation in neoliberal Mumbai, *Cities*, 25: 73–85.

Ribeiro, L. C. (2014) *The Metropolis of Rio de Janeiro: A Space in Transition*, Rio de Janeiro: IPPUR/UFRJ.

Ribeiro, L. C. and Santos Junior, O. (2014) Mega sporting events in Brazil: Transformation and commodification of the cities, in L. C. Ribeiro (ed.), *The Metropolis of Rio de Janeiro: A Space in Transition*, Rio de Janeiro: IPPUR/UFRJ: 249–62.

Robinson, J. (2011). Cities in a world of cities: The comparative gesture, *International Journal of Urban and Regional Research*, 35: 1–23.

Robinson, J. (2016) Thinking cities through elsewhere: Comparative tactics for a more global urban studies, *Progress in Human Geography*, 40: 3–29.

Rolnik, R. (2011) Democracy on the edge: Limits and possibilities in the implementation of an urban reform agenda in Brazil, *International Journal of Urban and Regional Research*, 35, 2: 239–55.

68
"The urban revolution"

from *The Urban Revolution* (2003, originally published in 1968)

Henri Lefebvre

EDITORS' INTRODUCTION

The French theorist Henri Lefebvre (1901–1991) is perhaps the most important critical urban thinker of the past five decades. His work on space (1991a), everyday life (1991b), and urbanism (1996; 2003) has influenced generations of critical urban scholars and activists. Although Lefebvre never wrote about global cities in the specific sense in which they are understood by most contributors to this volume, his approach to urbanization has considerable salience for the contemporary study of globalized urbanization. Lefebvre's masterful book, *The Urban Revolution*, was completed in the midst of the May 1968 revolt in Paris, when students and workers were engaged in a massive uprising against the French state and the capitalist order. Beyond the national political crisis that framed the uprising, colonial and postcolonial revolutions were never far afield: for it was partly the historic defeat of French colonialism in Algeria a few years previously that had set the stage for this metropolitan revolt. It is against this background that Lefebvre developed his famous claim that "society is completely urbanized." Rather than adopting then-fashionable notions of 'industrial' and 'post-industrial' society, Lefebvre argued that a new phase of human development was unfolding that was based on the worldwide (spatial) extension and (temporal) acceleration of urbanization processes. The multifaceted sociospatial transformations Lefebvre described as the "urban revolution" must be viewed as an essential precondition for the emergence of a specialized world city network, as analyzed by the contributors to this Reader. Moreover, the notion of "complexification," as developed by Lefebvre below, usefully underscores the multilayered sociospatial fabric of contemporary global cities.

Lefebvre's contribution is also relevant to attempts to define the global city concept itself. While French students and workers revolted in Paris, Mao Zedong's Cultural Revolution was taking place in China. Influenced by leading intellectual Lin Biao, China's leaders maintained that the global system was now being divided into a "world city" and a "world countryside"—a rather crude version of the core-periphery model that was being developed during this same period by Western theorists of the capitalist world-system. Lefebvre, however, was unconvinced that the "world countryside" could succeed in resisting the urbanized core, particularly in light of the dramatic forms of capitalist (and state socialist) urbanization that were unfolding during the second half of the 20th century; consequently, he takes issue with Mao's concept of a "world city." In the selection below, Lefebvre argues forcefully against the notion that large-scale urbanization—the consolidation of the "world city"—could somehow be halted through revolutionary praxis based in the rural peripheries. Interestingly, versions of this notion continue to reappear, for instance in debates on postcolonialism and, in Michael Hardt and Antonio Negri's influential musings on "multitude" (2004: 123–4). In both cases, the showdown of the revolutionary rural masses with the urbanized core of the world-system is viewed as a key aspect of resistance to imperialism and empire. Other writers, however, have rejected such arguments and supported Lefebvre's viewpoint. John Friedmann (2002: 1–2), for example, suggests that "the urban transition will not be reversed" and that "[w]illful attempts at ruralization, such as in Mao Zedong's China or Pol Pot's Kampuchea, were thus never more than temporary reversals." Lefebvre's insistence on the impossibility of suppressing the complete urbanization of human existence thus provides a provocative, if very much open-ended, conclusion to the debates and analyses surveyed in this volume.

FROM THE CITY TO URBAN SOCIETY

I'll begin with the following hypothesis: Society has been completely urbanized. This hypothesis implies a definition: An *urban society* is a society that results from a process of complete urbanization. This urbanization is virtual today, but will become real in the future.

Here, I use the term "urban society" to refer to the society that results from industrialization, which is a process of domination that absorbs agricultural production. This urban society cannot take shape conceptually until the end of a process during which the old urban forms, the end result of a series of *discontinuous* transformations, burst apart. An important aspect of the theoretical problem is the ability to situate the discontinuities and continuities with respect to one another. How could any absolute discontinuities exist without an underlying continuity, without support, without some inherent process? Conversely, how can we have continuity without crises, without the appearance of new elements or relationships?

Instead of the term "postindustrial society" – the society that is born of industrialization and succeeds it – I will use "urban society," a term that refers to tendencies, orientations, and virtualities, rather than any preordained reality. Such usage in no way precludes a critical examination of contemporary reality, such as the analysis of the "bureaucratic society of controlled consumption."

Economic growth and industrialization have become self-legitimating, extending their effects to entire territories, regions, nations, and continents. As a result, the traditional unit typical of peasant life, namely the village, has been transformed. Absorbed or obliterated by larger units, it has become an integral part of industrial production and consumption. The concentration of the population goes hand in hand with that of the mode of production. The *urban fabric* grows, extends its borders, corrodes the residue of agrarian life. This expression, "urban fabric," does not narrowly define the built world of cities but all manifestations of the dominance of the city over the country. In this sense, a vacation home, a highway, a supermarket in the countryside are all part of the urban fabric. Of varying density, thickness, and activity, the only regions untouched by it are those that are stagnant or dying, those that are given over to "nature." As this global process of industrialization and urbanization was taking place, the large cities exploded, giving rise to growths of dubious value: suburbs, residential conglomerations, and industrial complexes, satellite cities that differed little from urbanized towns. Small and midsize cities became dependencies, partial colonies of the metropolis. In this way my hypothesis serves both as a point of arrival for existing knowledge and a point of departure for a new study and new projects: complete urbanization. The hypothesis is anticipatory. It prolongs the fundamental tendency of the present. Urban society is gestating in and through the "bureaucratic society of controlled consumption." The expression "urban society" meets a theoretical need. It is more than simply a literary or pedagogical device, or even the expression of some form of acquired knowledge; it is an elaboration, a search, a conceptual formulation. A movement of thought toward a certain *concrete,* and perhaps toward *the* concrete, assumes shape and detail. This movement, if it proves to be true, will lead to a practice, *urban practice,* that is finally or newly comprehended.

Similarly, by "urban revolution" I refer to the transformations that affect contemporary society, ranging from the period when questions of growth and industrialization predominate (models, plans, programs) to the period when the urban problematic becomes predominant, when the search for solutions and modalities unique to urban society are foremost. Some of these transformations are sudden; others are gradual, planned, determined. But which ones? This is a legitimate question. It is by no means certain in advance that the answer will be clear, intellectually satisfying, or unambiguous. The words "urban revolution" do not in themselves refer to actions that are violent. Nor do they exclude them.

We can draw an axis as follows:

0 ————————————————— 100%

The axis runs from the complete absence of urbanization ("pure nature," the earth abandoned to the elements) on the left to the completion of the process on the right. A signifier for this signified – the *urban* (the urban reality) – this axis is both spatial and temporal: spatial because the process extends through space, which it modifies; temporal because it develops over time. Temporality, initially of secondary importance, eventually becomes the predominant aspect of practice and history. This schema presents no more than an aspect of this history, a division of time that is both abstract and arbitrary and gives rise to operations (periodizations) that have no absolute privilege but are as necessary (relative) as other divisions.

The rise of the mercantile city, which was grafted onto the political city but promoted its own ascendancy, was soon followed by the appearance of industrial capital and, consequently, the *industrial city.* This requires further explanation. Was industry associated with the city? One would assume it to be associated with the *non-city,* the absence or rupture of urban reality. We know that industry initially developed near the sources of energy (coal and water), raw materials (metals, textiles), and manpower reserves. Industry gradually made its way into the city in search of capital and capitalists, markets, and an abundant supply of low-cost labor. It could locate itself anywhere, therefore, but sooner or later made its way into existing cities or created new cities, although it was prepared to move elsewhere if there was an economic advantage in doing so. Just as the political city resisted the conquest – half-pacific, half-violent – of the merchants, exchange, and money, similarly the political and mercantile city defended itself from being taken over by a nascent industry, industrial capital, and capital itself. But how did it do this? Through corporatism, by establishing relationships. Historical continuity and evolution mask the effects and ruptures associated with such transitions. Yet something strange and wonderful was also taking place, which helped renew dialectical thought: the non-city and the anti-city would conquer the city, penetrate it, break it apart, and in so doing extend it immeasurably, bringing about the urbanization of society and the growth of the urban fabric that covered what was left of the city prior to the arrival of industry. This extraordinary movement has escaped our attention and has been described in piecemeal fashion because ideologues have tried to eliminate dialectical thought and the analysis of contradictions in favor of logical thought – that is, the

identification of coherence and nothing but coherence. Urban reality, simultaneously amplified and exploded, thus loses the features it inherited from the previous period: organic totality, belonging, an uplifting image, a sense of space that was measured and dominated by monumental splendor. It was populated with signs of the urban within the dissolution of urbanity; it became stipulative, repressive, marked by signals, summary codes for circulation (routes), and signage. It was sometimes read as a rough draft, sometimes as an authoritarian message. It was imperious. But none of these descriptive terms completely describes the historical process of implosion-explosion (a metaphor borrowed from nuclear physics) that occurred: the tremendous concentration (of people, activities, wealth, goods, objects, instruments, means, and thought) of urban reality and the immense explosion, the projection of numerous, disjunct fragments (peripheries, suburbs, vacation homes, satellite towns) into space.

The *industrial city* (often a shapeless town, a barely urban agglomeration, a conglomerate, or conurbation like the Ruhr Valley) serves as a prelude to a *critical zone.* At this moment, the effects of implosion-explosion are most fully felt. The increase in industrial production is superimposed on the growth of commercial exchange and multiplies the number of such exchanges. This growth extends from simple barter to the global market, from the simple exchange between two individuals all the way to the exchange of products, works of art, ideas, and human beings. Buying and selling, merchandise and market, money and capital appear to sweep away all obstacles. During this period of generalization, the effect of the process – namely the urban reality – becomes both cause and reason. Induced factors become dominant (inductors). The *urban problematic* becomes a global phenomenon. Can urban reality be defined as a "superstructure" on the surface of the economic structure, whether capitalist or socialist? The simple result of growth and productive forces? Simply a modest marginal reality compared with production? Not at all. Urban reality modifies the relations of production without being sufficient to transform them. It becomes a productive force, like science. Space and the politics of space "express" social relationships but react against them. Obviously, if an urban reality manifests itself and becomes dominant, it does so only through the urban problematic. What can be done to change this? How can we build cities or "something" that replaces what was formerly the City? How can we reconceptualize the urban phenomenon? How can we formulate, classify, and order the innumerable questions that arise, questions that move, although not without considerable resistance, to the forefront of our awareness? Can we achieve significant progress in theory and practice so that our consciousness can comprehend a reality that overflows it and a possible that flees before its grasp? We can represent this process as follows: *implosion-explosion* (urban concentration, rural exodus, extension of the urban fabric, complete subordination of the agrarian to the urban).

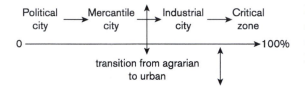

The onset of urban society and the modalities of urbanization depend on the characteristics of society as it existed during the course of industrialization (neocapitalist or socialist, full economic growth or intense automation). The onset of urban society at different times, the implications and consequences of these initial differences, are part of the problematic associated with the urban phenomenon, or simply the "urban." These terms are preferable to the word "city," which appears to designate a clearly defined, definitive *object,* a scientific object and the immediate goal of action, whereas the theoretical approach requires a critique of this "object" and a more complex notion of the virtual or possible object. Within this perspective there is no science of the city (such as urban sociology or urban economy), but an emerging understanding of the overall process, as well as its term (goal and direction).

The urban (an abbreviated form of urban society) can therefore be defined not as an accomplished reality, situated behind the actual in time, but, on the contrary, as a horizon, an illuminating virtuality. The *virtual object* is nothing but planetary society and the "global city," and it stands outside the global and planetary crisis of reality and thought, outside the old borders that had been drawn when agriculture was dominant and that were maintained during the growth of exchange and industrial production. Nevertheless, the urban problematic can't absorb every problem. There are problems that are unique to agriculture and industry, even though the urban reality modifies them. Moreover, the urban problematic requires that we

exercise considerable caution when exploring the realm of the possible. It is the analyst's responsibility to identify and describe the various forms of urbanization and explain what happens to the forms, functions, and urban structures that are transformed by the breakup of the ancient city and the process of generalized urbanization. Until now the critical phase was perceived as a kind of black box. We know what enters the box, and sometimes we see what comes out, but we don't know what goes on inside. This makes conventional procedures of forecasting and projection useless, since they extrapolate from the actual, from a set of facts. Projections and forecasts have a determined basis only in the fragmentary sciences: demography, for example, or political economy. But what is at stake here, "objectively," is a totality.

URBAN SOCIETY

During this exploration, the urban phenomenon appears as something other than, as something more than, a superstructure (of the mode of production). I say this in response to a form of Marxist dogmatism that manifests itself in a variety of ways. The urban problematic is worldwide. The same problems are found in socialism and in capitalism – along with the failure to respond. Urban society can only be defined as global. Virtually, it covers the planet by recreating nature, which has been wiped out by the industrial exploitation of natural resources (material and "human"), by the destruction of so-called natural particularities.

Moreover, the urban phenomenon has had a profound effect on the methods of production: productive forces, relationships of production, and the contradictions between them. It both extends and accentuates, on a new plane, the social character of productive labor and its conflict with the ownership (private) of the means of production. It continues the "socialization of society," which is another way of saying that the urban does not eliminate industrial contradictions. It does not resolve them for the sole reason that it has become dominant. What's more, the conflicts inherent in production (in the relationships of production and capitalist ownership as well as in "socialist" society) hinder the urban phenomenon, prevent urban development, reducing it to growth. This is particularly true of the action of the state under capitalism and state socialism.

To summarize then: Society becomes increasingly complex with the transition from the rural to the industrial and from the industrial to the urban. This multifaceted complexification affects space as well as time, for the complexification of space and the objects that occupy space cannot occur without a complexification of time and the activities that occur over time. This space is occupied by interrelated networks, relationships that are defined by interference. Its homogeneity corresponds to intentions, unified strategies, and systematized logics, on the one hand, and reductive, and consequently simplifying, representations, on the other. At the same time, differences become more pronounced in populating this space, which tends, like any abstract space, toward homogeneity (quantitative, geometric, and logical space). This, in turn, results in conflict and a strange sense of unease. For this space tends toward a unique code, an absolute system, that of exchange and exchange value, of the logical thing and the logic of things. At the same time, it is filled with subsystems, partial codes, messages, and signifiers that do not become part of the unitary procedure that the space stipulates, prescribes, and inscribes in various ways.

The theory of complexification anticipates the revenge of development over growth. The same is true for the theory of urban society. This revenge is only just beginning. The basic proposition, that growth cannot continue indefinitely and that the means can remain an end without a catastrophe occurring, still seems paradoxical.

These considerations evoke the prodigious extension of the urban to the entire planet, that is, urban society, its virtualities and potential. It goes without saying that this extension-expansion is not going to be problem-free. Indeed, it has been shown that the urban phenomenon tends to overflow borders, while commercial exchange and industrial and financial organizations, which once seemed to abolish those territorial limits (through the global market, through multinationals), now appear to reaffirm them. In any event, the effects of a possible rupture in industry and finance (a crisis of overproduction, a monetary crisis) would be accentuated by an extension of the urban phenomenon and the formation of urban society.

I have already introduced the idea of the "global city," generally attributed to Maoism, if not Mao Zedong himself. I would now like to develop this idea. The global city extends the traditional concept and image of the city to a global scale: a political center for the administration, protection, and operation of a vast territory. This is appropriate for the oriental city within the framework of an Asian mode of production. However, urban society cannot be constructed on the ruins of the classical city alone. In the West, this city has already begun to fragment. This fragmentation (explosion-implosion) may appear to be a precursor of urban society. It is part of its problematic and the critical phase that precedes it. However, a known strategy, which specifically makes use of urbanism, tends to view the political city as a decision-making center. Such a center is obviously not limited to collecting information upstream and distributing it downstream. It is not just a center of abstract decision making but a center of power. Yet power requires wealth, and vice versa. That is, the decision-making center, in the strategy being analyzed here, will serve as a point of attachment to the soil for a hyperorganized and rigidly systematized state. Formerly, the entire metropolitan land area played a central role with respect to the colonies and semicolonies, sucking up wealth, imposing its own order. Today, domination is consolidated in a physical locale, a capital (or a decision-making center that does not necessarily coincide with the capital). As a result, control is exercised throughout the national territory, which is transformed into a semicolony.

Part of my analysis may appear at first glance to correspond to the so-called Maoist interpretation of the "global city," but this interpretation raises a number of objections. There is nothing that prevents emerging centers of power from encountering obstacles and failing. What's more, any contradictions that occur no longer take place between city and country. The principal contradiction is shifted to the urban phenomenon itself: between the centrality of power and other forms of centrality, between the "wealth-power" center and the periphery, between integration and segregation.

Is the urban phenomenon the *total social phenomenon* long sought for by sociologists? Yes and no. Yes, in the sense that it tends toward totality without ever achieving it, that it is essentially totalizing (centrality) but that this totality is never effected. Yes, in the sense that no partial determinism, no fragmentary knowledge can exhaust it; it is simultaneously historical, demographic, geographic, economic, sociologic, psychologic, semiologic, and so on. It "is" that and more (thing or non-thing) besides: *form*, for example. In other words, a void, but one that demands or calls forth a content. If the urban is total, it is not total in the way a thing can be, as content that has been amassed, but in the way that thought is, which continues its activity of concentration endlessly but can never hold or maintain that state of concentration, which assembles elements continuously and discovers what it has assembled through a new and different form of concentration. Centrality defines the u-topic (that which has no place and searches for it). The u-topic defines centrality.

But neither the separation of fragment and content nor their confused union can define (and therefore express) the urban phenomenon. For it incorporates a *total reading,* combining the vocabularies (partial readings) of geographers, demographers, economists, sociologists, semiologists, and others. These readings take place on different levels. The phenomenon cannot be defined by their sum or synthesis or super-position. In this sense, it is not a totality. Similarly, it overcomes the separation between accident and necessity, but their synthesis doesn't determine it, assuming such synthesis can be determined. This is simply a repetition of the paradox of the urban pheno-menon, a paradox that in no way gives it precedence over the fundamental paradox of thought and awareness. For it is undoubtedly the same. The urban is specific: it is localized and focused. It is locally intensified and doesn't exist without that localization, or center. Thought and thinking don't take place unless they are themselves localized. The specificity of the fact, the event, is a given. And, consequently, a requirement. Near order occurs around a point, taken as a (momentary) center, which is produced by practice and can be grasped through analysis. This defines an isotopy. At the same time, the urban phenomenon is colossal; its prodigious extension-expansion cannot be constrained. While encompassing near order, a *distant order* groups distinct specificities, assembles them according to their differences (heterotopies). But isotopy and heterotopy clash everywhere and always, engendering an *elsewhere.* Although initially indispensable, the *transformed* centrality that results will be reabsorbed into the fabric of space-time. In this way the dialectical movement of the specific and the colossal, of place and non-place (elsewhere), of urban order and urban disorder assumes form (reveals itself as form).

The urban is not produced like agriculture or industry. Yet, as an act that assembles and distributes,

it does create. Similarly, manufacturing at one time became a productive force and economic category simply because it brought together labor and tools (technology), which were formerly dispersed. In this sense, the urban phenomenon contains a praxis (urban practice). Its form, as such, cannot be reduced to other forms (it is not isomorphous with other forms and structures), but it absorbs and transforms them.

The procedure for accessing urban reality as a form is reversed once the process is complete. In this way we can use linguistics to define isotopy and hete-rotopy. Once they have been identified in the urban text, these concepts assume a different meaning. Isn't it because human habitations assume the form that they do that they can be recognized in discourse? The urban is associated with a discourse and a route, or pathway. And it is for this reason, or formal cause, that there are different discourses and pathways in language. One cannot be separated from the other. Although different, language and dwelling are indis-solubly combined. Is it surprising then that there is a *paradigm of* the urban (high and low, private and public), just as there is for habiting (open and closed, intimate and public), although neither the urban nor habiting can be defined by a simple discourse or by a system? If there is any logic inherent in the urban and the habiting it implies, it is not the logic of a system (or a subject or an object). It is the logic of thought (subject) that looks for a content (object). It is for this reason that our understanding of the urban requires that we simultaneously abandon our illusions of subjectivity (representation, ideology) and objectivity (causality, partial determinism).

The urban consolidates. As a form, the urban trans-forms what it brings together (concentrates). It con-sciously creates difference where no awareness of difference existed: what was only distinct, what was once attached to particularities in the field. It consoli-dates everything, including determinisms, heteroge-neous materials and contents, prior order and disorder, conflict, preexisting communications and forms of communication. As a transforming form, the urban destructures and restructures its elements: the mes-sages and codes that arise in the industrial and agrar-ian domains.

The urban also contains a negative power, which can easily appear harmful. Nature, a desire, and what we call culture (and what the industrial era dissociated from nature, while during predominately agrarian periods, nature and culture were indissoluble) are reworked and combined in urban society. Heterogeneous, if not heteroclite, these contents are put to the test. Thus, by way of analogy, agricultural exploitation (the farm) and the enterprise (which came into existence with the rise of manufacturing) are put to the test, are transformed, and are incorporated in new forms within the urban fabric. We could consider this a form of second-order creativity *(poiesis),* agricultural and industrial production being forms of first-order creativity. This does not mean that the urban phenomenon can be equated with second-order discourse, metalanguage, exegesis, or commentary on industrial production. No, second-order creation and the secondary naturality of the urban serve to *multiply* rather than reduce or reflect creative activity. This raises the issue of an activity that produces (creates) meanings from elements that already possess signification (rather than units similar to phonemes, sounds, or signs devoid of signification). From this point of view,

the urban would create situations and acts just as it does objects.

REFERENCES FROM THE READING

Friedmann, J. (2002) *The Prospect of Cities,* Minneapolis, MN: University of Minnesota Press.

Hardt, M. and Negri, A. (2004) *Multitude,* New York: Penguin Press.

Lefebvre, H. (1991a [originally published 1974]) *The Production of Space,* trans. D. N. Smith, Cambridge, MA: Blackwell.

Lefebvre, H. (1991b [originally published 1947]) *Critique of Everyday Life, Volume 1,* Trans. J. Moore, London: Verso.

Lefebvre, H. (1996) *Writings on Cities,* Trans. E. Kofman and E. Lebas, Cambridge, MA: Blackwell.

Lefebvre, H. (2003 [originally published 1968]) *The Urban Revolution,* Trans. R. Bononno. Minneapolis, MN: University of Minnesota Press.

Index

Page numbers in italics refer to figures. Page numbers in bold refer to tables.